Advances in Intelligent Systems and Computing

Volume 400

Series editor

Janusz Kacprzyk, Polish Academy of Sciences, Warsaw, Poland
e-mail: kacprzyk@ibspan.waw.pl

About this Series

The series "Advances in Intelligent Systems and Computing" contains publications on theory, applications, and design methods of Intelligent Systems and Intelligent Computing. Virtually all disciplines such as engineering, natural sciences, computer and information science, ICT, economics, business, e-commerce, environment, healthcare, life science are covered. The list of topics spans all the areas of modern intelligent systems and computing.

The publications within "Advances in Intelligent Systems and Computing" are primarily textbooks and proceedings of important conferences, symposia and congresses. They cover significant recent developments in the field, both of a foundational and applicable character. An important characteristic feature of the series is the short publication time and world-wide distribution. This permits a rapid and broad dissemination of research results.

More information about this series at http://www.springer.com/series/11156

Troels Andreasen · Henning Christiansen
Janusz Kacprzyk · Henrik Larsen
Gabriella Pasi · Olivier Pivert
Guy De Tré · Maria Amparo Vila
Adnan Yazici · Sławomir Zadrożny
Editors

Flexible Query Answering Systems 2015

Proceedings of the 11th International Conference FQAS 2015, Cracow, Poland, October 26–28, 2015

Springer

Editors
Troels Andreasen
Department of Communication, Business and Information Technologies
Roskilde University
Roskilde, Denmark

Henning Christiansen
Department of Communication, Business and Information Technologies
Roskilde University
Roskilde, Denmark

Janusz Kacprzyk
Systems Research Institute
Polish Academy of Sciences
Warszawa, Poland

Henrik Larsen
Department of Electronic Systems
Aalborg University
Esbjerg, Denmark

Gabriella Pasi
Computer Science Department
University Milano-Biccoca
Milano, Italy

Olivier Pivert
IRISA
University of Rennes I
Rennes, France

Guy De Tré
Department of Telecommunications and Information Processing
Ghent University
Gent, Belgium

Maria Amparo Vila
Department of Computation and Artificial Intelligence Sciences
University of Granada
Granada, Spain

Adnan Yazici
Department of Computer Engineering
Middle East Technical University
Ankara, Turkey

Sławomir Zadrożny
Systems Research Institute
Polish Academy of Sciences
Warsaw, Poland

ISSN 2194-5357 ISSN 2194-5365 (electronic)
Advances in Intelligent Systems and Computing
ISBN 978-3-319-26153-9 ISBN 978-3-319-26154-6 (eBook)
DOI 10.1007/978-3-319-26154-6

Library of Congress Control Number: 2015952986

Springer Cham Heidelberg New York Dordrecht London

Printed on acid-free paper

Springer International Publishing AG Switzerland is part of Springer Science+Business Media (www.springer.com)

Foreword

This volume contains the papers presented at the Eleventh Flexible Query Answering Systems 2015 (FQAS-2015) held on October 26–28, 2015 in Cracow, Poland. The international conferences on Flexible Query Answering Systems (FQAS) are a series of premier conferences focusing on the key issue in the information society of providing easy, flexible, and intuitive access to information and knowledge to everybody, even people with a very limited computer literacy. In targeting this issue, the Conference draws on several research areas, such as information retrieval, database management, information filtering, knowledge representation, soft computing, management of multimedia information, and human-computer interaction. The Conference provides a unique opportunity for researchers, developers and practitioners to explore new ideas and approaches in a multidisciplinary forum. The previous FQAS conferences, which always attracted a large audience from all parts of the world, include: FQAS 2013 (Granada, Spain), FQAS 2011 (Ghent, Belgium), FQAS 2009 (Roskilde, Denmark), FQAS 2006 (Milano, Italy), FQAS 2004 (Lyon, France), FQAS 2002 (Copenhagen, Denmark), FQAS 2000 (Warsaw, Poland), FQAS 1998 (Roskilde, Denmark), FQAS 1996 (Roskilde, Denmark), FQAS 1994 (Roskilde, Denmark).

An important contribution of the Conference has also been the fact that has greatly facilitated, and often made possible, a deeper discussion on papers presented which as a rule has resulted in new collaborative works and a further progress in the areas.

The Workshop has been partially supported, financially and technically, by many organizations, notably: Systems Research Institute, Polish Academy of Sciences; Department IV of Engineering Sciences, Polish Academy of Sciences; Cracow Branch, Polish Academy of Sciences; Academia Europaea – The Hubert Curien Initiative Fund; Ghent University; Polish Association of Artificial Intelligence, and Polish Operational and Systems Research Society. Their support is acknowledged and highly appreciated.

We hope that the collection of main contributions presented at the Conference, completed with plenary talks by leading experts in the field, will provide a source of much needed information and inspiration on recent trends in the topics considered.

We wish to thank all the authors for their excellent contributions and their collaboration during the editing process of the volume. We are looking forward to the same fruitful collaboration during the next FQAS Conferences of this series that are planned for the years to come. Special thanks are due to the peer reviewers whose excellent and timely work has significantly contributed to the quality of the volume.

And last but not least, we wish to thank Dr. Tom Ditzinger, Dr. Leontina di Cecco and Mr. Holger Schaepe for their dedication and help to implement and finish this large publication project on time maintaining the highest publication standards.

August 2015

Troels Andreasen
Henning Christiansen
Janusz Kacprzyk
Henrik Larsen
Gabriella Pasi
Olivier Pivert
Guy De Tré
Maria Amparo Vila
Adnan Yazici
Sławomir Zadrożny

Contents

Part I
Preferences, Desires, Perception of Time, and Logical Foundations

Revising Desires – A Possibility Theory Viewpoint

Didier Dubois , Emiliano Lorini and Henri Prade

Abstract As extensively studied in the artificial intelligence literature, agents may have to revise their beliefs when they receive a new piece of information, for avoiding inconsistency in their epistemic states, since one cannot believe p and believe $\neg p$ at the same time. Similarly desiring p and $\neg p$ simultaneously does not sound reasonable, since this would amount to be pleased by anything. This motivates an approach for revising desires, a topic remained largely untouched. Desires do not behave as beliefs. While beliefs are closed under conjunction, one may argue that the disjunction of desires reflects the endorsement of each desire. In a possibility theory modeling setting, desires are expressed by a strong possibility set function, while beliefs are encoded by means of a necessity function. The paper outlines an original approach to the revision, the expansion, and the contraction of desires in the framework of possibility theory, and contrasts it with belief revision.

Introduction

Desires, goals, preferences, often used interchangeably (see, e.g., [23]), are all members of the family of *motivational* attitudes. This family is traditionally opposed to the family of *epistemic* attitudes including knowledge and belief. The distinction epistemic vs. motivational is in terms of the *direction of fit* of mental attitudes to the world. While epistemic attitudes aim at being true and their being true is their fitting the world, motivational attitudes aim at realization and their realization is the world fitting them [22, 25]. The philosopher J. Searle [26] calls "mind-to-world" the first kind of *direction of fit* and "world-to-mind" the second one.

While the word 'preferences' seems to have a generic meaning, *goals* are intended, they are like intentions, which is not the case for *desires*. Indeed a goal usually refers to a desire that has been selected by an agent in order to try to reach it. This distinction

D. Dubois · E. Lorini · H. Prade(✉)
IRIT – CNRS, 118, route de Narbonne, 31062 Toulouse Cedex 09, France
e-mail: {dubois,lorini,prade}@irit.fr

T. Andreasen et al. (eds.), *Flexible Query Answering Systems 2015*,
Advances in Intelligent Systems and Computing 400,
DOI: 10.1007/978-3-319-26154-6_1

between desires and goals is supported by the possibilistic setting, where they are modeled differently. A goal p with a priority level α is translated by a constraint of the form $N(p) \geq \alpha$ where N is necessity measure. This fits with the idea that having $p \wedge q$ as a goal is the same as having p as a goal and having q as a goal. In that respect, see [19] for a possibilistic logic view of flexible querying in terms of prioritized goals.

As suggested in [6], and advocated in [9], a desire p is properly represented by a constraint of the form $\Delta(p) \geq \alpha$ which stands for "the agent desires p with strength at least α", where Δ is a strong possibility measure [13].

In the context of possibility theory, this concept of desire is also contrasted with the concept of belief that is properly represented by a constraint of the form $N(p) \geq \alpha$ which stands for "the agent believes p with strength at least α", where N is a necessity measure. However, the fact that beliefs and goals are represented by constraints of the same type does not mean at all that beliefs and goals are the same. Since they respectively represent what it is known about the current state of the world, and how the agent would like the state of the world becomes, it is crucial to keep beliefs and goals separated in two different possibilistic logic bases in decision under uncertainty problems; see [7].

The use here of necessity measures and strong possibility measures for modeling the beliefs and the desires of agents, should not be confused with their use for modeling bipolar queries where we need to distinguish between what is compulsory to reach (since the opposite should be avoided), and what would be really satisfactory to get [14, 17]. Such a later bipolar view is left aside in the following, where one considers positive desires only (namely those that it would be really satisfactory to concretize). Modeling desires of endusers is a task ahead the expression of definite queries whose purpose is to check if intended desires are feasible and how.

"It makes no sense to want everything and its opposite at the same time" says the wisdom of mankind. Otherwise, one is led to indetermination. In other words, one cannot desire p and desire $\neg p$ at the same time. This parallels the fact that one cannot believe p and believe $\neg p$ at the same time, without being led to inconsistency. Revising beliefs copes with this constraint. Similarly, revising desires should cope with the previous constraint. But beliefs and desires obey different principles.

Beliefs, modeled by means of necessity measures, satisfy

$$N(p \wedge q) = \min(N(p), N(q))$$

i.e., believing p *and* q amounts to believing p and to believing q. Thus we have

$$\min(N(p), N(\neg p)) = N(\bot) = 0$$

where $\bot$ denotes contradiction. This expresses that one cannot believe in p and in $\neg p$ in the same time while remaining consistent. We also have

$$N(p \vee q) \geq \max(N(p), N(q))$$

which is nothing but the increasingness of N with respect to entailment (i.e., if $r \models s$ then $N(r) \leq N(s)$), and fits with the fact that one may believe $p \vee q$ without believing p or believing q more specifically.

Desires rather obey the principle

$$\Delta(p \vee q) = \min(\Delta(p), \Delta(q))$$

i.e., desiring p *or* q amounts to desiring p and to desiring q. Thus we have

$$\min(\Delta(p), \Delta(\neg p)) = \Delta(\top) = 0$$

where$\top$ denotes tautology. Moreover $\Delta(p \wedge q) \geq \max(\Delta(p), \Delta(q))$. This indicates that Δ is decreasing with respect to entailment (i.e., if $r \models s$ then $\Delta(r) \geq \Delta(s)$).

Since desires do not behave as beliefs – N increases while Δ decreases with respect to entailment – belief revision does not straightforwardly apply to desire revision. One needs a slightly different theory for desire revision.

After restating and explaining the modeling of desires in terms of Δ functions, and providing a refresher on belief revision in the setting of possibility theory, the paper introduces and discusses the revision of desires.

Modeling Desire Using Δ Function

We here assume that in order to determine how much a proposition p is desirable an agent takes into consideration the worst situation in which p is true. Let $\mathcal{L}(ATM)$ be the propositional language built out of the set of atomic formulas ATM and let Ω the set of all interpretations of this language, corresponding to the different possible states of the world that can be described by means of the language. Moreover, let $||p|| \subseteq \Omega$ denote the set of interpretations where the propositional formula p is true.

Let π and δ be two possibilistic functions with domain Ω and codomain a linearly ordered scale S with 1 and 0 as top and bottom elements. Functions π and δ capture, respectively, the degree of (epistemic) possibility and the degree of desirability of a given interpretation $\omega \in \Omega$. We assume that π and δ satisfy the following normality constraints: (i) there exists $\omega \in \Omega$ such that $\pi(\omega) = 1$ (i.e., at least one state of the world is fully possible), and (ii) there exists $\omega \in \Omega$ such that $\delta(\omega) = 0$ (i.e., at least one state of the world is not desired at all).

As suggested in [6] and advocated in [9], for a given formula p, we can interpret

$$\Delta(p) = \min_{\omega \in ||p||} \delta(\omega)$$

as the extent to which the agent desires p to be true. This can be contrasted with

$$N(p) = 1 - \max_{\omega \in ||\neg p||} \pi(\omega)$$

which estimates the extent to which the agent believes p to be true, all the more as $\neg p$ is found impossible. Indeed, the measure of necessity N is the dual of the possibility measure Π, namely $\Pi(p) = 1 - N(\neg p)$ (where $1 - (\cdot)$ denotes the order-reversing map of S). Let us justify the following two properties for desires:

$$\Delta(p \vee q) = \min(\Delta(p), \Delta(q)); \quad \Delta(p \wedge q) \geq \max(\Delta(p), \Delta(q)).$$

According to the first property, an agent desires p to be true with a given strength α and desires q to be true with a given strength β if and only if the agent desires p or q to be true with strength equal to $\min(\alpha, \beta)$. A similar intuition can be found in [6] about the min-decomposability of disjunctive desires, where however it is emphasized that it corresponds to a pessimistic view. Notice that in the case of epistemic states, this property would not make any sense because the plausibility of $p \vee q$ should be clearly *at least* equal to the maximum of the plausibilities of p and q. For the notion of desires, it seems intuitively satisfactory to have the opposite, namely the level of desire of $p \vee q$ should be *at most* equal to the minimum of the desire levels of p and q. Indeed, we only deal here with "*positive*"[1] desires (i.e., desires to reach something with a given strength).

Under the proviso that we deal with positive desires, the level of desire of $p \wedge q$ cannot be less than the maximum of the levels of desire of p and q. According to the second property, the joint occurrence of two desired events p and q is more desirable than the occurrence of one of the two events. This is the reason why in the right side of the equality we have the operator max. This latter property does not make any sense in the case of epistemic attitudes like beliefs, as the joint occurrence of two events p and q is epistemically less plausible than the occurrence of a single event. On the contrary it makes perfect sense for motivational attitudes likes desires, as suggested by the following example.

Example 1. Suppose Peter wishes to go to the cinema in the evening with strength α (i.e., $\Delta(\mathit{goToCinema}) = \alpha$) and, at the same time, he wishes to spend the evening with his girlfriend with strength β (i.e., $\Delta(\mathit{stayWithGirlfriend}) = \beta$). Then, according to the preceding property, Peter wishes to to go the cinema with his girlfriend with strength at least $\max(\alpha, \beta)$ (i.e., $\Delta(\mathit{goToCinema} \wedge \mathit{stayWithGirlfriend}) \geq \max(\alpha, \beta)$). This is a reasonable conclusion because the situation in which Peter achieves his two desires is (for Peter) at least as pleasant as the situation in which he achieves only one desire.

One might object that if it is generally the case that satisfying simultaneously two desires is at least as good as satisfying one of them, there may exist exceptional situations where it is not the case. Just imagine, in the above example, the case where Peter's

---

[1] The distinction between positive and negative desires is a classical one in psychology. Negative desires correspond to state of affairs the agent wants to avoid with a given strength, and then desires the opposite to be true. However, we do not develop this bipolar view [16] here.

girlfriend would be laughing aloud or crying all the time during movies, and so Peter would not like to go with her to the cinema. This is a situation of non monotonic desires that can be coped with in this setting; see [9]. It is a counterpart of non monotonic reasoning about beliefs where, while $N(p \vee q) \geq \max(N(p), N(q))$ expresses increasing monotonicity, this should be remedied by an appropriate prioritization as in the example "penguins (p) are birds (b), birds fly (f)", where one should block the consequences of $p \models b \Rightarrow N(\neg b \vee f) \leq N(\neg p \vee f))$; see [1] for details on non monotonic reasoning about beliefs in the possibilistic reasoning setting.

Besides, from the normality constraint of δ ($\exists \omega, \delta(\omega) = 0$, expressing that not everything is desired), we can deduce that if $\Delta(p) > 0$ then $\Delta(\neg p) = 0$. This means that if an agent desires p to be true – i.e., with some strength $\alpha > 0$ – then he does not desire at all p to be false. In other words, an agent's desires must be consistent. Note also that the operator Δ, which is monotonically decreasing, satisfies $\Delta(\bot) = 1$ by convention. There is no harm to desire $\bot$, which by nature is unreachable.

From Belief Revision to Desire Revision

It has been recognized early that the epistemic entrenchment relations underlying any well-behaved belief revision process obeying Gärdenfors' postulates [20] are qualitative necessity relations [10], thus establishing a link between belief revision and possibility theory [13]. In the possibility theory view of belief revision, the epistemic entrenchment is explicit and reflects a confidence-based priority ranking between pieces of information. This ranking is revised when a new piece of information is received. We restate the possibilistic expression of belief revision, before considering the revision of desires.

Belief Revision

Uncertain beliefs are represented in possibility theory by constraints of the form $N(p) \geq \alpha$, corresponding to possibilistic logic [15, 18] formulas (p, α), expressing that p is believed to be true, with a certainty level at least equal to $\alpha > 0$, where α belongs to a linearly ordered scale S with 1 and 0 as top and bottom elements. Necessity measures satisfy the characteristic axiom $N(p \wedge q) = \min(N(p), N(q))$ (one believes $p \wedge q$ at the extent to what the less believed propositions of p and q is believed). They are the dual of possibility measures Π such that $\Pi(p) = 1 - N(\neg p)$ $(1 - (\cdot)$ is the order-reversing map of S).

A set B of possibilistic logic formulas (p_i, α_i) (for $i = 1, \ldots, n$) is semantically associated to a possibility distribution

$$\pi_B(\omega) = \min_{i=1,\ldots,n} \max(||p_i||(\omega), 1 - \alpha_i),$$

where $||p_i||(\omega) = 1$ if $\omega \models p_i$ and $||p_i||(\omega) = 0$ otherwise [15]. π_B is the largest possibility distribution (minimum specificity principle) such that $N(p_i) \geq \alpha_i$ for $i = 1, \ldots, n$. The distribution π_B rank-orders the interpretations of the language induced by the p_i's according to their plausibility on the basis of the strength of the beliefs in B.

In qualitative possibility theory [13], conditioning is defined by means of equation

$$\Pi(p \wedge q) = \min(\Pi(q|p), \Pi(p)).$$

Applying the minimum specificity principle, we get the possibility distribution $\pi(\cdot|p)$ associated with the possibility measure $\Pi(\cdot|p)$:

$$\pi(\omega|p) = \left\{ \begin{array}{cc} 1 & \text{if } \pi(\omega) = \Pi(p) \text{ and } \omega \vDash p \\ \pi(\omega) & \text{if } \pi(\omega) < \Pi(p) \text{ and } \omega \vDash p \\ 0 & \text{if } \omega \vDash \neg p \end{array} \right\}.$$

Then the *revision* B_p^* of the belief base B revised by input p, is defined as:

$$\pi_{B_p^*}(\omega) = \pi_B(\omega|p).$$

It includes the *expansion* B_p^+ of B by p (where $core(\pi) = \{\omega \mid \pi(\omega) = 1\}$) as a particular case:

$$\pi_{B_p^+}(\omega) = \min(\pi(\omega), ||p||(\omega)) \text{ provided that } core(\pi) \cap ||p|| \neq \emptyset.$$

Besides, the *contraction* B_p^- of B by p is defined by [11, 12]:

$$\pi_{B_p^-}(\omega) = \left\{ \begin{array}{cc} 1 & \text{if } \pi(\omega) = \Pi(\neg p) \text{ and } \omega \vDash \neg p \\ \pi(\omega) & \text{otherwise} \end{array} \right\}.$$

In particular, if $\Pi(p) = \Pi(\neg p) = 1$, we have $\pi_{B_p^-}(\omega) = \pi(\omega)$.

Harper's and Levi's identities, which respectively relate expansion and contraction to revision, remain valid [11]. It can be checked that counterparts of Gärdenfors' postulates for expansion, contraction, and revision [20] hold in the possibilistic setting [11]. See [3] [4] for more thorough studies, and the syntactic counterpart in possibilistic logic of the above revision process.

In particular, the possibilistic base B_p^*, can be obtained syntactically as $\{(p_i, \alpha_i) \in B \text{ s.t. } \alpha_i > \lambda\} \cup \{(p, 1)\}$, where $\lambda = inc(B \cup \{(p, 1)\}) = 1 - \max_\omega \pi_{B \cup \{(p,1)\}}(\omega)$ is the degree of inconsistency [15] of $B \cup \{(p, 1)\}$. Let us illustrate the approach by a small example.

Example 2. Let $B = \{(p, \alpha), (p \vee q, \beta)\}$, with $\alpha < \beta$.
We have $\pi_B(pq) = \pi_B(p\neg q) = 1$; $\pi_B(\neg pq) = 1 - \alpha$; $\pi_B(\neg p \neg q) = 1 - \beta$.
Then, assume the input $\neg p$ is received.
It gives $\pi_{B^*_{\neg p}}(pq) = \pi_{B^*_{\neg p}}(p\neg q) = 0$; $\pi_{B^*_{\neg p}}(\neg pq) = 1$; $\pi_{B^*_{\neg p}}(\neg p \neg q) = 1 - \beta$
(since $\Pi_B(\neg p) = \max_{\omega \vDash \neg p} \pi_B(\omega) = 1 - \alpha$).
The syntactic counterpart is $B^*_{\neg p} = \{(\neg p, 1), (p \vee q, \beta)\}$
(where $inc(B \cup \{(\neg p, 1)\}) = \alpha$).

Desire Revision

As explained at the beginning of the paper, desires can be represented in terms of strong (or guaranteed) possibility measures (denoted by Δ). A desire for p is expressed by a constraint of the form $\Delta(p) \geq \alpha$. A desire p with strength α will be denoted $[p, \alpha]$. Strong possibility measures are governed by the characteristic property $\Delta(p \vee q) = \min(\Delta(p), \Delta(q))$. This implies that Δ is decreasing with respect to entailment; in particular $\Delta(\top) = 0$ and $\Delta(\bot) = 1$.

A set D of desires $[p_i, \alpha_i]$ (for $i = 1, \ldots, n$) is semantically associated to a possibility distribution

$$\delta_D(\omega) = \max_{i=1,\ldots,n} \min(||p_i||(\omega), \alpha_i).$$

δ_D is the smallest possibility distribution (maximum specificity principle) such that $\Delta(p_i) \geq \alpha_i$ for $i = 1, \ldots, n$. The distribution δ_D rank-orders the interpretations of the language induced by the p_i's according to their satisfaction level on the basis of the strength of the desires in D. Because we should have $\Delta(\top) = 0$, $\min_\omega \delta_D(\omega) = 0$ should hold. More generally,

$$una(D) = \min_\omega \delta_D(\omega)$$

may be viewed as a level of *unacceptability* of D. The larger $una(D)$, the more unacceptable the set of desires D.

The *contraction* of D by p amounts to no longer desire p at all after contraction:

$$\delta_{D_p^-}(\omega) = \left\{ \begin{array}{ll} 0 & \text{if } \delta(\omega) = \Delta(p) \text{ and } \omega \vDash p \\ \delta(\omega) & \text{otherwise} \end{array} \right\}.$$

In particular, we have $\delta_{D_p^-} = \delta(\omega)$ if $\Delta(p) = \Delta(\neg p) = 0$.

The *expansion* of a set of desires D by p amounts to perform the cumulate desire p with the desires in D, providing that the result is not the desire of everything to some extent (due to the postulate $\Delta(\top) = 0$). Thus, we have

$$\delta_{D_p^+}(\omega) = \max(\delta(\omega), ||p||(\omega))$$

under the proviso that $support(\delta) \cup ||p|| \neq \Omega$, where $support(\delta) = \{\omega | \delta(\omega) > 0\}$, Ω denoting the set of all possible interpretations.

The conditioning of a strong possibility measure Δ obeys the equation [2]:

$$\Delta(p \wedge q) = \max(\Delta(q|p), \Delta(p)). \quad (*)$$

Applying the *maximum* specificity principle, we get the smallest (i.e., corresponding to the least committed conditional desires) possibility distribution $\delta(\omega|p)$ obeying $(*)$:

$$\delta(\omega|p) = \left\{ \begin{array}{cc} 0 & \text{if } \delta(\omega) = \Delta(p) \text{ and } \omega \models p \\ \delta(\omega) & \text{if } \delta(\omega) > \Delta(p) \text{ and } \omega \models p \\ 1 & \text{if } \omega \models \neg p \end{array} \right\}.$$

As can be seen, what is no longer reachable is fully desirable by default ($\Delta(\neg p|p) = 1$), while what we have is no longer desired since $\Delta(p|p) = 0$, but still preserving what is strictly above $\Delta(p)$.

While the *revision* of a set of beliefs B by p exactly corresponds to the conditioning of π_B by p, this is no longer the case with respect to δ_D for the revision of a set of desires D by p. Indeed, while a belief input $(p, 1)$, i.e. $N(p) = 1$, really means that all the models of $\neg p$ should be impossible, i.e., $\Pi(\neg p) = \max_{\omega \models \neg p} \pi_B(\omega) = 0$, a desire input $[p, 1]$ means $\Delta(p) = \min_{\omega \models p} \pi_B(\omega) = 1$, which says that all the models of p are satisfactory after revision. Due to this change of focus from $\neg p$ to p, when moving from beliefs to desires, we state:

$$\delta_{D^*_p}(\omega) = \delta_D(\omega|\neg p) = \left\{ \begin{array}{cc} 0 & \text{if } \delta(\omega) = \Delta(\neg p) \text{ and } \omega \models \neg p \\ \delta(\omega) & \text{if } \delta(\omega) > \Delta(\neg p) \text{ and } \omega \models \neg p \\ 1 & \text{if } \omega \models p \end{array} \right\}$$

As easily seen, $\Delta_{D^*_p}(p) = 1$, which shows that the *success postulate* for desire revision is satisfied, in the sense that an agent desires p to be true after revising his desire base by p. The latter may be found too strong and weakened into $\Delta_{D^*_p}(p) > 0$. It can be defined by taking lesson of what is done in belief revision; see [3].

Let us illustrate the approach by some examples.

Example 3. Let $D = \{[p \wedge q, \alpha], [r, \beta]\}$, with $\alpha > \beta$.
We have $\delta_D(pqr) = \delta_D(pq\neg r) = \alpha$; $\delta_D(p\neg qr) = \delta_D(\neg pqr) = \delta_D(\neg p\neg qr) = \beta$; $\delta_D(p\neg q\neg r) = \delta_D(\neg pq\neg r) = \delta_D(\neg p\neg q\neg r) = 0$.
Clearly, $una(D) = 0$.
Now, assume we want to add desire $[\neg p, 1]$. Let us compute $\delta_{D^*_{\neg p}}$. We get:
$\delta_{D^*_{\neg p}}(pqr) = \delta_{D^*_{\neg p}}(pq\neg r) = \alpha$; $\delta_{D^*_{\neg p}}(p\neg qr) = \beta$; $\delta_{D^*_{\neg p}}(p\neg q\neg r) = 0$, which remain unchanged, while it gives
$\delta_{D^*_{\neg p}}(\neg pqr) = \delta_{D^*_{\neg p}}(\neg p\neg qr) = \delta_{D^*_{\neg p}}(\neg pq\neg r) = \delta_{D^*_{\neg p}}(\neg p\neg q\neg r) = 1$.
Observe that $una(D \cup \{[\neg p, 1]\}) = 0$,
which means that after addition of the new desire, the set of desires remains acceptable. In fact, we have just performed an expansion here.
As can be checked, we have $D^*_{\neg p} = D^+_{\neg p}$.
The syntactic counterpart is $D^*_{\neg p} = \{[p \wedge q, \alpha], [r, \beta], [\neg p, 1]\}$.

Let us now consider two other examples where the unacceptability level becomes positive.

Example 4. Let $D' = \{[p, \alpha], [r, \beta]\}$ with $\alpha > \beta$.
Then $\delta_{D'}(pr) = \delta_{D'}(p\neg r) = \alpha$; $\delta_{D'}(\neg pr) = \beta$; $\delta_{D'}(\neg p\neg r) = 0$,
and $una(D') = 0$.
Now, let us add desire $[\neg p, 1]$.
We get $\delta_{D'^*_{\neg p}}(pr) = \delta_{D'^*_{\neg p}}(p\neg r) = 0$; $\delta_{D'^*_{\neg p}}(\neg pr) = \delta_{D'^*_{\neg p}}(\neg p\neg r) = 1$.
We have $una(D' \cup \{[\neg p, 1]\}) = \alpha$ and $D'^*_{\neg p} = \{[\neg p, 1]\}$.

Example 5. Now consider $D'' = \{[p, \beta], [r, \alpha]\}$ (always with $\alpha > \beta$),
then $\delta_{D''}(pr) = \alpha$; $\delta_{D''}(p\neg r) = \beta$; $\delta_{D''}(\neg pr) = \alpha$; $\delta_{D''}(\neg p\neg r) = 0$, and $\delta_{D''^*_{\neg p}}(pr) = \alpha$; $\delta_{D''^*_{\neg p}}(p\neg r) = 0$; $\delta_{D''^*_{\neg p}}(\neg pr) = \delta_{D''^*_{\neg p}}(\neg p\neg r) = 1$.
Finally, $una(D'' \cup \{[\neg p, 1]\}) = \beta$, and $D''^*_{\neg p} = \{[r, \alpha], [\neg p, 1]\}$.

Generally speaking, it can be shown that only the desires strictly above the level of unacceptability are saved (the others are drown, as $[r, \beta]$ in D'):

$$D^*_p = \{[p_i, \alpha_i] \in D \text{ s.t. } \alpha_i > una(D \cup \{[p, 1]\}) \cup \{[p, 1]\}.$$

Conclusion

The paper has outlined a formal approach to the revision of desires. Much remains to be done: providing the postulates characterizing this type of revision, laying bare the counterparts of Harper's and Levi's identities for desires, studying iterated desire revision. Moreover, the success postulate is translated by $\Delta_{D^*_p}(p) = 1$, which may be found too strong; this may be weakened into $\Delta_{D^*_p}(p) > 0$, which corresponds to the idea of *natural* revision in the sense of Boutilier [5].

Besides, it is known that belief revision and nonmonotonic reasoning are two sides of the same coin [21]. This remains to be checked for nonmonotonic desires [9] and desires revision. Finally, we plan to extend the static modal logic of belief and desire we proposed in [8] by dynamic operators of belief revision and desire revision. This will provide a unified modal logic framework based on possibility theory dealing with both the static and the dynamic aspects of beliefs and desires, to be compared with the proposal made in [24].

References

1. Benferhat, S., Dubois, D., Prade, H.: Practical handling of exception-tainted rules and independence information in possibilistic logic. Applied Intelligence **9**(2), 101–127 (1998)
2. Benferhat, S., Dubois, D., Kaci, S., Prade, H.: Bipolar possibilistic representations. In: Darwiche, A., Friedman, N. (ed.) Proc. 18th Conf. in Uncertainty in Artificial Intelligence (UAI 2002), Edmonton, Alberta, August 1–4, pp. 45–52. Morgan Kaufmann (2002)
3. Benferhat, S., Dubois, D., Prade, H.: A computational model for belief change and fusing ordered belief bases. In: Williams, M.-A., Rott, H. (eds.) Frontiers in Belief Revision, pp. 109–134. Publ., Kluwer Acad. (2001)
4. Benferhat, S., Dubois, D., Prade, H., Williams, M.-A.: A practical approach to revising prioritized knowledge bases. Studia Logica **70**, 105–130 (2002)

5. Boutilier, C.: Revision sequences and nested conditionals. In: Proc. 13th Int. Joint Conf. on Artificial Intelligence (IJCAI 1993), Chambéry, August 28 - September 3, pp. 519–525. Morgan Kaufmann (1993)
6. Casali, A., Godo, L., Sierra, C.: A graded BDI agent model to represent and reason about preferences. Artificial Intelligence **175**, 1468–1478 (2011)
7. Dubois, D., Le Berre, D., Prade, H., Sabbadin, R.: Logical representation and computation of optimal decisions in a qualitative setting. In: Proc. 15th National Conf. on Artificial Intelligence (AAAI 1998), July 26–30, Madison, Wisc, pp. 588–593. AAAI Press / The MIT Press (1998)
8. Dubois, D., Lorini, E., Prade, H.: Bipolar possibility theory as a basis for a logic of desires and beliefs. In: Liu, W., Subrahmanian, V.S., Wijsen, J. (eds.) SUM 2013. LNCS, vol. 8078, pp. 204–218. Springer, Heidelberg (2013)
9. Dubois, D., Lorini, E., Prade, H.: Nonmonotonic desires - A possibility theory viewpoint. In: Booth, R., Casini, G., Klarman, S., Richard, G., Varzinczak, I.J. (eds.) Proc. Int. Workshop on Defeasible and Ampliative Reasoning (DARe@ECAI 2014), Prague, August 19, vol. 1212. CEUR Workshop Proc. (2014)
10. Dubois, D., Prade, H.: Epistemic entrenchment and possibilistic logic. Art. Int. **50**, 223–239 (1991)
11. Dubois, D., Prade, H.: Belief change and possibility theory. In: Głrdenfors, P. (ed.) Belief Revision, pp. 142–182. University Press, Cambridge (1992)
12. Dubois, D., Prade, H.: A synthetic view of belief revision with uncertain inputs in the framework of possibility theory. Int. J. Approx. Reasoning **17**(2–3), 295–324 (1997)
13. Dubois, D., Prade, H.: Possibility theory: qualitative and quantitative aspects. In: Gabbay, D., Smets, P. (eds.) Quantified Representation of Uncertainty and Imprecision, Handbook of Defeasible Reasoning and Uncertainty Management Systems, vol. 1, pp. 169–226. Kluwer (1998)
14. Dubois, D., Prade, H.: Bipolarity in flexible querying. In: Andreasen, T., Motro, A., Christiansen, H., Larsen, H.L. (eds.) FQAS 2002. LNCS (LNAI), vol. 2522, pp. 174–182. Springer, Heidelberg (2002)
15. Dubois, D., Prade, H.: Possibilistic logic: a retrospective and prospective view. Fuzzy Sets and Systems **144**, 3–23 (2004)
16. Dubois, D., Prade, H.: An introduction to bipolar representations of information and preference. Int. J. of Intelligent Systems **23**(8), 866–877 (2008)
17. Dubois, D., Prade, H.: An overview of the asymmetric bipolar representation of positive and negative information in possibility theory. Fuzzy Sets and Systems **160**(10), 1355–1366 (2009)
18. Dubois, D., Prade, H.: Possibilistic logic - An overview. In: Gabbay, D.M., Siekmann, J., Woods, J. (eds.) Handbook of the History of Logic. Computational Logic, vol. 9, pp. 283–342. Elsevier (2014)
19. Dubois, D., Prade, H., Touazi, F.: A possibilistic logic approach to conditional preference queries. In: Larsen, H.L., Martin-Bautista, M.J., Vila, M.A., Andreasen, T., Christiansen, H. (eds.) FQAS 2013. LNCS, vol. 8132, pp. 376–388. Springer, Heidelberg (2013)
20. Gärdenfors, P.: Knowledge in Flux. The MIT Press (1990)
21. Gärdenfors, P.: Belief revision and nonmonotonic logic: Two sides of the same coin? In: Proc. 9th Europ. Conf. on Artificial Intelligence (ECAI 1990), Stockholm, pp. 768–773 (1990)
22. Humberstone, I.L.: Direction of fit. Mind **101**(401), 59–83 (1992)

23. Lang, J.: Conditional desires and utilities: An alternative logical approach to qualitative decision theory. In: Wahlster, W. (ed.) Proc. 12th Eur. Conf. Artif. Intellig. (ECAI 1996), Budapest, August 11–16, pp. 318–322. J. Wiley (1996)
24. Lang, J., van der Torre, L.W.N., Weydert, E.: Hidden uncertainty in the logical representation of desires. In: Gottlob, G., Walsh, T. (eds.) Proc. 18th Int. Joint Conf. on Artificial Intelligence (IJCAI 2003), Acapulco, August 9–15, pp. 685–690. Morgan Kaufmann (2003)
25. Platts, M.: Ways of meaning. Routledge and Kegan Paul (1979)
26. Searle, J.: Expression and meaning. Cambridge University Press (1979)

Discovering Consumers' Purchase Intentions Based on Mobile Search Behaviors

Mingyue Zhang, Guoqing Chen and Qiang Wei

Abstract Search activity is an essential part for gathering useful information and supporting decision making. With the exponential growth of mobile e-commerce, consumers often search for products and services that are closely relevant to the current context such as location and time. This paper studies the search behaviors of mobile consumers, which reflect their customized purchase intentions. In light of machine learning, a probabilistic generative model is proposed to discover underlying search patterns, i.e., *when to search,where to search* and *in what category*. Furthermore, the predicting power of the proposed model is validated on the dataset released by Alibaba, the biggest e-commerce platform in the world. Experimental results show the advantages of the proposed model over the classical content-based methods, and also illustrate the effectiveness of integrating contextual factors into modeling consumers search patterns.

Keywords Search patterns · Context-aware · Probabilistic model · Recommendation

1 Introduction

Nowadays, the Internet acts as a core information source for human worldwide, and many information gathering activities take place online [5, 10]. Search activities widely exist in daily life, from which irrelevant information is filtered and useful one is extracted to support decision making. For example, 'information retrieval' and 'database querying' are two common activities to extract relevant documents or other types of information. According to the CNNIC (China Internet Network Information Center) report, the number of search engine users in China had reached 522 million in 2014, with a growth rate of 6.7% compared to last year. In addition,

M. Zhang · G. Chen · Q. Wei(✉)
Department of Management Science and Engineering,
School of Economics and Management, Tsinghua University, Beijing 100084, China
e-mail: {zhangmy.12,chengq,weiq}@sem.tsinghua.edu.cn

T. Andreasen et al. (eds.), *Flexible Query Answering Systems 2015*,
Advances in Intelligent Systems and Computing 400,
DOI: 10.1007/978-3-319-26154-6_2

mobile search engines also attract 429 million people and had an even higher growth rate of 17.6%. The search services have been extended to the combined presentation of pictures, applications, products and other types instead of just text and links. On one hand, search becomes an essential part for consumers to find useful information in their decision making processes. On the other hand, by keeping track of the search patterns of the consumers, online merchants can have a better understanding of the consumers' behaviors and intentions [5].

In mobile e-commerce, potential consumers search for product information before making purchasing decisions due to the overwhelming information [8, 12, 16]. Since search could reflect consumers' purchase intentions and affect their choices online [9], it is worthy of deep exploration and has attracted a lot of interest from both academia and practitioners. Moreover, as Bhatnagar & Ghose (2004) [3] indicated, consumers exhibit differences in their search patterns, i.e, time spent per search episode and search frequency, which are attributed to product categories and consumer characteristics.

In marketing practice, clickstream data are commonly used to quantify the customers' search behaviors [8, 9, 16]. Usually, the clickstream data provide information about the sequence of pages or the path viewed by the consumers as they navigate websites [18]. With the prevalence of smart mobile devices [6], the consumers' clickstream data have been enriched with various contextual information [25], such as geographical information, which poses significant new opportunities as well as challenges [13]. Some data mining techniques have been employed to extract consumers' context-aware preferences [12, 20]. However, these research efforts mostly focused on purchase records while ignoring the search activities. While purchasing indicates consumers' final preferences over different products in the same category, search is an essential reflection of their purchase intentions towards a specific category. Therefore, a more precise model is needed to capture each consumer's search behavior relating to the particular context.

In this paper, we aim to understand the mobile e-commerce consumers' potential purchase intentions by studying their search patterns. That is, because the examination and inspection of products/services come at the cost of the consumers' time and effort, search outcomes become informative about what the consumers want [12]. We start by analyzing the search history of each consumer and then examine whether there is a relationship between search activities and the contextual factors (i.e., time and location). Based on the assumption that search patterns are time and location dependent, a probabilistic generative process is proposed to model each consumer's search history, in which the latent context variable is introduced to capture the simultaneous influence from both time and location. By identifying the search patterns of the consumers, we can predict their click decisions in specific contexts and recommend the products/services with the maximum clicking probabilities of the consumers.

The remaining part of the paper is organized as follows. Section 2 reviews related research from three aspects: consumer information search, clickstream data, and context-aware preference. Section 3 presents the consumer search model and parameter estimation process. Section 4 demonstrates the experimental results on a real-world dataset and Section 5 concludes the paper.

2 Related Work

This section discusses the existing work that is related to this study, consisting of three aspects: consumer information search, clickstream data, and context-aware preference.

Consumer Information Search. Consumer information search is an important part of purchase decision making [4, 8, 19], which attracts continuous attention from researchers. By using page-to-page clickstream data, Moe (2003) [16] examined in-store navigational behavior in terms of the pattern and content of pages viewed, and classified consumers into four categories according to their underlying objectives, namely, direct buying, search and deliberation, hedonic browsing, and knowledge building. This helps understand the objectives of the consumers better, thereby providing some insights into purchasing behaviors. Huang et al. (2009) [8] investigated the differences in consumer search patterns between search goods and experience goods in the online context based on clickstream data. In the empirical examination, they found the type of information that the consumers seek, and the way they search and make choices, was different for the two types of goods. Further, these differences affected the amount of time spent per page of information, the number of pages searched, the likelihood of free riding, and the relative importance of interactive mechanisms. Branco et al. (2012) [4] discussed the optimal stopping strategy for consumer search. Specially, they provided a parsimonious model of consumer search for gradual information that captured the benefits and costs of search, resulting in the optimal stopping threshold where the marginal costs outweighed the benefits. Similarly, Kim et al. (2010) [12] introduced the optimal sequential search process into a model of choice and estimated consumer information search and online demand for durable goods based on dynamic programming framework.

Clickstream Data. The widespread availability of Internet clickstream data has contributed greatly to marketing research [16], which allows both practitioners and academics to examine consumer search behaviors in a large-scale field setting. For example, Banerjee & Ghosh (2001) [2] clustered users based on a function of the longest common subsequence of their clickstreams, considering both the trajectory taken through a website and the time spent at each page. Montgomery et al. (2004) [18] used a dynamic multinomial probit model to extract information from consumers' navigation path, which is helpful in predicting their future movements. Kim et al. (2004) [11] used the clickstream data as implicit feedback to design a hybrid recommendation model, which resulted in better performance. Moe (2006) [17] proposed an empirical two-stage choice model based on clickstream data to capture observed choices for two stages: products viewed and products purchased. They found that the product attributes evaluated in Stage 1 differred from those evaluated in Stage 2. Overall, the clickstream data provided a great opportunity for researchers to dig into consumer search and purchase behaviors. Nevertheless, very little research has been conducted to describe the generative process of consumer search, especially the search behaviors in specific contexts.

Context-Aware Preference. Due to the exponential growth of mobile e-commerce, large volume of contextual information is available, which enables researchers to study the problem of personalized context-aware recommendation [13, 21, 24]. Shabib & Krogstie (2011) [21] proposed a step-by-step approach to assessing the context-aware preferences, which consists of four phases: product classification, interest matrix formation, clustering similar users and making recommendation. Zheng et al. (2010) [24] discovered useful knowledge from GPS trajectories based on users' partial location and activity annotations to provide targeted collaborative location and activity recommendations together. Specifically, they modeled the user-location-activity relations in a tensor and designed the algorithm based on regularized tensor and matrix decomposition. Liu et al. (2015) [13] considered users' check-in behavior in mobile devices to provide personalized recommendations of places, which integrated the effect of geographical factor and location based social network factor. Different from previous research, our study aims to formalize consumers' search behaviors as (*when, where, what*) patterns through a probabilistic generative model, which has better explanation ability and can be used to design appropriate recommendation strategies.

3 Consumer Search Model

3.1 Search Behavior Analysis

In the context of e-commerce, consumers commonly search for product information online before making purchase decisions [4]. Since more and more people access the Internet with their smart phone, the mobile search activities have distinctive features where contextual information can be captured by the search logs, including time and location [22, 25]. While consumers are often overwhelmed by excessive number of products in the platform, the specific category that products belong to is a good reflection of consumers' purchase intentions. In addition, the data of search log will be very sparse if each product is treated as an individual item, which may hardly be able to discover common search patterns. Therefore, we focus on the searched 'category' instead of 'product' in the following analysis. Predicting the category that consumers are most likely to click in a given timeslot and location is of great importance. If we can model a consumer's purchase intentions toward a specific category, we can recommend the appropriate products/services taking into account both his/her preferences and location information.

To analyze the possible factors that affect mobile consumers' search patterns, the relationships between the number of clicks and contextual factors will be explored. The clickstream data were released by Alibaba.com, which will be described in detail later. Figure 1 displays the click times in a range of 24 hours and Figure 2 shows the distribution of clicks under different geographic areas. Note that locations were clustered into different geographic areas without overlap.

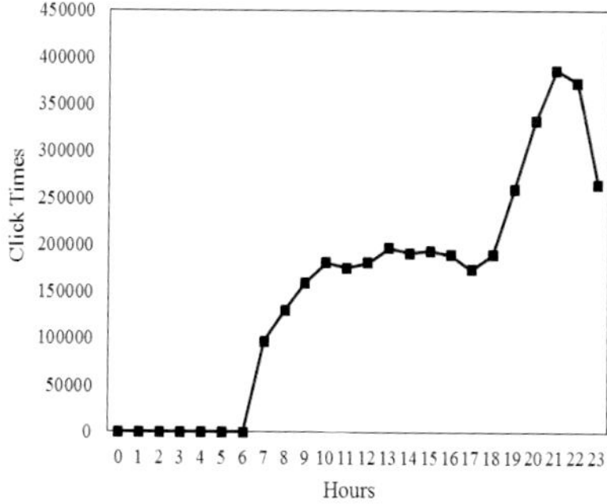

Fig. 1 Click times in different hours

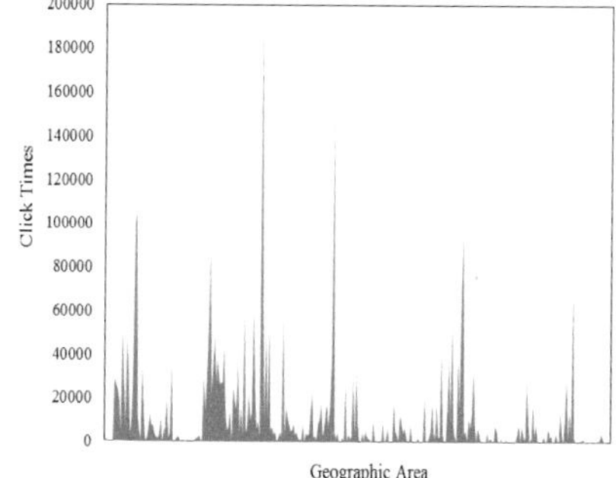

Fig. 2 Click times in different geographic areas

Here, the search events mostly occurred in the evening and reached the peak at 21:00. It can be seen that the consumers were more active in certain geographic areas, which illustrates the importance of location information. Thus, it is reasonable to assume that consumers search/click patterns are dependent on contextual factors including time and location.

Problem Definition: Suppose that we have consumers' clicking histories in mobile terminal, which reflect their search behaviors. For a consumer u, its clicking trace can be represented with a tuple (l, t, g) where a consumer u clicks category g at time t in location l. The problem is to model all the consumers' search histories and discover the patterns, in terms of *when to search, where to search*, and *in what category*. In this way, appropriate recommendation strategies can be designed so as to present the right product/service categories to right people at right time and right locations.

3.2 Generative Process

The advantage of probabilistic generative model is that it can mimic the process of a consumer's purchase behavior. Inspired by the classic topic model "Probabilistic Latent Semantic Analysis" (PLSA) [7, 15], the process of consumer search could be considered as an extension [23].

Specifically, each search is composed of particular search time, location and category, which are correlated with each other. In order to discover the common search patterns that underly these tuples, we introduce a latent factor, i.e., *shared search context* c. Thus, each value of context c is assumed to be the mixture of time, location and category, representing some common patterns among all the consumers, hence consumers' preferences towards time, location, and category are simultaneously captured by the shared search context.

For example, in the case that u wants to choose a category, he/she may choose a context according to his/her personal preference distribution, which is commonly a multinomial distribution. After that, the selected context in turn 'generates' the specific search tuple (l, t, g) following the context's generative distribution. Thus, this model simulates the process of how u picks the category g at the specific time t and location l. To simplify the analysis, we assume that these specific tuples, including time, location and category, are conditionally independent given the latent context. Therefore, the generative process for each consumer's clicked category can be summarized as follows:

- For each consumer u:
  - For each search action (l, t, g) of consumer u:
    (1) Generate a shared search context $c \sim p(c|u)$, which is a multinomial distribution;
    (2) Generate a search location $l \sim p(l|c)$ conditional on the latent context;
    (3) Generate a search timeslot $t \sim p(t|c)$ conditional on the latent context;
    (4) Generate a search category $g \sim p(g|c)$ conditional on the latent context.

Note that $p(l|c)$, $p(t|c)$ and $p(g|c)$ are also assumed to be multinomial distributions. The above Bayesian generative process can be represented as a graphical model in which a directed graph is used to describe probability distributions (see Figure 3).

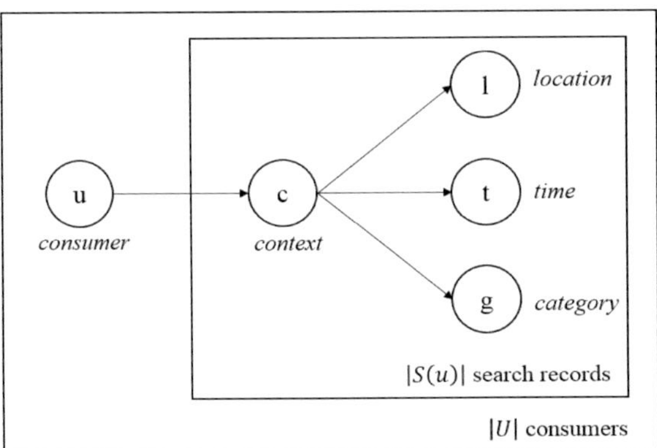

Fig. 3 A graphical representation of the probabilistic model

Thus, a consumer's search history is regarded as a sample of the following mixture model.

$$p(l, t, g|u) = \sum_{c} p(l, t, g|c) \cdot p(c|u) \quad (1)$$

Since l, t, g are assumed to be conditional independent with each other given the latent shared search context c, the above equation can be transformed to:

$$p(l, t, g|u) = \sum_{c} p(c|u) \cdot p(l|c) \cdot p(t|c) \cdot p(g|c) \quad (2)$$

The proposed model is a latent class statistical mixture model, which discovers: 1) a consumer's personal preference distribution over latent search context; 2) a category generative distribution for each latent context; 3) a consumer's preference distribution over locations/timeslots.

3.3 Parameter Estimation

To estimate parameters, the MLE (Maximum Likelihood Estimation) method is used to maximize the log likelihood of the collected search history for all consumers U that are generated by this model. The log likelihood function is given by:

$$\log p(U;\theta) = \sum_{u} \sum_{<l,t,g>\in S(u)} \{\log \sum_{c} p(c|u)\cdot p(l|c)\cdot p(t|c)\cdot p(g|c)\} \tag{3}$$

where $S(u)$ denotes the search history for consumer u; θ denotes all the parameters in the model including $p(c|u)$, $p(l|c)$, $p(t|c)$ and $p(g|c)$. Since it is difficult to directly optimize the above equation due to the log calculation being out of a summation, the EM (Expectation Maximization) algorithm is employed here to estimate these parameters.

In the **E-step**, the posterior distribution of hidden variable (i.e., context c) is computed, given the observed data and the current values of parameters according to Bayesian rule:

$$p(c|u,l,t,g) = \frac{p(c|u)\cdot p(l|c)\cdot p(t|c)\cdot p(g|c)}{\sum_{c'} p(c'|u)\cdot p(l|c')\cdot p(t|c')\cdot p(g|c')} \tag{4}$$

In the **M-step**, the new optimal values for parameters are obtained given the current settings of hidden variables calculated in E-step. By maximizing the log likelihood function, the parameters can be updated as follows:

$$p(c|u) = \frac{\sum_{<l,t,g>\in S(u)} p(c|u,l,t,g)}{\sum_{c'}\sum_{<l,t,g>\in S(u)} p(c'|u,l,t,g)} \tag{5}$$

$$p(l|c) = \frac{\sum_{u}\sum_{<l,t,g>\in S(u)} p(c|u,l,t,g)}{\sum_{l'}\sum_{u}\sum_{<l,t,g>\in S(u)} p(c|u,l',t,g)} \tag{6}$$

$$p(t|c) = \frac{\sum_{u}\sum_{<l,t,g>\in S(u)} p(c|u,l,t,g)}{\sum_{t'}\sum_{u}\sum_{<l,t,g>\in S(u)} p(c|u,l,t',g)} \tag{7}$$

$$p(g|c) = \frac{\sum_{u}\sum_{<l,t,g>\in S(u)} p(c|u,l,t,g)}{\sum_{g'}\sum_{u}\sum_{<l,t,g>\in S(u)} p(c|u,l,t,g')} \tag{8}$$

Table 1 Data description

# consumers	# categories	# click events	# locations
7,079	7,618	3,680,662	561,259

When implementing the EM algorithm, some intial values are randomly assigned to the parameters, then the iterations between E-step and M-step are conducted until the log likelihood gets converged.

4 Experiments

With the proposed consumer search model (CSM), the context-aware recommendations for mobile consumers can be implemented. In this section, the effectiveness of the CSM are demonstrated on real-world dataset, and the performances are compared with several baseline methods.

4.1 Data Description

The dataset used in the experiments is from a public data set released by Alibaba mobile recommendation challenge[1]. It contains the clickstream data in a month of 10,000 consumers on 8,898 categories which consist of 1,200,000 items. Specially, the collected records were created by smart phones instead of laptops, where location information is available. The given items mainly belong to experience goods, that is, consumers trade online and experience the service offline, such as restaurants service, hotel service and photography service. We sampled a collection of click records which reflect consumers' search behavior from the dataset, and each record was formatted as (*consumerId, categoryId, clickLocation, clickTime*). To be specific, consumers with less than 30 click records in the period were removed and the data description after preprocessing is presented in Table 1.

The dataset was split according to the timestamp of the event, that is, click records occurred in the first 26 days were treated as training set while the remaining ones were treated as test set. Then the CSM model was fitted with the training set and its performance was evaluated on test set.

4.2 Context-Aware Recommendation

Considering the effect of different contextual factors on consumers' search patterns, we can design a context-aware recommendation approach based on the CSM model. The objective of our recommendation is to present the right item category to the right people at right time and right location, hoping that the recommended category will

[1] `http://tianchi.aliyun.com/competition/index.htm`

be clicked by the target consumer. Therefore, given all the parameters inferred from the model, the categories can be ranked according to the probability that consumer u would click the category g at time t and location l. Formally, the key procedure for recommendation is to estimate $p(g|u, t, l)$. Based on the Bayesian rule, it can be calculated as:

$$p(g|u,t,l) = \frac{\sum_c p(c|u) \cdot p(t|c) \cdot p(l|c) \cdot p(g|c)}{\sum_{g'} \sum_c p(c|u) \cdot p(t|c) \cdot p(l|c) \cdot p(g'|c)} \tag{9}$$

In the above formula, the time is divided into 48 timeslots, which corresponds to different hours in weekends and weekdays.

(1) Baseline Methods

Several baseline recommendation methods were selected to compare with the CSM. First, since the click records in the dataset are implicit feedback from consumers, it is not feasible to directly apply collaborative filtering techniques that often use rating data. Thus, we adopted the content-based recommendation method [1] where items that were similar to consumers' click history were recommended. Second, in order to study the effectiveness of different contextual factors (i.e., time and location information), we simplified the CSM by removing the factors one by one and evaluated their recommendation performances. Concretely, CSM_time is the model without time factor and $CSM\_location$ is the model without location factor. Finally, $random$ is also served as a baseline method where item categories are randomly recommended to consumers.

(2) Evaluation Metrics

In the experiments, the click record in the test set were used as the ground truth and Mean Average Precision (MAP) at position K was applied to evaluate the performances of context-aware recommendation methods. The size of recommendation set K ranged from 5 to 15, that is, $K \in \{5, 6, 7, 8, 9, 10, 11, 12, 13, 14, 15\}$. Thus, $MAP@K$ represents the mean value of average precision at top K recommendation results over all click/search records in test set. Formally, $MAP@K = \frac{\sum_{s \in S} AP^{(s)}@K}{|S|}$ where S is the collection of records in test set and $AP^{(s)}@K$ can be computed by:

$$AP^{(s)}@K = \frac{\sum_{r=1}^{K} P(r) \times rel(r)}{N_s} \tag{10}$$

where r is the rank in the sequence of recommended categories, K is the size of recommendation set, $P(r)$ is the precision at cut-off r in the rank list, $rel(r)$ is an indicator function equaling one if the category at rank r is clicked by the target user and zero otherwise, and N_s denotes the real click count in record s which always equals to 1 in our experiment.

4.3 Results and Discussion

(1) Parameter Selection

Since there was a hidden variable (i.e., the latent context c) in the model, we first investigated the different recommendation performances with various values of c and chose the most appropriate parameter in the remaining experiments. Specially, the model with different number of latent contexts were trained where $c \in \{5, 10, 15, 20, 25, 30, 35, 40, 45, 50\}$ and the recommendation performances were evaluated regarding to these numbers. Figure 4 provides an illustration of the MAP at top five recommendations with different parameter settings. It is clear that the performance was best when the number of context was set to 20. Thus, we set $c = 20$ in the remaining experiments.

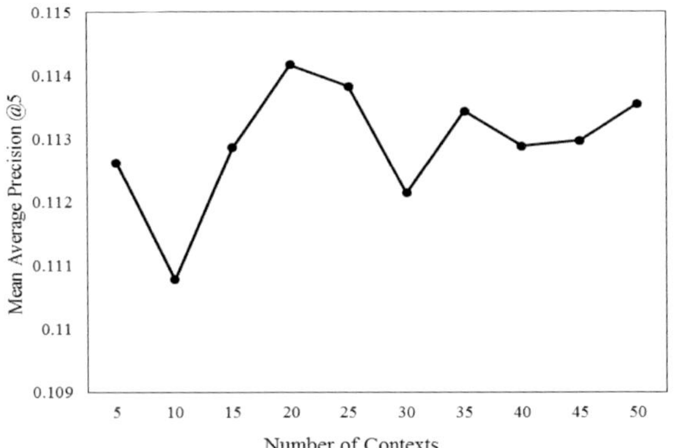

Fig. 4 Recommendation results with different parameter settings

(2) Recommendation Results

Figure 5 shows the comparing results of recommendation performances between CSM and other baseline methods. Overall, CSM outperformed other methods with different number of recommendations, indicating the advantage of proposed method. Moreover, CSM was much more explicable for consumers' search behaviors than the heuristic method (i.e., content-based method). The values of $MAP@K$ for CSM are between 0.1 to 0.2, this is because it's still a very challenging recommendation task without explict feedback, especially when the number of categories is extremely large.

Additionally, from Figure 5(b) and Figure 5(c) we can see that the $MAP@K$ in CSM is higher than that in CSM_time and $CSM\_location$, illustrating the effectiveness of contextual factors when modeling consumers' search patterns. This also verifies the assumption that consumers' search behaviors are time and location dependent. It is worth mentioning that the improvement of CSM over $CSM\_location$ is higher than that of CSM over CSM_time, which demonstrates that location factor is more influential in mobile e-commerce environment.

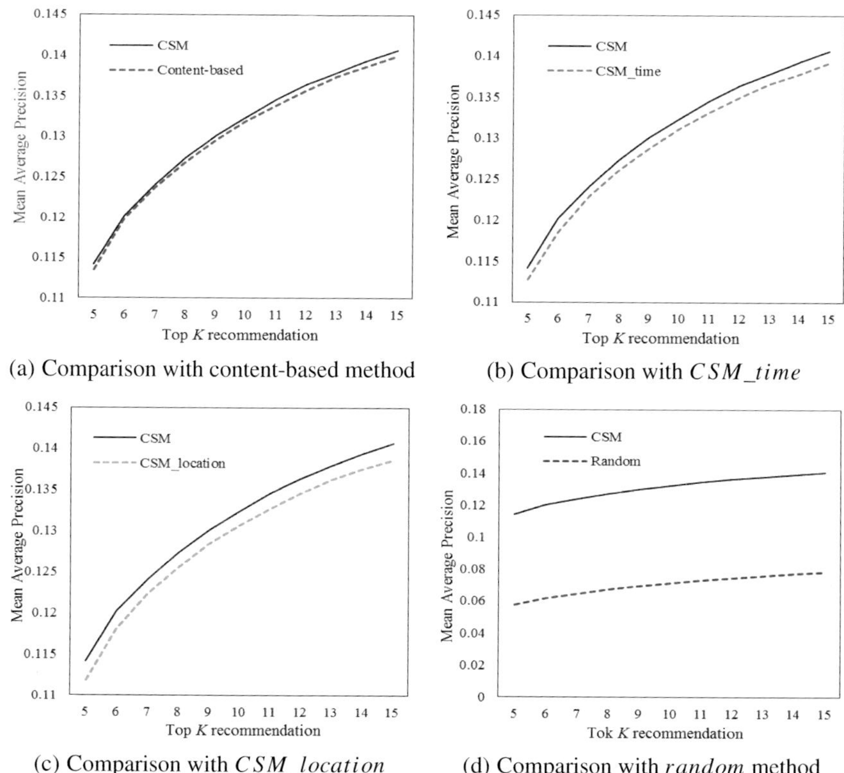

(a) Comparison with content-based method

(b) Comparison with *CSM_time*

(c) Comparison with *CSM_location*

(d) Comparison with *random* method

Fig. 5 Performances for top K recommendations

(3) Discussion

As mentioned above, the timeslot was set to 48 corresponding to different hours in weekdays and weekends. This partition standard is similar to that in [14] where it was applied to the retweeting time in Twitter. Unlike retweeting patterns, consumers' shopping behaviors may not be so sensitive to whether the day is weekday or weekend. Figure 6 shows the average click times in each hour on weekdays and weekends, separately in our experimental dataset. It can be found that there was no apparent difference between weekdays and weekends, which confirms our speculation.

With this observation, we re-trained the CSM with 24 timeslots and the performances of top K recommendations were shown in Table 2. It can be observed that there was some small improvements after adjusting the model, which indicates that the model can be further improved if consumers' search behaviors can be better understood.

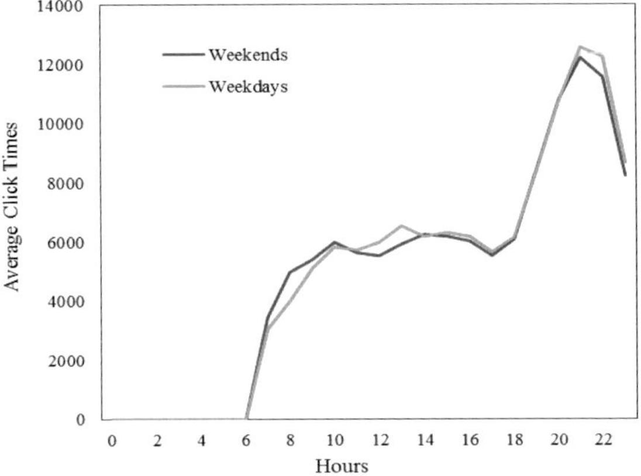

Fig. 6 Click patterns in weekdays and weekends

Table 2 $MAP@K$ for CSM with 24 timeslots

K	$Timeslots = 48$	$Timeslots = 24$	K	$Timeslots = 48$	$Timeslots = 24$
5	0.1142	0.1146	10	0.1324	0.1331
6	0.1202	0.1206	11	0.1346	0.1353
7	0.1241	0.1247	12	0.1364	0.1372
8	0.1274	0.1279	13	0.1379	0.1387
9	0.1301	0.1306	14	0.1393	0.1401

5 Conclusion

Consumer information search has long been recognized as an important procedure before consumers making purchase decisions, which reflects their purchase intentions. In this paper, we have studied the search behaviors of mobile consumers. In doing so, the click times in terms of different timeslots and geographical areas have been analyzed, where the contextual information including time and location has great influence on consumer search behaviors. Then, the problem has been formalized as to model all the consumers' search histories so as to discover underlying patterns, in terms of *when to search*, *where to search* and *in what category*. A probabilistic generative model (i.e., the consumer search model — CSM) has been developed to identify search patterns for its good explanation. In CSM, a latent variable (i.e., *shared search context*) has been introduced to capture the simultaneous effects among different factors. With the inferred parameters of the model, a context-aware recommendation method has been designed where categories are ranked according to the probability that the target consumer would click the category at a particular time and location. Real-world data experiments from Alibaba.com have revealed the advantages of our proposed method over others in related metrics.

In future explorations, the proposed model can be extended by incorporating other contextual factors, such as companions and e-retailers. Another future effort may apply the proposed method to other mobile e-commerce platforms such as the ones in western countries so as to test and find the similarities and differences in consumers' behaviors.

Acknowledgments The work was partly supported by the National Natural Science Foundation of China (grant numbers 71490724/71110107027) and the MOE Project of Key Research Institute of Humanities and Social Sciences at Universities of China (grant number 12JJD630001).

References

1. Adomavicius, G., Tuzhilin, A.: Toward the next generation of recommender systems: A survey of the state-of-the-art and possible extensions. IEEE Transactions on Knowledge and Data Engineering **17**(6), 734–749 (2005)
2. Banerjee, A., Ghosh, J.: Clickstream clustering using weighted longest common subsequences. In: Proc of the Workshop on Web Mining SIAM Conference on Data Mining, pp. 33–40 (2001)
3. Bhatnagar, A., Ghose, S.: Online information search termination patterns across product categories and consumer demographics. Journal of Retailing **80**(3), 221–228 (2004)
4. Branco, F., Sun, M., Villas-Boas, J.M.: Optimal search for product information. Management Science **58**(11), 2037–2056 (2012)
5. Curme, C., Preis, T., Stanley, H.E., Moat, H.S.: Quantifying the semantics of search behavior before stock market moves. Proceedings of the National Academy of Sciences **111**(32), 11600–11605 (2014)
6. Einav, L., Levin, J., Popov, I., Sundaresan, N.: Growth, adoption, and use of mobile E-commerce. American Economic Review **104**(5), 489–494 (2014)
7. Hong, L.: Probabilistic Latent Semantic Analysis. Science, Computer **2**, 1–13 (2012)
8. Huang, P., Lurie, N.H., Mitra, S.: Searching for experience on the web: an empirical examination of consumer behavior for search and experience goods. Journal of Marketing **73**(2), 55–69 (2009)
9. Jabr, W., Zheng, E.: Know Yourself and Know Your Enemy: An Analysis of Firm Recommendations and Consumer Reviews in a Competitive Environment. MIS Quarterly **38**(3), 635–654 (2014)
10. Kamvar, M., Kamvar, M., Baluja, S., Baluja, S.: A large scale study of wireless search behavior: Google mobile search. In: Proceedings of the SIGCHI Conference on Human Factors in Computing Systems, pp. 701–709 (2006)
11. Kim, D., Atluri, V., Bieber, M., Adam, N.: A clickstream-based collaborative filtering personalization model: towards a better performance. In: 6th Annual ACM International Workshop on Web Information and Data Management, pp. 88–95 (2004)
12. Kim, J.B., Albuquerque, P., Bronnenberg, B.J.: Online Demand Under Limited Consumer Search. Marketing Science **29**(6), 1001–1023 (2010)
13. Liu, B., Xiong, H., Papadimitriou, S., Fu, Y., Yao, Z.: A General Geographical Probabilistic Factor Model for Point of Interest Recommendation. IEEE Transactions on Knowledge and Data Engineering **27**(5), 1167–1179 (2015)

14. Liu, G., Fu, Y., Xu, T., Xiong, H., Chen, G.: Discovering temporal retweeting patterns for social media marketing campaigns. In: IEEE International Conference on Data Mining, pp. 905–910 (2014)
15. Lu, Y., Mei, Q., Zhai, C.: Investigating task performance of probabilistic topic models: An empirical study of PLSA and LDA. Information Retrieval **14**(2), 178–203 (2011)
16. Moe, W.W.: Buying, searching, or browsing: Differentiating between online shoppers using in-store navigational clickstream. Journal of Consumer Psychology **13**(1), 29–39 (2003)
17. Moe, W.W.: An empirical two-stage choice model with varying decision rules applied to internet clickstream data. Journal of Marketing Research **43**(4), 680–692 (2006)
18. Montgomery, A.L., Li, S., Srinivasan, K., Liechty, J.C.: Modeling Online Browsing and Path Analysis Using Clickstream Data. Marketing Science **23**(4), 579–595 (2004)
19. Putrevu, S., Ratchford, B.T.: A model of search behavior with an application to grocery shopping. Journal of Retailing **73**(4), 463–486 (1998)
20. Rendle, S., Gantner, Z., Freudenthaler, C., Schmidt-Thieme, L.: Fast context-aware recommendations with factorization machines. In: Proceedings of the 34th International ACM SIGIR Conference on Research and Development in Information, pp. 635–644 (2011)
21. Shabib, N., Krogstie, J.: The use of data mining techniques in location-based recommender system. In: Proceedings of the International Conference on Web Intelligence, Mining and Semantics, pp. 28:1–28:7 (2011)
22. Wen, W.H., Huang, T.Y., Teng, W.G.: Incorporating localized information with browsing history for on-demand search. In: Proceedings of the International Symposium on Consumer Electronics, ISCE, pp. 14–17 (2011)
23. Ye, M., Liu, X., Lee, W.C.: Exploring social influence for recommendation: A probabilistic generative model approach. In: Proceedings of the 35th International ACM SIGIR Conference on Research and Development in Information Retrieval, p. 671 (2012)
24. Zheng, V.W., Cao, B., Zheng, Y., Xie, X., Yang, Q.: Collaborative filtering meets mobile recommendation: a user-centered approach. In: Proceedings of the Twenty-Fourth AAAI Conference on Artificial Intelligence, pp. 236–241 (2010)
25. Zhu, H., Chen, E., Yu, K., Cao, H., Xiong, H., Tian, J.: Mining personal context-aware preferences for mobile users. In: Proceedings - IEEE International Conference on Data Mining, ICDM, pp. 1212–1217 (2012)

Monotonization of User Preferences

M. Kopecky, L. Peska, P. Vojtas and M. Vomlelova

Abstract We consider the problem of user-item recommendation as a multiuser instance ranking learning. A user-item preference is monotonizable if the learning can restrict to monotone models. A preference model is monotone if it is a monotone composition of rankings on domains of explanatory attributes (possibly describing user behavior, item content but also data aggregations). Target preference ordering of users on items is given by a preference indicator (e.g. purchase, rating).

In this paper we focus on learning the (partial) order of vectors of rankings of user-item attribute values. We measure degree of agreement of comparable vectors with ordering given by preference indicators for each user. We are interested in distribution of this degree across users. We provide sets of experiments on user behavior data from an e-shop and on a subset of the semantically enriched Movie Lens 1M data.

Keywords User modeling · Preference learning · Decision support · Recommender systems · Generalized monotonicity constraint · Distribution of quality of learning along users

1 Introduction, Related Work, Data Examples

Main motivation of this paper is the huge and always growing number of users as well as the growing amount of e-shops. Main task is easy, efficient, flexible, human-friendly and intuitive recommender systems personalized to each single user specifically.

Our work can be considered as a work on a multi-user monotone prediction problems, in which the target variable is non-decreasing given an increase of the explanatory variables (for each user separately). In our preference setting if an

---

M. Kopecky · L. Peska · P. Vojtas(✉) · M. Vomlelova
Faculty of Mathematics and Physics, Charles University in Prague,
Malostranske namesti 25, Prague, Czech Republic
e-mail: {kopecky,peska,vojtas}@ksi.mff.cuni.cz, marta@ktiml.mff.cuni.cz

T. Andreasen et al. (eds.), *Flexible Query Answering Systems 2015*,
Advances in Intelligent Systems and Computing 400,
DOI: 10.1007/978-3-319-26154-6_3

object is better in all explanatory attributes (depending on user and giving his/her preference on attribute domains) as another object, then it should not be preferred lower (in the sense of target preference indicator).

A usual approach to monotone prediction considers linear ordering naturally given on attribute domains. In [4] we tried to find an ordering better fitting the task by simple attribute value-transformations. Nevertheless tests in [4] were evaluated on benchmark datasets from the UCI machine learning repository [10] – where the ordering of the domain of the target variable was (in fact) considered as a single-user preference ordering. Noticeable is also the fact that the data table was full – no missing data, no null values.

In this paper we would like to continue in this research contributing to

- multi-user setting and very sparse data
- a new monotone model and metrics for evaluation
- experiments listed by users (not only averages over all users)

We introduce MUP – monotone user preference model. Tests were provided on two sorts of data sets – one private and second coming from [6].

To be more intuitive and human-friendly we consider only preferences expressible as generalized annotation programs rules (GAP) of Kifer-Subrahmanian ([5]) to explain user's preference. Our model is equivalent to Fagin-Lotem-Naor (FLN) data model and optimal threshold algorithm ([2]) – this makes the use of learned model for top-k computation effective. Our model can be considered as instance ranking in the terminology of Furnkranz-Hullermeier ([3]). Last but not least we are motivated by recommender systems, several challenges and our previous research (e.g. [1, 7, 8 and 9]).

In this paper we focus on learning the (partial) order of vectors of rankings of user-item attribute values. We measure degree of agreement of comparable vectors with ordering given by preference indicators for each user. We are interested in distribution of this degree across users.

Of course, not all users behave monotone (see e.g. [1]), nevertheless the task is to increase the number of users which were recommended in a satisfactory way.

In general, our approach can be seen in the framework of multiple objectives decision support, see e.g. [11]. Nevertheless it differs in learning item ranking and a new way of evaluation of agreement of ground truth and computed estimation.

In what follows we describe data for experiments, in Chapter 2 we present our formal model, next we describe building the model and in Chapter 4 results of experiments. We finish with conclusions and future work.

1.1 Example – Travel Agency and Implicit User Behavior

We have collected usage data from one of the major Czech travel agencies. Data were collected from August 2014 to May 2015. Travel agency is a typical e-commerce enterprise, where customers buy products only once in a while (most typically once a year). The site does not force users to register and so we can track unique users only with cookies stored in the browser. User typically either land straight on the intended object via search engine (less interesting case), or browses through several categories, compares objects (possibly on more websites) and eventually buys an object. Table 1 contains full description of used implicit indicators. Note that we aggregate feedback by Count and Sum, we do not use here LogFile of sessions. Note that indicators are stored on user×item bases. Feedback dataset contains approx. 185000 user×item records with 0.07% density of user×item matrix and in average 1.6 visited objects per user (119 000 distinct users and 2300 objects). For the purpose of the experiment, the dataset was then restricted to only those users with at least one purchase (some outliers were removed) and at least 4 visited objects, leaving over 8400 strictly rated pairs from 380 users with 450 purchases. So in our prediction we know that there is a purchase, we would like to distinguish between visited and non-purchased and visited and purchased objects. More general setting of this experiment is out of the scope of this paper.

Table 1 Description of the considered implicit feedbacks for user visiting an object.

	Factor	***Description***
1	*PageView*	Count(*OnLoad()* event on object page)
2	*Mouse*	Count(*OnMouseOver()* events on object page)
3	*Scroll*	Count(OnScroll() events on object page)
4	*Time*	Sum(time spent on object page – in seconds)
5	*Purchase*	1 IFF user bought the item, 0 OTHERWISE

1.2 Example – Movie Recommendation and Explicit Rating

In the other case we got restricted MovieLens 1M collection of movie ratings from the MovieLens[1] site that concerns non-commercial, personalized movie recommendations.

Table 2 Sample of movie flags extracted and aggregated from URI.

	DBPedia movie flag	***MovieID 50***
1	*SOUNDMIX*	*1*
2	*SOUNDEDIT*	*1*
3	*SPIELBERG*	*1*
4	*ORIGINAL*	*0*
5	*VISUAL*	*0*
6	*CAMERON*	*0*
7	*WILLIAMS*	*0*
8	*MAKEUP*	*0*
9	*NOVELS*	*0*

[1] `https://movielens.org/`

Data cover ratings of 3 156 movies from 1 000 users, where the five star rating matrix is sparse. Only 145 419 ratings (4.6%) were available. The goal was to predict and recommend 5 best movies for each of users.

Each user and movie was described by a same set of attributes as in original MovieLens 1M data. Additionally, each movie was labeled by 5 003 additional binary flags taken from database described in [6]. Each flag said whether the movie belongs to certain category or not, see Table 2.

2 Formal Model of MUP - Monotone User Preferences

In following examples we describe our use-case and data. We always assume we have a set of users U, a set of items I and a set of attributes which (or at least some of them) can after appropriate ordering serve as explanation for user's preference.

2.1 Ground Truth $<_r^u$

Ground truth assumes a (very sparse) user-item rating matrix with values from some preference indicator from a linearly ordered set (L_r ,$<_r$). L_r can be {1 = purchased, 0 = not purchased} as in 1.1 or a five star hierarchy as in 1.2.

This matrix can be also viewed as a partial mapping on the set of all rated pairs (denoted as ARP)

$$r: \text{Rated} \subseteq \text{UxI} \rightarrow L_r$$

This induces user ground truth preference ordering $<_r^u$ on the set Rated (u) ={i∈I: (u,i) ∈Rated}, where $i_1 <_r^u i_2$ if $r(u, i_1) <_r r(u, i_2)$.

For organization of our experiments (without repetition of pairs of items) we divide pairs of all rated pairs of items (denoted ARP = Rated(u)^2) to those which are strictly rated (pairs, denoted SRP) and equally rated pairs, formally ARP = SRP ∪ EQ. For a fixed user u this gives $ARP^u{}_r = LT^u{}_r \cup EQ^u{}_r$ to pairs strictly $<_r^u$ ordered by rating and with equal rating.

2.2 MUP Monotone User Preferences $<_e^u$ - Our Estimation to be Computed

Formally, we assume we have attributes A_1,... A_m (explanatory variables) and their domains D_1,... D_m . These can be item properties (as in 1.2) and/or user behavior on item (as in 1.1). We illustrate our approach with user 4741 from 1.2 example depicted in Figure 1.

Assume our attributes are as in Table 2 and each movie gets assigned a nine coordinates 0-1 vector, e.g. g(M44) = [010000000], g(M608) = [000000000], g(M50) = [111000000], g(M904) = [001000000], g(M1918) = [010000000]. This generates coordinate wise ordering $<_e^u$ as depicted on Fig.1.

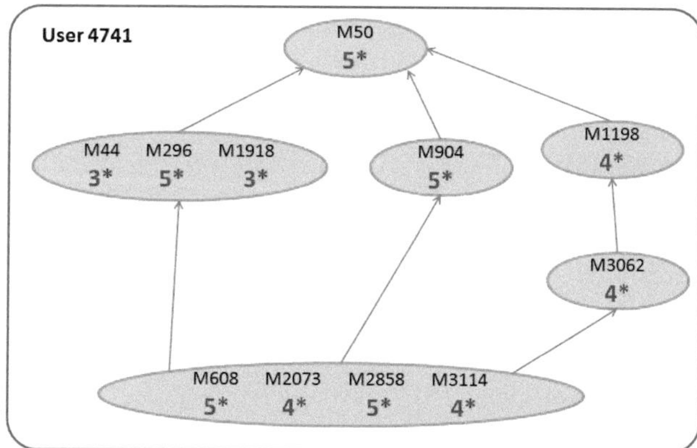

Fig. 1 Agreement of estimated partial ordering of movies of a user with his/her rating

In general, for each user and item we have a vector of attribute values from respective domains

$$(u, i, x_1^{u,i}, \ldots, x_j^{u,i}, \ldots, x_m^{u,i})$$

Our task is to find those attributes and user specific orderings $<_j^u$ on domains of attributes. In practice, we assume they are generated by ranking functions g_j^u: $D_j \rightarrow L_e$. Our estimation of user preference is the ordering $<_e^u$ that user u prefers item i_1 above item i_2 defined by

$$g^u(i_1) = (g_1^u(x_1^{u,i1}), \ldots, g_m^u(x_m^{u,i1})) <_e^u g^u(i_2) = (g_1^u(x_1^{u,i2}), \ldots, g_m^u(x_m^{u,i2}))$$

in the sense of coordinate wise (partial) ordering of vectors. We divide pairs of vectors $(\Pi D_j)^2 = LT^u_e \cup EQ^u_e \cup NC^u_e$ to strictly ordered, equal and non-comparable in the sense of the ordering $<_e$. Note that this ordering is on the whole Cartesian product of attributes domain as we want our estimation has predictive function and can be compared with test set ordering.

In Figure 1, note that pair M1918$<_e^u$ M50 is ordered correctly, pair M608$<_e^u$ M44 is incorrect, pair (M904, M44) is $<_e^u$ non-comparable

2.3 Comparison – Quality of Models

We compare this estimation ordering $<_e^u$ with ground truth ordering $<_r^u$. We consider following approaches and respective notations (used in Figures 2 and 3 with results).

Notation. Our notation has form

MethodExpression.TypeOfComparison.SetRestriction

MethodExpression denotes the estimation method and will be described in modelling part (if empty then it is our vector comparison ordering). In general can take values {empty String (denoting vector coordinate wise comparison, direct, distance, k-NN, bAVG)}. *Set restrictions* will take values APR (all rated pairs), SRP (strictly rated pairs), 50 – users which rated at most 50 movies, a real number describing}. Main part of experiments is measurement of agreement between $<_e^u$ and $<_r^u$ ordering restricted to SetRestricton as the ratio of $<_e^u$ and $<_r^u$ pairs fulfilling *TypeOfComparison*.

TypeOfComparison is an expression of form SEOP_SROP. Here SEOP is a specification of a set of pairs ordered by (our computed) estimation ordering, and SROP in rating ordering resp.

$LT^u_e\_LT^u_r$ denotes the ratio of pairs where $i_1 <_e^u i_2$ iff $i_1 <_r^u i_2$ in LT^u_r (ratio of correct estimations). In a more formal and detailed formula

$$LT^u_e\_LT^u_r\,.\,ARP = \#(\{(i_1, i_2): i_1 <_e^u i_2\} \cap \{(i_1, i_2): i_1 <_r^u i_2\}) / \#ARP$$

And for instance $LT^{4741}_e\_LT^{4741}_r$. ARP = 0,18

$LT^u_e\_GT^u_r$ denotes the ratio of pairs where $i_1 >_e^u i_2$ iff $i_1 <_r^u i_2$ in LT^u_r (ratio of incorrect estimations)

$NC^u_e\_LT^u_r$ denotes the ratio of pairs where i_1 is incomparable with i_2 in $<_e^u$ ordering and $i_1 <_r^u i_2$ (hence our estimation does not give answer).

NC^u_e_All denotes the ratio of pairs where i_1 is $>_e^u$ incomparable to i_2 for all pairs as above (here the estimator is "fair" and cannot order vectors, still final aggregation can give right answer)

EQ^u_e_All denotes the case when L_e vectors of i_1 and i_2 are equal (here the estimator is "cheating" and aggregation cannot correct it, in Fig. 3 we denote this by ?)

$LT^u_e\_EQ^u_r$ denotes the ratio of pairs where $i_1 <_e^u i_2$ iff $i_1 =_r^u i_2$ for all pairs as above (here the estimator tries to decide pairs which were not decided by user). Note that ARP = LT_LT+LT_EQ+LT_GT (we omit subscripts).

Already now, without knowing methods and all restrictions we can observe that:

- Correct estimation cannot be broken by any monotonicity preserving aggregation
- Incorrectness and equality cannot be repaired by any monotonicity preserving aggregation
- Only $<_e$ incomparable objects will be decided by an aggregation (which makes estimated ordering linear).

3 Learning Models

In this section we describe how did we choose explanatory attributes and their ranking to describe user's preference.

Main interest is in the ratio of movies rated where the principle is kept (correct) and the ratio of movies rated where the principle is broken (incorrect).

Note that here we are not dealing with final aggregation which makes our estimated preference ordering linear.

So our aim is to keep agreement (correct estimation) as high as possible and contradiction (incorrect estimation) as small as possible. In our Figures 2 and 3 the desired ideal value is denoted by positioning a ● sign. A smaller portion of incomparable object would be nice but it is not a strict requirement, we denote the desired value by ●.

Note that in higher dimensions the overall ratio of comparable objects vanishes with dimension d as fast as

$$1/2^{d-1}$$

We have to keep this in mind when reading results of our experiments.

3.1 Travel Agency and Implicit User Behavior

The user preference is defined as binary purchase indicator. For other types of feed-back we used several orderings. First its original value in a way "the more the better" (denoted as direct comparison in the experiments in Fig.2). Second, we use two types of average of purchase indicator of other users with similar values of feedback. We distinguish two types of similarity: we either use all entries from a neighborhood defined as ε* feedback value (denoted as distance in the experiments) or we use k-nearest neighbors, where K = ε* percent of total number of records (denoted as KNN in the experiments). All approaches have some (dis)advantages. The direct comparison is the computationally easiest method, however based on a heuristics which might not hold perfectly in the real-world application. Both KNN and Distance overcomes this difficulty by using other user's purchasing behavior to predict it for the current one and expects that similarity is encoded in the feedback value. However behavior patterns of various users might be different too. Furthermore the Distance method might suffer from an empty neighborhood especially in the long-tail, and the KNN might use too broad neighborhood to reach the K-th neighbor in the same situation.

3.2 RuleML Enriched Movie Recommendation

Due to large number of properties in movie dataset of [6], our first guess was to take most frequent properties and check their influence on our prediction task (see Table 3). We saw that these do not influence our achievements on the task significantly. Our second attempt was based on observation that certain types of properties repeat (Table 4).

Nevertheless, this did not help. Last approach was to create set of explanatory attributes of movies and set them to 1, respectively 0 according to appearance of some word or phrase in the movie flag description. So we created attribute Spielberg and set it to 1 for each movie directed, produced or any other way connected to Steven Spielberg, similarly we created attribute LA and assigned it to all movies connected to LA etc. Then we tried to compare model with truth. We obtained results (initial segment) shown in Table 5.

Table 3 Some of most frequent properties, up to last with more than 100 occurrences (out of 5 003 DBPedia properties)

Order number	*Category name*	*Number of movies*
1	English-language_films	2104
2	American_films	1188
3	Directorial_debut_films	371
4	1990s_drama_films	344
…	…	…
51	Miramax_Films_films	103

Table 4 Many properties repeat, forming clusters

Repeated part of property	Number of properties
Films_directed_by_...	995
1757-2032…	443
Films_set_in_...	364
Films_shot_in_...	244
Films_about_...	220

Table 5 Description of importance of the movie attributes we used for estimation.

	Explanatory attribute	*#of movie pairs*	*Sharp Agree*	*Agree*
1	*SOUNDMIX*	1.726.816	0,4947	0,7812
2	*SOUNDEDIT*	1.475.594	0,4887	0,7787
3	*SPIELBERG*	833.012	0,4913	0,7719
4	*ORIGINAL*	1.798.581	0,4751	0,766
5	*VISUAL*	1.636.696	0,4607	0,7498
6	*CAMERON*	516.333	0,4348	0,7276
7	*WILLIAMS*	1.218.578	0,4301	0,7157
8	*MAKEUP*	820.031	0,399	0,696
9	*NOVELS*	3.420.889	0,3814	0,6875

Models based on attributes from Table 5 (and some data mining based on data aggregate values) can be expressed as GAP rules ([5], [9]).

4 Experiments

4.1 Travel Agency Data Experiments

Figure 2 depicts values (LT_LT), incorrect (LT_GT) and incomparable (NC) pairs for *Direct* comparison, Distance and KNN methods with several values of ε. The results are aggregated according to the users. For the space reasons we do not provide results of all experimental settings, only the interesting values of ε.

The results of Direct comparison method were quite surprising. Direct comparison was the best for Time feedback and close to best also for other feedback types. How-ever in aggregation it produces more LT_GT errors instead of more NC errors for *Distance and KNN* methods (with proper settings of ε), which might be an important disadvantage based on the aggregation method. The boxplots on Fig. 2 also illustrates that preferences are well monotonizable not just in average, but also for the majority of the users. However on the other hand, there are always a few "difficult" users for whom the volume of incorrectly ordered pairs reaches 100%.

Quite interesting was also the optimal value of ε. Both Distance and KNN methods requires large portion of data to produce good results (i.e. the best results for KNN is ε=0.5, so the resulting preference is determined according to a half of the dataset). We think that it is caused by low volume of purchases, so a large portion of data is needed to separate signal from a noise.

Table 6 shows how the pairs of objects were evaluated according to each type of feedback. The best results across all experimented methods are shown. One interesting observation is that each type of feedback requires different processing

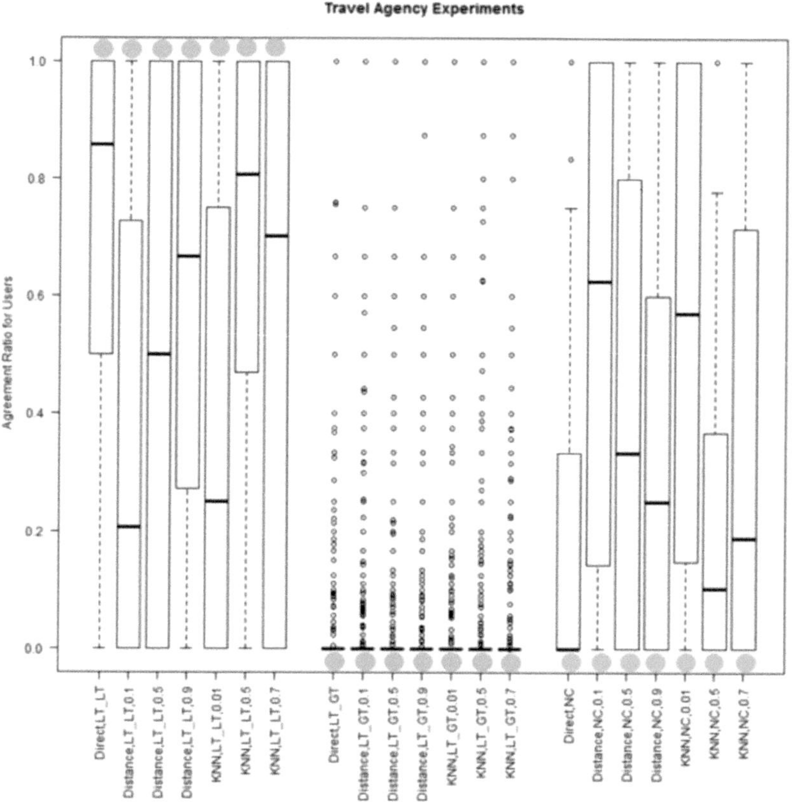

Fig. 2 Experiments on 1.1 data - user-wise distribution of results on sharply rated object with equality modified by a multiplicative constant ranging from 0 to 0.5.

in order to produce optimal results. Thus we cannot conclude with one “best for all” method for processing user feedback. We can conclude that from the examined types of feed-backs, Mouse and Page View indicators have the best ordering capability for the majority of experimental settings. The Scrolling event was not triggered so often as others (only 1/3 of pairs has some non-zero scrolling feedback) and thus ends up as equal often. Time on page has in general higher rate of incorrectly ordered pairs (for some settings of ε was the number of incorrectly ordered pairs even higher than the correct ones).

Table 6 Experiments on Travel agency dataset. SRP evaluated according to each type of feedback. The results of the best method for each feedback are displayed.

Feedback	Total	Correct	Incorrect	Equal
Time (*Direct comparison*)	8405	6418	1528	459
Scroll (*KNN*, ε=0.01)	8405	2035	648	5722
Mouse (*KNN*, ε=0.7)	8405	6088	663	1654
PageView (*KNN*, ε=0.7)	8405	6182	707	1516

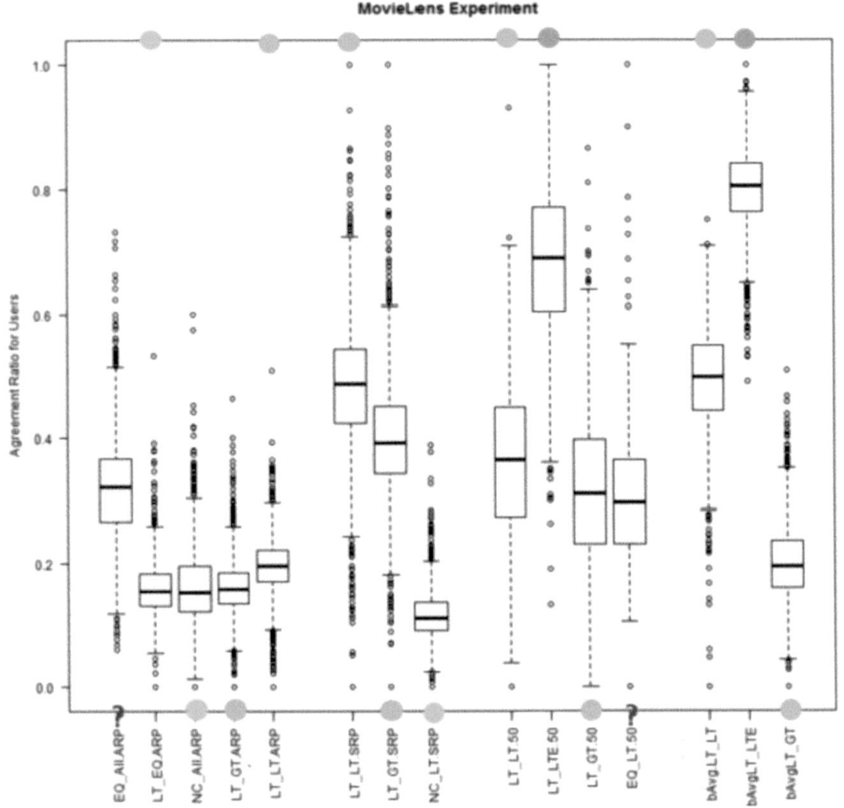

Fig. 3 Distribution of values on 1.2 data experiments, in notation from 2.3.

4.2 *Movie Rating Data Experiments*

User distribution on movie data from 3.2 gives a totally different picture – results are much worse than for 3.1 (probably because the dimension of data is much higher).

- A group of ARP experiments (5 plots) and SRP experiments (3 plots)
- A group of four plots on user which rated at most 50 movies
- A group of experiments which uses machine learning (biased movie average) – only 3 plots because AVG makes $<_e$ a linear order. Note that ideal value where an overlap with counting item pairs with previous boxplot can appear is denoted with dark green bullet ●.

Surprisingly best are results with biased movie average. Although the predicate AVG (in fact linear ordering) is defined in the second order predicate calculus, we can adopt data as facts to our rule system (we use the equivalence of GAP and monotone model mentioned in the beginning).

Users which rated at most 50 movies are harder to predict than in global average. This is a surprise, because in data there are users who rated more than 1000 movies – so a question arose whether these data are a good description of user preferences.

Experiments were more demanding on computer power, as we have 27,428,634 rated pairs in total for 1000 users (in comparison to 8400 in 3.1 experiments).

There are 10,921,798 strictly ordered pairs. This can be viewed as in a 5 star $<_r$ there 10 possible ordered pairs (whereas in purchase yes/no there is only one). Overall results give LT_LT/SRP = 0,482; GT_LT/SRP = 0,401; NC_LT/SRP = 0,116. It seems that results with SRP are better than with ARP. It is a question, in a practical situation, what has to be avoided – whether only conflicts with sharp claims or all. Still SRP results for 3.2 data are much worse in conflict LT_GT than for 1.1 data.

5 Discussion, Comparison, Conclusion and Future Work

In this paper we proposed methods for approximation of user rating with a (partial) order of vector of explanatory variables with domains ordered in monotone (positive) way. We provided experiments

- On two data sets with orthogonal characteristics
- In a multi-user setting, very sparse data and computationally demanding
- We introduced a new monotone model and several metrics for evaluation
- Experiments provide not only average values but are listed by users and their agreement of computed estimation and user's rating – this show that some users are "difficult" but most of them behave better when concerning monotonization

In general experiments with only strictly rated pairs and smaller dimension give better results (in our case travel agency data).

Experiments with all rated pairs and data with higher dimension give significantly worse results (in our case semantically enriched MovieLens data). Although a simple data mining approach gives here best results.

For future work we left cases when monotonization does not work. Using 2CP-regression from [1] would be a first try. We will also try to estimate "difficult users" in order to be able automatically switch the choice of estimation method. We plan also to make more extensive experiments on other sorts of movie rating data collected in a business process.

Acknowledgement The work was supported by the Czech grants SVV-2015-260222, GAUK-126313, GACR-P103-15-19877S and P46.

References

1. Eckhardt, A., Vojtáš, P.: Learning user preferences for 2CP-regression for a recommender system. In: van Leeuwen, J., Muscholl, A., Peleg, D., Pokorný, J., Rumpe, B. (eds.) SOFSEM 2010. LNCS, vol. 5901, pp. 346–357. Springer, Heidelberg (2010)
2. Fagin, R., Lotem, A., Naor, M.: Optimal aggregation algorithms for middleware. Journal of Computer and System Sciences **66**, 614–656 (2003)
3. Fürnkranz, J., Hüllermeier, E.: Preference learning: An introduction. In: Preference Learning, pp. 1–17. Springer (2011)
4. Horvath, T., Eckhardt, A., Buza, K., Vojtas, P., Schmidt-Thieme, L.: Value-transformation for monotone prediction by approximating fuzzy membership functions. In: 2011 IEEE 12th CINTI, IEEE 2011, pp. 367–372 (2012)
5. Kifer, M., Subrahmanian, V.S.: Theory of generalized annotated logic programming and its ap-plications. Journal of Logic Programming **12**(4), 335–367 (1992)
6. Kuchar, J.: Augmenting a Feature Set of Movies Using Linked Open Data. CEUR Proceedings Vol-1417 Rule Challenge and Doctoral Consortium @ RuleML 2015, Track 2: Rule-based Recommender Systems for the Web of Data. http://ceur-ws.org/Vol-1417/paper16.pdf Dataset. http://nbviewer.ipython.org/urls/s3-eu-west-1.amazonawscom/recsysrules2015/RecSysRules2015-Dataset.ipynb
7. Peska, L., Vojtas, P.: Modelling user preferences from implicit preference indicators via compensational aggregations. In: Hepp, M., Hoffner, Y. (eds.) EC-Web 2014. LNBIP, vol. 188, pp. 138–145. Springer, Heidelberg (2014)
8. Vojtas, P., Peska, L.: e-Shop user preferences via user behavior. In: Obaidat, M.S., et al (eds.) ICE-B 2014, pp. 68–75. SciTePress (2014)
9. Vomlelova, M., Kopecky, M., Vojtas, P.: Transformation and aggregation preprocessing for top-k recommendation GAP rules induction. CEUR Proceedings Vol-1417 Rule Challenge and Doctoral Consortium @ RuleML 2015, Track 2: Rule-based Recommender Systems for the Web of Data. http://ceur-ws.org/Vol-1417/paper18.pdf
10. Asuncion, A., Newman, D.J.: UCI Machine Learning Repository, Irvine, CA (2007). http://www.ics.uci.edu/~mlearn/MLRepository.html
11. Keeney, R.L., Raiffa, H.: Decisions with multiple objectives–preferences and value tradeoffs, 569 p. Cambridge University Press, Cambridge (1993)

The Role of Computational Intelligence in Temporal Information Retrieval: A Survey of Imperfect Time in Information Systems

Christophe Billiet and Guy De Tré

Abstract In existing literature, many proposals have concerned the modeling or handling of imperfection in time in information systems. However, although reviews, surveys and overviews about either imperfection or time in information systems exist, no reviews, surveys or overviews about imperfection in time in information systems seem to exist. The main contribution of the work presented in this paper is to attempt to fill this void by presenting a survey of some existing scientific contributions dealing with time in information systems. A more modest contribution is an attempt at identifying some open research challenges or opportunities concerning imperfection in time in information systems.

1 Introduction

Generally, an information system (IS) may have two main purposes. One is to preserve (and possibly guard) data, information or knowledge and the other is to allow (some) users to retrieve its contents or data, information or knowledge derived from its contents. For example, an IS containing medical patient history data is typcally used to preserve information about former afflictions and to retrieve patient condition information.

To both purposes, imperfection is an important issue. Indeed, it is the experience of the authors that, in many cases, the contents of IS are subject to imperfections such as uncertainties [26, 32, 46, 61], imprecisions [26, 61], vaguenesses [61], ... and that users want to retrieve imperfect data, information or knowledge or use flexible or even imperfect ways of specifying their retrieval preferences [26, 53]. Moreover, it is the experience of the authors that, in many cases, IS may improve their quality of service significantly by correctly and comprehensively modeling and handling

C. Billiet(✉) · G. De Tré
Department of Telecommunications and Information Processing, Ghent University, Sint-Pietersnieuwstraat 41, 9000 Ghent, Belgium
e-mail: {Christophe.Billiet,Guy.DeTre}@UGent.be

T. Andreasen et al. (eds.), *Flexible Query Answering Systems 2015*,
Advances in Intelligent Systems and Computing 400,
DOI: 10.1007/978-3-319-26154-6_4

these imperfections. Hence, many proposals in existing literature have concerned the modeling and handling of imperfections in IS. Moreover, many surveys or reviews concerning imperfections in IS exist. Some present an overview of research [27, 88], while others try to identify open challenges and opportunities [7, 26, 61]. Among many others, these reviews or surveys focus on imperfections in IS in general [26, 61], imperfections in database systems [27], in expert systems [88] or in data mining [7].

Additionally, it is the experience of the authors that the contents of many IS are related to time and that useful ways of retrieving data, information or knowledge from such IS require the possibility to express time-related preferences. Hence, many proposals in existing literature have concerned the modeling and handling of time in IS. Moreover, many surveys or reviews concerning time in IS exist. Some bundle existing work [74, 86], some present an overview of research [23, 66, 84] and others try to identify open challenges and opportunities [5]. Among many others, these reviews or surveys focus on time in database systems [74, 86], in artificial intelligence [66, 84], in information retrieval [5] or in information processing [23].

Surprisingly, to the knowledge of the authors, no reviews, surveys or overviews on the topic of imperfection in time in IS exist. Therefore, the main contribution of the work presented in this paper is to attempt to fill this void by presenting a survey of a collection of existing scientific proposals dealing with imperfection in time in IS. Due to page restrictions, both the extent of this collection and the extents of the analyses of the works in it are rather limited. Moreover, the focus of the work presented in this paper is not so much on natural language understanding or processing, but more limited to dealing with imperfections. However, the authors of this paper have tried to include those works they believe to be more or less influential or groundbreaking in this collection and they have tried to include those aspects into the works' analyses that may answer both questions a reader with rather practical needs and a reader with a more academic point of view might have. A more modest contribution of the work presented in this paper is an attempt to identify some open challenges or opportunities concerning the research on imperfection in time in IS.

This paper is structured as follows. In Section 2, some preliminary concepts and terms about time and imperfection are presented. In Section 3, some specificalities about the research presented in this paper are explained and in Section 4 this research is presented. Finally, in Section 5, some conclusions are drawn.

2 Preliminary Concepts and Terminology

2.1 Basic Time Units

Time as Perceived by Information Systems. Usually, time is perceived by IS as linear, flowing at a constant pace and in one direction (from the past to the future). Therefore, in the context of IS, time is often thought of as following an axis (a *time axis*), where the points on this axis are infinitesimally short 'moments' in time.

Thus, time is often seen as a totally ordered set of such points (the time axis), which are called *instants*.

Definition 1. *Instant [22, 45]. An* instant *is a time point on an underlying time axis.*

For example, the point on an unspecified time axis described by the words '... December 6th at 10h exactly ...' is an instant.

An interval subset of a time axis can be defined using two of its instants. Such interval subset is called a *time interval*.

Definition 2. *Time interval [22, 45]. In the presented work, a subset of a time axis, containing all instants of this time axis between two given instants of this time axis, is called a* time interval.

For example, the subset of an unspecified time axis described by the words '... from December 6th at 10h exactly until December 6th at noon exactly ...' is a time interval, bounded by the instants described by the words '... December 6th at 10h exactly ...' and '... December 6th at noon exactly ...'.

Definition 3. *Duration [22, 45]. A* duration *is an amount of time with known length, but no specific starting or ending instants.*

A time interval is bounded by two instants, whereas a duration is not. For example, the amount of time described by the words 'a month' is a duration: its length is known, but its bounding instants are not.

Time as Modeled by Information Systems. IS may model time intervals and instants following a *time model*. As IS are usually finitely precise and instants have an infinitesimally short duration, such time models usually model an underlying time axis using *chronons*.

Definition 4. *Chronon [22, 45]. In a data model, a* chronon *is a non-decomposable time interval of some fixed, minimal duration.*

If using a time model, an IS usually introduces a datatype to represent instants and time intervals. Such a datatype is called a *time datatype*. Generally, a time datatype has a time domain at its disposal, which is a set of values allowed by the datatype. For example, such a datatype could use $\mathbb{Z}$ as time domain. A time domain used by an IS is usually related to a sequence of consecutive chronons. Every element of the time domain then corresponds to exactly one chronon, with the ordering of the consecutive time domain elements reflecting the temporal ordering of these consecutive chronons. These chronons are the smallest time intervals an IS using the time domain can distinguish, which is generally why they are used [22, 45]. An instant is usually modeled as a single *time domain element* (which is an element in the used time domain) corresponding to the single chronon containing the instant somehow. A time interval can be modeled as a *time domain interval* (which is an interval in the used time domain) corresponding to the set of consecutive chronons together containing all instants of the time interval.

Definition 5. *Time domain element.* A time domain element *is an element of a time domain.*

Definition 6. *Time domain interval.* A time domain interval *is an interval subset of a time domain.*

Time models can be categorized based on how they perceive and thus model time, among others. One of these categorizations partitions time models in *discrete models*, *dense models* and *continuous models* [22, 45]. In a discrete model, every time domain element has a single successor. In a dense model, there is always a time domain element between every two given time domain elements. In a continuous model, the time domain is a continuous set.
For example, the set $\mathbb{Z}$ of integers could be used as the time domain of a discrete time model, the set $\mathbb{Q}$ of rational numbers could be used as the time domain of a dense time model and the set $\mathbb{R}$ of real numbers could be used as the time domain of a continuous time model.

In general, IS may or may not use a time model. Thus, they may represent and reason with instants, time intervals and durations an sich or they may represent instants, time intervals and durations using time domain elements, time domain intervals and amounts of time domain elements and reason with those. In this paper, the concepts any examined work uses as fundamental units to consider or reason with, will be called 'basic time units'.

2.2 *Temporal Relationships*

Human interaction with time in IS often considers or makes use of so-called *temporal relationships.* Temporal relationships are relationships between instants or time intervals or between time domain elements or time domain intervals, where these relationships have a clear temporal nature.

In the context of IS, in recent literature, most attention has gone to *qualitative* temporal relationships. Qualitative temporal relationships usually only rely on the ordering of the underlying time axis and don't consider any metric. For example, a qualitative temporal relationship may express that a time interval lies before another, but not what duration both are apart. Early proposals considering qualitative temporal relationships attempted to define sets of qualitative temporal relationships in which instants/time intervals or time domain elements/intervals could be. One of the most groundbreaking contributions in this area is the framework proposed by Allen [2, 3, 4, 49]. Following this approach, two arbitrary time intervals or two arbitrary time domain intervals are always in one of the thirteen mutually exclusive qualitative temporal relationships which Allen proposed as part of this framework.

Other proposals consider *quantitative* temporal relationships. Quantitative temporal relationships usually both rely on the ordering of the underlying time axis and take into account a metric (usually, a metric related to the underlying time axis). For example, a quantitative temporal relationship may both express that a time interval lies before another and give a quantification of the duration both are apart.

2.3 *Types of Imperfection*

Although the work presented in this paper didn't exclude any types of imperfection, the most prevalent proved to be uncertainty, imprecision and vagueness.

When a datum like a basic time unit or temporal relationship is subject to uncertainty, this means it is known that one otherwise perfect datum is intended, but it is unkown which one this is. In some cases, this uncertainty is caused by a (partial) lack of knowledge about the factors influencing the intended datum. In this case, existing scientific literature suggests the usage of possibility theory [43, 87] to model confidence about the intended datum. In other cases, this uncertainty is caused by variability in the candidates for the intended datum. In this case, existing scientific literature suggests the usage of probability theory [43, 56] to model such confidence. Other causes exist, but seem to be less prevalent.

When a datum is subject to imprecision, this may be interpreted in one of two ways [44]. On one hand, this imprecision may be seen as a form of uncertainty caused by the fact that the datum is not described with a sufficient level of precision [21, 67]. On the other hand, a datum may be seen as imprecise if it is impossible to decide to which category or collection it belongs, due to its clear gradual nature [13, 44]. Usually, vagueness is considered to be (very similar to) imprecision described using linguistic terms [37, 53].

In the next section, some aspects of the research presented in this paper are discussed.

3 Research Methodology

In what follows, the term 'a work' will be used to refer to one of the scientific contributions which are the subjects of the study presented in this paper. Moreover, the terminology as presented in Section 2 will always be used.

The authors of this paper have chosen over 70 of the scientific contributions which deal with uncertainty, imprecision and/or vagueness in time in IS and which they deem somehow influential or groundbreaking. The amount of covered works is limited due to page restrictions and the selection is rather oriented towards fuzzy logic due to the authors' interests, which doesn't mean other works cannot be considered influential or groundbreaking. In what follows, the chosen works are categorized and discussed based on the following of their properties. The highest-level categorization is made based on whether the works consider imperfection to be located in basic time units, in temporal relationships or in both. A lower-level categorization is made based on the nature of the basic time units chosen by the works. For every group of works considered similar by these categorizations, the type(s) of time domains and the type(s) of temporal relationships are presented and the nature of the considered imperfection is discussed. More properties were studied, but page restrictions inhibit their mentioning.

4 Imperfection, Time Models and Temporal Relationships

In tables 1, 2a and 2b, an overview of the amounts of works in each category is shown.

Table 1 Amounts of papers considering basic time units and imperfection locations

location of imperfection	basic time units: instant	time interval	duration	instant and time interval	instant and duration	time interval and duration	time domain element	time domain interval	time domain element and interval	time domain element and duration	unclear	total
basic time units	0	16	3	6	0	1	5	1	9	1	6	48
relationships	0	5	0	0	0	0	0	0	0	0	0	5
basic time units and relationships	1	4	0	3	3	0	4	1	1	0	0	17
unclear	0	1	0	1	0	0	0	0	0	0	2	4
total	1	26	3	10	3	1	9	2	10	1	8	74

Table 2 Amounts of papers considering time models and temporal relationships

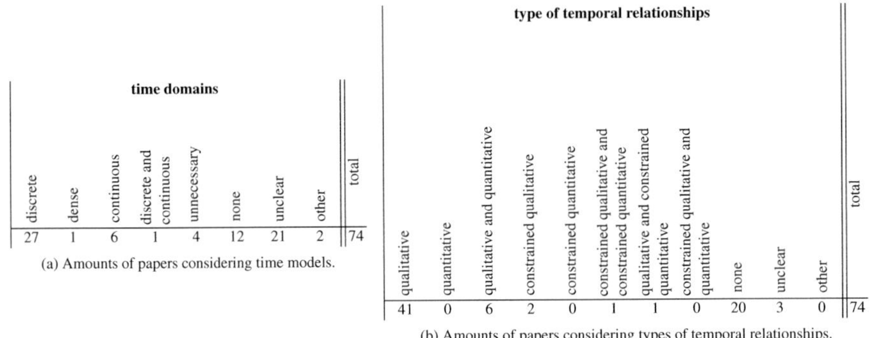

time domains: discrete	dense	continuous	discrete and continuous	unnecessary	none	unclear	other	total
27	1	6	1	4	12	21	2	74

(a) Amounts of papers considering time models.

type of temporal relationships: qualitative	quantitative	qualitative and quantitative	constrained qualitative	constrained quantitative	constrained qualitative and constrained quantitative	qualitative and constrained quantitative	constrained qualitative and quantitative	none	unclear	other	total
41	0	6	2	0	1	1	0	20	3	0	74

(b) Amounts of papers considering types of temporal relationships.

4.1 *Imperfection in Basic Time Units*

48 works consider imperfection only in basic time units. These works are discussed below.

For 16 Works, the Basic Time Units Are Time Intervals. Some works only consider a *single* type of imperfection. Works [16, 17, 18, 29, 68, 70] consider situations in which time intervals are subject to uncertainty caused by a (partial) lack of knowledge. Hence, possibility theory is used to model confidence in the context of this uncertainty. As usually done, fuzzy set theory is used to define the possibility distributions involved. All works seem to consider the uncertainty about the time intervals to be located in the interval boundaries and consider the uncertainty about the time intervals to be a consequence of this. In [1, 8, 47, 72], time intervals subject to uncertainty, where the uncertainty is localised in the interval boundaries, are considered, but the source of this uncertainty, and hence the theory used to model confidence in the context of this uncertainty, is not made explicit. In [1, 8, 72] rough set theory is used to extend the concept of time intervals to account for this uncertainty. In [47], no such theory is used. Of these works, only [1] considers no temporal relationships, all others consider qualitative temporal relationships.

Other works simultaneously consider several *different* types of imperfections. In [13], both time intervals subject to uncertainty and time intervals subject to imprecision or vagueness are considered. As this uncertainty is assumed to be caused by a (partial) lack of knowledge, possibility theory is used to model confidence in this context. Both this confidence and the gradual nature of data or information, which is seen as the effect of imprecision or vagueness, are supported through the usage of fuzzy sets. While in [15], both uncertainty with an unspecified source and imprecision are considered and in [73] both uncertainty with an unspecified source and vagueness are considered, in [62] uncertainty with an unspecified source, imprecision and vagueness are considered. Both [15] and [62] use fuzzy set theory to support modeling confidence in the context of uncertainty and the gradual nature of data or information inherent to imprecision and vagueness, while in [73], rough set theory is used to handle uncertainty and fuzzy set theory is used to support vagueness. Of these works, only [13] considers no temporal relationships, whereas all others consider qualitative relationships.

Finally, in both [34] and [71], the considered imperfection is not (always) specified. However, in [34], fuzzy set theory is used to support this unspecified type of imperfection, while in [71], both rough set theory and fuzzy set theory are used to the same end. Both of these works consider qualitative temporal relationships.

For 4 Works, the Basic Time Units Are Durations. In [38] and [51], durations are assumed to be subject to uncertainty caused by a (partial) lack of knowledge. Therefore, confidence about a duration is modeled using possibility theory, supported by fuzzy set theory. In [57], durations are assumed to be subject to uncertainty with an unspecified source. Confidence in the context of this uncertainty seems not to be modeled, but the cardinality of the set of candidates to be the intended duration is considered an important indication. In [52], IS users are allowed to describe vagueness in a duration through a conjunctively-interpreted fuzzy set over existing durations. The membership degrees of this set reflect to what degree these users consider the existing durations similar. As these works consider their basic time units to be durations, they obviously don't consider temporal relationships.

For 6 Works, the Basic Time Units Are Both Time Intervals and Instants. These works only consider a *single* type of imperfection. In [19, 32, 50, 69], both instants and time intervals are explicitely considered to be subject to uncertainty caused by a (partial) lack of knowledge, and confidence in the context of uncertainty is modeled using possibility theory, supported by fuzzy set theory. In [83], both instants and time intervals are explicitely considered to be subject to uncertainty. However, the source of this uncertainty is not made explicit, but probability theory is used to model confidence in this context. In [33], fuzzy set theory is used to support an unspecified type of imperfection. Of these works, only [33, 83] consider no temporal relationships, while all others consider qualitative relationships.

For 1 Work, the Basic Time Units Are Both Time Intervals and Durations. In [59], both time intervals and durations are assumed to be subject to uncertainty with an unspecified source, for which no modeling of confidence is mentioned, although an intuitive form of rough set theory seems to be used to support the imperfection. No temporal relationships are considered.

Of the works referred to up to this point, [32, 33, 34, 38, 50, 52, 57] mention a discrete time model, [62, 73] mention a continuous time model, [51] mentions that the set of basic time units is chosen to be $\mathbb{Q}$ and all other works don't (explicitly) introduce a time model. None of these use a time model to the extent that time domain elements, intervals or amounts become basic time units.

The authors of this paper find it remarkable that none of the examined works consider imperfection in both instants and durations. Indeed, theoretically, time intervals could be derived using instants and durations. This could lead to another cause for imperfection in time intervals.

For 5 Works, the Basic Time Units Are Time Domain Elements. As could be expected from works that explicitely consider a time model, many of these works concern contributions with a stronger applied character or which fit in a rather applied context. In [35, 36, 89], time domain elements are considered to be subject to uncertainty caused by variability. Hence, confidence in this context is modeled using probability theory. In [6], time domain elements are also considered to be subject to uncertainty, but the cause of the uncertainty is not made explicit. Accordingly, confidence in the context of this uncertainty is not modeled, but a time domain element subject to uncertainty is replaced by a disjunction of time domain elements that might be intended. In [85], time dependency in standard binary logic is considered, where basic logical statements are allowed to be expressed through fuzzy sets, without explicit interpretation. Of these works, only in [89], a very constrained form of temporal relationships are considered. All of these works employ discrete time models.

For 1 Work, the Basic Time Units Are Time Domain Intervals. In [14], time domain intervals subject to uncertainty caused by a (partial) lack of knowledge are considered. As usual, confidence is modeled using possibility theory supported by fuzzy set theory. Qualitative temporal relationships are considered and a discrete time model is employed.

For 9 Works, the Basic Time Units Are Both Time Domain Elements and Time Domain Intervals. In [30, 46, 63, 75, 76, 82], time domain elements and intervals subject to uncertainty caused by variability are considered and confidence in the context of this uncertainty is modeled using probability theory. In [31], the basic time units are considered to be subject to vagueness, where this vagueness is interpreted as uncertainty caused by an insufficient level of precision in the descriptions of the respective basic time units. Fuzzy sets are used to support the modeling of this vagueness. In [55], time domain elements and intervals are considered to be subject to uncertainty of which the source doesn't seem to be clearly specified. Probability theory seems to be used to model confidence in this context, although fuzzy set theory is used to support this modeling. In [58], time domain elements and intervals are clearly subject to uncertainty, but no modeling of confidence in this context seems to be made explicit. In [46, 75, 76], qualitative temporal relationships are considered, while none of the other works seem to consider temporal relationships. In [58], a dense time model is employed, in [82], a continuous time model is employed and in [31], no time model seems to be employed. In all other works, discrete time models are employed.

Overviewing the works mentioned above, the authors of this paper find it remarkable that almost all of these works consider uncertainty about intervals to be an effect of uncertainty about their boundaries whereas almost none of them consider uncertainty about an interval as uncertainty about which interval is intended, given a universe of candidate intervals. Moreover, none of the covered works elaborate on the possible relationships between intervals subject to uncertainty in one of these two approaches with respect to intervals subject to uncertainty in the other. Only some very short suggestions in this context can be found in [16, 17, 19, 32, 70], in which uncertainty is always assumed to be caused by a (partial) lack of knowledge.

The authors of this paper find it remarkable that rather few works specifically consider their basic time units to be time domain elements, intervals or amounts. Indeed, the choice of such basic time units relates more closely to applications, which would suggest a higher popularity. Moreover, the authors believe transforming techniques or approaches based on instants, time intervals or durations to similar techniques or approaches based on time domain elements, intervals or amounts is not always evident.

Other Works. For 6 works, the used basic time units seem not to be clearly specified.

4.2 Imperfection in Temporal Relationships

Only 5 works consider imperfection located only in temporal relationships. In [20], a discrete time model is considered, in all others no time model is considered. In [60] and [78], uncertainty about the intended qualitative temporal relationship(s) between two intervals with clearly temporal interpretations, is considered. It seems not to be mentioned whether these intervals are time intervals or time domain intervals. Although both works contain very few information about the source of the uncertainty, in [60], possibility theory supported by fuzzy set theory is used to model confidence in the context of this uncertainty, whereas in [78], probability theory is used to model this confidence. In [9], an attempt is made to extend Allen's interval algebra [2] to allow soft constraints, which may be satisfied to a degree, using the graduality inherent to fuzzy sets, and to allow prioritizing between constraints, using possibility theory. In [10], a similar attempt allows soft constraints, using the graduality inherent to fuzzy sets, in Allen's interval algebra but additionally allows the membership degrees in these fuzzy sets to be interpreted as possibility degrees, as to support confidence in the context of uncertainty caused by a (partial) lack of knowledge. Finally, in [20], temporal relationships subject to imperfection are considered, but the exact type of imperfection is not made clear.

4.3 Imperfection in Both Basic Time Units and Temporal Relationships

17 works consider imperfection located in both basic time units and temporal relationships. These works are discussed below.

For 1 Work, the Basic Time Units Are Instants. In [24], a clear distinguishment is made between instants subject to uncertainty caused by a (partial) lack of knowledge, for which confidence is modeled using possibility theory, and vagueness used to allow a flexible specification of user preferences, both supported by fuzzy set theory. The imperfection in the qualitative temporal relationships is seen as an effect of the imperfections in the basic time units. No time model is considered.

For 4 Works, the Basic Time Units Are Time Intervals. In [79, 80, 81], continuous time models are considered and time intervals are considered to be subject to vagueness and/or imprecision which is interpreted as them having a gradual nature. Fuzzy sets on the time axis, with a conjunctive interpretation, are used to model this imperfection. Remarkably, in these works, both qualitative and quantitative relationships are considered. In [80] and [81], temporal relationships are explicitely assumed or allowed to be subject to vagueness or imprecision, as to highten their expressive power by allowing them to have a gradual nature. In [79], temporal relationships are subject to vagueness or imprecision as a consequence of the vagueness or imprecision in time intervals. In [11], time intervals subject to uncertainty seem to be considered, but no theory for modeling confidence in this context is specified. A constricted form of rough sets seems to be used to support this uncertainty. The imperfection in the qualitative temporal relationships seems to be an effect of the imperfection in the basic time units. No time model is specified.

For 3 Works, the Basic Time Units Are Both Instants and Time Intervals. In [41, 77], qualitative temporal relationships are considered to be subject to uncertainty, which is an effect of uncertainty in the basic time units. Whereas in [41] this uncertainty is caused by a (partial) lack of knowledge and confidence is modeled using possibility theory supported by fuzzy set theory, in [77] this uncertainty is caused by variability and confidence is modeled using probability theory. Finally, in [12], although not explicitly specified, uncertainty seems to be assumed in the basic time units, resulting in uncertainty in the qualitative temporal relationships. Confidence doesn't seem to be modeled, but the uncertainty seems to be covered by including all existing options. In none of these works, a time model is specified.

For 3 Works, the Basic Time Units Are Both Instants and Durations. In [64] and [65], no time model is specified and instants are considered as basic time units, but the durations between them are subject to uncertainty caused by a (partial) lack of knowledge and confidence is modeled using possibility theory, supported by fuzzy set theory. Imperfection in the qualitative temporal relationships is not explicitly specified, but inherently assumed to be the consequence of the imperfection in the basic time units. In [54], instants and durations are assumed to be subject to a form of imperfection not clearly identified. A discrete time model is considered and constrained forms of both qualitative and quantitative relationships are considered.

For 4 Works, the Basic Time Units Are Time Domain Elements. In [28, 42], time domain elements are considered to be subject to uncertainty caused by a (partial) lack of knowledge and confidence is modeled using possibility theory, supported by fuzzy set theory. The temporal relationships between these elements are subject to the same uncertainty as a result of this. However, whereas in [42] only qualitative relationships are considered, in [28] attention is given to the quantitative nature of temporal relationships too. In [40], elementary temporal relationships between time domain elements are assumed to be subject to uncertainty caused by a (partial) lack of knowledge and confidence is modeled using possibility theory. Although this is not explicitly specified, the authors of this paper assume the time domain elements themselves to be subject to the same uncertainty (indeed, if two time domain elements are perfectly known, there could be no uncertainty about the temporal relationship between them). In [25], imperfection in the time domain elements is supported using fuzzy set theory. Qualitative temporal relationships are considered, which may be subject to vagueness as to allow system users to flexibly interact with the IS. In all works, discrete time models are employed.

For 1 Work, the Basic Time Units Are Time Domain Intervals. In [48], fuzzy set theory is used to support imperfection in time domain intervals and in qualitative temporal relationships. Both uncertainty caused by a (partial) lack of knowledge and vagueness are supported. However, although a time model is clearly used, it is not clearly specified.

For 1 Work, the Basic Time Units Are Both Time Domain Elements and Intervals. In [39], on one hand, a situation is covered in which time domain elements and intervals may be subject to uncertainty caused by a (partial) lack of knowledge or to imprecision interpreted as uncertainty caused by an insufficient level of precision in the description of the respective elements and intervals. On the other hand, a situation is covered in which qualitative and a constrained form of quantitative temporal relationships are needed to be subject to imperfection, to allow the flexible expression of preferences.

5 Conclusions

In this paper, over 70 scientific contributions dealing with imperfection in time in information systems are cathegorized based on where they consider imperfection and what units of time they consider basic to their approach. Each contribution is concisely analysed with special attention going to the type of imperfection they consider, the types of temporal relationships they cover and the time model they employ. Moreover, some open research challenges or opportunities concerning imperfection in time in information systems are identified.

References

1. Aigner, W., Miksch, S., Thurnher, B., Biffl, S.: Planninglines: Novel glyphs for representing temporal uncertainties and their evaluation. In: Proceedings of the 9th International Conference on Information Visualisation, pp. 457–463 (2005)
2. Allen, J.F.: Maintaining knowledge about temporal intervals. Communications of the ACM **26**(11), 832–843 (1983)
3. Allen, J.F.: Towards a general theory of action and time. Artificial Intelligence **23**(2), 123–154 (1984)
4. Allen, J.F.: Time and time again: The many ways to represent time. International Journal of Intelligent Systems **6**(4), 341–355 (1991)
5. Alonso, O., Strötgen, J., Baeza-Yates, R., Gertz, M.: Temporal information retrieval: challenges and opportunities. In: Proceedings of the 1st International Temporal Web Analytics Workshop, vol. i, pp. 1–8, Hyderabad, India (2011)
6. Anselma, L., Terenziani, P., Snodgrass, R.T.: Valid-time indeterminacy in temporal relational databases: A family of data models. In: 17th International Symposium on Temporal Representation and Reasoning, pp. 139–145, Paris, France (2010)
7. Arotaritei, D., Mitra, S.: Web mining: a survey in the fuzzy framework. Fuzzy Sets and Systems **148**(1), 5–19 (2004)
8. Asmussen, K., Qiang, Y., De Maeyer, P., Van De Weghe, N.: Triangular models for studying and memorising temporal knowledge. In: Proceedings of the International Conference on Education, Research and Innovation, pp. 1849–1859. IATED, Madrid (2009)
9. Badaloni, S., Giacomin, M.: A fuzzy extension of allen's interval algebra. In: Lamma, E., Mello, P. (eds.) AI*IA 1999. LNCS (LNAI), vol. 1792, pp. 155–165. Springer, Heidelberg (2000)
10. Badaloni, S., Giacomin, M.: The algebra ia fuz: a framework for qualitative fuzzy temporal reasoning. Artificial Intelligence **170**, 872–908 (2006)
11. Bassiri, A., Malek, M.R., Alesheikh, A.A., Amirian, P.: Temporal relationships between rough time intervals. In: Gervasi, O., Taniar, D., Murgante, B., Laganà, A., Mun, Y., Gavrilova, M.L. (eds.) ICCSA 2009, Part I. LNCS, vol. 5592, pp. 543–552. Springer, Heidelberg (2009)
12. van Beek, P.: Temporal query processing with indefinite information. Artificial Intelligence in Medicine **3**(6), 325–339 (1991)
13. Billiet, C., De Tré, G.: Combining uncertainty and vagueness in time intervals. In: Angelov, P. (ed.) Intelligent Systems 2014. AISC, vol. 322, pp. 353–364. Springer, Heidelberg (2014)

14. Billiet, C., De Tré, G.: Twodimensional visualization of discrete time domain intervals subject to uncertainty. In: Proceedings of the 6th International Conference on Fuzzy Computation Theory and Applications, pp. 137–145. Scitepress, Rome (2014)
15. Billiet, C., Pons, J.E., Matthé, T., De Tré, G., Pons Capote, O.: Bipolar fuzzy querying of temporal databases. In: Christiansen, H., De Tré, G., Yazici, A., Zadrozny, S., Andreasen, T., Larsen, H.L. (eds.) FQAS 2011. LNCS, vol. 7022, pp. 60–71. Springer, Heidelberg (2011)
16. Billiet, C., Pons, J.E., Pons Capote, O., De Tré, G.: Evaluating possibilistic valid-time queries. In: Greco, S., Bouchon-Meunier, B., Coletti, G., Fedrizzi, M., Matarazzo, B., Yager, R.R. (eds.) IPMU 2012, Part I. CCIS, vol. 297, pp. 410–419. Springer, Heidelberg (2012)
17. Billiet, C., Pons Frias, J.E., Pons, O., De Tré, G.: Bipolarity in the Querying of Temporal Databases, pp. 21–37. SRI PAS/IBS PAN, new trends edn. (2013)
18. Billiet, C., Pons, J.E., Pons, O., De Tré, G.: Bipolar querying of valid-time intervals subject to uncertainty. In: Larsen, H.L., Martin-Bautista, M.J., Vila, M.A., Andreasen, T., Christiansen, H. (eds.) FQAS 2013. LNCS, vol. 8132, pp. 401–412. Springer, Heidelberg (2013)
19. Billiet, C., Pons Frias, J.E., Pons Capote, O., De Tré, G.: A comparison of approaches to model uncertainty in time intervals. In: Advances in Intelligent Systems Research, pp. 626–633. Atlantis Press, Milano (2013)
20. Bittner, T.: Approximate qualitative temporal reasoning. Annals of Mathematics and Artificial Intelligence **36**(1–2), 39–80 (2002)
21. Black, M.: Vagueness. an exercise in logical analysis. Philosophy of Science **4**(4), 427–455 (1937)
22. Jensen, C.S., et al.: The consensus glossary of temporal database concepts - february 1998 version. In: Etzion, O., Jajodia, S., Sripada, S. (eds.) Dagstuhl Seminar 1997. LNCS, vol. 1399, pp. 367–405. Springer, Heidelberg (1998)
23. Bolour, A., Anderson, T.L., Dekeyser, L.J., Wong, H.K.T.: The role of time in information processing: A survey. ACM SIGART Bulletin **80**, 28–46 (1982)
24. Bordogna, G., Bucci, F., Carrara, P., Pagani, M., Rampini, A.: Extending inspire metadata to imperfect temporal descriptions. International Journal of Spatial Data Infrastructures Research, Special Issue GSDI-11 (2009)
25. Bordogna, G., Carrara, P., Pagani, M., Pepe, M., Rampini, A.: Managing imperfect temporal metadata in the catalog services of spatial data infrastructures compliant with inspire. In: Proceedings of the Joint 2009 International Fuzzy Systems Associations World Congress and 2009 European Society of Fuzzy Logic and Technology Conference, pp. 915–920, Lisbon, Portugal (2009)
26. Bosc, P., Kraft, D., Petry, F.: Fuzzy sets in database and information systems: Status and opportunities. Fuzzy Sets and Systems **156**(3), 418–426 (2005)
27. Bosc, P., Pivert, O.: Modeling and querying uncertain relational databases: A survey of approaches based on the possible worlds semantics. International Journal of Uncertainty, Fuzziness and Knowledge-Based Systems **18**(5), 565–603 (2010)
28. Bosch, A., Torres, M., Marín, R.: Reasoning with disjunctive fuzzy temporal constraint networks. In: Proceedings of the 9th International Symposium on Temporal Representation and Reasoning, pp. 36–43. IEEE (2002)
29. Bronselaer, A., Pons, J.E., De Tré, G., Pons, O.: Possibilistic evaluation of sets. International Journal of Uncertainty, Fuzziness and Knowledge-Based Systems **21**(3), 325–346 (2013)
30. Chountas, P., Petrounias, I.: Modelling and representation of uncertain temporal information. Requirements Engineering **5**(3), 144–156 (2000)

31. De Caluwe, R., Devos, F., Maesfranckx, P., De Tré, G., Van Der Cruyssen, B.: The Semantics and Modelling of Flexible Time Indications. Studies in Fuzziness and Soft Computing, chap. 11, pp. 229–256. Springer (1999)
32. De Tré, G., Bronselaer, A.J., Billiet, C., Qiang, Y., Van De Weghe, N., De Maeyer, P., Pons, J.E., Pons, O.: Visualising and handling uncertain time intervals in a two-dimensional triangular space. In: Proceedings of the Second World Conference on Soft Computing, pp. 585–592, Baku, Azerbaijan (2012)
33. De Tré, G., De Caluwe, R., Van Der Cruyssen, B.: Dealing with time in fuzzy and uncertain object-oriented database models. In: Proceedings of the EUFIT 97 Conference, pp. 1157–1161, Aachen, Germany (1997)
34. De Tré, G., Van de Weghe, N., de Caluwe, R., De Maeyer, P.: Towards a flexible visualization tool for dealing with temporal data. In: Larsen, H.L., Pasi, G., Ortiz-Arroyo, D., Andreasen, T., Christiansen, H. (eds.) FQAS 2006. LNCS (LNAI), vol. 4027, pp. 109–120. Springer, Heidelberg (2006)
35. Dekhtyar, A., Ozcan, F., Ross, R., Subrahmanian, V.S.: Probabilistic temporal databases, ii: Calculus and query processing (2001)
36. Dekhtyar, A., Ross, R., Subrahmanian, V.S.: Probabilistic temporal databases, i: Algebra. ACM Transactions on Database Systems **26**(1), 41–95 (2001)
37. Dubois, D., Esteva, F., Godo, L., Prade, H.: An Information-based Discussion of Vagueness: Six Scenarios Leading to Vagueness, chap. 40, pp. 891–909. Elsevier (2005)
38. Dubois, D., Fargier, H., Galvagnon, V.: On latest starting times and floats in activity networks with ill-known durations. European Journal Of Operational Research **147**, 266–280 (2003)
39. Dubois, D., HadjAli, A., Prade, H.: Fuzziness and uncertainty in temporal reasoning. Journal of Universal Computer Science **9**(9), 1168–1194 (2003)
40. Dubois, D., Hadjali, A., Prade, H.: A possibility theory-based approach to the handling of uncertain relations between temporal points. International Journal of Intelligent Systems **22**(2), 157–179 (2007)
41. Dubois, D., Lang, J., Prade, H.: Timed possibilistic logic. Fundamenta Informaticae **15**(3–4), 211–234 (1991)
42. Dubois, D., Prade, H.: Processing fuzzy temporal knowledge. IEEE Transactions on Systems, Man and Cybernetics **19**(4), 729–744 (1989)
43. Dubois, D., Prade, H.: Formal representations of uncertainty. In: Decision-Making Process: Concepts and Methods, chap. 3, p. 59. Wiley, London (2009)
44. Dubois, D., Prade, H.: Gradualness, uncertainty and bipolarity: Making sense of fuzzy sets. Fuzzy Sets and Systems **192**(1), 3–24 (2012)
45. Dyreson, C.E., et al.: A consensus glossary of temporal database concepts. SIGMOD Record **23**(1), 52–64 (1994)
46. Dyreson, C.E., Snodgrass, R.T.: Supporting valid-time indeterminacy. ACM Transactions on Database Systems **23**(1), 1–57 (1998)
47. Freksa, C.: Temporal reasoning based on semi-intervals. Artificial Intelligence **54**(1–2), 199–227 (1992)
48. Galindo, J., Medina, J.M.: Ftsql2: Fuzzy time in relational databases*. In: Proceedings of the 2nd International Conference in Fuzzy Logic and Technology, pp. 47–50. Leicester (2001)
49. Galton, A.: A critical examination of allen's theory of action and time. Artificial Intelligence **42**(2–3), 159–188 (1990)
50. Garrido, C., Marín, N., Pons, O.: Fuzzy intervals to represent fuzzy valid time in a temporal relational database. International Journal of Uncertainty, Fuzziness and Knowledge-Based Systems **17**(Suppl. 1), 173–192 (2009)

51. Godo, L., Vila, L.: Possibilistic temporal reasoning based on fuzzy temporal constraints. In: Proceedings of the 14th International Joint Conference on Artificial Intelligence, pp. 1916–1922. Morgan Kaufmann, Montréal (1995)
52. Guil, F., Bosch, A., Bailón, A., Marín, R.: A fuzzy approach for mining generalized frequent temporal patterns. In: Proceedings of the 4th IEEE International Conference on Data Mining, Workshop on Alternative Techniques for Data Mining and Knowledge Discovery, p. 6. IEEE Computer Society, Washington, DC (2004)
53. Kacprzyk, J., Zadrozny, S.: Computing with words in intelligent database querying: Standalone and internet-based applications. Information Sciences **134**(1–4), 71–109 (2001)
54. Kahn, K., Gorry, G.A.: Mechanizing temporal knowledge. Artificial Intelligence **9**(1), 87–108 (1977)
55. Kalczynski, P.J., Chou, A.: Temporal document retrieval model for business news archives. Information Processing & Management **41**, 635–650 (2005)
56. Kolmogorov, A.: Grundbegriffe der Wahrscheinlichkeitsrechnung. Julius Springer, Berlin (1933)
57. Kopetz, H., Kim, K.H.K.: Temporal uncertainties in interactions among real-time objects. In: Proceedings of the 9th Symposium on Reliable Distributed Systems, pp. 165–174, Huntsville, U.S.A (1990)
58. Koubarakis, M.: Database models for infinite and indefinite temporal information. Information Systems **19**(2), 141–173 (1994)
59. Long, W.: Temporal reasoning for diagnosis in a causal probabilistic knowledge base. Artificial Intelligence in Medicine **8**(3), 193–215 (1996)
60. Mitra, D., Gerard, M.L., Srinivasan, P., Hands, A.E.: A possibilistic interval constraint problem: Fuzzy temporal reasoning. In: Proceedings of the 3rd IEEE Conference on Fuzzy Systems, pp. 1434–1439 (1994)
61. Motro, A., Smets, P.: Uncertainty Management in Information Systems: from Needs to Solutions. Kluwer Academic Publishers (1997)
62. Nagypál, G., Motik, B.: A fuzzy model for representing uncertain, subjective, and vague temporal knowledge in ontologies. In: Meersman, R., Schmidt, D.C. (eds.) CoopIS 2003, DOA 2003, and ODBASE 2003. LNCS, vol. 2888, pp. 906–923. Springer, Heidelberg (2003)
63. O'Connor, M.J., Tu, S.W., Musen, M.A.: Representation of temporal indeterminacy in clinical databases. In: Proceedings of the AMIA Symposium, pp. 615–619 (2000)
64. Palma, J., Juarez, J.M., Campos, M., Marin, R.: A fuzzy approach to temporal model-based diagnosis for intensive care units. In: Proceedings of the 16th European Conference on Artificial Intelligence, pp. 868–872. Amsterdam, The Netherlands (2004)
65. Palma, J., Juarez, J.M., Campos, M., Marin, R.: Fuzzy theory approach for temporal model-based diagnosis : An application to medical domains. Artificial Intelligence in Medicine (2006)
66. Pani, A.K., Bhattacharjee, G.P.: Temporal representation and reasoning in artificial intelligence: A review. Mathematical and Computer Modelling **34**, 55–80 (2001)
67. Parsons, S.: Current approaches to handling imperfect information in data and knowledge bases. IEEE Transactions on Knowledge and Data Engineering **8**(3), 353–372 (1996)
68. Pons, J.E., Billiet, C., Pons Capote, O., De Tré, G.: A possibilistic valid-time model. In: Greco, S., Bouchon-Meunier, B., Coletti, G., Fedrizzi, M., Matarazzo, B., Yager, R.R. (eds.) IPMU 2012, Part I. CCIS, vol. 297, pp. 420–429. Springer, Heidelberg (2012)
69. Pons, J.E., Marín, N., Pons, O., Billiet, C., Tré, G.D.: A relational model for the possibilistic valid-time approach. International Journal of Computational Intelligence Systems **5**(6), 1068–1088 (2012)

70. Pons Frias, J.E., Billiet, C., Pons, O., De Tré, G.: Aspects of dealing with imperfect data in temporal databases. In: Pivert, O., Zadrożny, S. (eds.) Flexible Approaches in Data, Information and Knowledge Management. ACI, vol. 497, pp. 189–220. Springer, Heidelberg (1991)
71. Qiang, Y.: Modelling Temporal Information in a Two-dimensional Space - A Visualization Perspective. Phd thesis, Ghent University (2012)
72. Qiang, Y., Asmussen, K., Delafontaine, M., Stichelbaut, B., De Tré, G., De Maeyer, P., Van De Weghe, N.: Visualising rough time intervals in a two-dimensional space. In: Proceedings of IFSA World Congress/EUSFLAT Conference, pp. 1480–1485 (2009)
73. Qiang, Y., Delafontaine, M., Asmussen, K., Stichelbaut, B., De Maeyer, P., Van De Weghe, N.: Modelling imperfect time intervals in a two-dimensional space*. Control And Cybernetics **39**(4), 983–1010 (2010)
74. Researchers, T.D.S.: Summaries of current work. In: Temporal Databases: Research and Practice, pp. 414–428. Springer (1998)
75. Ryabov, V.: Uncertain relations between indeterminate temporal intervals. In: Proceedings of The 10th International Conference on Management of Data, pp. 87–95. Tata McGraw-Hill Publishing Company, New Delhi (2000)
76. Ryabov, V.: Probabilistic estimation of uncertain temporal relations. Revista Colombiana de Computacion **2**(2) (2001)
77. Ryabov, V., Terziyan, V.: Industrial diagnostics using algebra of uncertain temporal relations. In: Proceedings of the 21st International Multi-Conference on Applied Informatics, pp. 351–356. ACTA Press, Innsbruck (2003)
78. Ryabov, V., Trudel, A.: Probabilistic temporal interval networks. In: Proceedings of the 11th International Symposium on Temporal Representation and Reasoning, pp. 64–67 (2004)
79. Schockaert, S., Cock, M.D.: Temporal reasoning about fuzzy intervals. Artificial Intelligence **172**, 1158–1193 (2008)
80. Schockaert, S., De Cock, M., Kerre, E.E.: Imprecise temporal interval relations. In: Bloch, I., Petrosino, A., Tettamanzi, A.G.B. (eds.) WILF 2005. LNCS (LNAI), vol. 3849, pp. 108–113. Springer, Heidelberg (2006)
81. Schockaert, S., De Cock, M., Kerre, E.E.: Fuzzifying allens tempora interval relations. IEEE Transactions on Fuzzy Systems **16**(2), 517–533 (2008)
82. Tossebro, E., Nygard, M.: Uncertainty in spatiotemporal databases. In: Yakhno, T. (ed.) ADVIS 2002. AIS, vol. 2457, pp. 43–53. Springer, Heidelberg (2002)
83. Trajcevski, G.: Probabilistic range queries in moving objects databases with uncertainty. In: Proceedings of the 3rd ACM International Workshop on Data Engineering for Wireless and Mobile Access, pp. 39–45. ACM (2003)
84. Vila, L.: A survey on temporal reasoning in artificial intelligence. AI Communications **7**(1), 4–28 (1994)
85. Virant, J., Zimic, N.: Attention to time in fuzzy logic. Fuzzy Sets and Systems **82**(1), 39–49 (1996)
86. Wu, Y., Jajodia, S., Wang, X.S.: Temporal database bibliography update. In: Etzion, O., Jajodia, S., Sripada, S. (eds.) Dagstuhl Seminar 1997. LNCS, vol. 1399, pp. 338–366. Springer, Heidelberg (1998)
87. Zadeh, L.A.: Fuzzy sets as a basis for a theory of possibility. Fuzzy Sets and Systems **1**(1), 3–28 (1978)
88. Zadeh, L.A.: The role of fuzzy logic in the management of uncertainty in expert systems. Fuzzy Sets and Systems **11**, 199–227 (1983)
89. Zhang, H., Diao, Y., Immerman, N.: Recognizing patterns in streams with imprecise timestamps. Proceedings of the VLDB Endowment **3**(1–2), 244–255 (2010)

Specification of Underspecified Quantifiers via Question-Answering by the Theory of Acyclic Recursion

Roussanka Loukanova

Abstract This paper introduces a technique for specifying quantifier scope distribution in formal terms that represent underspecified quantifier scopes. For this purpose, we extend the higher-order theory of acyclic recursion, by adding generalized quantifiers and terms for multiple quantifiers with underspecified scope. The specification of the quantifier scopes is by using interactive questions and answers that are also rendered into formal terms of the theory of acyclic recursion.

Keywords Semantics · Algorithms · Recursion · Type-theory · Underspecified scope · Quantifiers · Question-answers · Specifications

1 Introduction

Moschovakis [6] initiated development of a new approach to the mathematical notion of algorithm. The theory in [6] concerns full, untyped recursion. In a next stage of development, in Moschovakis [7] the new theory of recursion was, in one direction, limited to acyclic recursion, and on another direction, extended to higher-order. We refer to the classes of formal languages and theory of higher-order, acyclic recursion by the notation $\mathrm{L}^{\lambda}_{\mathrm{ar}}$. The formal languages $\mathrm{L}^{\lambda}_{\mathrm{ar}}$ represent the concept of language meaning as the process for computation of denotation in typed models. Moschovakis recursion system $\mathrm{L}^{\lambda}_{\mathrm{ar}}$ is a higher-order type theory, which strictly extends Gallin TY_2, see Gallin [1], and thus, of Montague Intensional Logic (IL), see Montague [9]. This is achieved by using several means, in particular, (1) two kinds of variables: *recursion variables*, which we call also *memory variables*, and *pure variables*; (2) by formation of an additional set of *recursion terms*; (3) systems of rules that form various reduction calculi, e.g., the reduction calculus of $\mathrm{L}^{\lambda}_{\mathrm{ar}}$ and the induced referential synonymy. In the next section, we give the formal definitions of the reduction rules of L^{λ}_{ar}, including

R. Loukanova(✉)
Department of Mathematics, Stockholm University, Stockholm, Sweden
e-mail: rloukanova@gmail.com

T. Andreasen et al. (eds.), *Flexible Query Answering Systems 2015*,
Advances in Intelligent Systems and Computing 400,
DOI: 10.1007/978-3-319-26154-6_5

the new (γ)-rule, since they are directly used in this paper. Here, we briefly outline the syntax of L^{λ}_{ar}. For more detailed introduction to L^{λ}_{ar}, see Moschovakis [7] and Loukanova [3]. The γ-rule and γ-reduction are introduced in [3].

1.1 Syntax of L^{λ}_{ar}

Types of L^{λ}_{ar}**:** The set *Types* is the smallest set defined recursively (by a BNF notation used in computer science): $\tau :\equiv e \mid t \mid s \mid (\tau_1 \to \tau_2)$.

The vocabulary of L^{λ}_{ar} consists of pairwise disjoint sets of: typed constants, $K = \bigcup_{\tau \in Types} K_{\tau}$; typed pure variables, $PureVars = \bigcup_{\tau \in Types} PureVars_{\tau}$; and typed recursion (memory) variables, $RecVars = \bigcup_{\tau \in Types} RecVars_{\tau}$.

The Terms of L^{λ}_{ar}**:** (by using a notational variant of "typed" BNF, with the assumed types given as superscripts)

$$A :\equiv \mathsf{c}^{\tau} : \tau \mid x^{\tau} : \tau \mid B^{(\sigma\to\tau)}(C^{\sigma}) : \tau \mid \lambda(v^{\sigma})(B^{\tau}) : (\sigma \to \tau)$$
$$\mid A_0^{\sigma} \text{ where } \{p_1^{\sigma_1} := A_1^{\sigma_1}, \ldots, p_n^{\sigma_n} := A_n^{\sigma_n}\} : \sigma$$

where $\{p_1^{\sigma_1} := A_1^{\sigma_1}, \ldots, p_n^{\sigma_n} := A_n^{\sigma_n}\}$ is a set of assignments that satisfies the *acyclicity condition*: For any terms $A_1 : \sigma_1$, ..., $A_n : \sigma_n$, and recursion variables $p_1 : \sigma_1$, ..., $p_n : \sigma_n$, there is a function $\mathsf{rank} : \{p_1, \ldots, p_n\} \longrightarrow \mathbb{N}$ such that, for all $p_i, p_j \in \{p_1, \ldots, p_n\}$, if p_j occurs free in A_i then $\mathsf{rank}(p_j) < \mathsf{rank}(p_i)$.

The terms of the form A_0^{σ} where $\{p_1^{\sigma_1} := A_1^{\sigma_1}, \ldots, p_n^{\sigma_n} := A_n^{\sigma_n}\}$ are called *recursion terms*.

1.2 Two Kinds of Semantics of L^{λ}_{ar}

Denotational Semantics of L^{λ}_{ar}**:** An L^{λ}_{ar} *structure* is a tuple $\mathfrak{A} = \langle \mathbb{T}, \mathcal{I}, \mathsf{den} \rangle$, satisfying the following conditions (S1)–(S4): (S1) $\mathbb{T}$ is a set, called a *frame*, of sets $\mathbb{T} = \{\mathbb{T}_{\sigma} \mid \sigma \in Types\}$, where $\mathbb{T}_e \neq \varnothing$ is a nonempty set of entities, $\mathbb{T}_t = \{0, 1, er\} \subseteq \mathbb{T}_e$ is the set of the *truth values*, $\mathbb{T}_s \neq \varnothing$ is a nonempty set of objects called *states*. (S2) $\mathbb{T}_{(\tau_1\to\tau_2)} = \{ p \mid p \colon \mathbb{T}_{\tau_1} \longrightarrow \mathbb{T}_{\tau_2} \}$. (S3) $\mathcal{I}$ is a function $\mathcal{I} \colon K \longrightarrow \mathbb{T}$, called the *interpretation function* of $\mathfrak{A}$, such that for every $\mathsf{c} \in K_{\tau}$, $\mathcal{I}(\mathsf{c}) = c$ for some $c \in \mathbb{T}_{\tau}$. (S4) Given that $G = \{ g \mid g \colon PureVars \cup RecVars \longrightarrow \mathbb{T}$ and $g(x) \in \mathbb{T}_{\sigma}$, for every $x : \sigma \}$ is the set G of the variable assignments, the *denotation function* $\mathsf{den} \colon Terms \longrightarrow \{ f \mid f \colon G \longrightarrow \mathbb{T} \}$ is defined, for each $g \in G$, by recursion on the structure of the terms:

(D1) $\mathsf{den}(x)(g) = g(x)$; $\mathsf{den}(\mathsf{c})(g) = \mathcal{I}(\mathsf{c})$;

(D2) $\mathsf{den}(A(B))(g) = \mathsf{den}(A)(g)(\mathsf{den}(B)(g))$;

(D3) $\mathsf{den}(\lambda(x)(B))(g) : \mathbb{T}_{\tau} \longrightarrow \mathbb{T}_{\sigma}$, where $x : \tau$ and $B : \sigma$, is the function such that, for every $t \in \mathbb{T}_{\tau}$, $[\mathsf{den}(\lambda x(B))(g)](t) = \mathsf{den}(B)(g\{x := t\})$;

(D4) $\mathsf{den}(A_0 \text{ where } \{p_1 := A_1, \ldots, p_n := A_n\})(g) =$
$$\mathsf{den}(A_0)(g\{p_1 := \overline{p}_1, \ldots, p_n := \overline{p}_n\}),$$

where for all $i \in \{1, \ldots, n\}$, $\overline{p}_i \in \mathbb{T}_{\tau_i}$ are defined by recursion on $\mathsf{rank}(p_i)$, so that: $\overline{p_i} = \mathsf{den}(A_i)(g\{p_{k_1} := \overline{p}_{k_1}, \ldots, p_{k_m} := \overline{p}_{k_m}\})$, where $p_{k_1}, \ldots, p_{k_m}$ are all the recursion variables $p_j \in \{p_1, \ldots, p_n\}$ such that $\mathsf{rank}(p_j) < \mathsf{rank}(p_i)$.

Intensional Semantics of $\mathrm{L}^{\lambda}_{\mathrm{ar}}$: The notion of intension in the languages of recursion covers the most essential, computational aspect of the concept of meaning. The *referential intension*, $\mathsf{Int}(A)$, of a meaningful term A is the tuple of functions (a recursor) that is defined by the denotations $\mathsf{den}(A_i)$ ($i \in \{0, \ldots n\}$) of the parts (i.e., the head sub-term A_0 and of the terms A_1, ..., A_n in the system of assignments) of its canonical form $\mathsf{cf}(A) \equiv A_0$ where $\{p_1 := A_1, \ldots, p_n := A_n\}$. Intuitively, for each meaningful term A, the intension of A, $\mathsf{Int}(A)$, is the *algorithm* for computing its denotation $\mathsf{den}(A)$. Two meaningful expressions are synonymous iff their referential intensions are naturally isomorphic, i.e., they are the same algorithms. Thus, the languages of recursion offer a formalisation of central computational aspects: denotation, with at least two semantic "levels": *referential intensions (algorithms)* and *denotations*. The terms in canonical form represent the algorithmic steps for computing semantic denotations.

Semantic Underspecification: By using linguistic motivations, we present arguments for the new kind of recursion variables and the distinctions between λ-calculus terms, recursion terms, subject to week β-reduction, and terms in canonical forms. One of the distinctive characteristics of the algorithmic theory of $\mathrm{L}^{\lambda}_{\mathrm{ar}}$ is the possibility to formalize the concept of semantic underspecification, at the object level of the language $\mathrm{L}^{\lambda}_{\mathrm{ar}}$. We give renderings of NL expressions into $\mathrm{L}^{\lambda}_{\mathrm{ar}}$ terms. that represent computational patterns with potentials for further specifications depending on context.

2 Reduction Rules of $\mathrm{L}^{\lambda}_{\mathrm{ar}}$

Congruence: If $A \equiv_{\mathsf{c}} B$, then $A \Rightarrow B$ (cong)

Transitivity: If $A \Rightarrow B$ and $B \Rightarrow C$, then $A \Rightarrow C$ (trans)

Compositionality:

If $A \Rightarrow A'$ and $B \Rightarrow B'$, then $A(B) \Rightarrow A'(B')$ (comp-ap)

If $A \Rightarrow B$, then $\lambda(u)(A) \Rightarrow \lambda(u)(B)$ (comp-λ)

If $A_i \Rightarrow B_i$, for $i = 0, \ldots, n$, then (comp-rec)

A_0 where $\{\, p_1 := A_1, \ldots, p_n := A_n\} \Rightarrow B_0$ where $\{\, p_1 := B_1, \ldots, p_n := B_n\}$

The head rule:

$$\big(A_0 \text{ where } \{\overrightarrow{p} := \overrightarrow{A}\}\big) \text{ where } \{\overrightarrow{q} := \overrightarrow{B}\} \qquad \text{(head)}$$
$$\Rightarrow A_0 \text{ where } \{\overrightarrow{p} := \overrightarrow{A},\ \overrightarrow{q} := \overrightarrow{B}\}$$

given that no p_i occurs free in any B_j, for $i = 1, \ldots, n$, $j = 1, \ldots, m$.

The Bekič-Scott rule:

$$A_0 \text{ where } \{p := \big(B_0 \text{ where } \{\overrightarrow{q} := \overrightarrow{B}\}\big), \overrightarrow{p} := \overrightarrow{A}\} \qquad \text{(B-S)}$$
$$\Rightarrow A_0 \text{ where } \{p := B_0, \overrightarrow{q} := \overrightarrow{B}, \overrightarrow{p} := \overrightarrow{A}\}$$

given that no q_i occurs free in any A_j, for $i = 1, ..., n$, $j = 1, ..., m$.

The recursion-application rule:

$$\big(A_0 \text{ where } \{\overrightarrow{p} := \overrightarrow{A}\}\big)(B) \Rightarrow A_0(B) \text{ where } \{\overrightarrow{p} := \overrightarrow{A}\} \qquad \text{(recap)}$$

given that no p_i occurs free in B for $i = 1, ..., n$.

The application rule:

$$A(B) \Rightarrow A(p) \text{ where } \{p := B\} \qquad \text{(ap)}$$

given that B is a proper term and p is a fresh location

The λ-rule:

$$\lambda(u)(A_0 \text{ where } \{ p_1 := A_1, \ldots, p_n := A_n \}) \qquad (\lambda)$$
$$\Rightarrow \lambda(u)A'_0 \text{ where } \{ p'_1 := \lambda(u)A'_1, \ldots, p'_n := \lambda(u)A'_n \}$$

where for all $i = 1, ..., n$, p'_i is a fresh location and

$$A'_i :\equiv A_i\{p_1 :\equiv p'_1(u), \ldots, p_n :\equiv p'_n(u)\} \qquad (5)$$

The γ-rule

$$A \equiv \quad A_0 \text{ where } \{ \overrightarrow{a} := \overrightarrow{A}, \; p := \lambda(v)P, \; \overrightarrow{b} := \overrightarrow{B} \}$$
$$\Rightarrow^*_\gamma A'_0 \text{ where } \{ \overrightarrow{a} := \overrightarrow{A'}, \; p' := P, \quad \overrightarrow{b} := \overrightarrow{B'} \} \qquad (\gamma)$$

where

- the term $A \in \mathsf{Terms}$ satisfies the γ-condition (in Definition 1) for the assignment $p := \lambda(v)P$
- $p' \in \mathsf{RV}_\tau$ is a fresh recursion variable
- $\overrightarrow{A'} \equiv \overrightarrow{A}\{p(v) :\equiv p'\}$ is the result of the replacements $A_i\{p(v) :\equiv p'\}$ of all occurrences of $p(v)$ by p', in all parts A_i in $\overrightarrow{A}$ $(i \in \{0, \ldots, n\})$
- $\overrightarrow{B'} \equiv \overrightarrow{B}\{p(v) :\equiv p'\}$ is the result of the replacements $B_j\{p(v) :\equiv p'\}$ of all occurrences of $p(v)$ by p', in all parts B_j in $\overrightarrow{B}$ $(j \in \{1, \ldots, k\})$

Definition 1 **(γ-condition).** *We say that a term $A \in$ Terms satisfies the γ-condition for an assignment $p := \lambda(v)P$ if and only if A is of the form:*

$$A \equiv A_0 \text{ where } \{ \overrightarrow{a} := \overrightarrow{A}, \; p := \lambda(v)P, \; \overrightarrow{b} := \overrightarrow{B} \} \qquad (7)$$

for some $v \in \mathsf{PV}_{\sigma}$, $p \in \mathsf{RV}_{(\sigma \to \tau)}$, $\lambda(v)P \in \mathsf{Terms}_{(\sigma \to \tau)}$, $\overrightarrow{A} \equiv A_1, \ldots, A_n \in \mathsf{Terms}$, $\overrightarrow{a} \equiv a_1, \ldots, a_n \in \mathsf{RV}$ $(n \geq 0)$, $\overrightarrow{B} \equiv B_1, \ldots, B_k \in \mathsf{Terms}$, $\overrightarrow{b} \equiv b_1, \ldots, b_k \in \mathsf{RV}(k \geq 0)$, *of corresponding types and such that the following clauses (1) and (2) hold:*

1. *The term* $P \in \mathsf{Terms}_{\tau}$ *does not have any (free) occurrences of* v *(and of* p*, by the acyclicity) in it, i.e.,* $v \notin \mathsf{FV}(P)$.
2. *All the occurrences of* p *in* A_0, $\overrightarrow{A}$ *and* $\overrightarrow{B}$ *are occurrences in a sub-term* $p(v)$ *that are in the scope of* $\lambda(v)$ *(modulo congruence with respect to renaming the scope variable* v*).*

Theorem 1 (Referential γ-Synonymy Theorem). *Two terms* A, B *are referentially* γ*-synonymous,* $A \approx_{\gamma} B$*, if and only if there are explicit, irreducible terms of corresponding types,* $A_i : \sigma_i$, $B_i : \sigma_i$ $(i = 0, \ldots, n)$, $(n \geq 0)$*, such that:*

$$A \Rightarrow_{\mathsf{gcf}} A_0 \text{ where } \{ p_1 := A_1, \ldots, p_n := A_n \} \quad (\gamma\text{-irreducible}) \tag{8a}$$

$$B \Rightarrow_{\mathsf{gcf}} B_0 \text{ where } \{ p_1 := B_1, \ldots, p_n := B_n \} \quad (\gamma\text{-irreducible}) \tag{8b}$$

and for all $i = 0, \ldots, n$,

$$\mathsf{den}(A_i)(g) = \mathsf{den}(B_i)(g), \quad \text{for all } g \in G. \tag{9}$$

For more on the (γ)-rule and the Referential γ-Synonymy Theorem 1, see Loukanova [5].

3 Specific Readings of Sentences Involving Quantification

3.1 Specific Distributions of Two Quantifiers

We represent the general problem with a sentence like (10) that represents a specific instance of a pattern sentence that has a transitive head verb. I.e., the head verb requires two syntactic arguments, which are NPs.

$$\text{Every professor reads some paper} \tag{10}$$

There are two different interpretations of the sentence (10), corresponding to which quantifier has its scope over the other. We call these two different scope distributions, respectively, *de dicto quantifier scope* in (11b), and *de re quantifier scope* in (12b). Without following the detailed reduction steps, these two terms, (11b) and (12b), are reduced to their canonical and γ-canonical forms as follows, by using the reduction rules given in Section 2.

$$\text{Every professor reads some paper} \xrightarrow{\text{render}} \tag{11a}$$

$$\begin{aligned} &every(professor)(\lambda(x_2)some(paper) \\ &\qquad (\lambda(x_1)read(x_1)(x_2))) \qquad \text{(de dicto)} \end{aligned} \tag{11b}$$

$$\Rightarrow_{\text{cf}} every(p)(R_2) \text{ where } \{R_2 := \lambda(x_2)some(b'(x_2))(R_1(x_2)), \tag{11c}$$

$$R_1 := \lambda(x_2)\,\lambda(x_1)read(x_1)(x_2), \tag{11d}$$

$$p := professor,\ b' := \lambda(x_2)paper\} \tag{11e}$$

$$\Rightarrow_{\text{gcf}} every(p)(R_2) \text{ where } \{R_2 := \lambda(x_2)some(b)(R_1(x_2)), \tag{11f}$$

$$R_1 := \lambda(x_2)\,\lambda(x_1)read(x_1)(x_2), \tag{11g}$$

$$p := professor,\ b := paper\} \tag{11h}$$

And respectively, for the de re quantifier scope, we have the following reductions to canonical and γ-canonical forms.

$$\text{Every professor reads some paper} \xrightarrow{\text{render}} \tag{12a}$$

$$\begin{aligned} &some(paper)(\lambda(x_1)every(professor) \\ &\qquad (\lambda(x_2)read(x_1)(x_2))) \qquad \text{(de re)} \end{aligned} \tag{12b}$$

$$\Rightarrow_{\text{cf}} some(b)(R_1) \text{ where } \{R_1 := \lambda(x_1)every(p'(x_1))(R_2(x_1)), \tag{12c}$$

$$R_2 := \lambda(x_1)\,\lambda(x_2)read(x_1)(x_2), \tag{12d}$$

$$p' := \lambda(x_1)professor,\ b := paper\} \tag{12e}$$

$$\Rightarrow_{\text{gcf}} some(b)(R_1) \text{ where } \{R_1 := \lambda(x_1)every(p)(R_2(x_1)), \tag{12f}$$

$$R_2 := \lambda(x_1)\,\lambda(x_2)read(x_1)(x_2), \tag{12g}$$

$$p := professor,\ b := paper\} \tag{12h}$$

By the Referential Synonymy Theorem, see Moschovakis [7] and Loukanova [3], the term in (11b) is algorithmically (i.e., referentially) synonymous to its canonical form in (11c)–(11e), and, respectively, the term (12b) to its canonical form in (12c)–(12e). While the terms in (11b) and (12b) have the same denotations as their respective canonical forms, the canonical forms represent the algorithmic semantics of the terms, i.e., the computational steps for computing the corresponding denotations. Furthermore, the recursion assignments in the canonical forms are reduced to the most basic steps and facts, so that the computations, by following the ranking of the recursion (memory) variables, do not involve any complex recursive "backtracking".

The reductions to canonical forms from (11b) to (11c)–(11e), and from (12b) to (12c)–(12e), use the (λ) rule, which produces the λ-abstractions in the assignments. These λ-abstractions are important since they provide the correct dependencies on arguments, with corresponding applications, and binding links to the corresponding argument roles including for the quantifiers' arguments and scopes. However, the λ-abstracts in the assignments $b' := \lambda(x_2)paper$, in (11e), and $p' := \lambda(x_1)professor$, in (12e), create unnecessary constant functions with redundant applications in $b'(x_2)$ and $p'(x_1)$, respectively. The (γ) rule reduces these vacuous λ-abstracts and applications.

Strictly, the new terms are not algorithmically synonymous to the ones from which they are γ-reduced, but they are algorithmically γ-equivalent, by preserving all algorithmic steps, except removing the vacuous ones.

Now, there is one more positive effect by having the γ-canonical forms from (11b) to (11f)–(11h), and from (12b) to (12f)–(12h). The γ-canonical forms (11f)–(11h) and eqrefdere to (12f)–(12h) reveal the common pattern in their different scope distributions. This shared pattern becomes more explicit if we move the head parts of the γ-canonical forms into the corresponding sets of assignments, as in the terms (13b)–(13d) and (14b)–(14d).

$$R_3 \text{ where } \{R_3 := every(p)(R_2), \tag{13a}$$
$$R_2 := \lambda(x_2) some(b)(R_1(x_2)), \tag{13b}$$
$$R_1 := \lambda(x_2)\, \lambda(x_1) read(x_1)(x_2), \tag{13c}$$
$$p := professor,\ b := paper\} \tag{13d}$$

$$R_3 \text{ where } \{R_3 := some(b)(R_1), \tag{14a}$$
$$R_1 := \lambda(x_1) every(p)(R_2(x_1)), \tag{14b}$$
$$R_2 := \lambda(x_1)\, \lambda(x_2) read(x_1)(x_2), \tag{14c}$$
$$p := professor,\ b := paper\} \tag{14d}$$

The variables bound by recursion operator where and by the λ-abstractions can be renamed by using arbitrary variables that do not cause variable collisions. However, we used indexing the bound variables to visualize the patterns in the scope bindings. Thus, for $i = 1, 2$, in each assignment to R_i, the argument slot of R_i created by the innermost λ-abstraction, i.e., $\lambda(x_i)$, in the right hand side of its assignment, is (via binding) the i-th argument slot of the head verbal relation $read$, recursively across all assignments. This linking is preserved by renaming the bounding variables.

3.2 *Underspecification of Quantifier Scopes*

The underspecified term (15a)–(15f) represents the common pattern in (13b)–(13d) and (14b)–(14d).

$$U \equiv R_3 \text{ where } \{l_1 := Q_1(R_1),\ l_2 := Q_2(R_2), \tag{15a}$$
$$Q_1 := q_1(d_1),\ Q_2 := q_2(d_2), \tag{15b}$$
$$q_1 := some,\ q_2 := every, \tag{15c}$$
$$h := read,\ d_1 := paper,\ d_2 := professor\} \tag{15d}$$
$$\text{s.th. } \{ Q_i \text{ binds the } i\text{-th argument of } h, \tag{15e}$$
$$R_3 \text{ is assigned to a closed subterm with fully scope specified } Q_i, \tag{15f}$$
$$R_3 \text{ dominates each } Q_i \text{ (for } i = 1, 2)\} \tag{15g}$$

Note that the term (15a)–(15f) has an additional constraint operator denoted by s.th., for restrictions over recursion variables in $\mathrm{L}^{\lambda}_{\mathrm{ar}}$ terms, as in Loukanova [5]. (In this paper, we use the abbreviated notation s.th. of the operator such that in [5].) The constraint that Q_i binds the i-th argument of h, requires that the argument slot of R_i, via the sub-terms $Q_i(R_i)$, is the i-th argument of h, i.e., links to it across the recursion assignments.

A term for the de dicto interpretation can be obtained from the underspecified (15a)–(15f), by adding assignments and modifications to the existing ones to satisfy the constraints in it. E.g., the added λ-abstractions reflect the constraints on linking the corresponding argument slots. The superscripts in l_1^1 and R_1^1 are renaming of the corresponding recursion variables to satisfy typing constraints, i.e., $l_1 : \widetilde{\mathsf{t}}$, $R_1 : (\widetilde{\mathsf{e}} \to \widetilde{\mathsf{t}})$ while $l_1^1 : (\widetilde{\mathsf{e}} \to \widetilde{\mathsf{t}})$, $R_1^1 : (\widetilde{\mathsf{e}} \to (\widetilde{\mathsf{e}} \to \widetilde{\mathsf{t}}))$.

$$
\begin{aligned}
U_{21} \equiv R_3 \text{ where } \{\, & R_3 := l_2,\ l_2 := Q_2(R_2), && \text{(16a)}\\
& R_2 := \lambda(x_2) l_1^1(x_2),\ l_1^1 := \lambda(x_2) Q_1(R_1^1(x_2)), && \text{(16b)}\\
& R_1^1 := \lambda(x_2)\,\lambda(x_1) h(x_1)(x_2), && \text{(16c)}\\
& Q_1 := q_1(d_1),\ Q_2 := q_2(d_2), && \text{(16d)}\\
& q_2 := \mathit{every},\ d_2 := \mathit{professor}, && \text{(16e)}\\
& h := \mathit{read},\ q_1 := \mathit{some},\ d_1 := \mathit{paper}\,\} && \text{(16f)}
\end{aligned}
$$

Each pair of the first two assignments in (16a) and (16b) can be merged. We get the following simplified term (17a)–(17f), which is algorithmically close to (16a)–(16f), except for removing the unnecessary computational steps.

$$
\begin{aligned}
S_{21} \equiv R_3 \text{ where } \{\, & R_3 := Q_2(R_2), && \text{(17a)}\\
& R_2 := \lambda(x_2) Q_1(R_1^1(x_2)), && \text{(17b)}\\
& R_1^1 := \lambda(x_2)\,\lambda(x_1) h(x_1)(x_2), && \text{(17c)}\\
& Q_1 := q_1(d_1),\ Q_2 := q_2(d_2), && \text{(17d)}\\
& q_2 := \mathit{every},\ d_2 := \mathit{professor}, && \text{(17e)}\\
& h := \mathit{read},\ q_1 := \mathit{some},\ d_1 := \mathit{paper}\,\} && \text{(17f)}
\end{aligned}
$$

Algorithmically, the term (16a)–(16f) is closer to the underspecified term (15a)–(15f), by preserving the recursion (memory) slots l_2 and the corresponding l_1^1. The term (17a)–(17f) is simpler, and carries the clear pattern of specifications, including the added λ-abstractions for linking the argument slots and bindings. The recursion variables in (15a)–(15f) can be considered as playing the roles of the corresponding linking, and thus identified, respectively, R_3 by l_2, R_2 by $\lambda(x_2) l_1^1(x_2)$. The remaining R_1 is renamed for the λ-abstractions and linked, via assignment to h, as $R_1^1 := \lambda(x_2)\,\lambda(x_1) h(x_1)(x_2)$. Note the important in this assignment, for a general pattern of specification: the order of the λ-abstractions corresponds to the quantification order.

The other order of quantifier scope is obtained by specifications in a similar pattern, from the underspecified term (15a)–(15f), resulting S_{12} in (18a)–(18f).

$$
\begin{aligned}
S_{12} \equiv R_3 \text{ where } \{\, & R_3 := Q_1(R_1), && \text{(18a)}\\
& R_1 := \lambda(x_1) Q_2(R_2^1(x_1)), && \text{(18b)}\\
& R_2^1 := \lambda(x_1)\,\lambda(x_2) h(x_1)(x_2), && \text{(18c)}\\
& Q_1 := q_1(d_1),\ Q_2 := q_2(d_2), && \text{(18d)}\\
& q_2 := every,\ d_2 := professor, && \text{(18e)}\\
& h := read,\ q_1 := some,\ d_1 := paper \,\} && \text{(18f)}
\end{aligned}
$$

4 Specifications via Questions

4.1 Rules of Existential Instantiation and Introduction

In this section, we consider how an underspecified term, such as (15a)–(15f), can be specified via answering questions. We consider tentative interactions between a tentative system that can handle automatically the reduction calculi of L_{ar}^{λ}, and theoretical rules, including the rules Existential Instantiation (ExistInst) defined in (19a)–(23h), and Existential Introduction (ExistIntro) defined, respectively, in Table 1 and Table 2. We consider that the system has a database of informational facts represented by L_{ar}^{λ} terms and rules. The implementation of an actual question-answering system based on the theory of acyclic recursion L_{ar}^{λ} is future work, as described in the Section 5.

Table 1 Existential Instantiation (ExistInst) rule

Existential Instantiation (ExistInst) For any fresh $c \in \mathsf{RV}_{\tilde{\mathsf{e}}}$

$$
\begin{aligned}
& E \text{ where } \{\, E := Q(R), && \text{(19a)}\\
& \qquad R := \lambda(x)A,\ Q := q(d),\ q := some,\ d := D, && \text{(19b)}\\
& \qquad \overrightarrow{a} := \overrightarrow{A},\ \} && \text{(19c)}\\
& \text{s.th. } \{\, \overrightarrow{C} \,\} && \text{(19d)}\\
& \therefore \\
& R' \text{ where } \{\, R' := A\{x :\equiv c\},\ d := D, && \text{(19e)}\\
& \qquad \overrightarrow{a} := \overrightarrow{A} \,\} && \text{(19f)}\\
& \text{s.th. } \{ d(c),\ \overrightarrow{C} \,\} && \text{(19g)}
\end{aligned}
$$

Table 2 Existential Introduction (ExistIntro) rule

Existential Introduction (ExistIntro)

For any $c \in \mathsf{RV}_{\tilde{\mathsf{e}}}$, such that c does not occur in $\overrightarrow{A}$, $\overrightarrow{C}$, D

$$R' \text{ where } \{\, R' := A\{\, x :\equiv c \,\},\ d := D, \tag{20a}$$

$$\overrightarrow{a} := \overrightarrow{A} \,\} \tag{20b}$$

$$\text{s.th. } \{\, d(c),\ \overrightarrow{C} \,\} \tag{20c}$$

$\therefore$

$$E \text{ where } \{\, E := Q(R), \tag{20d}$$

$$R := \lambda(x)A,\ Q := q(d),\ q := \mathit{some},\ d := D, \tag{20e}$$

$$\overrightarrow{a} := \overrightarrow{A} \,\} \tag{20f}$$

$$\text{s.th. } \{\, \overrightarrow{C} \,\} \tag{20g}$$

4.2 *Specifications via Existential Instantiation*

Assume that a user, such as a human, robot, or other computational system, is prompted by the system with the term (21a)–(21g), as available it its database.

$$U \equiv R_3 \text{ where } \{\, l_1 := Q_1(R_1),\ l_2 := Q_2(R_2), \tag{21a}$$

$$Q_1 := q_1(d_1),\ Q_2 := q_2(d_2), \tag{21b}$$

$$q_1 := \mathit{some},\ q_2 := \mathit{every}, \tag{21c}$$

$$h := \mathit{read},\ d_1 := \mathit{paper},\ d_2 := \mathit{professor} \,\} \tag{21d}$$

$$\text{s.th. } \{\, Q_i \text{ binds the } i\text{-th argument of } h, \tag{21e}$$

$$R_3 \text{ is assigned to a closed subterm with fully scope specified } Q_i, \tag{21f}$$

$$R_3 \text{ dominates each } Q_i \text{ (for } i = 1, 2) \,\} \tag{21g}$$

$$U \text{ specified-to } ? \tag{21h}$$

The user can respond either by providing a specification term like (17a)–(17f) or (18a)–(18f) in case he has information that is sufficient to provide an answer, without further query from his perspective, e.g., as follows:

U specified-to ?	prompt	(22a)
U specified-to S_{12}	answer	(22b)

where S_{12} is the term in (18a)–(18f). The user can respond by a query depending on his perspective, knowledge, or a guess. The respond can be providing a specification by requesting further information. Significantly, the query-answer system can be equipped with inference rules such as Existential Introduction (ExistIntro) and Existential Instantiation (ExistInst).

$$U \text{ specified-to } ? \tag{23a}$$
$$\text{Which is this paper?} \xrightarrow{\text{render}} \tag{23b}$$
$$R_3 \text{ where } \{ R_3 := Q_1(R_1), \tag{23c}$$
$$R_1 := \lambda(x_1) Q_2(R_2^1(x_1)), \tag{23d}$$
$$R_2^1 := \lambda(x_1)\, \lambda(x_2) h(x_1)(x_2), \tag{23e}$$
$$Q_1 := q_1(d_1),\ Q_2 := q_2(d_2), \tag{23f}$$
$$q_2 := \mathit{every},\ d_2 := \mathit{professor}, \tag{23g}$$
$$h := \mathit{read},\ q_1 := \mathit{some},\ d_1 := \mathit{paper} \} \tag{23h}$$

$$\therefore \tag{24a}$$
$$R_1' \text{ where } \{ R_1' := Q_2(R_2^1(c)), \tag{24b}$$
$$R_2^1 := \lambda(x_1)\, \lambda(x_2) h(x_1)(x_2), \tag{24c}$$
$$Q_2 := q_2(d_2), \tag{24d}$$
$$q_2 := \mathit{every},\ d_2 := \mathit{professor}, \tag{24e}$$
$$h := \mathit{read},\ d_1 := \mathit{paper} \} \tag{24f}$$
$$\text{s.th. } \{ d_1(c) \} \tag{24g}$$
$$c \text{ specified-to?} \tag{24h}$$

In response to the prompt (24h) issued by the user, the system searches through the database for a term T, which matches the term (24a)–(24f), where a sub-term C occurs instead of the occurrences of c.

1. In case there is such a term C, the system responds with $c := C$:

$$\therefore \tag{25a}$$
$$R_1' \text{ where } \{ R_1' := Q_2(R_2^1(c)), \tag{25b}$$
$$R_2^1 := \lambda(x_1)\, \lambda(x_2) h(x_1)(x_2), \tag{25c}$$
$$Q_2 := q_2(d_2), \tag{25d}$$
$$q_2 := \mathit{every},\ d_2 := \mathit{professor}, \tag{25e}$$
$$h := \mathit{read},\ d_1 := \mathit{paper},\ c := C \} \tag{25f}$$
$$\text{s.th. } \{ d_1(c) \} \tag{25g}$$
$$c \text{ specified-to } C \tag{25h}$$

2. Otherwise, the system answers:

$$\text{There is no such } c. \tag{26a}$$

$$U \text{ specified-to } S_{21} \qquad \text{answer} \tag{26b}$$

where S_{21} is the term in (17a)–(17f).

Other specifications of terms with underspecified quantifier scopes, via Existential Introduction, can be explored similarly to using Existential Instantiation. Such exploration is outside the subject and space of this paper, and we leave that for extended work.

5 Conclusions and Future Work

In this paper, we concentrate on introducing a technique for specifying quantifier scope distribution based on available underspecified scopes that are followed by question-answers. We demonstrated the technique on pattern sentences as the one in (10), having as components an NP that is interpreted as a universal quantifier and another NP as an existential quantifier. Further development of the technique of interactive, dynamic question-answer steps, for specification of underspecified scope distribution of more varieties of quantifiers, and combinations of more than two mixed quantifiers, is forthcoming.

In forthcoming work, we extend L^{λ}_{ar} by adding introduction and elimination rules for the universal quantifier every, corresponding to the ones for the existential quantifier some in Tables 1–2. Furthermore, along the quantifiers some, every, which are abundant in human language, we add analysis of other quantifiers, e.g., few, a few, most, and numerical quantifiers, in the type-theory L^{λ}_{ar}, with respective rules. In effect, we define them as generalized quantifiers, i.e., “lifting” the syntactic and semantic definitions of the generalized quantifiers, introduced by Mostowski [8], into L^{λ}_{ar} terms in γ-canonical forms, by using the γ-reduction introduced in Loukanova [3]. Furthermore, we extend the technique by adding generalized quantifiers that bind arbitrary numbers of variables over respective sequences of propositional terms, i.e., adding definitions of generalized quantifiers, as introduced by Lindström [2], into L^{λ}_{ar}. Our conjecture is that there are algorithmic procedures for mapping Mostowski and Lindström quantifiers into γ-canonical terms, by using the γ-reduction from [3].

Furthermore, the technique can be generalized to sentences with more varieties of NPs, such as quantifier NPs mixed with indefinite or definite referential NPs, and numerical NPs, e.g., such as in (27a)–(27d).

Every man built a (some) house. (27a)

Thirty four men built a (the) house. (27b)

Thirty four men built (the) two houses. (27c)

The thirty four men built a (the) house. (27d)

In order to render a sentence, which includes referential NPs, such as the man, mixed with quantifier NPs, into a single L_{ar}^{λ} term representing underspecification over the alternatives, we work on extending the formal system L_{ar}^{λ} in several directions. E.g., Loukanova [4] extends the type system of L_{ar}^{λ} with underspecified types. In other upcoming work, we add analysis for indefinite and definite numerals and definite descriptions other than numerical ones.

Clearly, in the contemporary intelligent computational systems, it is necessary to computerize exploration of data involving quantifiers with underspecified scopes relative to each other, and possible scope specifications depending on context, computerized systems and users (which can be humans or computing systems, e.g., robots). In this paper, we have considered the higher-order type theory of acyclic recursion as the formal foundation of such tasks. The implementation of an actual question-answering system based on the theory of acyclic recursion L_{ar}^{λ}, and using the rules presented in this paper, including the rules of Existential Instantiation (ExistInst) and Existential Introduction (ExistIntro) is future work.

References

1. Gallin, D.: Intensional and Higher-Order Modal Logic. North-Holland (1975)
2. Lindström, P.: First order predicate logic with generalized quantifiers. Theoria **32**(3), 186–195 (1966)
3. Loukanova, R.: γ-Reduction in Type Theory of Acyclic Recursion (to appear)
4. Loukanova, R.: Introduction to the Type Theory of Acyclic Recursion with Underspecified Types (to appear)
5. Loukanova, R.: Algorithmic granularity with constraints. In: Imamura, K., Usui, S., Shirao, T., Kasamatsu, T., Schwabe, L., Zhong, N. (eds.) BHI 2013. LNCS, vol. 8211, pp. 399–408. Springer, Heidelberg (2013)
6. Moschovakis, Y.N.: Sense and denotation as algorithm and value. In: Oikkonen, J., Vaananen, J. (eds.) Lecture Notes in Logic. Lecture Notes in Logic, No. 2, pp. 210–249. Springer (1994)
7. Moschovakis, Y.N.: A logical calculus of meaning and synonymy. Linguistics and Philosophy **29**, 27–89 (2006)
8. Mostowski, A.: On a generalization of quantifiers. Fundamenta Mathematicae **1**(44), 12–36 (1957)
9. Thomason, R.H. (ed.): Formal Philosophy: Selected Papers of Richard Montague. Yale University Press, New Haven (1974)

Part II
Aspects of Data Representation, Processing and Mining

A Certainty-Based Approach to the Cautious Handling of Suspect Values

Olivier Pivert and Henri Prade

Abstract In this paper, we consider the situation where a database may contain suspect values, i.e. precise values whose validity is not certain but whose attached uncertainty level is unknown. We propose a database model based on the notion of possibilistic certainty to deal with such values. The operators of relational algebra are extended in this framework. A crucial aspect is that queries have the same data complexity as in a classical database context.

1 Introduction

The need to handle uncertain values in a database context has been recognized a long time ago, and a variety of uncertain database models have been proposed to this aim. In these models, an ill-known attribute value is generally represented by a probability distribution (see, e.g. [4, 10]) or a possibility distribution [1], i.e. a set of weighted candidate values. However, in many situations, it may be very problematic to quantify the level of uncertainty attached to the different candidate values. One may not even know the set of (probable/possible) alternative candidates. Then, using a probabilistic model in a rigorous manner appears quite difficult, not to say impossible. In this work, we assume that all one knows is that a given precise value is suspect, i.e. not totally certain, and we show that a database model based on the notion of possibilistic certainty is a suitable tool for representing and handling suspect data. An introductory and abridged version can be found in [8].

O. Pivert(✉)
Irisa, University of Rennes 1, Lannion, France
e-mail: pivert@enssat.fr

H. Prade
IRIT–CNRS, University of Toulouse 3, Toulouse, France
e-mail: prade@irit.fr

T. Andreasen et al. (eds.), *Flexible Query Answering Systems 2015*,
Advances in Intelligent Systems and Computing 400,
DOI: 10.1007/978-3-319-26154-6_6

The remainder of the paper is structured as follows. Section 2 presents the uncertain database model that we advocate for representing tuples that may i) involve suspect attribute values, ii) be themselves uncertain (due to the fact that some operations generate "maybe tuples"). Section 3 gives the definitions of the algebraic operators in this framework. Query equivalences are studied in Section 4. In Section 5, we discuss a way to make selection queries more flexible, which makes it possible to discriminate the uncertain answers to a query. Finally, Section 6 recalls the main contributions and outlines perspectives for future work.

2 The Model

In possibility theory [3, 11], each event E — defined as a subset of a universe Ω — is associated with two measures, its possibility $\Pi(E)$ and its necessity $N(E)$. Π and N are two dual measures, in the sense that $N(E) = 1 - \Pi(\overline{E})$ (where the overbar denotes complementation). This clearly departs from the probabilistic situation where $Prob(E) = 1 - Prob(\overline{E})$. In possibility theory, being somewhat certain about E ($N(E)$ has a high value) forces you to have $\overline{E}$ rather impossible ($1 - \Pi$ is impossibility), but it is allowed to have no certainty neither about E nor about $\overline{E}$. Generally speaking, possibility theory is oriented towards the representation of epistemic states of information, while probabilities are deeply linked to the ideas of randomness, and of betting in case of subjective probability.

In the following, we assume that the certainty degree associated with the uncertain events considered (that concern the actual value of an attribute in a tuple, for instance) is unknown. Thus, we use a fragment of possibility theory where three values only are used to represent certainty : 1 (completely certain), α (somewhat certain but not totally), 0 (not at all certain). The fact that one uses α for every somewhat certain event does not imply that the certainty degree associated with these events is the same; α is just a conventional symbol that means "a certainty degree in the open interval (0, 1)". Notice that this corresponds to using three symbols for representing possibility degrees as well: 0, β ($= 1 - \alpha$), and 1 (but we are not interested in qualifying possibility here). In other words, we are representing pieces of information of the form $N(E) \geq \alpha$, which might be regarded as "known unknowns" [5] in the sense that their certainty level α cannot be precisely assessed, while "unknown unknowns" remain out of reach (states of acknowledged ignorance corresponding to $\Pi(E) = \Pi(\overline{E}) = 1$, or equivalently $N(E) = N(\overline{E}) = 0$, cannot be captured either).

The model that we introduce hereafter is a simplified — thus "lighter" — version of that introduced in [2] and detailed in [9], where a certainty level is attached to each ill-known attribute value (by default, an attribute value has certainty 1). As we will see, representing suspect attribute values also leads us to representing the fact that the existence of some tuples (in the result of some queries) may not be totally certain either. Let us first discuss the philosophy of the approach, before describing the model and the operations more precisely.

Let us consider a database containing suspect values. In the following, a suspect value will be denoted using a star, as in 17*. A value a^* means that it is somewhat certain (thus completely possible) that a is the actual value of the considered attribute for the considered tuple, but not totally certain (otherwise we would use the notation a instead of a^*). When evaluating a selection query Q based on a condition ψ (made of atomic conjuncts ψ_i), one may distinguish between three groups of answers:

- the *completely certain* answers. A tuple t is a completely certain answer to Q iff t does not contain any suspect value concerned by ψ, and every attribute value of t concerned by a conjunct ψ_i satisfies ψ_i. For instance, $t = \langle$John, 35*, Paris$\rangle$ is a completely certain answer to $Q = \sigma_{city=\text{'Paris'}}(r)$.
- the *somewhat certain* (thus completely possible) answers. A tuple t is a somewhat certain answer to Q iff i) t contains at least one suspect value concerned by a conjunct ψ_i, and every suspect value concerned by a conjunct ψ_i satisfies this conjunct, ii) every nonsuspect attribute value of t concerned by a conjunct ψ_i satisfies ψ_i. For instance, $t = \langle$John, 35*, Paris$\rangle$ is a somewhat certain answer to $Q = \sigma_{city=\text{'Paris' and } age=35}(r)$.
- the *somewhat possible* (but not at all certain) answers. A tuple t is a somewhat possible answer iff i) t contains at least one suspect value that does not satisfy the corresponding conjunct ψ_i from ψ, ii) every nonsuspect attribute value of t concerned by a conjunct ψ_i satisfies ψ_i. For instance, $t = \langle$John, 35*, Paris$\rangle$ is a somewhat possible answer to $Q = \sigma_{city=\text{'Paris' and } age=40}(r)$.

As to the other tuples (those that contain at least one nonsuspect value that does not satisfy the associated conjunct ψ_i from ψ), they are of course discarded.

In fact, in the model we propose, we restrict ourselves to the computation of the two first groups (i.e., the completely or somewhat certain answers), since dealing with the answers that are only somewhat possible raises important difficulties. Namely, in order to have a sound compositional framework (and to preserve the possible worlds semantics), one would have to be able to represent not only values of the type a^* but one would need to maintain a complete representation of attribute values in terms of possibility distributions. Moreover, we are then faced with the problem of handling intertuple dependencies generated by the join operation in particular [9].

Notice that the framework we propose is compatible with the use of NULL values for representing attribute values that exist but are currently (totally) unknown. If a tuple includes a NULL for an attribute concerned by a selection condition, it will simply be discarded since we are only interested in answers that are somewhat certain.

Uncertain tuples are denoted by α/t where α has the same meaning as above. α/t means that the existence of the tuple in the considered relation is only somewhat certain (thus, it is also possible to some extent that it does not exist). It is mandatory to have a way to represent such uncertain tuples since some operations of relational algebra (selection, in particular) may generate them. The tuples whose existence is completely certain are denoted by $1/t$. A relation of the model will thus involve an extra column denoted by N, representing the certainty attached to the tuples.

3 Algebraic Operators

In this section, we give the definition of the different operators of relational algebra in the certainty-based model defined above.

3.1 Selection

Let us denote by $c(t.A)$ the certainty degree associated with the value of attribute A in tuple t: $c(t.A)$ equals 1 if $t.A$ is a nonsuspect value, and it takes the (conventional) value α otherwise (with $\alpha < 1$). It is the same thing for the certainty degree N associated with a tuple (the notation is then N/t).

Case of a condition of the form $A\ \theta\ q$ where A is an attribute, θ is a comparison operator, and q is a constant:

$$\sigma_{A\,\theta\,q}(r) = \{N'/t \mid N/t \in r \text{ and } t.A\ \theta\ q \text{ and } N' = \min(N,\ c(t.A))\} \tag{1}$$

Example 1. Let us consider the relation *Emp* represented in Table 1 (left) and the selection query $\sigma_{job=\text{'Engineer'}}(Emp)$. Its result appears in Table 1 (right). ◇

Table 1 Relation *Emp* (left), result of the selection query (right)

#id	*name*	*city*	*job*	*N*
37	John	Newton*	Engineer*	1
53	Mary	Quincy*	Clerk*	1
71	Bill	Boston	Engineer	1

#id	*name*	*city*	*job*	*N*
37	John	Newton*	Engineer*	α
71	Bill	Boston	Engineer	1

Case of a condition of the form $A_1\ \theta\ A_2$ where A_1 and A_2 are two attributes and θ is a comparison operator:

$$\begin{aligned}\sigma_{A_1\,\theta\,A_2}(r) = \{N'/t \mid N/t \in r \text{ and } t.A_1\ \theta\ t.A_2 \text{ and}\\ N' = \min(N,\ c(t.A_1),\ c(t.A_2))\}.\end{aligned} \tag{2}$$

Case of a conjunctive condition $\psi = \psi_1 \wedge \ldots \wedge \psi_m$:

$$\begin{aligned}\sigma_{\psi_1 \wedge \ldots \wedge \psi_m}(r) = \{N'/t \mid N/t \in r \text{ and } \psi_1(t.A_1) \text{ and } \ldots \text{ and } \psi_m(t.A_m)\\ \text{and } N' = \min(N,\ c(t.A_i),\ \ldots,\ c(t.A_m))\}.\end{aligned} \tag{3}$$

Case of a disjunctive condition $\psi = \psi_1 \vee \ldots \vee \psi_m$:

$$\begin{aligned}\sigma_{\psi_1 \vee \ldots \vee \psi_m}(r) = \{N'/t \mid N/t \in r \text{ and } (\psi_1(t.A_1) \text{ or } \ldots \text{ or } \psi_m(t.A_m))\\ \text{and } N' = \min(N,\ \max_{i \text{ such that } \psi_i(t.A_i)} (c(t.A_i)))\}.\end{aligned} \tag{4}$$

3.2 Projection

Let r be a relation of schema (X, Y). The projection operation is straightforwardly defined as follows:

$$\pi_X(r) = \{N/t.X \mid N/t \in r \text{ and } \nexists N'/t' \in r \text{ such that } sbs(N'/t'.X,\ N/t.X)\}.$$

The only difference w.r.t. the definition of the projection in a classical database context concerns duplicate elimination, which is here based on the concept of "possibilistic subsumption". Let $X = \{A_1, \ldots, A_n\}$. The predicate sbs, which expresses subsumption, is defined as follows:

$$\begin{aligned} sbs((N'/t'.X,\ N/t.X) \equiv \\ &\forall i \in \{1, \ldots, n\},\ t.A_i = t'.A_i \text{ and} \\ &c(t.A_i) \leq c(t'.A_i) \text{ and } N \leq N' \text{and} \\ &((\exists i \in \{1, \ldots, n\},\ c(t.A_i) < c(t'.A_i)) \text{ or } N < N'). \end{aligned} \tag{5}$$

Example 2. Let us consider relation *Emp* represented in Table 2 (left) and the projection query $\pi_{\{city,\ job\}}$ (*Emp*). Its result is represented in Table 2 (right). ◇

Table 2 Relation *Emp* (left), result of the projection query (right)

#id	*name*	*city*	*job*	*N*
35	Phil	Newton	Engineer*	1
52	Lisa	Quincy*	Clerk*	α
71	Bill	Newton	Engineer	α
73	Bob	Newton*	Engineer*	α
84	Jack	Quincy*	Clerk	α

city	*job*	*N*
Newton	Engineer*	1
Newton	Engineer	α
Quincy*	Clerk	α

3.3 Join

The definition of the join in the context of the model considered is:

$$\begin{aligned} r_1 \bowtie_{A=B} r_2 = \{ &\min(N_1,\ N_2,\ c(t_1.A),\ c(t_2.B))/t_1 \oplus t_2 \mid \\ &\exists\, N_1/t_1 \in r_1,\ \exists\, N_2/t_2 \in r_2 \text{ such that } t_1.A = t_2.B \end{aligned} \tag{6}$$

where $\oplus$ denotes concatenation.

Example 3. Consider the relations *Person* and *Lab* from Table 3 and the query:

$$PersLab = Person \bowtie_{Pcity=Lcity} Lab$$

Table 3 Relations *Person* (left), *Lab* (right), result of the join query (bottom)

#Pid	*Pname*	*Pcity*	*N*
11	John	Boston*	1
12	Mary	Boston	α
17	Phil	Weston*	α
19	Jane	Weston	1

#Lid	*Lname*	*Lcity*	*N*
21	BERC	Boston*	α
22	IFR	Weston	1
23	AZ	Boston	1

#Pid	*Pname*	*Pcity*	*#Lid*	*Lname*	*Lcity*	*N*
11	John	Boston*	21	BERC	Boston*	α
11	John	Boston*	23	AZ	Boston	α
12	Mary	Boston	21	BERC	Boston*	α
12	Mary	Boston	23	AZ	Boston	α
17	Phil	Weston*	22	IFR	Weston	α
19	Jane	Weston	22	IFR	Weston	1

which looks for the pairs (p, l) such that p (somewhat certainly) lives in a city where (somewhat certainly) a research center l is located. Its result appears in Table 3 (bottom). ◇

In the case of a natural join (i.e., an equijoin on all of the attributes common to the two relations), one keeps only one copy of each join attribute in the resulting table. Here, this "merging" keeps the more uncertain value for each join attribute. This behavior is illustrated in Table 4.

Table 4 Result of the natural join query (assuming a common attribute *City*)

#Pid	*Pname*	*City*	*#Lid*	*Lname*	*N*
11	John	Boston*	21	BERC	α
11	John	Boston*	23	AZ	α
12	Mary	Boston*	21	BERC	α
12	Mary	Boston	23	AZ	α
17	Phil	Weston*	22	IFR	α
19	Jane	Weston	22	IFR	1

3.4 Intersection

For the sake of readability of the following definition, we denote a suspect value v^* by $(v,\ \alpha)$ and a totally certain value v by $(v,\ 1)$. The intersection $r_1 \cap r_2$ is defined as follows:

$$
\begin{aligned}
& r_1 \cap r_2 = \\
& \quad \{\min(N_1,\ N_2)/\langle(t.A_1,\ \min(\rho_{1,1},\ \rho_{2,1})),\ \ldots,(t.A_n,\ \min(\rho_{1,n},\ \rho_{2,n}))\rangle \\
& \qquad \text{such that } N_1/\langle(t.A_1,\ \rho_{1,1}),\ldots,\ (t.A_n,\ \rho_{1,n})\rangle \in r_1 \\
& \qquad \text{and } N_2/\langle(t.A_1,\ \rho_{2,1}),\ldots,\ (t.A_n,\ \rho_{2,n})\rangle \in r_2\}.
\end{aligned} \tag{7}
$$

Recall that $\rho_{i,j}$ equals either 1 (certain value) or α (suspect value).

Example 4. Consider the relations from Table 5 (left and middle). The result of the intersection query $r_1 \cap r_2$ is given in Table 5 (right). ◇

Table 5 Relations r_1 (left), r_2 (middle), and $r_1 \cap r_2$ (right)

Job	*City*	*N*
Engineer*	Boston	1
Engineer	Newton*	α
Technician	Boston	1
Technician	Quincy*	1

Job	*City*	*N*
Engineer*	Boston*	1
Engineer	Newton	1
Clerk*	Boston	α
Technician	Boston	1

Job	*City*	*N*
Engineer*	Boston*	1
Engineer	Newton*	α
Technician	Boston	1

3.5 Union

Union is defined as usual (and has the same data complexity), except that duplicate elimination is based on the notion of "possibilistic subsumption" (see Subsection 3.2) :

$$\begin{aligned} & r_1 \cup r_2 = \\ & \quad \{N/t \mid (N/t \in r_1 \text{ and } N'/t \notin r_2) \text{ or } (N/t \in r_2 \text{ and } N'/t \notin r_1) \text{ or} \\ & \quad (N/t \in r_1 \text{ and } N/t \in r_2 \text{ or} \\ & \quad (N/t \in r_1 \text{ and } N'/t' \in r_2 \text{ and } V(t) = V(t') \text{ and } sbs(N/t,\ N'/t')) \text{ or} \\ & \quad (N/t \in r_2 \text{ and } N'/t' \in r_1 \text{ and } V(t) = V(t') \text{ and } sbs(N/t,\ N'/t'))\} \end{aligned} \tag{8}$$

where $V(t)$ is the tuple formed with the attribute values of t made certain. For instance, if $t = \langle 17, \text{John}, \text{Engineer*}, \text{Boston}, 35^* \rangle$, then $V(t) = \langle 17, \text{John}, \text{Engineer}, \text{Boston}, 35 \rangle$.

Example 5. Consider the relations from Table 6 (left and middle). The result of the union query $r_1 \cup r_2$ is given in Table 6 (right). ◇

3.6 Difference

Let us now consider the difference $r_1 - r_2$. Its definition is as follows:

$$\begin{aligned} r_1 - r_2 = \{N'/t \mid N/t \in r_1 \text{ and } N' = \min(N,\ \delta) \text{ with} \\ \delta = \min_{t' \in r_2} \max_{i=1..n} N(t.A_i \neq t'.A_i)\} \end{aligned} \tag{9}$$

where

$$N(t.A_i \neq t'.A_i) = \begin{cases} 1 \text{ if } t.A_i \neq t'.A_i \text{ and } c(t.A_i) = c(t'.A_i) = 1, \\ 0 \text{ if } t.A_i = t'.A_i, \\ \alpha \text{ otherwise.} \end{cases}$$

Table 6 Relations r_1 (left), r_2 (middle), and $r_1 \cup r_2$ (right)

Job	*City*	*N*
Engineer*	Boston	1
Engineer*	Quincy*	α
Manager	Newton	1

Job	*City*	*N*
Manager*	Boston	α
Manager	Newton	1
Engineer*	Boston*	α

Job	*City*	*N*
Engineer*	Boston	1
Engineer*	Quincy*	α
Manager	Newton	1
Manager*	Boston	α

In Formula 9, δ corresponds to the certainty degree associated with the event "all the tuples t' from r_2 differ from t" or, in other words, "for every tuple t' that appears in r_2, one of the attribute values of t' differs from the corresponding attribute value in t". The universal quantifier ("for every tuple") is interpreted by a min whereas the existential one ("one of the attribute values") is interpreted by a max.

Example 6. Consider the relations from Table 5 (left and middle). The result of $r_1 - r_2$ is given in Table 7 (right). Note that the tuple ⟨Manager, Chicago⟩ is uncertain in the result because of the tuple ⟨Cashier*, Boston*⟩ from r_2 which has a $(1-\alpha)$-possible interpretation equal to ⟨Manager, Chicago⟩. ◇

Table 7 Relations r_1 (left), r_2 (middle), and $r_1 - r_2$ (right)

Job	*City*	*N*
Engineer*	Boston*	1
Engineer	Quincy*	1
Manager	Chicago	1
Clerk	Springfield	1

Job	*City*	*N*
Engineer	Newton*	α
Clerk	Quincy*	α
Clerk	Springfield	α
Cashier*	Boston*	α

Job	*City*	*N*
Engineer*	Boston*	α
Engineer	Quincy*	α
Manager	Chicago	α

A crucial point is that the join operation does not induce intertuple dependencies in the result, due to the semantics of certainty. This is not the case when a probabilistic or a full possibilistic [1] model is used, and one then has to use a variant of c-tables [6] to handle these dependencies, which implies a non-polynomial complexity. On the other hand, since none of the operators of relational algebra induces intertuple dependencies in our model, the queries have the same data complexity as in a classical database context; see [9] for a more complete discussion.

4 About Query Equivalences

Let us recall that relational algebraic queries can be represented as a tree where the internal nodes are operators, leaves are relations, and subtrees are subexpressions. The primary goal of query optimization is to transform expression trees into equivalent ones, where the average size of the relations yielded by subexpressions in the tree are smaller than they were before the optimization. This transformation process uses a set of properties (query equivalences), and the question arises whether these properties remain valid in the certainty-based model. The most common query equivalences are:

1. $\pi_X(\pi_{XY}(r)) = \pi_X(r)$,
2. $\sigma_{\psi_2}(\sigma_{\psi_1}(r)) = \sigma_{\psi_1}(\sigma_{\psi_2}(r)) = \sigma_{\psi_1 \wedge \psi_2}(r)$,
3. $\sigma_{\psi_1 \vee \psi_2}(r) = \sigma_{\psi_1}(r) \cup \sigma_{\psi_2}(r)$,
4. $\sigma_{\psi}(\pi_X(r)) = \pi_X(\sigma_{\psi}(r))$ if ψ concerns X only,
5. $\sigma_{\psi}(r_1 \times r_2) = \sigma_{\psi_1 \wedge \psi_2 \wedge \psi_3}(r_1 \times r_2) = \sigma_{\psi_3}(\sigma_{\psi_1}(r_1) \times \sigma_{\psi_2}(r_2))$ where $\psi = \psi_1 \wedge \psi_2 \wedge \psi_3$ and ψ_1 concerns only attributes from r_1, ψ_2 concerns only attributes from r_2, and ψ_3 is the part of ψ that concerns attributes from both r_1 and r_2,
6. $\sigma_{\psi}(r_1 \cup r_2) = \sigma_{\psi}(r_1) \cup \sigma_{\psi}(r_2)$,
7. $\sigma_{\psi}(r_1 \cap r_2) = \sigma_{\psi}(r_1) \cap \sigma_{\psi}(r_2) = \sigma_{\psi}(r_1) \cap r_2 = r_1 \cap \sigma_{\psi}(r_2)$,
8. $\sigma_{\psi}(r_1 - r_2) = \sigma_{\psi}(r_1) - \sigma_{\psi}(r_2) = \sigma_{\psi}(r_1) - r_2$,
9. $\pi_X(r_1 \cup r_2) = \pi_X(r_1) \cup \pi_X(r_2)$,
10. $\pi_Z(r_1 \times r_2) = \pi_X(r_1) \times \pi_Y(r_2)$ if X (resp. Y) denotes the subset of attributes of Z present in r_1 (resp. r_2).

It is straightforward to prove that all of these equivalences remain valid in the model we propose (they are direct consequences of the definitions of the operators given above). As an illustration, let us demonstrate Property 3.

Proof. Let us assume that condition ψ_1 (resp. ψ_2) concerns attribute A_1 (resp. A_2). Let us consider a tuple N/t from r such that $\psi_1(t.A_1) \vee \psi_2(t.A_2)$ holds (which is a necessary condition for t to belong to $(\sigma_{\psi_1 \vee \psi_2}(r))$ on the one hand, and to $(\sigma_{\psi_1}(r) \cup \sigma_{\psi_2}(r)))$ on the other hand. Four cases have to be considered:

- $c(t.A_1) = 1$ and $c(t.A_2) = 1$: then, N/t generates a tuple N/t in $\psi_1(t.A_1) \vee \psi_2(t.A_2)$ (indeed $\min(N,\ \max(1,\ 1)) = \min(N,\ 1) = N$) and a tuple N/t both in σ_{ψ_1} and in σ_{ψ_2}, thus in $(\sigma_{\psi_1}(r) \cup \sigma_{\psi_2}(r))$.
- $c(t.A_1) = 1$ and $c(t.A_2) = \alpha$: then, N/t generates a tuple N/t in $\psi_1(t.A_1) \vee \psi_2(t.A_2)$ (indeed $\min(N,\ \max(1,\ \alpha)) = \min(N,\ 1) = N$). It also generates a tuple N/t in σ_{ψ_1} and a tuple α/t in σ_{ψ_2}. Thus, it generates a tuple N/t in $(\sigma_{\psi_1}(r) \cup \sigma_{\psi_2}(r))$ since N/t subsumes (or is equal to) α/t.
- $c(t.A_1) = \alpha$ and $c(t.A_2) = 1$: this case is similar to the previous one.
- $c(t.A_1) = \alpha$ and $c(t.A_2) = \alpha$: then N/t generates a tuple α/t in $\psi_1(t.A_1) \vee \psi_2(t.A_2)$ (indeed $\min(N,\ \max(\alpha,\ \alpha)) = \alpha$). It also generates a tuple α/t in σ_{ψ_1} and in σ_{ψ_2}. Thus, it generates a tuple α/t in $(\sigma_{\psi_1}(r) \cup \sigma_{\psi_2}(r))$.

5 Making Selection Queries More Flexible

If one assumes that the relation concerned by a selection condition is a base relation (i.e., where all the tuples have a degree $N = 1$), a tuple in the result is uncertain iff it involves at least one suspect value concerned by the selection condition. If such a tuple involves several such suspect values, it will be no more uncertain ($N = \alpha$) than if it involves only one. However, one may find it desirable to distinguish between these situations. For instance, considering the query

$$\sigma_{job=\text{`}Engineer' \text{ and } city=\text{`}Boston' \text{ and } age=30}(Emp)$$

the tuple ⟨John, Engineer*, Boston, 30⟩ could be judged more satisfactory (less risky) than, e.g., ⟨Bill, Engineer*, Boston*, 30⟩, itself more satisfactory than ⟨Paul, Engineer*, Boston*, 30*⟩.

For a selection condition $\psi = \psi_1 \wedge \ldots \psi_m$ and a tuple t, this amounts to saying that "every attribute value (certain and suspect) of t must satisfy the condition ψ_i that concerns it, and the less there are suspect values concerned by a ψ_i in t, the more t is preferred". In other words, the selection condition becomes:

$\psi_1 \wedge \ldots \wedge \psi_m$ and *as many* $(t.A_1, \ldots, t.A_m)$ *as possible* are totally certain.

In a user-oriented language based on the algebra described above, one may then introduce an operator IS CERTAIN (meaning "is totally certain"), in the same way as there exists an operator IS NULL in SQL.

The fuzzy quantifier [12] *as many as possible* (*amap* for short) corresponds to a function from [0, 1] to [0, 1]. Its associated membership function μ_{amap} is such that: i) $\mu_{amap}(0) = 0$, ii) $\mu_{amap}(1) = 1$, iii) $\forall x, y, x > y \Rightarrow \mu_{amap}(x) > \mu_{amap}(y)$. Typically, we shall take $\mu_{amap}(x) = x$.

The selection condition as expressed above is made of two parts: a "value-based one" — that may generate uncertain answers —, and a "representation-based" one that generates gradual answers. A tuple of the result is assigned a satisfaction degree μ (seen as the complement to 1 of a suspicion degree), on top of its certainty degree N. For a conjunctive query made of m atomic conjuncts ψ_i, the degree μ associated with a tuple t is computed as follows:

$$\mu(t) = \mu_{amap}\left(\frac{\sum_{i=1}^{m} certain(t, i)}{m}\right) \tag{10}$$

where

$$certain(t, i) = \begin{cases} 1 \text{ if } \psi_i \text{ if of the form } A\ \theta\ q \text{ and } c(t.A) = 1, \\ 1 \text{ if } \psi_i \text{ if of the form } A_1\ \theta\ A_2 \text{ and } \min(c(t.A_1), c(t.A_2)) = 1, \\ 0 \text{ otherwise.} \end{cases}$$

In order to display the result of the query, one rank-orders the answers on N first, then on μ (in a decreasing way in both cases).

Example 7. Let us consider the relation represented in Table 1 (top) and the selection query $\sigma_\psi(Emp)$ where ψ is the condition

job = 'Engineer' and *city* = 'Boston' and *age* > 30 and
amap(*job* IS CERTAIN, *city* IS CERTAIN, *age* IS CERTAIN)

Let us assume that the membership function associated with the fuzzy quantifier *amap* is $\mu_{amap}(x) = x$. The result of the selection query is represented in Table 8 (bottom). ◇

Table 8 Relation *Emp* (top), result of the selection query (bottom)

#id	name	city	job	age	N
38	John	Boston*	Engineer*	32	1
54	Mary	Quincy*	Engineer*	35	1
72	Bill	Boston	Engineer	40	1
81	Paul	Boston*	Engineer*	31*	1
93	Phil	Boston	Engineer	52*	1

#id	name	city	job	age	N	μ
72	Bill	Boston	Engineer	40	1	1
93	Phil	Boston	Engineer	52*	α	0.67
38	John	Boston*	Engineer*	32	α	0.33
81	Paul	Boston*	Engineer*	31*	α	0

This extended framework, where two degrees (N and μ) are associated with each tuple in the relations, can be easily made compositional. One just has to manage the degrees μ, in the definition of the algebraic operators, as in a gradual (fuzzy) relation context, see [7]. In base relations, it is assumed that $\mu(t) = 1\ \forall t$.

Notice that an alternative solution, of a more qualitative nature, is also possible for discriminating the tuples that (somewhat certainly) satisfy a conjunctive selection condition. It consists in using the lexicographic ordering (denoted by $>_{lex}$ hereafter) as follows. For every tuple t of the result, one builds a vector $V(t)$ of m values (1 or α):

$$V(t)[i] = \begin{cases} 1 \text{ if } \psi_i \text{ if of the form } A\ \theta\ q \text{ and } c(t.A) = 1, \\ 1 \text{ if } \psi_i \text{ if of the form } A_1\ \theta\ A_2 \text{ and } \min(c(t.A_1),\ c(t.A_2)) = 1, \\ \alpha \text{ otherwise.} \end{cases}$$

Then $V(t)$ is transformed into $V'(t)$ by ordering the i components in decreasing order (first the 1's, then the α's if any). Finally,

$$t \succ t' \Leftrightarrow V'(t) >_{lex} V'(t') \tag{11}$$

where $t \succ t'$ means that t is considered a more satisfactory answer than t'.

A refinement consists in taking into account that atomic conditions in ψ may have different importance levels. Let us denote by $w(\psi_i)$ the weight (importance level) associated with the atomic condition ψ_i and let us assume that the greater $w(\psi_i)$, the more important ψ_i. With the first method (based on scores), the idea consists in using a weighted mean instead of an arithmetic mean in Formula 10 for computing $\mu(t)$. With the second method (based on lexicographic ordering), the idea would be, in case of ties, to use the complete preorder on the importance levels of the attributes for introducing another lexicographic refinement in order to break these ties.

6 Conclusion

In this paper, we have proposed a database model and defined the associated algebraic operators for dealing with the situation where some attribute values in a dataset are suspect, i.e., have an uncertain validity, in the absence of further information about the precise levels of uncertainty attached to such suspect values. The framework used is that of possibility theory restricted to a certainty scale made of three levels. It is likely that the idea of putting some kind of tags on suspect values/tuples/answers is as old as information systems. However, the benefit of handling such a symbolic tag in the framework of possibility theory is to provide a rigorous setting for this processing.

A crucial point is that the data complexity of all of the algebraic operations is the same as in the classical database case, i.e., it is either linear or polynomial, which makes the approach perfectly tractable. Moreover, the definitions of both the model and the operators are quite simple and do not raise any serious implementation issues.

It is worth mentioning that this model could be rather straightforwardly extended in order to handle disjunctive suspect values — of the form $(v_1 \vee \ldots v_n)^*$ — instead of singletons. The definitions of the operators would then become a bit more complicated, but the data complexity would not change. Another extension would be to use a small set of suspicion levels rather than one. This would bring us closer to the general setting of the certainty-based model described in [9].

References

1. Bosc, P., Pivert, O.: About projection-selection-join queries addressed to possibilistic relational databases. IEEE Trans. on Fuzzy Systems **13**(1), 124–139 (2005)
2. Bosc, P., Pivert, O., Prade, H.: A model based on possibilistic certainty levels for incomplete databases. In: Godo, L., Pugliese, A. (eds.) SUM 2009. LNCS, vol. 5785, pp. 80–94. Springer, Heidelberg (2009)
3. Dubois, D., Prade, H.: Possibility Theory. Plenum, New York (1988)
4. Haas, P.J., Suciu, D.: Special issue on uncertain and probabilistic databases. VLDB J. **18**(5), 987–988 (2009)
5. Hastings, D., McManus, H.: A framework for understanding uncertainty and its mitigation and exploitation in complex systems. In: Proc. of Engineering Systems Symposium MIT (2004)
6. Imielinski, T., Lipski, W.: Incomplete information in relational databases. J. of the ACM **31**(4), 761–791 (1984)
7. Pivert, O., Bosc, P.: Fuzzy Preference Queries to Relational Databases. Imperial College Press, London (2012)
8. Pivert, O., Prade, H.: Database querying in the presence of suspect values. In: Morzy, T., Valduriez, P., Bellatreche, L. (eds.) ADBIS 2015. CCIS, vol. 539, pp. 44–51. Springer, Heidelberg (2015)

9. Pivert, O., Prade, H.: A certainty-based model for uncertain databases. IEEE Transactions on Fuzzy Systems (2015) (to appear)
10. Suciu, D., Olteanu, D., Ré, C., Koch, C.: Probabilistic Databases. Synthesis Lectures on Data Management. Morgan & Claypool Publishers (2011)
11. Zadeh, L.: Fuzzy sets as a basis for a theory of possibility. Fuzzy Sets and Systems **1**(1), 3–28 (1978)
12. Zadeh, L.: A computational approach to fuzzy quantifiers in natural languages. Computing and Mathematics with Applications **9**, 149–183 (1983)

Representation and Identification of Approximately Similar Event Sequences

T.P. Martin and B. Azvine

Abstract The MARS (Modelling Autonomous Reasoning System) project aims to develop a collaborative intelligent system combining the processing powers and visualisation provided by machines with the interpretive skills, insight and lateral thinking provided by human analysts. There is an increasing volume of data generated by online systems, such as internet logs, transaction records, communication records, transport network monitors, sensor networks, etc. Typically, these logs contain multiple overlapping sequences of events related to different entities. Information that can be mined from these event sequences is an important resource in understanding current behaviour, predicting future behaviour and identifying non-standard patterns. In this paper, we describe a novel approach to identifying and storing sequences of related events, with scope for approximate matching. The event sequences are represented in a compact and expandable *sequence pattern* format, which allows the addition of new event sequences as they are identified, and subtraction of sequences that are no longer relevant. We present an algorithm enabling efficient addition of a new sequence pattern. Examination of the sequences by human experts could further refine and modify general patterns of events.

1 Introduction

With the increasing volume of data generated by online systems (internet logs, transaction records, communication records, transport networks, sensor networks, etc.) there is a clear requirement for machine assistance in filtering, fusing and finding causal relations in data flows. Current AI approaches rely heavily on statistical and

T.P. Martin(✉)
Machine Intelligence & Uncertainty Group, University of Bristol, Bristol BS8 1UB, UK
e-mail: trevor.martin@bristol.ac.uk

T.P. Martin · B. Azvine
BT TSO Security Futures, Adastral Park, Ipswich IP5 3RE, UK

T. Andreasen et al. (eds.), *Flexible Query Answering Systems 2015*,
Advances in Intelligent Systems and Computing 400,
DOI: 10.1007/978-3-319-26154-6_7

machine learning algorithms which require large amounts of data and rest on an assumption that the previous behaviour of a system is a good guide to future behaviour.

The aim of the MARS[1] (Modeling Autonomous Reasoning System) project is to develop a collaborative intelligent system able to process real-time (or near real-time) unstructured data feeds by combining the processing powers and visualisation provided by machines with the interpretive skills, insight and lateral thinking provided by human analysts. The overall product of this research is a cognitive software system that exhibits highly flexible and autonomous behaviour, and that can provide insights into the functioning of a broad range of complex social-technical systems monitored by data flows. An early version of MARS was successful in helping to combat cable theft[2], leading to significant savings for UK telecoms and rail networks.

A key aspect in the analysis of lower level data such as event logs is the identification of sequences of linked events. For example, internet logs, physical access logs, transaction records, email and phone records contain multiple overlapping sequences of events related to different users of a system. Information that can be mined from these event sequences is an important resource in understanding current behaviour, predicting future behaviour and identifying non-standard patterns and possible security breaches.

An event sequence is a time-stamped series of observations or measurements of a fixed set of attributes. An example could be a sequence of records containing:
access card no., date/time, building and entrance identifier, direction of access (in/out), result (access granted / refused)
or
source IP address/port, destination IP address/port, date/time, amount of data transferred.

Specific problems in extracting sequences of related events include determination of what makes events "related", finding groups of "similar" sequences, identification of typical sequences, and detection of sequences that deviate from previous patterns. In this paper we describe a novel approach to identifying sequences of related events, with scope for assistance from human experts. The event sequences are represented in a compact and expandable sequence pattern format, which allows the addition of new event sequences as they are identified, and subtraction of sequences that are no longer relevant. Examination of sequences can be used to refine and modify general patterns of events. The specific focus of this short paper is the algorithm to incrementally add a new sequence pattern.

2 Background

Humans are adept at acquiring knowledge of sequential patterns, even when the knowledge is acquired and used in an implicit manner, i.e. without conscious aware-

[1] UK Technology Strategy Board Reference: 1110_CRD_TI_ICT_120182

[2] www.newscientist.com/article/dn21989-ai-system-helps-spot-signs-of-copper-cable-theft.html

ness [1]. In an effort to reproduce this learning behaviour in machines, investigators have focused on statistical methods such as time-series analysis and machine learning approaches. These are not ideally suited to the task, because they generally require a large sample of data, usually represented as numerical features, and because they typically seek to fit data to known distributions. There is evidence that human behaviour sequences can differ significantly from such distributions - for example, in sequences of asynchronous events such as sending emails or exchanging messages. Barabasi [2] showed that many activities do not obey Poisson statistics, and consist instead of short periods of intense activity followed by longer periods in which there is no activity.

A related problem is that statistical machine learning methods generally require a significant number of examples to form meaningful models, able to make predictions. Where we are looking for a new behaviour pattern (for example, in network intrusion) it may be important to detect it quickly, before a statistically significant number of incidents have been seen. A malicious agent may even change the pattern before it can be detected. Finally, it is highly advantageous to allow human experts to specify patterns. This is difficult or impossible with data-driven methods.

2.1 *Fuzzy Indiscernibility of Events*

A fundamental aspect of human information processing and intelligence is the ability to identify a group of entities e.g. physical objects, events, abstract ideas) and subdivide them into smaller subgroups at an appropriate level of granularity for the task at hand. Early humans needed to determine whether an animal was dangerous or not; for the non-dangerous case, a second level of analysis might distinguish between edible and non-edible (or uncatchable). From the perspective of decision-making (stay or run, hunt or ignore), the precise identification of different species is not relevant - there is no need for discernibility within the broad sub-categories.

The concept of information granulation was introduced by Zadeh [3] to formalise the process of dividing a group of objects into sub-groups (granules) based on "indistinguishability, similarity, proximity or functionality". In this view, a granule is a fuzzy set whose members are (to some degree) equivalent to each other. In a similar manner, humans are good at dividing events into related groups, both from the temporal perspective (event A occurred a few minutes before event B but involves the same entities) and from the perspective of non-temporal event properties (event C is very similar to event D because both involve entities which are similar).

2.2 *The X-Mu Approach to Fuzzy*

Standard fuzzy approaches typically require modification of crisp algorithms to allow set-valued variables. This is most apparent in fuzzifications of arithmetic, where a single value is replaced by a (so-called) fuzzy number, which actually represents an *interval*. For example, calculating the average age of four employees known to be 20, 30, 50 and 63 is inherently simpler than the same calculation when the ages are given as *young, quite young, middle-aged* and *approaching retirement*. In the latter case, we must handle interval arithmetic as well allowing for membership grades. In a similar fashion, querying a database to find employees who are aged over 60 is simpler than finding employees who are *approaching retirement age*.

In the classical fuzzy approach, for any predicate on a universe U, we introduce a membership function

$\mu : U \rightarrow [0,\ 1]$

representing the degree to which each value in U satisfies the predicate; since there is generally a set of values which satisfy the predicate to some degree, we must modify algorithms to handle sets of values rather than single values. These sets represent equivalent values - that is, values which cannot be distinguished from each other. In this work, we are dealing with events that are equivalent because their attributes are indiscernible - however, these sets of events may vary according to membership, interpreted as the degree to which elements can be distinguished from each other.

The $X - \mu$ method [4] recasts the fuzzy approach as a mapping from membership to universe, allowing us to represent a set, interval or a single value that varies with membership. An example of a single value is the mid-point of an interval. This is a natural idea which is difficult to represent in standard fuzzy theory, even though it arises in common aspects such as the cardinality of a discrete fuzzy set or the number of answers returned in response to a fuzzy query.

3 Algorithm for Converting a Stream of Events into a Minimal Directed Graph

3.1 Directed Graph Representation of Event Sequences

For any sequence of events, we create a directed graph representation of the events in which each edge represents a set of indiscernible events. Clearly for reasons of storage and searching efficiency it is desirable to combine event sequences with common sub-sequences, as far as possible, whilst only storing event sequences that have been observed. This problem is completely analogous to dictionary storage, where we are dealing with single letters rather than sets of events, and we can utilise efficient solutions that have been developed to store dictionaries. In particular, we adopt the notion of a directed acyclic word graph, or DAWG [5]. Words with common letters (or events) at the start and/or end are identified and the common paths are merged to give a minimal graph, in the sense that it has the smallest number of nodes for a DAWG representing the set of words (event sequences). Several algorithms for creating minimal DAWGs have been proposed. In the main, these have been applied to creation of dictionaries and word checking, efficient storage structure for lookup of key-value pairs and in DNA sequencing (which can be viewed as a variant of dictionary storage). Most methods (e.g. [6, 7]) operate under the assumption that all words (letter sequences) are available and can be presented to the algorithm in a specific order. Sgarbas [5] developed an incremental algorithm which allowed additional data to be added to a DAWG structure, preserving the minimality criterion (i.e. assuming the initial DAWG represented the data in the most compact way, then the extended DAWG is also in the most compact form).

We assume that the data is in a time-stamped tabular format (for example, as comma separated values with one or more specified fields storing date and time information) and that data arrives in a sequential manner, either row by row or in larger groups which can be processed row-by-row. Each row represents an event; there may be several unrelated event sequences within the data stream but we assume events in a single sequence arrive in time order. It is not necessary to store the data once it has been processed, unless required for later analysis.

Each column i in the table has a domain D_i and a corresponding attribute name A_i . There is a special domain O whose elements act as unique identifiers (e.g. row number or event id).

Formally, the data can be represented by a function

$$f : O \to D_1 \times D_2 \times \cdots \times D_n \text{ or a relation } R \subset O \times D_1 \times D_2 \times \cdots \times D_n$$

where any given identifier $o_i \in O$ appears at most once. We use the notation $A_k(o_i)$ to denote the value of the kth attribute for object o_i. Our aim is to find ordered sequences of events (and subsequently, groups of similar sequences, where the grouping is based on indiscernibility of attributes).

3.2 Sequence-Extending Relations

Event sequences obey

- each event is in at most one sequence
- events in a sequence are ordered by date and time
- an event and its successor are linked by relations between their attributes, such as equivalence, tolerance, and other relations.

These are referred to as *sequence-extending relations*. Note that it is possible to have different sequence-extending relations for different sequences; also it is possible to change the sequence-extending relations dynamically. In the graph structure described below, the sequence-extending relations are associated with nodes in the graph. It is also possible to identify sequences when an event is missing, although this is not covered here. Any event that is not part of an existing sequence is the start of a new sequence. For any attribute A_i we can define a tolerance relation R_i where

$$R_i : D_i \times D_i \to [0,\ 1]$$

is a reflexive and symmetric fuzzy relation and $\forall j : \ R_i\left(o_j,\ o_j\right) = 1$ Then the tolerance class of objects linked through attribute A_i is

$$T\left(A_i,\ o_m\right) = \left\{o_j / \chi_{mj} \left| R_i\left(A_i\left(o_m\right),\ A_i\left(o_j\right)\right) = \chi_{mj} \right.\right\}$$

Note that this set includes (with membership 1) all objects with the attribute value $A_i(o_m)$. The tolerance class can be expressed equivalently as a set of pairs.

Finally we requiref a total order relation P_T, defined on a distinguished attribute (or small subset of attributes) representing a timestamp. We can then define sequences (in the obvious way) and projected sequences

$$\begin{array}{l}\forall i : \ P_T\left(A_T\left(o_i\right),\ A_T\left(o_i\right)\right)\\ \forall i \neq j : \ P_T\left(A_T\left(o_i\right),\ A_T\left(o_j\right)\right) > 0 \ \rightarrow \ P_T\left(A_T\left(o_j\right),\ A_T\left(o_i\right)\right) = 0\\ Q\left(o_i\right) = \left\{o_i / \chi_{ti} \ \middle| P_T\left(o_t,\ o_i\right) = \chi_{ti}\right\}\end{array}$$

where A_T is the timestamp attribute (or attributes) and the ordering of events models the obvious temporal ordering. The time attribute t_i obeys $t_i \leq t_{i+1}$ for all i. It is treated as a single attribute although it could be stored as more than one (such as date, time of day). We assume that a number of sequence-extending relations $R_1 \ldots R_n$ are defined on appropriate domains.

Two events o_i and o_j are potentially linked in the same sequence if

$$\min\left(Q_T\left(o_i,\ o_j\right),\ \min_m R_m\left(o_i,\ o_j\right)\right) \geq \mu$$

i.e. all required attributes satisfy the specified sequence-extending relations to a degree greater than some threshold μ. We write

$$potential - link\left(o_i\ ,\ o_j\ ,\mu\right) \leftrightarrow \min\left(Q_T\left(o_i,\ o_j\right),\ \min_m R_m\left(o_i,\ o_j\right)\right) \geq \mu$$

and

$$\begin{array}{c}linked\left(o_i\ ,\ o_j\ ,\mu\right) \leftrightarrow potential - link\left(o_i\ ,\ o_j\ ,\mu\right)\\ AND\\ \nexists o_k : potential - link\left(o_i\ ,\ o_k\ ,\mu\right)\ AND\ potential - link\left(o_{ki}\ ,\ o_j\ ,\mu\right)\end{array}$$

i.e. two events are linked if they satisfy the specified tolerance and equivalence relations to a degree μ or greater, and there is no intermediate event.

3.3 Event Categorisation Relations

We also define equivalence classes on some of the domains, which are used to compare and categorise events from different sequences. An equivalence class on one or more domains is represented by a value from each domain - for times denoted by *day* and *hour* values, we might define equivalence for weekday rush hour (*day=Mon-Fri, hour=8,9,17,18*), other-weekday (*day=Mon-Fri, hour≠8,9,17,18*) and weekend (*day=Sat,Sun*).

These can easily be extended to fuzzy equivalence classes. The important feature is that the equivalence classes partition the objects - i.e. each object belongs to exactly one equivalence class for each domain considered. In the $X - \mu$ fuzzy case, the equivalence classes can vary with membership but always partition the objects. Where necessary operations can be incrementally extended to cover different memberships.

Formally, for a specified attribute A_i we define

$$S\left(A_i, o_m\right) = \left\{o_j \,\middle|\, A_i\left(o_j\right) = A_i\left(o_m\right)\right\}$$

and the set of associated equivalence classes (also called elementary concepts) is

$$C_i = \left\{S\left(A_i, o_m\right) \,\middle|\, o_m \in O\right\}$$

(as an example consider time or elapsed time below). In the propositional case C_i contains just one set, whose elements are the objects for which attribute i is true. In the fuzzy case, elements are equivalent to some degree. Specifying a membership threshold gives a nested set of equivalence relations, so that once a membership threshold is known, we can proceed as in the crisp case. The operation can be extended to multiple attributes.

We use the selected attributes to find the *EventCategorisation*. This is an ordered set of equivalence classes from one or more attributes (or $n-$tuples of attributes):

$$\begin{aligned}&B_k \in \{A_1, \ \ldots \ , \ A_n\}\\&EventCategorisation\left(o_i\right) = \left[B_k\left(o_i\right) \,\middle|\, k = 1, \ \ldots \ , \ m\right]\end{aligned}$$

i.e. each B_k is an attribute, and the event categorisation of some object o_i is given by the equivalence classes corresponding to its attribute values. The result is not dependent on the order in which the attributes are processed. This order can be optimised to give fastest performance when deciding which edge to follow from given node.

For any set of sequences, we create a minimal representation using a DASG (directed acyclic sequence graph) as shown in Fig. 1. The graph is a deterministic finite automaton, with no loops and a unique end node. Each event is represented by a labelled edge. The edge label shows the equivalence classes applicable to the event, referred to as the event categorisation. The source node S is a single starting point for all sequences; to ensure we have a unique end node, F, we append a dummy "end of sequence" ($\#END$) event to all sequences.

3.4 Worked Example Based on the VAST Challenge 2009 Dataset

We illustrate the algorithm using a small subset (Table 1) of benchmark data taken from mini-challenge 1 of the VAST 2009 dataset[3]. This gives swipecard data showing

[3] `http://hcil2.cs.umd.edu/newvarepository/benchmarks.php`

Table 1 Sample Data from the VAST 2009 MC1 Dataset

eventID	Date	Time	Emp	Entrance	Direction
1	jan-2	7:30	10	b	in
2	jan-2	13:30	10	b	in
3	jan-2	14:10	10	c	in
4	jan-2	14:40	10	c	out
5	jan-2	9:30	11	b	in
6	jan-2	10:20	11	c	in
7	jan-2	13:20	11	c	out
8	jan-2	14:10	11	c	in
9	jan-2	15:00	11	c	out
10	jan-3	9:20	10	b	in
11	jan-3	10:40	10	c	in
12	jan-3	14:00	10	c	out
13	jan-3	14:40	10	c	in
14	jan-3	16:50	10	c	out
15	jan-3	9:00	12	b	in
16	jan-3	10:20	12	c	in
17	jan-3	13:00	12	c	out
18	jan-3	14:30	12	c	in
19	jan-3	15:10	12	c	out

employee movement into a building and in and out of a classified area within the building. No data is provided on exiting the building. In this data,

Emp = set of employee ids = {10, 11, 12}
Date,Time = date / time of event
Entry points = {B - building, C - classified section}
Access direction = {in, out}

We have selected three employees for illustration purposes; rows in the initial table were ordered by date/time, but have been additionally sorted by employee here to make the sequences obvious. We first define the sequence-extending relations, to detect candidate sequences. Here, for a candidate sequence of n events:

$$S_1 = (o_{11}, o_{12}, o_{13}, \ \dots \ , o_{1n})$$

we define the following computed quantities :

$$\begin{aligned} ElapsedTime \quad \Delta T_{ij} &= Time\left(o_{ij}\right) - Time\left(o_{ij-1}\right) \\ with \quad \Delta T_{i1} &= Time\left(o_{i1}\right) \end{aligned}$$

and restrictions (for $j > 1$) :

Table 2 Allowed Actions (row = first action, column = next action)

	b,in	c,in	c,out
b,in	x	x	
c.in			x
c,out	x	x	

$$\begin{aligned} Date\left(o_{ij}\right) &= Date\left(o_{ij-1}\right) \\ 0 < Time\left(o_{ij}\right) - Time\left(o_{ij-1}\right) &\leq T_{thresh} \\ Emp\left(o_{ij}\right) &= Emp\left(o_{ij-1}\right) \\ \left(Action\left(o_{ij-1}\right), Action\left(o_{ij}\right)\right) &\in AllowedActions \\ \text{where} \quad Action\left(o_{ij}\right) &= \left(Entrance\left(o_{ij}\right),\ Direction\left(o_{ij}\right)\right) \end{aligned}$$

where the relation *AllowedActions* is specified in Table 2.

These constraints can be summarised as

- events in a single sequence refer to the same employee
- successive events in a single sequence conform to allowed transitions between locations and are on the same day, within a specified time (T_{thresh}) of each other

We choose a suitable threshold e.g. $T_{thresh} = 8$, ensuring anything more than 8 hours after the last event is a new sequence. We identify candidate sequences by applying the sequence-extending relations. Any sequence has either been seen before or is a new sequence. In Table 1, candidate sequences are made up of the events:
$1 - 2 - 3 - 4$,
$5 - 6 - 7 - 8 - 9$,
$10 - 11 - 12 - 13 - 14$,
$15 - 16 - 17 - 18 - 19$

We also define the *EventCategorisation* equivalence classes used to compare events in different sequences. Here,

$$\begin{aligned} EquivalentAction &= I_{Action} \\ \text{For direction } In, \quad EquivalentEventTime &= \{[7], [8], \ldots\} \\ \text{For direction } Out, \quad EquivalentElapsedTime &= \{[0], [1], [2], \ldots\} \end{aligned}$$

where I is the identity relation and the notation [7] represents the set of start times from 7:00-7:59. With this definition, events 5 and 10 are equivalent since both have *Entrance=b, Direction=In* and *Time* in 7:00-7:59. Formally,

$$\begin{aligned} EventCategorisation\left(o_5\right) &= \left([b, in], [7]\right) \\ EventCategorisation\left(o_{10}\right) &= \left([b, in], [7]\right) \end{aligned}$$

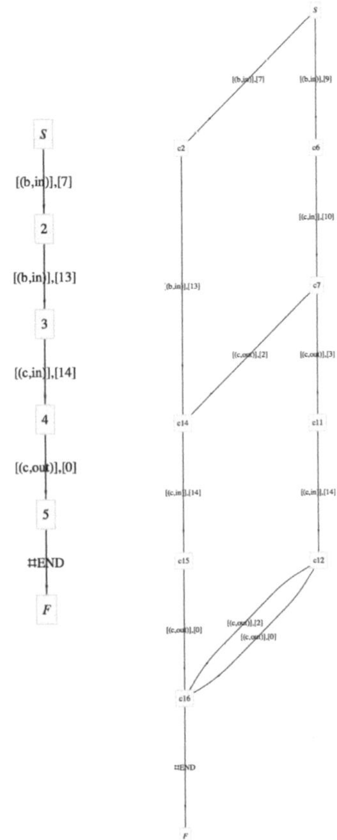

Fig. 1 (left) Event sequence 1-2-3-4 from Table 1. The label [(b,in)],[7] represents all events whose entrance and direction is (b, in) and time is equivalent to 7 (at a given membership). (right) DASG representing the 4 distinct sequences in Table 1

Similarly, events 7 and 12 are equivalent, as both have *Entrance =c, Direction = Out* and *ElapsedTime* in 3:00-3:59.

We represent each identified sequence as a path labelled by its event categorisations (Fig. 1 left) and combine multiple sequences into a minimal DASG representing all sequences seen so far (Fig. 1 right). Nodes are labelled by unique numbering; since the graph is deterministic, each outgoing edge is unique. An edge can be specified by its start node and event categorisation. Below, we also refer to an edge by its event categorisation if there is no ambiguity about its start node.

Standard definitions are used for *InDegree, OutDegree, IncomingEdges* and *OutgoingEdges* of a node, giving respectively the number of incoming and outgoing edges, the set of incoming edges and the set of outgoing edges. We also apply

functions *Start* and *End* to an edge, to find or set its start and end nodes respectively and *EdgeCategorisation* to find its categorisation class.
Finally, let the function $ExistsSimilarEdge(edge, endnode)$ return true when:
edge has end node *endnode*, event categorisation *L* and start node *S1*
AND
a second, distinct, edge has the same end node and event categorisation *L* but a different start node $S2$
AND
$S1$ and $S2$ have $OutgoingEdges(S1) == OutgoingEdges(S2)$

If such an edge exists, its start node is returned by the function
$StartOfSimilarEdge(edge, endnode)$

The function $MergeNodes(Node1, Node2)$ deletes $Node2$ and merges its incoming and outgoing edges with those of $Node1$.

The function $CreateNewNode(Incoming, Outgoing)$ creates a new node with the specified sets of incoming and outgoing edges.

The graph can be used to identify sequences of events that have already been seen. If a new sequence is observed (i.e. a sequence which differs from each sequence in the graph by at least one event categorisation), it can be added to the graph using the algorithm below.

The algorithm proceeds in three stages. In the first and second parts, we move step-by-step through the event sequence and graph, beginning at the start node S. If an event categorisation matches an outgoing edge, we follow that edge to the next node and move to the next event in the sequence. If the new node has more than one incoming edge, we copy it; the copy takes the incoming edge that was just followed, and the original node retains all other incoming edges. Both copies have the same set of output edges. This part of the algorithm finds other sequences with one or more common starting events.

If we reach a node where there is no outgoing edge matching the next event's categorisation, we create new edges and nodes for the remainder of the sequence, eventually connecting to the end node F. As the sequence is new, we must reach a point at which no outgoing edge matches the next event's categorisation; if this happens at the start node S then the first stage is (effectively) missed.

Finally, in the third stage, we search for sequences with one or more common ending events. Where possible, the paths are merged. As shown in Sgarbas [5], adding a new sequence using the incremental algorithm takes time roughly in the order of $|S|$, the number of unique sequences stored.

```
Algorithm ExtendGraph
Input :Graph G (minimal current sequence), start node S, end node F
CandidateSequence Q[0 - NQ] representing the candidate sequence;
        each element is an event identifier. The sequence is not
         already present in the graph and is terminated by #END
Output : updated minimal graph, incorporating the new sequence
Local variables :  Node startNode, newNode, endNode, matchNode
        Edge currentEdge, matchEdge
        Categorisation currentCategorisation
        integer seqCounter;

startNode = S
seqCounter = 0
WHILE  EventCategorisation(S[seqCounter]) ∈ OutgoingEdges(StartNode)
    currentEdge = (startNode, EventCategorisation(Q[seqCounter] )
    endNode = End (currentEdge)
    IF InDegree (endNode) > 1
    THEN
        newNode =CreateNewNode({currentEdge}, OutgoingEdges(endNode))
        IncomingEdges(endNode) = IncomingEdges (endNode) -
    currentEdge
        startNode = newNode
    ELSE
        startNode = endNode
    seqCounter++
ENDWHILE

WHILE seqCounter < NQ                          // create new path
    currentEdge = (startNode, EventCategorisation (S[seqCounter]) )
    startNode = CreateNewNode({currentEdge}, { })
    seqCounter++
ENDWHILE

currentCategorisation = #END
currentEdge = (startNode, #END )      // last edge, labelled by #END

IncomingEdges(F) = IncomingEdges (F) + currentEdge
endNode = F
nextEdgeSet = {currentEdge}

WHILE  nextEdgeSet contains exactly one element (i.e currentEdge)
                AND ExistsSimilarEdge(currentEdge, endNode)

    matchNode = StartOfSimilarEdge(currentEdge, endNode)
    startNode = Start (currentEdge)
    IncomingEdges(endNode) = IncomingEdges (endNode) - {currentEdge}
    nextEdgeSet = IncomingEdges (startNode)
    IncomingEdges (matchNode)= nextEdgeSet∪IncomingEdges (matchNode)
    endNode = matchNode
    currentEdge ∈ edgeSet              // choose any element,
END WHILE                              //  "while" loop terminates if >1
```

Fig. 2 Extending a minimal graph by incremental addition of a sequence of edges

4 Summary

The paper outlines a way of storing event sequences in a compact directed graph format, and gives an efficient incremental algorithm to update the graph with an unseen sequence. A human expert can easily add sequence patterns, even if these have not been seen in the data yet. An algorithm to remove sequence patterns has also been developed and will be presented in a future paper, together with experimental results showing the efficiency and effectiveness of the approach.

Acknowledgments This work was partly supported by the UK Technology Strategy Board, grant reference: 1110_CRD_TI_ICT_120182.

References

1. Cleeremans, A., James, L., McClelland, J.: Learning the structure of event sequences. J. Experimental Psychology **120**, 235–253 (1991)
2. Barabasi, A.L.: The origin of bursts and heavy tails in human dynamics. Nature, 207–211 (2005)
3. Zadeh, L.A.: Toward a theory of fuzzy information granulation and its centrality in human reasoning and fuzzy logic. Fuzzy Sets and Systems **90**, 111–127 (1997)
4. Martin, T.P.: The x-mu representation of fuzzy sets. Soft Computing **19**, 1497–1509 (2015)
5. Sgarbas, K.N., Fakotakis, N.D., Kokkinakis, G.K.: Optimal insertion in deterministic dawgs. Theoretical Computer Science **301**(1–3), 103–117 (2003)
6. Revuz, D.: Minimization of acyclic deterministic automata in linear time. Theoretical Computer Science **92**, 181–189 (1992)
7. Hopcroft, J.E., Ullman, J.D.: Introduction to Automata Theory, Languages, and Computation. Addison-Wesley, Reading (1979)

Detection of Outlier Information Using Linguistic Summarization

Agnieszka Duraj, Piotr S. Szczepaniak and Joanna Ochelska-Mierzejewska

Abstract The main goal of automatic summarization of databases is usually to characterize the collection of data in terms of the dominant information involved. In complement to this task, the present paper shows the use of linguistic summarization for the characterization of databases containing textual records through detection of outlier information involved. The method applies a fuzzy measure of similarity between sentences to the summarization result.Certain level of standadization of textual records is assumed.

Keywords Detection of outlier information · Textual records comparison · Summarization of textual data bases · Fuzzy similarity · Information granularity

1 Introduction

Today, the number and size of documents contained in databases may be, and in fact frequently are, too large to be handled and interpreted effectively without the use of computer systems. Information is manageable for people only if it is condensed (granulated) and given in a natural language. Frequently, for practical reasons, compression is of more value than the accuracy of formulation (although wrong statements are obviously unacceptable).

In this context, two concepts (or definitions) need to be considered: *summarization* and *information granularity*. Summarization is the process of distilling the most important information from a source (or sources) to produce an abridged version for a particular user (or users) and task (or tasks) [1].

A. Duraj(✉) · P.S. Szczepaniak · J. Ochelska-Mierzejewska
Institute of Information Technology, Lodz University of Technology, ul. Wolczanska 215, 90-924 Lodz, Poland
e-mail: {agnieszka.duraj,joanna.ochelska-mierzejewska}@p.lodz.pl

T. Andreasen et al. (eds.), *Flexible Query Answering Systems 2015*,
Advances in Intelligent Systems and Computing 400,
DOI: 10.1007/978-3-319-26154-6_8

The term *information granularity* carries various meanings. In general, an *information granule* is an intuitively appealing construct which is expected to be useful in human perceptive, interpretative, cognitive and finally decision-making activity. Moreover, one expects that the granulation is justifiable, in the sense that the intuitively motivated requirements for the information granule are meaningful [2]. The granular approach is related to intelligent analysis of all kinds of data.

In particular, the term "granule" is understood as a group, cluster, or class of clusters in a universe of discourse. Within the "granular computing", theory, methodologies and techniques which deal with the processing of information granules are conceptually located [3], [4]. Usually, the granulation of information is considered when an intelligent analysis of numerical data is performed, and the vehicle is proximity-based fuzzy clustering, e.g. [2], [5].

In practice, we are usually satisfied with two kinds of information extracted: general summarization on the one hand, and detection of important abnormalities on the other. Frequently, at the same time, two kinds of information are expected to be generated: general and detailed.

An outlier often contains useful information about the characteristics of the system. The recognition of such unusual characteristics provides useful application-specific insights. An abnormality (outlier) is also referred to as *discordant, deviant*, or *anomaly* in the data mining and statistics literature. For example, the formal proposition of Hawkins [6] is as follows: *"An outlier is an observation which deviates so much from the other observations as to arouse suspicions that it was generated by a different mechanism*".

In the natural language there are a lot of imprecise expressions. A measure of the similarity of n-grams indicating the similarity of two sentences gives a numerical value which provides a powerful tool for the modelling of expression in the natural language. The concept of fuzzy quantifiers was introduced by Zadeh [9], [10]. The applications of linguistic summaries have been studied by many authors. Different approaches to fuzzy quantification and inaccuracies are presented in [12]. In the work of Delgado et al. [14], the authors focused on discussing fuzzy and generalized quantification for the linguistic summaries. Works related to the generation of linguistic summaries can be divided into those that deal with the creation of new summaries and those that seek to extend the existing ones to include more complex expressions. The starting point for the present work is the summarization of databases involving numerical data, as defined by Yager [8]. The operating instrument is the concept of similarity based on fuzzy sets, which serves to estimate the similarity of textual records. The merging of the two concepts (i.e. Yager's summarization and the extended Zadeh's fuzziness) enables the construction of a method capable of summarizing databases containing textual records, e.g. words, sentence words or complete sentences. New forms of linguistic summaries were introduced in [11], [18]. The summaries can be also extended to include complex expressions, as presented in [16], [17], [12]. Outlier detection in textual data were also considered by Aggarwal [7]. Other works deal with the application of linguistic summaries to various domains, e.g., the statistical exchange, financial, medical, or weather data [15].

As stated above, both summarization and detection of outliers are considered as useful characterizations of data collections. In the sequel, we focus on the detection of information which is not common for many records, i.e. can be labelled as having a specific, outlier character. The use of outlier detection in the context of linguistic summarization is a novelty. The method proposed in the present work, rooted in the theory of fuzzy sets, is based on concepts used for the comparison of natural language words and sentences. In particular, it employs the concept of fuzzy relation.

In Section 2, we introduce the definitions of the similarity of words and sentences based on definitions of fuzzy relationship. Section 3 focuses on the defining of an exception in linguistic summaries, which is of key importance for the present study. In Section 4, we describe the process of outlier detection, which is performed on the basic of two queries. The definition of an exception is illustrated using the example of a database containing 98 records. The database contains numerical and textual records. We take into account such attributes as the age of the person and the degree of unemployment. The unemployment rate is a text attribute. Finally, Section 5 presents a brief summary of the method used and the results obtained.

2 Fuzzy Text Similarity

2.1 Fuzzy Set and Fuzzy Relation

The concept of a fuzzy set was introduced by Zadeh in 1965 [9]. To each element of a considered space, he added a positive real number from the [0,1] interval interpreted as a "membership level (or degree)".
Formally, a fuzzy set A in a non-empty space X is a set of ordered pairs

$$A = \{< x, \mu_A(x) >: x \epsilon X\} \tag{1}$$

where $\mu_A : X \to [0, 1]$ the membership function.
The fuzzy relation R in the Cartesian product of two non-empty spaces X and Y is a set of pairs:

$$R = \{< (x, y), \mu_R(x, y) >: x \epsilon X, y \epsilon Y\} \tag{2}$$

where $\mu_R : X \times Y \to [0, 1]$ is the membership function. The positive real number $\mu_R(x, y)$ is usually interpreted as the degree of relationship between x and y.
It has the following properties:

a) reflexivity, if and only if

$$\mu_R(x, x) = 1 \qquad \forall x \epsilon X \tag{3}$$

b) symmetry, if and only if

$$\mu_R(x, y) = \mu_R(y, x) \qquad \forall x, y \epsilon X \tag{4}$$

The fuzzy relation called a "neighborhood relation" can be interpreted as a model of non-transitive similarity.

2.2 Comparision of Words and Sentences

The fuzzy relation RW is proposed as a useful instrument for the comparison of single words [11], [13]:

$$RW = \{(< w_1, w_2 >, \mu_{RW}(w_1, w_2)) : w_1, w_2 \epsilon W\} \tag{5}$$

where W is the set of all words within the universe of discourse, for example, a language or a dictionary.
A possible form of the membership function $\mu_{RW} : W \times W \rightarrow [0, 1]$ may be [11], [13].

$$\mu_{RW}(w_1, w_2) = \frac{2}{(N^2 + N)} \sum_{i=1}^{N(w_1)} \sum_{j=1}^{N(w_1)-i+1} h(i, j) \qquad \forall w_1, w_2 \epsilon W \tag{6}$$

where:
$N(w_1), N(w_2)$ the number of letters in words w_1, w_2, respectively;
$N = max\{N(w_1), N(w_2)\}$ – the maximal length of the considered words;
$h(i, j)$ – the value of the binary function,
i.e. $h(i, j) = 1$ if a subsequence, containing i letters of word w_1 and beginning at the $j - th$ position in w_1, appears at least once in word w_2;
otherwise $h(i, j) = 0$;
$h(i, j) = 0$ also if $i > N(w_2)$ or $i > N(w_1)$.
Note that 0.5 $(N^2 + N)$ is the number of possible subsequences to be considered. Obviously, on the basis of word comparison, one can also analyze the whole text. The membership function (6) is asymmetric. Thus, the result of the similarity between two words depends on the order of words compared. This inconvenience can be removed quite arbitrarily as follows:

$$\forall w_1, w_2 \in W \qquad \mu_{\overset{\circ}{RW}} = min\{\mu_{RW}(w_1, w_2)\}, \mu_{RW}(w_2, w_1)\} \tag{7}$$

Such a construction of the membership function μ_{RW} reflects the human intuition: "the more common the subsequences two given words contain, the more similar they are". Note that the method determines the syntactical similarity only for a given pair of words. The value of the membership function contains no information on the sense or semantics of the arguments.

In a natural way, the comparison of sentences is based on the following definition of the word similarity measure.
The fuzzy relation RZ on the set of all sentences Z, is of the form:

$$RZ = \{(< z_1, z_2 >, \mu_{RZ}(z_1, z_2)) : z_1, z_2 \epsilon Z\} \tag{8}$$

with the membership function $\mu_{RZ} : Z \times Z \rightarrow [0, 1]$.

$$\mu_{RZ}(z_1, z_2) = \frac{1}{N} \sum_{i=1}^{N(z_1)} \max_{j \epsilon 1,\dots,N(z_2)} \mu_{RW}(w_i, w_j) \tag{9}$$

where:
w_i – the word of number i in sentence z_1,
w_j – the word of number j in sentence z_2,
$\mu_{RW}(w_i, w_j)$ – the value of function μ_{RW} for the pair (w_i, w_j) - determined by formula (6)
$N(z_1)$, $N(z_2)$ – the number of words in sentences z_1, z_2, respectively,
$N = max\{N(z_1), N(z_2)\}$ – the number of words in the longer of the two sentences under comparison.
The symmetry can be achieved in the following way:

$$\mu_{\overset{\circ}{RZ}}(z_1, z_2) = min\{\mu_{RZ}(z_1, z_2), \mu_{RZ}(z_2, z_1)\} \tag{10}$$

The similarity measures can be applied to the summarization of structured textual databases in Yager's standard. This proposition is obviously only one of many possible ways to define a fuzzy relation.

3 Application of Yager's Linguistic Summaries to Outlier Detection

Outlier detection aims to find small groups of objects that are exceptional in comparison with the order objects.

Hawkins [6]:
For any object $x \epsilon X$, if x has some **abnormal characteristics** as compared to the other objects in X, we may consider x as an outlier.

Def.1. Linguistic summary
The linguistic database summaries proposed by Yager [8] have been extended to the following form:
Q P being R are S (have property S) [and correctness of this is of degree T].

The symbols are interpreted as follows:
Q – amount determination,
P – subject of the summary,
R – subject's description of the summary,
S – property,
T – quality of the summary.

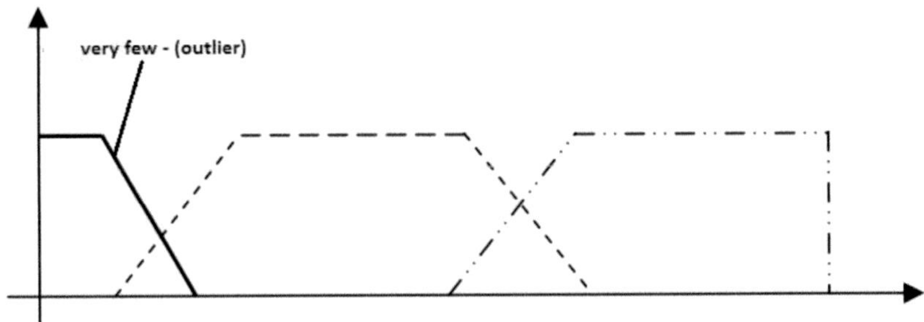

Fig. 1 Graphical interpretation of linguistic values for an exception

It has been assumed that the value of a summarized attribute is a crisp number, while the amount determination Q is a linguistic variable, for example: "*very few*", "*many*", "*almost each*", which defines how many objects have a given property. The subject of the summary P is determined as an object described with the values from its record. The quality of the summary T is a real number from the interval [0, 1] and can be interpreted as the level of truth (confidence) for the given summary.

Def. 2. Outlier
If for any subject P being R and having property S, the amount determination Q is **small** and T is **big** the P is considered **to be abnormal**. The **"smallness"** of Q and **"big value"** (or **"satisfactory grade")** of confidence T need to be defined (by the user or the system designer).
In other words, one has to define a proper linguistic variable (Fig. 1), for example "*very few*", which will classify a number of objects as belonging to a small atypical group. The search for outliers may be performed using the following queries:

F1) How many P being R are S?
F2) Are Q P being R which are S?
where Q is defined as the "*smallness*" variable describing the objects in the database.
For example, if we have the set of linguistic variables describing Q as "*very few*", "*few*", "*many*", "*almost all*", Q is "*very few*".

The procedure for determining outliers is as follows:

1) We define a set of linguistic variables
$X = \{Q_1, \ldots, Q_N\}$
2) Determination of the **big value** of T.
3) The degree of correctness T is determined in the following way:

   a) Calculation of the membership function of the features S, R for each tuple
   b) Calculation of r for all tuples according to the formula (11)

$$r = \frac{\sum_{i=1}^{n} (\mu_R(x_i) * \mu_S(x_i))}{\sum_{i=1}^{n} \mu_R(x_i)} \tag{11}$$

c) Calculation of the membership function Q for r, which determines the quality of the summary T with the equation (12).

$$T = \mu_{Q_i}(r) \qquad for \qquad i = 1, ..., N \tag{12}$$

d) The final form of the linguistic summary for
F1the first query takes the form
Q_1 P being R are S $[T]$
Q_2 P being R are S $[T]$
.....
Q_N P being R are S $[T]$

F2the second query takes the form
Q_1 P being R are S $[T]$

4) We verify if the value of T is compatible with the "big value" defined by the user in step 2. If the value of T corresponds to the "big value" for the linguistic variable Q_1, the outliers have been found.

In the case when a big number of diverse summaries performed for the same data base is quantified as a vey few means that:

- there is a small number of dominating clusters or there are no distinct clusters, and the records are splited over the space; or
- querries where formulated inappropriately to the information contained in the considered data base; or
- the textual records of data base are written in a very free fashion the data base is not standardized.

4 Example

Let us consider a database containing 98 records. Each record contains information such as the age of the person and the degree of unemployment. The database contains numerical and textual records. The unemployment rate is a text attribute. Table 1 presents some records extracted from a database containing information about 98 persons related to the age and unemployment in the form of both numerical and textual data. In order to illustrate the summary, we present the calculation performed for query (F1) only. A similar illustration can be easily provided also for query (F2). Outliers are always sought for the smallest linguistic variable. The term smallest

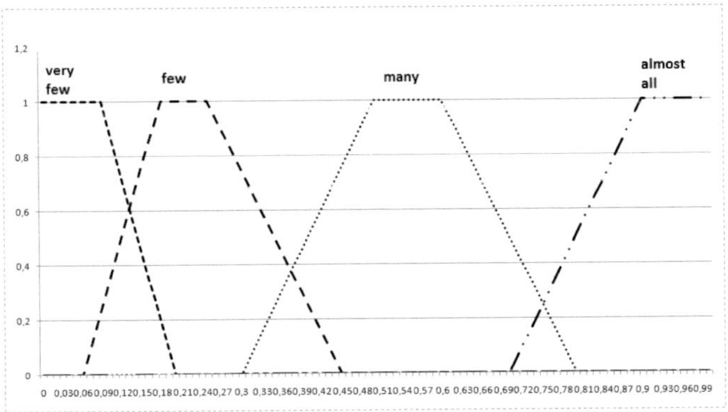

Fig. 2 Graphical interpretation of the membership function of the linguistic variable$X$.

can be defined as a fuzzy variable; for practical reasons it can also be taken in a non-standard way.

Let us define a set of linguistic variables X (13).

$$X = \{Q_1, Q_2, Q_3, Q_4\} = \{\text{“}\textit{very few}\text{”}, \text{“}\textit{few}\text{”}, \text{“}\textit{many}\text{”}, \text{“}\textit{almost all}\text{”}\} \quad (13)$$

For definitions see, respectively: “very few” – (14), “few” – (15), “many” – (16), “almost all” – (17). The graphical representation is given in Fig. 2.

$$\mu_{very\,few}(r) = \begin{cases} 1 & r < 0.09 \\ \frac{0.2-r}{0.1} & 0.09 \le r < 0.20 \\ 0 & r \ge 0.20 \end{cases} \quad (14)$$

$$\mu_{few}(r) = \begin{cases} \frac{r-0.06}{0.12} & 0.06 \le r < 0.18 \\ 1 & 0.18 \le r < 0.30 \\ \frac{0.45-r}{0.05} & 0.30 \le r < 0.45 \end{cases} \quad (15)$$

$$\mu_{many}(r) = \begin{cases} \frac{r-0.3}{0.2} & 0.3 \le r < 0.5 \\ 1 & 0.5 \le r < 0.6 \\ \frac{0.8-r}{0.2} & 0.6 \le r < 0.8 \end{cases} \quad (16)$$

$$\mu_{almost\,all}(r) = \begin{cases} 0 & r < 0.7 \\ \frac{r-0.7}{0.2} & 0.7 \le r < 0.9 \\ 1 & r \ge 0.9 \end{cases} \quad (17)$$

We are looking for outliers for the variable ***“very few”***. Next, the concept of **big value** of T needs to be defined.

Table 1 Sample record. Sentences describe the level of unemployment and the age

ID	Age	Sentences - unemployment rate
1	45	Very high unemployment
2	26	Unemployment is very high
3	37	Unemployment is high
...	...	...
98	41	Unemployment rather high

Table 2 The results of the membership function calculated for the records contained in Table 1: MFvAge - Membership function values for young people; SlevelCopm - Similarity level for the comparison of the two sentences; MVcomp - Membership values for the comparison of the two sentences

ID	MFvAge	SlevelComp	MVcomp
1	0	0.73	0.8
2	1	0.55	0.4
3	0	0.6752	0.8
...	...	...	...
98	0.8	0.3809	0

Let us consider the query no. 1:
How many young people are in the group of high unemployment?

Young people are qualified according to the membership function defined by the formula (18), where x is the age. The graphical representation of the specific age groups is given in Figure 3.

$$\mu_{young}(x) = \begin{cases} 0 & x < 20 \\ \frac{x-20}{2} & 20 \le x < 25 \\ 1 & 25 \le x < 30 \\ \frac{35-x}{5} & 30 \le x < 35 \\ 0 & x \ge 35 \end{cases} \tag{18}$$

The values of the membership function (18) calculated for the records contained in Table 1 are given in Table 2.

The next step is to determine the level of similarity between two sentences. Let the basic sentence be *'High unemployment rate'*. Now, we determine the membership function for the similarity of two sentences (the basic sentence and one of the

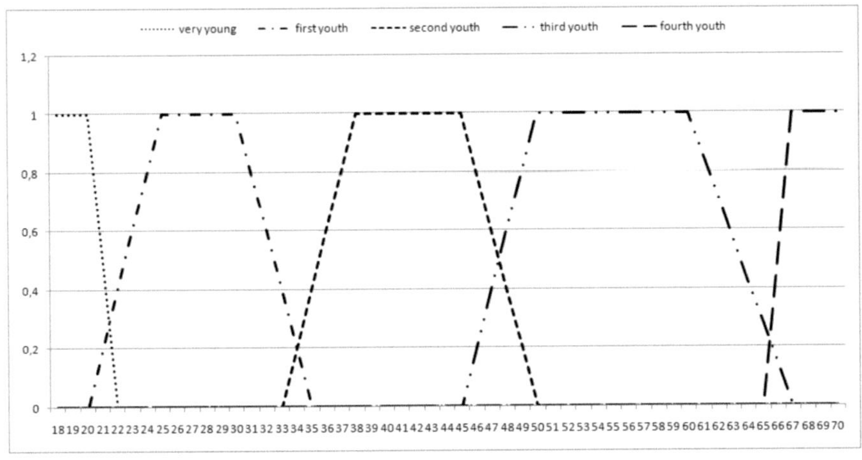

Fig. 3 Graphical interpretation of the membership function of each age group.

sentences from the record), see the SlevelComp column in Table 2. The membership function of a pair of compared sentences is set on the basis of the formulas (6) and (9). The fuzzy similarity measure of textual records is determined by the equation (8). Since the value of similarity between the two sentences is almost always close to 0.9, we introduce the membership function $t(z)$, determined by equation (19), where z is the value representing the similarity of the sentences. The values of the function $t(z)$ are given in Table 2 in the MVcomp column, where z is the similarity of the two sentences.

$$t(z) = \begin{cases} 0 & 0.00 \leq z < 0.45 \\ 0.40 & 0.45 \leq z < 0.65 \\ 0.80 & 0.65 \leq z < 0.80 \\ 1 & 0.80 \leq z < 1.0 \end{cases} \tag{19}$$

The last step is to determine the quality of the summary for the given query. We computed r using the equation (20).

$$r = \frac{\sum_{i=1}^{n} [\mu_{age}(x_i) * t(z_i)]}{\sum_{i=1}^{n} \mu_{age}(x_i)} = 0.19 \tag{20}$$

Now, the value of $T = \mu_{Q_i}(r)$ for $i = 1, 2, 3, 4$ can be computed. The quality of the summary T is determined for every linguistic variable $\{Q_1, Q_2, ..., Q_N\}$. The graphical interpretation of the summary is illustrated in Fig. 4.

Hence, the complete summary expressed as follows:

Very few young people are in the group of high unemployment [0.09].
Few young people are in the group of high unemployment [0.91].
Many young people are in the group of high unemployment [0].
Almost all people are in the group of high unemployment [0].

Now we return to the previously defined values **smallness** Q and **big value** of T. T needs to be defined (by the user or the system designer). For example, if T is defined (as "big value") at $T > 0.8$, we might receive outliers. However, if the user defines "smallness" Q as a linguistic variable "very few", outliers are not accepted because we receive $T = 0.91$ for Q="smallness"="few".
Now, let us consider another question:
How many people in the high unemployment group are young?
Analogously to the first question, we calculate r (21).

$$r = \frac{\sum_{i=1}^{n} [\mu_{age}(x_i) * t(z_i)]}{\sum_{i=1}^{n} \mu_t(z_i)} = 0.13 \tag{21}$$

Similarly as above, we verify r for every linguistic variable Q_i.
The complete summary expressed as follows:

Very few in the high unemployment group are young people [0.63].
Few in the high unemployment group are young people [0.41].
Many in the high unemployment group are young people [0].
Almost all in the high unemployment group are young people [0].

Similarly as above, we verify the **big value** of T defined by the user or the system designer If, for example, T is defined (as "big value") at $T > 0.8$, outliers are not accepted. If T is defined at $T > 0.6$, the obtained quality of the summary for query 2 indicates the presence of outliers, because $T = 0.63$ belongs to the user-defined outliers T.

Therefore, on the basis of the summary, we may conclude that the exceptions are detected by a factor of credibility of summary $T = [0.63]$ for the quantity *"very few"*. Therefore, the presence of outliers has been confirmed. In F1 in begin the search from the feature, ie. young age, then we compare the the senteces to the given reference sentence. In the F2 we begin the search by comparing texts, then we are looking for people at young age.

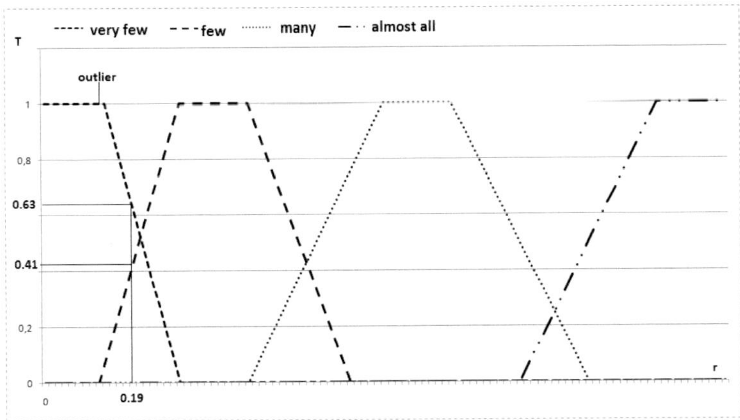

Fig. 4 Graphical interpretation of ouliers

5 Final Remarks

The paper presents a new approach to the summarization of databases containing both numerical and partly standardized textual records. The method described enables the detection of outlier information. In order to explain our idea, we used Yagers proposition. However, the present approach is universal, as it can be extended to more sophisticated forms of summarization. The method may be applied in computer tools used to assist managers in taking strategic decisions on the basis of qualitative description of an object.

References

1. Mani, I., Maybury, M.T.: Advances in Automatic Text Summarization. The MIT Press, Cambridge (1999)
2. Pedrycz, W., Al-Hamouz, R., Morfeq, R., Balamash, A.: The Design of Free Structure Granular Mappings: The Use of the Prniciple of Justifiable Granularity. IEEE Trans. on Cybernetics **43**(6), 2105–2113 (2013)
3. Pedrycz, W.: Granular computing in data mining. In: Last, M., Kandel, A. (eds.) Data Mining and Computational Intelligence. Springer, Singapore (2001)
4. Lin, T.Y., et al. (eds.): Data mining, rough sets and granular computing. Physica-Verlag, Berlin (2002)
5. Pedrycz, W., Loia, V.: P-FCM: A proximity-based fuzzy clustering. Fuzzy Sets and Systems **128**, 21–41 (2004)
6. Hawkins, D.: Identification of Outliers. Chapman and Hall (1980)
7. Aggarwal, C.: Outlier detection in categorical, text and mixed attribute data. In: Outlier Detection, Spronger Book, pp. 199–223 (2013)
8. Yager, R.R.: Linguistic summaries as a tool for databases discovery. In: Workshop on Fuzzy Databases System and Information Retrieval, Yokohama, Japa (1995)
9. Zadeh, L.A.: Fuzzy Sets. Information and Control **8**, 338–353 (1965)

10. Zadeh, L.A.: Toward a theory of fuzzy systems. In: Kalman, R.E., De Claris, N. (eds.) Aspects of Network and System Theory. Holt. Rinehart and Winston. New York (1971)
11. Niewiadomski, A.: Appliance of fuzzy relations for text documents comparing. In: Proceedings of the 5th Conference on Neural Networks and Soft Computing, Zakopane, Poland, pp. 347–352 (2002)
12. Niewiadomski, A.: Metods for the Linguistic Summarization of Data: Applications of Fuzzy Sets, EXIT, Warszawa (2008)
13. Szczepaniak, P.S., Niewiadomski, A.: Internet search based on text intuitionistic fuzzy similarity. In: Szczepaniak, P.S., Segovia, J., Kacprzyk, J., Zadeh, L. (eds.) Intelligent Exploration of the Web. Physica-Verlag. A Springer-Verlag Company, Heidelberg, New York (2003)
14. Delgado, M., Riuz, M., Sanchez, M.D., Vila, M.A.: Fuzzy quantification: a state of the art. Fuzzy Sets and Systems **242**, 1–30 (2014)
15. Ramos Soto, A.: Linguistic Descriptions for Automatic Generation of textual Short-Therm Weather Forecast on Real Prediction Data. IEEE Transaction on Fuzzy Systems **1–1**, 1–13 (2014)
16. Kacprzyk, J., Zadrozny, S.: Computing with words is an implementable paradigm: fuzzy queries, linguistic data summaries, and natural-language generation. IEEE Transactions on Fuzzy Systems (2010)
17. Kacprzyk, J., Zadrożny, S.: Bipolar queries: Some inspirations from intention and preference modeling. In: Trillas, E., Bonissone, P.P., Magdalena, L., Kacprzyk, J. (eds.) Combining Experimentation and Theory. STUDFUZZ, vol. 271, pp. 191–208. Springer, Heidelberg (2011)
18. Kacprzyk, J., Wilbik, A., Zadrozny, S.: An approach to the liguistic summarization of the series using a fuzzy quantifier driven aggregation. International Journal of Intelligent Systems **25**(5), 411–439 (2010)

Analysis of Fuzzy String Patterns with the Help of Syntactic Pattern Recognition

Mariusz Flasiński, Janusz Jurek and Tomasz Peszek

Abstract One of the main problems in the syntactic pattern recognition area concerns analysis of distorted/fuzzy string patterns. Classical methods developed to solve the problem are based on the error-correcting approach or the stochastic one. These methods are useful but have several limitations. Therefore, there is still the need to construct effective models of syntactic recognition of distorted/fuzzy patterns. The new approach to the problem is presented in the paper. It is based on the fuzzy primitives and the new class of fuzzy automata. The advantages of the approach are presented in the paper, as well as its comparison to classical approaches.

Keywords Syntactic pattern recognition · Fuzzy/distorted patterns · Error-correcting parsing · Stochastic grammars · Fuzzy automata

1 Introduction

Syntactic pattern recognition [3, 8, 12, 13, 19, 24] is one of the main approaches in the area of machine recognition. It is based on the idea of representing a pattern as a structure of the form of string, tree or graph and a set of structures as a formal language. A analysis and a recognition of an unknown structure is performed with a formal automaton (parser).

Syntactic pattern recognition is more suitable than other pattern recognition approaches (probabilistic, discriminant function-based, neural networks, etc.) when patterns considered can be characterized better with structural features than numerical features. Using this approach not only can we make a classification (in a sense

M. Flasiński(✉) · J. Jurek · T. Peszek
Information Technology Systems Department, Jagiellonian University in Cracow,
ul. prof. St. Łojasiewicza 4, 30-348 Cracow, Poland
e-mail: mariusz.flasinski@uj.edu.pl
http://www.ksi.uj.edu.pl/en/index.html

T. Andreasen et al. (eds.), *Flexible Query Answering Systems 2015*,
Advances in Intelligent Systems and Computing 400,
DOI: 10.1007/978-3-319-26154-6_9

of ascribing a pattern to a pre-defined category), but also a (structural) interpretation of an unknown pattern. Therefore, for structurally-oriented recognition problems, such as character recognition, speech recognition, scene analysis, chemical and biological structures analysis, texture analysis, fingerprint recognition, geophysics, this approach is particularly useful.

In the paper we consider a syntactic pattern recognition model based on *string* structures. String grammars and syntax analyzers of string languages have been a basic formal tool for syntactic pattern recognition since the very beginning of the research in this area [11, 21] and they are still being successfully used in many different applications [5, 6, 10, 18, 23].

The general scheme of a syntactic pattern recognition system in the case of string patterns is shown in Fig. 1. Firstly, an unknown pattern is preprocessed, i.e., segmented and decomposed. Then elementary patterns are recognized. These elementary patterns are called *primitives* and they are represented with symbols of a formal language alphabet. A string consisting of primitives, i.e., the symbolic representation of the pattern, is delivered to the parser module which is responsible for syntax analysis of the string on the basis of a pre-defined formal grammar. A classification and a description of the pattern in the form of a list of productions used during syntax analysis is the final result of the analysis.

One of the main practical problems with the use of syntactic pattern recognition methods concerns fuzzy/distorted structural components defining analyzed patterns [24]. If a pattern is distorted or fuzzy, then an identification of primitives can be

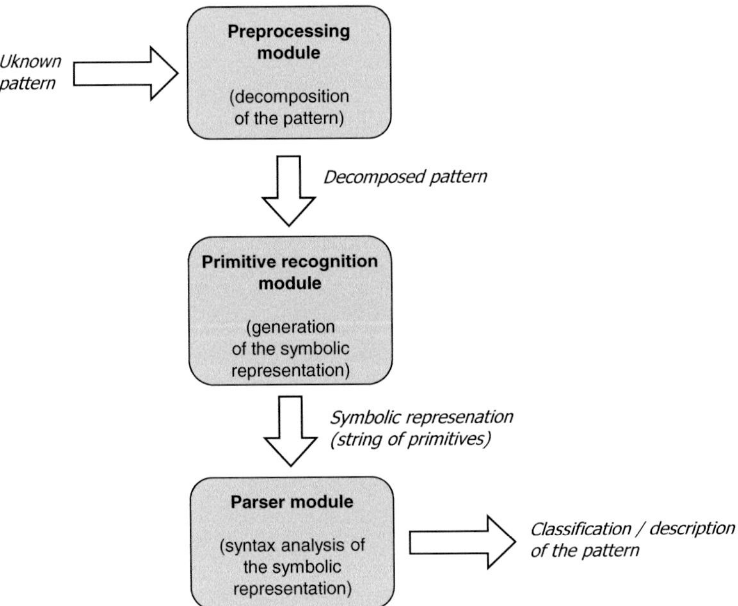

Fig. 1 The general scheme of a syntactic pattern recognition system

wrong. In this case parser module often does not bring a proper result of recognition either. Let us notice that even one misrecognized primitive could lead to an erroneous syntax analysis of the whole pattern.

There are two main classical approaches of solving the problem known in the literature: the error-correcting approach and the stochastic one [12]. We present both approaches in the case of the use of GDPLL(k) grammars for syntactic pattern recognition in sections 3 and 4.

It occurs that classical approaches have several limitations even if we use such strong descriptively and efficient computationally grammars as GDPLL(k) grammars. Therefore, research into a fuzzy model of GDPLL(k) grammars-based syntactic pattern recognition has been led recently. The model developed is presented in section 5. The comparison of this model with classical approaches is presented in section 6. Final remarks are included in section 7.

2 Basic Definitions

Let us introduce the formalism [5, 15] which will be used for discussing basic approaches to the problem of syntactic recognition of fuzzy/distorted patterns.

Definition 1. A *generalized dynamically programmed context-free grammar* is a six-tuple $G = (V, \Sigma, M, O, P, S)$, where V is a finite, nonempty alphabet; $\Sigma \subset V$ is a finite, nonempty set of terminal symbols ($N = V \setminus \Sigma$ is a set of nonterminal symbols); M is a memory; O is a set of basic operations on the values stored in the memory M; P is a finite set of productions in the form $p_i = (\mu_i, L_i, R_i, A_i)$, in which $\mu_i : M \longrightarrow \{TRUE, FALSE\}$ is the predicate of applicability of the production p_i defined over M, $L_i \in N$ and $R_i \in V^*$ are left- and right-hand sides of p_i, respectively, A_i is the sequence of operations ($\in O$) over M which are performed if the production is applied; $S \in N$ is the starting symbol. □

Definition 2. Let $G = (V, \Sigma, O, P, S, M)$ be a generalized dynamically programmed context-free grammar. The grammar G is called a *GDPLL(k) grammar* if the following two conditions are fulfilled.

1. Stearns's condition of LL(k) grammars. (The top-down left-hand side derivation is deterministic if this is allowed to look at k input symbols to the right of the current position of the input head in the string).
2. There exists a a certain number ξ such that after the application of ξ productions in a left-hand side derivation we get at the "left-hand side" of the sentence at least one new terminal symbol. □

A derivation in GDPLL(k) grammars is defined in the following way. Before an application of a production p_i we test whether L_i occurs in a sentential form derived. Then we check the predicate of applicability of the production. The predicate is defined as an expression defined with the variables stored in the memory. If the predicate is true then we replace L_i with R_i and we perform the sequence of operations over the memory. The execution of the operations changes the content of the memory. It is done with the help of arithmetical instructions.

The automaton for GDPLL(k) grammars (GDPLL(k) parser) has been described in [15]. We do not present it in this paper. We just notice here that the automaton performs the syntax analysis in the top-down manner, i.e., it simulates a process of the derivation made with the help of productions of the corresponding GDPLL(k) grammar.

GDPLL(k) grammars are characterized by three main features. Firstly, they are of very good discriminative properties, i.e., they are able to generate a large class of context-sensitive languages [5]. Secondly, the languages generated by them can be efficiently analyzed, i.e., this is possible to construct an automaton of the linear computational complexity [15]. Thirdly, the grammatical inference algorithm for GDPLL(k) grammars has been constructed [16]. Due to these advantages they have been used to solve many recognition problems in the various application areas [8].

3 The Error-Correcting Approach

The error-correcting approach in syntactic pattern recognition is based on the following idea. If some errors occur in the analyzed symbolic representation of a pattern, i.e. some primitives are wrongly identified, then a parsing algorithm should try to correct them.

The minimum-distance error-correcting parsing is the most popular strategy to achieve this goal. In [1] errors in a string are considered to be of the three types, i.e., substitution, deletion, and insertion. For each kind of an error the so-called *error transformation* is defined. Then the distance between two strings is defined as the smallest number of error transformations required to derive one string from another.

The extended context-free grammar is constructed by adding *error productions*, corresponding to each type of the transformations, to the *core (standard) grammar* [2]. The first minimum-distance error-correcting parsing algorithm has been defined for context free grammars according to the following rules [1]:

- if this is impossible to apply a *core production*, i.e., errors occurred in the input string (in), the proper error production is applied,
- each application of any error production increases the distance between the input string (in) and the corrected string (out),
- error productions are chosen in such a way that it ensures the minimum distance between in and out.

The complexity of this algorithm is of $O(n^3)$. However, the usefulness of the algorithm for solving some practical applications is questionable. The generative power of context-free grammars used in the method is too weak for solving, e.g., complex pattern recognition problems. On the other hand, if we try to apply the minimum-distance approach for more powerful grammars, e.g., for the programmed grammars [20], then the computational complexity increases significantly. For example, for the programmed grammars, the corresponding automaton is of the non-polynomial complexity [12].

Therefore, the error-correcting model has been used in the case of of GDPLL(k) grammars of the strong generative power which were introduced in section 2. Let us present the following definition [7].

Definition 3. An *extended generalized dynamically programmed context-free grammar* is a six-tuple $G_e = (V, \Sigma_e, M, O, P_e, S)$, where V is a finite, nonempty alphabet; $\Sigma_e \subset V$ is a finite, nonempty *extended* set of terminal symbols ($N = V \setminus \Sigma_e$ is a set of nonterminal symbols); M is a memory; O is a set of basic operations on the values stored in the memory M; P_e is a finite set of *extended* productions in the form $p_i = (\mu_i, L_i, R_i, A_i, ec_i)$, in which $\mu_i : M \longrightarrow \{TRUE, FALSE\}$ is the predicate of applicability of the production p_i defined over M, $L_i \in N$ and $R_i \in V^*$ are left- and right-hand sides of p_i, respectively, A_i is the sequence of operations ($\in O$) over M which are performed if the production is applied, ec_i is the error cost corresponding to the application of p_i;[1] $S \in N$ is the starting symbol. □

The difference between generalized dynamically programmed grammars and *extended* generalized dynamically programmed grammars consists in the use of *distorted terminal symbols* and *error productions* for correction of the distortions in a fuzzy string pattern. Distorted symbols represent most typical distortions of standard terminal symbols which can appear in a pattern. They have to be defined a'priori by a designer of a grammar according to his/her knowledge about analyzed phenomena. Error productions are added to the set of production to correct such distorted symbols. Each error production has its cost. The cost is not determined automatically, e.g., according to the type of the error like in standard context-free grammars, but it is set according to knowledge of a grammar designer. In this way we provide a flexible tool for distinguishing two patterns not only on the basis of the distance, but on the basis of complex context dependencies.

The definition of *extended* GDPLL(k) grammars [7] is analogous to Definition 2, so it will not be presented here. Let us notice that restrictions of the definition of *extended* GDPLL(k) grammars should be imposed to make a derivation process deterministic.

At the end of this section let us sum up our considerations concerning the error-correcting approach.

1. Any error-correcting syntactic pattern recognition method works according to the scheme shown in Fig. 1. This means that each primitive is identified as *strictly one symbol* of a terminal alphabet. A string of such symbols is delivered to a parser module which performs syntax analysis trying to correct errors.
2. In this approach this is possible to use GDPLL(k) grammars which are of the good discriminative and computational properties. In this way we can recognize the large class of quasi context-sensitive languages efficiently.
3. The error-correcting approach should be applied only when we are able to specify typical distortions *a'priori* and we know how these distortions should be corrected during a derivation process.

---

[1] If $ec_i > 0$ the production p_i is called an *error production*.

4 The Stochastic Approach

The second approach to analysis of fuzzy/distorted patterns in syntactic pattern recognition is based on the idea of *stochastic* grammars [12]. In the case of this approach we can also use GDPLL(k) grammars in order to enhance the generative power of context-free grammars. Let us introduce the following definition [7].

Definition 4. A *stochastic generalized dynamically programmed context-free grammar* is a six-tuple $G_s = (V, \Sigma, M, O, P_s, S)$, where V is a finite, nonempty alphabet; $\Sigma \subset V$ is a finite, nonempty set of terminal symbols ($N = V \setminus \Sigma$ is a set of nonterminal symbols); M is a memory; O is a set of basic operations on the values stored in the memory M; P_s is a finite set of *stochastic* productions in the form $p_i = (\mu_i, L_i, RS_i, A_i)$, in which $\mu_i : M \longrightarrow \{TRUE, FALSE\}$ is the predicate of applicability of the production p_i defined over M; $L_i \in N$ is the left-hand side of p_i; $RS_i = \{(R_{ij}, s_{ij}) : R_{ij} \in V^*, 0 < s_{ij} \leq 1, j = 1, ..., n_i, \sum_{j=1}^{n_i} s_{ij} = 1\}$ is the set of right-hand sides of p_i such that the right-hand side R_{ij} has assigned the probability s_{ij};[2] A_i is the sequence of operations ($\in O$) over M which are performed if the production is applied; $S \in N$ is the starting symbol. □

The difference between generalized dynamically programmed grammars and *stochastic* generalized dynamically programmed grammars consists in the definition of *stochastic productions*. Let us notice that there are many different right-hand sides with the probabilities assigned in a single stochastic production. The probability of the derivation is equal to the product of the probabilities of stochastic productions used in the derivation.

The definition of *stochastic* GDPLL(k) grammars is analogous to the definition of standard GDPLL(k) grammars, i.e., Definition 2. Hence, it will not be presented here.

The parser for stochastic GDPLL(k) grammars [7] performs syntax analysis in the top-down manner, i.e., it simulates a process of the derivation. The production to be applied is chosen on the basis of the computed probabilities. The procedure of the probabilities' computation is complex, since it analyzes all the paths in the derivation tree which lead to deriving the first terminal symbol of the lookahead string. As the result the procedure returns the maximum probabilities for all the analyzed paths. Let us notice that the number of iterations is limited by the constant according to Definition 2. Therefore, this procedure does not increase the computational complexity of the parsing algorithm.

The decision which approach, i.e., the error-correcting one or the stochastic one, is made according to the following premises. Stochastic grammars are used when it is difficult to specify typical distortions a'priori, e.g., we do not know the characteristics of analyzed phenomena. In this case we can try to compute the probabilities of possible distortions on the basis of the pattern frequencies belonging to the learning set. However, although the grammatical inference algorithm for *standard* GDPLL(k) grammars has been defined [14], it has not been constructed for stochastic GDPLL(k) grammars.

[2] For a given production p_i we have n_i different right-hand sides.

At the end of this section let us summarize the properties of the stochastic approach.

1. Any stochastic syntactic pattern recognition method works according to the scheme shown in Fig. 1. This means that each primitive is identified as *strictly one symbol* of a terminal alphabet. A string of such symbols is delivered to a parser module which performs syntax analysis choosing most probable paths of derivation.
2. In this approach this is possible to use GDPLL(k) grammars which are of the good discriminative and computational properties. In this way we can recognize the large class of quasi context-sensitive languages efficiently.
3. The stochastic approach should be applied when it is difficult to specify typical distortions *a'priori* and we are able to compute the probabilities of the grammar productions on the basis of the pattern frequencies belonging to the learning set.

5 The Novel Fuzzy-Pattern-Based Model

The novel model of the analysis of fuzzy/distorted patterns is the result of the recent research concerning the application of syntactic pattern recognition for electrical load forecasting [9]. The processing scheme of the model is different from the error-correcting model and the stochastic one. The general scheme of the system based on this model is shown in Fig. 2.

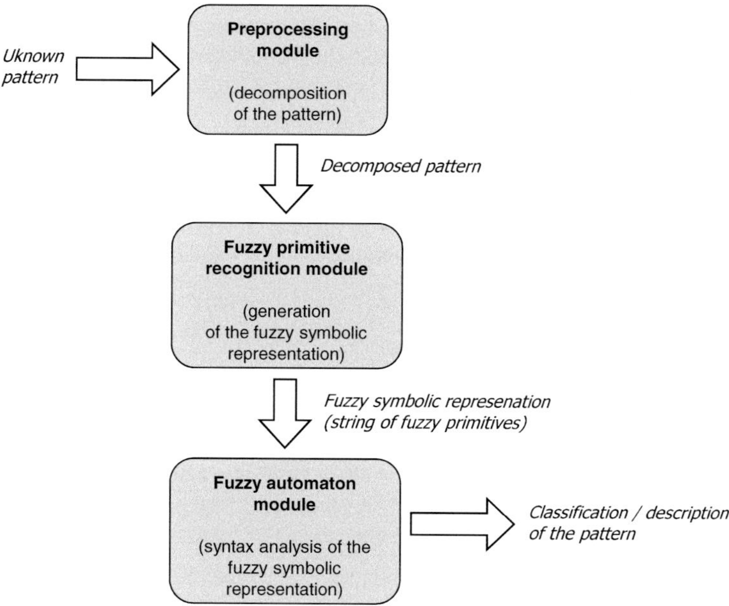

Fig. 2 The general scheme of the fuzzy-based syntactic pattern recognition system

The main idea of the model consists in constructing the so-called *fuzzy primitives*. They have been defined to describe distortions/uncertainty in the process of primitives recognition. A fuzzy primitive is represented by a vector of elements (instead of a one particular symbol). The elements of the vector correspond to the possible different recognitions of a primitive and their probabilities. Let us define a fuzzy primitive in a formal way.

Definition 5. Let $\Sigma = \{a_{i_1}, ..., a_{i_s}\}$ be a set of symbols. A *fuzzy primitive* is a vector

$$(a_{i_1} p_{i_1}, ..., a_{i_s} p_{i_s}), \text{ where}$$

$a_{i_1}, ..., a_{i_s} \in \Sigma$ are different symbols,
$p_{i_1}, ..., p_{i_s}$ are probabilistic measures corresponding to each symbol such that

- $p_{i_k} > 0, k = 1, ..., s,$
- $\Sigma_{k=1}^{s} p_{i_k} \leq 1,$
- $p_{i_1} \geq \cdots \geq p_{i_s}.$

Each pair (a_{i_k}, p_{i_k}) is called *a unit* of a fuzzy primitive, and each symbol a_{i_k} is called *an element* of a fuzzy primitive. The number of units of a fuzzy primitive is called its *dimension*. The set of fuzzy primitives over Σ is denoted by FP_Σ. □

The fuzzy primitive recognition module (cf. Fig. 2) of our system is implemented with the help of the probabilistic neural network, PNN, introduced by Specht in [22]. Apart from the classification, the network delivers an additional information about the distance of an unknown object to other classes[3]. In our case the PNN delivers strings of fuzzy primitives to their output.

Then the fuzzy representation of a pattern is analyzed by the fuzzy automaton module constructed on the basis of GDPLL(k) grammars. This automaton generates the classification/description of the pattern in the form of so-called *symbolic (variant) identification* of the fuzzy pattern, which is defined as follows.

Definition 6. Let Σ be a set of terminal symbols. A *symbolic (variant) identification* of a fuzzy pattern is a set of pairs (w_i, p_i), where $w_i \in \Sigma^*$ is one of the high-probability-recognized words describing the fuzzy pattern, p_i is the corresponding probability of the recognition of the fuzzy pattern as the word w_i. The symbolic identification of the pattern over Σ is denoted by SI_Σ. □

Now, let us introduce the definition of the fuzzy automaton.

Definition 7. An *FGDPLL(k) automaton* (a fuzzy GDPLL(k)-based automaton) is a seven-tuple $A = (Q, \Sigma, M_A, \delta, q_0, F, S_{GDPLL(k)})$, where Q is a finite set of states; Σ is a finite set of input symbols; M_A is an auxiliary memory; $\delta : Q \times FP_\Sigma \times M_A \longrightarrow Q \times M_A \times (SI_\Sigma \cup \emptyset)$ is the transition function[4], in which FP_Σ is the set of fuzzy primitives, SI_Σ is the symbolic identification; $q_0 \in Q$ is the initial state; $F \subseteq Q$ is the set of final states; $S_{GDPLL(k)} = \{A_{GDPLL(k)} : A_{GDPLL(k)}$ is a GDPLL(k) parser $\}$ is a pool of GDPLL(k) parsers. □

---

[3] That is other classes than the winner one.

[4] The empty set is generated to the output if a fuzzy primitive is not recognized.

An FGDPLL(k) automaton consists of three main elements as shown in Fig. 3. As we mentioned before, an input for the FGDPLL(k) automaton is in the form of a string of fuzzy primitives. The FGDPLL(k) automaton includes a pool of GDPLL(k) parsers which are needed to perform simultaneously several derivation processes for each possible value of an input primitive. The control of the automaton is responsible for verifying, which of the derivation processes should be continued. For this purpose it uses an auxiliary memory which contains computed probabilities for each derivation process.

Let us mention that FGDPLL(k) automaton is different than some fuzzy automata defined before [4]. They employed the idea of replacing classical *"transitions between states"* by appropriate probability distributions of current and next automaton state. Although FGDPLL(k) automaton is defined with the use of the classical *"state transition"* concept, it is able to analyze an input of a fuzzy form. What is more, FGDPLL(k) automaton, based on GDPLL(k) parsers, can efficiently (in polynomial time) recognize strings belonging to the large subclass of context-sensitive languages.

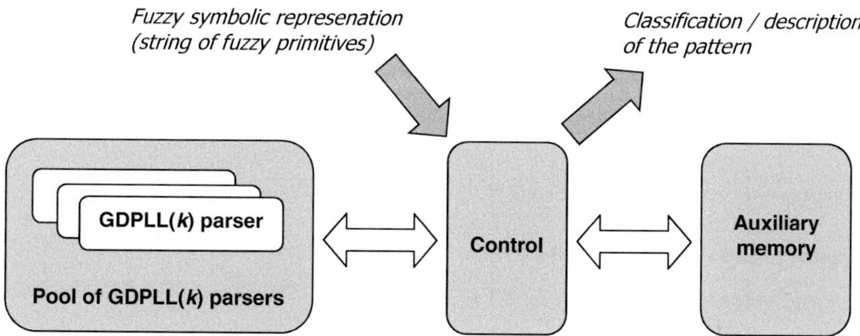

Fig. 3 The scheme of an FGDPLL(k) automaton

The novelty of the proposed approach consists in the use of the *complete* fuzzy information describing a recognized pattern in the phase of syntax analysis. In classical approaches presented in sections 3 and 4 as well as in approaches based on fuzzy automata a structural representation of a pattern is of the deterministic nature. (That is, in these approaches, after the primitive recognition phase, the fuzzy aspect of the pattern representation is, somehow, lost.) As a consequence of the possibility of the use of fuzzy primitives, there is no need to design special classes of grammars, such as the extended grammars or the stochastic ones. One can use standard grammars in this approach.

The approach proposed in the paper has been verified in practice. The system for improving the accuracy of the electrical load prediction, based on fuzzy primitives and FGDPLL(k) automata, has been constructed and verified with the use of real data delivered by one of Polish energy distribution companies [10].

Let us sum up advantages of the method based on fuzzy primitives and FGDPLL(k) automata.

1. The FGDPLL(k)-automata-based method works according to the scheme shown in Fig. 2, which differs from classical approaches to syntactic recognition of distorted/noisy patterns. The *complete* fuzzy information describing recognized patterns is used in the phase of syntax analysis.
2. The method is computationally efficient because an FGDPLL(k) automaton is of the polynomial complexity. At the same time it has good discriminative properties because of the strong generative power of GDPLL(k) grammars.
3. A syntactic pattern recognition system constructed on the basis of the method can be devised with a self-learning mechanism because the grammatical inference algorithm has been defined for GDPLL(k) grammars [14, 16] and one can train the probabilistic neural network in the phase of a recognition of fuzzy primitives.

6 Comparison of Approaches

In this section we present the comparison of approaches described in the paper. Let us make it with the help of the following table.

	Error-correcting approach	**Stochastic approach**	**FGDPLL(k)-based approach**
Primitive form	one strictly determined symbol	one strictly determined symbol	fuzzy symbolic representation
Grammar type	extended GDPLL(k) grammar	stochastic GDPLL(k) grammar	GDPLL(k) grammar
Language type	quasi-context sensitive	quasi-context sensitive	quasi-context sensitive
Analyzer type	extended GDPLL(k) parser	stochastic GDPLL(k) parser	FGDPLL(k) automaton equipped with a pool of GDPLL(k) parsers
Complexity of analysis	polynomial	polynomial	polynomial
Machine learning	partially assured (GDPLL(k) grammar can be inferred, but not extended GDPLL(k) grammar)	partially assured (GDPLL(k) grammar can be inferred, but not stochastic GDPLL(k) grammar)	assured (training of the probabilistic neural network and inferencing of GDPLL(k) grammars)
Application	only in case of possibility of specifying typical distortions a'priori	only in case of possibility of computing grammar productions' probabilities	no limitations

7 Concluding Remarks

The analysis of fuzzy/distorted string patterns, being one of the main problems in the syntactic pattern recognition area, is discussed in the paper. Two classical approaches to the problem, i.e., the error-correcting approach and the stochastic one, as well as the novel approach, which is based on the application of fuzzy primitives and a fuzzy automaton, have been presented.

The class of GDPLL(k) grammars has been used as the basis for defining syntactic pattern recognition methods within these approaches. These grammars are of the very good discriminative properties and the languages generated by them can be analyzed efficiently. Moreover, the grammatical inference algorithm for these grammars has been defined. These features make them useful in practical applications.

On the other hand, the advantages of the GDPLL(k) grammars do not eliminate all the limitations if the classical approaches to recognizing fuzzy/distorted patterns are concerned. The main drawback of these approaches is the loss of information about distortions/fuziness of patterns during the phase of primitive recognition, i.e., each primitive is identified as *strictly one symbol* and information about the similarity to other primitives is lost. The lack of learning mechanisms allowing one to construct the syntactic recognition system for fuzzy/distorted patterns automatically is another disadvantage of these approaches.

The novel approach is based on the application of fuzzy primitives. Each fuzzy primitive is represented as the sequence of most probable symbols (together with their probabilities). In this way additional information which improves the recognition process is provided. The fuzzy primitives can be analyzed efficiently by an FGDPLL(k) automaton which has been recently implemented for the practical application (improving the accuracy of the electrical load prediction for one of Polish energy distribution companies). In the new approach learning capabilities are much more effective than in the case of the classical approaches since this is possible to train the neural network of the system to recognize fuzzy primitives automatically and to infer the grammar on the basis of the learning set.

References

1. Aho, A.V., Peterson, T.G.: A Minimum Distance Error-correcting Parser for Context-free Languages. SIAM J. Comput. **4**, 305–317 (1972)
2. Aho, A.V., Ullman, J.D.: The Theory of Parsing, Translation, and Compiling. Prentice-Hall, Englewood Cliffs (1972)
3. Bunke, H.O., Sanfeliu, A. (eds.): Syntactic and Structural Pattern Recognition – Theory and Applications. World Scientific, Singapore (1990)
4. Doostfatemeh, M., Kremer, S.C.: New directions in fuzzy automata. International Journal of Approximate Reasoning **38**, 175–214 (2005)
5. Flasiński, M., Jurek, J.: Dynamically Programmed Automata for Quasi Context Sensitive Languages as a Tool for Inference Support in Pattern Recognition-Based Real-Time Control Expert Systems. Pattern Recognition **32**, 671–690 (1999)
6. Flasiński, M., Reroń, E., Jurek, J., Wójtowicz, P., Atıasiewicz, K.: On the construction of the syntactic pattern recognition-based expert system for auditory brainstem response analysis. In: Kurzyński, M., Puchaıa, E., Woźniak, M., Żoınierek, A. (eds.) Computer Recognition Systems. ASC, vol. 30, pp. 503–510. Springer, Heidelberg (2005)
7. Flasiński, M., Jurek, J.: On the Analysis of Fuzzy String Patterns with the Help of Extended and Stochastic GDPLL(k) Grammars. Fundamenta Informaticae **71**, 1–14 (2006)

8. Flasiński, M., Jurek, J.: Fundamental Methodological Issues of Syntactic Pattern Recognition. Pattern Analysis and Applications **17**, 465–480 (2014)
9. Flasiński, M., Jurek, J., Peszek, T.: Parallel processing model for syntactic pattern recognition-based electrical load forecast. In: Wyrzykowski, R., Dongarra, J., Karczewski, K., Waśniewski, J. (eds.) PPAM 2013, Part I. LNCS, vol. 8384, pp. 338–347. Springer, Heidelberg (2014)
10. Flasiński, M., Jurek, J., Peszek, T.: Application of Syntactic Pattern Recognition Methods for Electrical Load Forecasting. Advances in Intelligent Systems and Computing. Springer (in print)
11. Freeman, H.: On the Encoding of Arbitrary Geometric Configurations. IEEE Trans. Electron. Comput. **EC–10**, 260–268 (1961)
12. Fu, K.S.: Syntactic Pattern Recognition and Applications. Prentice Hall (1982)
13. Gonzales, R.C., Thomason, M.G.: Syntactic Pattern Recognition: An Introduction. Addison-Wesley, Reading (1978)
14. Jurek, J.: Towards grammatical inferencing of GDPLL(k) grammars for applications in syntactic pattern recognition-based expert systems. In: Rutkowski, L., Siekmann, J.H., Tadeusiewicz, R., Zadeh, L.A. (eds.) ICAISC 2004. LNCS (LNAI), vol. 3070, pp. 604–609. Springer, Heidelberg (2004)
15. Jurek, J.: Recent Developments of the Syntactic Pattern Recognition Model Based on Quasi-context Sensitive Languages. Pattern Recognition Letters **26**, 1011–1018 (2005)
16. Jurek, J.: Grammatical inference as a tool for constructing self-learning syntactic pattern recognition-based agents. In: Bubak, M., van Albada, G.D., Dongarra, J., Sloot, P.M.A. (eds.) ICCS 2008, Part III. LNCS, vol. 5103, pp. 712–721. Springer, Heidelberg (2008)
17. Jurek, J., Peszek, T.: Model of syntactic recognition of distorted string patterns with the help of GDPLL(k)-based automata. In: Burduk, R., Jackowski, K., Kurzynski, M., Wozniak, M., Zolnierek, A. (eds.) CORES 2013. AISC, vol. 226, pp. 101–110. Springer, Heidelberg (2013)
18. Ogiela, M.R., Ogiela, L., Tadeusiewicz, R.: Mathematical Linguistic in Cognitive Medical Images Interpretation Systems. Journal of Mathematical Imaging and Vision **34**, 328–340 (2009)
19. Pavlidis, T.: Structural Pattern Recognition. Springer, New York (1977)
20. Rosenkrantz, D.J.: Programmed Grammars and Classes of Formal Languages. J. ACM **16**, 107–131 (1969)
21. Shaw, A.C.: A Formal Picture Description Scheme as Basis for Picture Processing Systems. Information and Control **14**, 9–52 (1969)
22. Specht, D.F.: Probabilistic Neural Networks. Neural Networks **3**, 109–118 (1990)
23. Tadeusiewicz, R., Ogiela, M.R.: Medical Image Understanding Technology. Springer, Heidelberg (2004)
24. Tanaka, E.: Theoretical Aspects of Syntactic Pattern Recognition. Pattern Recognition **28**, 1053–1061 (1995)

Effective Multi-label Classification Method for Multidimensional Datasets

Kinga Glinka and Danuta Zakrzewska

Abstract Multi-label classification, contrarily to the traditional single-label one, aims at predicting more than one predefined class label for data instances. Multi-label classification problems very often concern multidimensional datasets where number of attributes significantly exceeds relatively small number of instances. In the paper, new effective problem transformation method which deals with such cases is introduced. The proposed Labels Chain (LC) algorithm is based on relationship between labels, and consecutively uses result labels as new attributes in the following classification process. Experiments conducted on several multidimensional datasets showed the good performance of the presented method, taking into account predictive accuracy and computation time. The obtained results are compared with those obtained by the most popular Binary Relevance (BR) and Label Power-set (LP) algorithms.

Keywords Multi-label classification · Labels chain · Machine learning · Problem transformation methods

1 Introduction

In machine learning, supervised classification is a task of automatic assignment of cases to predefined classes within the specified categories. A model for classifying and labelling new instances from a test set is built up based on known objects from a training set. This model also allows to analyze characteristics of available classes.

Multi-label classification, contrarily to the traditional single-label one, aims at predicting more than one predefined class label for data instances. This approach

K. Glinka(✉) · D. Zakrzewska
Institute of Information Technology, Lodz University of Technology,
Wólczańska 215, 90-924 Lodz, Poland
e-mail: 800559@edu.p.lodz.pl, zakrzewska@p.lodz.pl

T. Andreasen et al. (eds.), *Flexible Query Answering Systems 2015*,
Advances in Intelligent Systems and Computing 400,
DOI: 10.1007/978-3-319-26154-6_10

has a wide range of possible application domains – for text [1] [2] or music categorization [3], semantic scene classification [4], protein function classification [5] as well as medical data classification [6]. Multi-label classification seems to be better solution to present reality and characteristic of instances. A document or a song can be classified into two or more categories, and similarly, a patient may be suffering from more than one disease at the same time.

Applications of multi-label classification very often have to deal with multidimensional datasets with many attributes and relatively small number of instances. Such situation can take place in the case of medical records as well as specialised documents or images. The objects are described by many features what makes the process of multi-label classification more complex, and thus specialized methods to deal with that kind of data are required. There exist several universal techniques for multi-label classification that can be used for any dataset. However, they do not provide satisfactory accuracy in many cases, especially when sets of attributes are large.

In the paper, the universal problem transformation method, which deals with multi-label classification when the number of attributes significantly exceeds the number of instances, is proposed. The technique is validated by the experiments conducted on datasets of different number of instances and attributes. The results are compared with the ones obtained by application of the most commonly used methods: Binary Relevance and Label Power-set.

The reminder of the paper is organised as follows. In the next section some multi-label classification methods are shortly described. Then, the proposed algorithm is presented. In the following section the experiments and their results are depicted. Finally, some concluding remarks and future research are presented.

2 Multi-label Classification

Multi-label classification is intended to predict more than one class label for data instance. Thus, applications of multi-label classifications require specialized, dedicated methods in order to provide satisfactory accuracy and effectiveness.

There exist two main categories of multi-label classification methods: (i) algorithm adaptation methods and (ii) problem transformation methods (compare [7], [8]). Methods of the first category are the ones which extend specific algorithms in order to handle multi-label data directly. In turn, problem transformation methods, independently of the specific learning algorithms, transform multi-label classification problem into one or more traditional single-label tasks. Such approach allows to apply a variety of well-known classification algorithms to multi-label classification problems.

There exist several transformation techniques ([8]). As the most trivial ones, there should be mentioned the two of them which change the learning problem into traditional single-label classification (see [4], [7]). The first of the methods eliminates every multi-label instance from the dataset, while the second one randomly selects one of the multiple labels of each multi-label instance and eliminates the rest. These techniques are very simple but discard a lot of the information of the original multi-label dataset. Alternatively to these methods there were introduced Binary Relevance (BR) and Label Power-set (LP) techniques which became the most popular and widely used in multi-label classification. In the current research, the effectiveness of these two methods will be compared with the performance of the proposed LC technique.

Binary Relevance (BR) method constitutes a separate binary classifier for each existing label l from the label-set $\mathcal{L}$. The result of every single classification is the value l or $\neg l$, new instance has label l or not. In this way, the method divides set $\mathcal{L}$ into $|\mathcal{L}|$ two-elements sets, simplifying the multi-label problem to $|\mathcal{L}|$ binary classifications. For the final classification of a new instance x, this method outputs the union of the labels which are output of binary classifiers. As the main disadvantage of the method, one should mention about ignoring label correlations which may exist in a dataset, and thus losing some important information. The detailed description including experiments of *Binary Relevance* method can be found in [3], [8] and [9].

Label Power-set (LP) method creates new classes of all unique sets of labels which exist in the multi-label training data. Such approach allows to transform every complex task of multi-labelling into one single-label classification. Therefore, this method can be used regardless of number and variety of labels assigned to the instances. One of the negative aspects of creating new labels is that it may lead to datasets with a large number of classes and few instances representing them. The *Label Power-set* method has been investigated in [2], [4], [8] and [9].

3 Materials and Methods

3.1 The Proposed Approach

Multi-label classification very often concerns the datasets of the big attribute numbers. Such situation especially takes place when we classify documents, images or music – objects for which multi-label classification occurs the most often. However, the performance of the commonly used problem transformation methods seems to be not sufficient enough in such cases. We propose an alternative problem transformation method which gives promising effects, while dealing with multi-label classification for multidimensional datasets of relatively small number of instances comparing to the number of attributes.

Independent Labels (IL). Firstly, the approach called *Independent Labels (IL)*, where each label constitutes a separate single-label task, will be considered. The algorithm seems to be similar to *Binary Relevance (BR)* method, however, it requires to learn $|L|$ multiclass classifiers, instead of $|\mathcal{L}|$ binary classifiers. L is a set of relevant labels for an instance $x = (x_1, x_2, ..., x_m) \in \mathcal{X}$, subset of all label-set $\mathcal{L}$ (classes). It makes the method competitive in time and in computational complexity in the cases of small number of labels per instance. We also assume that numbers of labels for instances are known. Similarly to *BR*, during transformation process *IL* ignores label correlations which exist in the training data, that may result in losing some vital information and may provide poor prediction quality. The algorithm of *Independent*

Algorithm 1. Independent Labels (IL)

Require: training set $D_{tr} \neq \emptyset$, testing set $D_{ts} \neq \emptyset$, $D_{tr} \cap D_{ts} = \emptyset$,
single-label classifier $h(\cdot)$

Transform

$\chi' = \{\}$
for all $x \in D_{tr}$ **do**
    for all l_i **in** labels L of x **do**
        $\chi' \leftarrow x \cup \{l_i\}$ ▷ Create instances, each with one label
    end for
end for

Learning

$L' = \{\}$
for all $x \in \chi'$ **do**
    $L' = h(x)$ ▷ Classify instances
end for

Result

for all $x \in \langle D_{ts}, L' \rangle$ **do**
    if $x_{L'} = L_x$ **then** ▷ Check if all labels for instance are correct
        $result \leftarrow good$
    else
        $result \leftarrow bad$
    end if
end for

Labels is presented in Alg. 1. It is divided into three main parts: *Transform, Learning* and *Result*, and leads to the intended results in a finite number of steps. The number of steps needed to achieve final result of the classification depends on the number of instances in training and test datasets and primarily the size of relevant labels-set L, assigned to the instances.

Labels Chain (LC). *IL* method can be improved by using a view to map relationship between labels. New problem transformation method *Labels Chain (LC)* also requires to learn $|L|$ multiclass classifiers, but this one, in contrast to *IL*, consecutively uses result labels as new attributes in the following classification process, creating the classifications chain (the idea has been so far used only for binary classifications [10]). In the result, classifications in *Labels Chain* are not totally independent from themselves what enables providing better predictive accuracy. The fact should be observed especially in multi-label problems with small number of labels in $\mathcal{L}$, because in these cases the value of a new, added attribute is more significant for classification process. The *Labels Chain* method can be also applied taking into account different order of classifications, with the number of $|L|!$ available order combinations. Similarly to *IL*, *LC* algorithm also assumes that the number of labels for instances are known. The algorithm of *Labels Chain* is described in Alg. 2. As in the previous one, the number of required iterations to attain intended results is dependent on the size of training and test datasets and number of labels assigned to the instances.

In our further considerations, *Independent Labels* is considered to be only indirect method, improved by *Labels Chain* approach. We will compare the obtained results with those got by the most popular *Binary Relevance (BR)* and *Label Power-set (LP)* algorithms, taking into account predictive accuracy and computation time.

Algorithm 2. Labels Chain LC

```
Require: training set D_tr ≠ ∅, testing set D_ts ≠ ∅, D_tr ∩ D_ts = ∅,
   single-label classifier h(·)
   Transform
χ' = {}
for all x ∈ D_tr do
    χ' ← x ∪ {l_1}            ▷ Create instance with first label from set of relevant labels
end for
   Learning
L' = {}
for all l_i in labels L of x do
    for all x ∈ χ' do
        x ← x ∪ {L'_{i−1}}    ▷ Add result label from previous step as new attribute
        L'_i = h(x)           ▷ Classify instance
    end for
end for
   Result
for all x ∈ ⟨D_ts, L'⟩ do
    if x_{L'} = L_x then      ▷ Check if all labels for instance are correct
        result ← good
    else
        result ← bad
    end if
end for
```

3.2 Evaluation

Classification Accuracy. As the basic evaluation measure, *Classification Accuracy* will be considered, the measure which is also commonly used for single-label problems. This measure provides a very strict evaluation because it ignores partially correct sets of labels (considering them as incorrect), and requires the predicted set of labels to be an exact match of the true set of labels. *Classification Accuracy* for multi-label classification is defined as [11]:

$$CA = \frac{1}{N} \sum_{i=1}^{N} I\left(Y_i = F(x_i)\right) \tag{1}$$

where: x_i are instances, $i = 1..N$, N is their total number in the test set, Y_i denotes the list of labels and $F(x_i)$ is a sequence of predicted labels during classification process; $I(true) = 1$, $I(false) = 0$.

Table 1 Datasets used in the experiments

Dataset name	Domain	Instances	Attributes	Label set $\|\mathcal{L}\|$
scene176	image	176	294	5
scene90	image	90	294	5
emotions150	music	150	72	6
emotions61	music	61	72	6

Computation Time. As additional measure, *Computation Time* was examined. It is represented by the length of time required to perform the computational process of classification, and is counted as the average time in milliseconds for 10 repeated classification processes for every method. In that way, errors related to the Java Virtual Machine optimizations during application process are avoided. It is also worth mentioning that time is computed for all the process, including data transformation, learning and classifying.

3.3 Datasets

In order to evaluate the proposed method, the experiments were carried out on four datasets. We have taken into account the sets where the number of instances is relatively small comparing to the number of attributes. Thus, the datasets can be considered as multidimensional. In the three out of four datasets number of attributes significantly exceeds number of instances. All the characteristics of the considered datasets, such as the number of data instances, attributes or labels, are presented in Tab. 1.

The datasets come from A Java Library for Multi-Label Learning [12] and for experiments they were randomly limited to small number of instances, not exceeding 200, all with two labels assigned. The datasets were randomized by using Weka Open Source WEKA [13].

There were considered two original Mulan datasets: *scene* and *emotions*. From both of them there were created two new datasets of small number of instances, and one of them has about half of instances of the other. The *scene* dataset consists of images which may belong to multiple semantic classes: beach, fall foliage, field, mountain and urban. The second dataset *emotions* contains music (signal audio). It is designed for classification of music emotions and the labels concern genre of songs such as rock, classic, metal etc.

4 Experiment Results and Discussion

The experiments aimed at examining the performance of the proposed technique. They were carried out on the datasets described in Section 3.3. During the experiments classification accuracy and computation time were calculated and compared for distinct problem transformation methods of multi-label classification: *Binary Relevance (BR)*, *Label Power-set (LP)*, *Independent Labels (IL)* and *Labels Chain (LC)*. In the case of the *LC* techniques, the possible label orders were considered. The best values were taken into account as the final results of the *LC* method.

During experiments five well known classifiers were applied: k-nearest neighbours kNN, naive Bayes NB, support vector machine SVM, multilayer perceptron MLP and C4.5 decision tree [14]. They were conjuncted with the considered problem transformation methods.

For the experiment purpose the environment was created. The software was based on WEKA Open Source [13]. All algorithms were tested and evaluated by using, in most of the cases, the same default parameters of WEKA software. The system was running under Java JDK 1.8, on 64-bit machine with a dual-core processor Intel Core i5 2.27 GHz and 4 GB RAM. All the datasets: *scene176*, *scene90*, *emotions150*, *emotions61* were taken into account for all the considered techniques. Each dataset was randomized with the number of seed equal to 100 and divided into two parts – training set (60% of instances) and test set (40% of instances).

Results for the *scene176* are presented in Tab. 2, Tab. 3 and Fig. 1. It is easy to notice that for *scene176* dataset, in most of the cases, the best classification accuracy was obtained for *LC* technique, with the smallest Computation Time value. Only in the case of decision tree C4.5 the best results have been got for *LP* method. However, the Computation Time of *LC* and *LP* remains comparable.

Tab. 4, Tab. 5 and Fig. 2 present results for the dataset *scene90*. In most of the cases, together with decreasing the number of instances Computational Accuracy was getting worse, however still remained the best for *LC* method. The one exception

Table 2 Dataset *scene176*: Classification Accuracy (CA) [%]

	BR	LP	IL	LC
kNN	67.14	67.14	67.14	67.86
NB	44.29	50.00	50.00	67.86
SVM	41.43	41.43	41.43	46.43
MLP	61.43	71.43	68.57	71.43
C4.5	41.43	65.71	54.29	50.00

Table 3 Dataset *scene176*: Computation Time (T) [ms]

	BR	LP	IL	LC
kNN	670	446	174	130
NB	574	191	197	195
SVM	1815	452	709	512
MLP	300089	61867	122896	81502
C4.5	642	186	237	165

concerned SVM classifier for which Computational Accuracy has been grown for all the problem transformation techniques. Similarly to the set *scene176*, Computational Time was also the shortest for *LC* technique.

Tab. 6, Tab. 7 and Fig. 3 show results for the dataset *emotions150*. The results for *LC* do not differ significantly from the other problem transformation method. However, the best Classification Accuracy (71.43%) has been obtained for the conjunction of *LC* and SVM or MLP classifiers. The worst results can be noticed for the conjunction of *BR* and C4.5 classifier (41.67%).

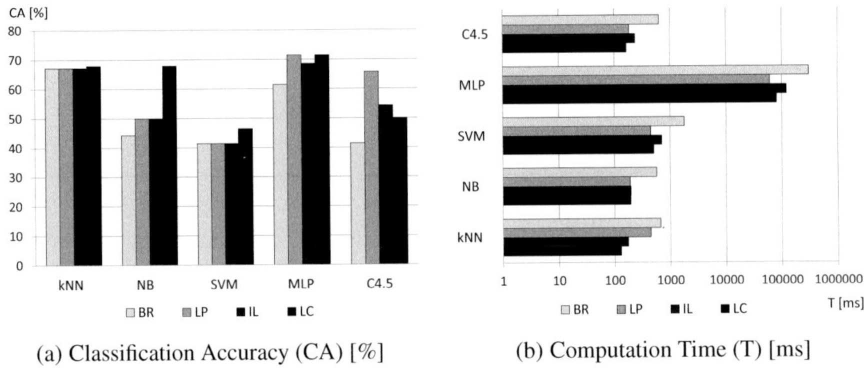

(a) Classification Accuracy (CA) [%]

(b) Computation Time (T) [ms]

Fig. 1 Dataset *scene176*

Table 4 Dataset *scene90*: Classification Accuracy (CA) [%]

	BR	LP	IL	LC
kNN	55.56	55.56	55.56	64.29
NB	63.89	61.11	63.89	57.14
SVM	47.22	47.22	69.44	71.43
MLP	66.67	63.89	69.44	71.43
C4.5	41.67	61.11	52.78	50.00

Table 5 Dataset *scene90*: Computation Time (T) [ms]

	BR	LP	IL	LC
kNN	457	373	89	65
NB	320	125	102	111
SVM	737	266	277	200
MLP	155560	31505	61382	43130
C4.5	328	75	107	86

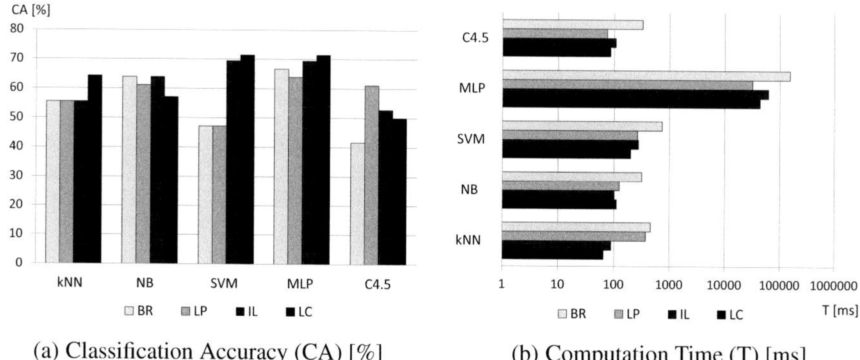

(a) Classification Accuracy (CA) [%]

(b) Computation Time (T) [ms]

Fig. 2 Dataset *scene90*

Table 6 Dataset *emotions150*: Classification Accuracy (CA) [%]

	BR	LP	IL	LC
kNN	35.00	35.00	35.00	41.67
NB	26.67	41.67	38.33	45.83
SVM	16.67	36.67	25.00	33.33
MLP	18.33	36.67	21.67	45.83
C4.5	25.00	33.33	28.33	45.83

Table 7 Dataset *emotions150*: Computation Time (T) [ms]

	BR	LP	IL	LC
kNN	494	27	47	46
NB	211	42	53	45
SVM	347	85	105	119
MLP	23363	4825	7787	4330
C4.5	179	40	67	67

Tab. 8, Tab. 9 and Fig. 4 present results for the dataset *emotions61*. They are the worst from all the ones obtained for the other datasets. However, again the best Classification Accuracy (50%) has been obtained for the conjunction of *LC* and MLP or C4.5 classifiers. The worst results can be noticed for the conjunction of *BR* and SVM classifier (8.33%).

Summing up the investigations concerning Computation Time, experiments on all the datasets gave the shortest, comparable times for *LP* and *LC* methods (Tab. 3, 5, 7 and 9, Fig. 1(b), 2(b), 3(b) and 4(b)). The worst in this criteria was *BR*, significantly exceeding times of other methods. Finally, we can conclude that the proposed

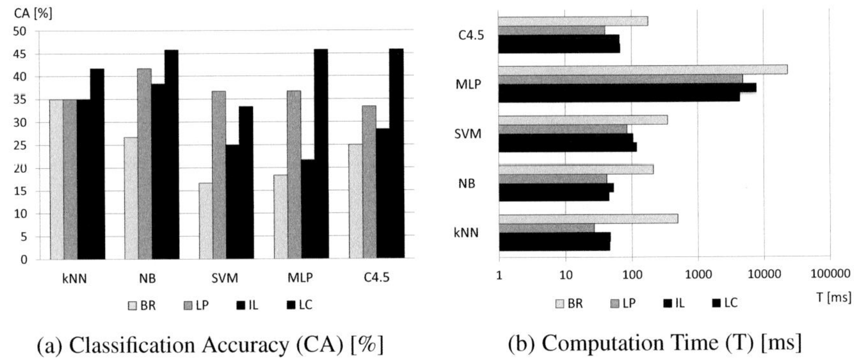

(a) Classification Accuracy (CA) [%] (b) Computation Time (T) [ms]

Fig. 3 Dataset *emotions150*

Table 8 Dataset *emotions61*: Classification Accuracy (CA) [%]

	BR	LP	IL	LC
kNN	33.33	33.33	33.33	40.00
NB	16.67	16.67	25.00	40.00
SVM	8.33	25.00	25.00	10.00
MLP	33.33	50.00	45.83	50.00
C4.5	16.67	33.33	29.17	50.00

Table 9 Dataset *emotions61*: Computation Time (T) [ms]

	BR	LP	IL	LC
kNN	458	20	27	23
NB	137	26	32	30
SVM	253	33	47	34
MLP	7917	2092	2633	1843
C4.5	103	24	35	25

method *Labels Chain* is effective not only in terms of Classification Accuracy but also Computation Time.

5 Concluding Remarks

In the paper, new effective problem transformation method of multi-label classification for multidimensional datasets is presented. The proposed *Labels Chain* algorithm is based on relationship between labels, and consecutively uses result labels as new attributes in the following classification process.

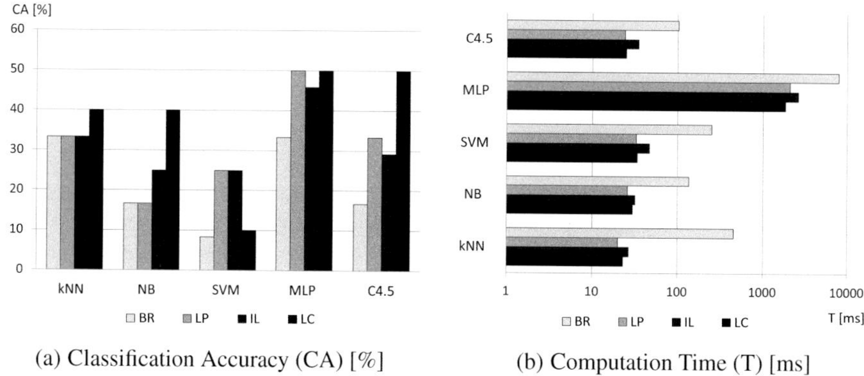

(a) Classification Accuracy (CA) [%]

(b) Computation Time (T) [ms]

Fig. 4 Dataset *emotions61*

Experiments carried out on datasets of small number of instances and significantly bigger number of attributes showed that the proposed *Labels Chain* integrated with certain classifiers performs better than the most popular problem transformation techniques, taking into account Classification Accuracy and Computation Time.

Future research will consist in further investigations of the proposed method, taking into account datasets of different sizes and characterised by bigger number of attributes and labels as well as different evaluation criteria. Worth considering is building the tool, based on the conjunction of problem transformation method and traditional classifier, which will effectively solve multi-label classification problems.

References

1. Fujino, A., Isozaki, H., Suzuki, J.: Multi-label text categorization with model combination based on f1-score maximization. In: Third International Joint Conference on Natural Language Processing, IJCNLP 2008, pp. 823–828, Hyderabad, India (2008)
2. Sajnani, H., Javanmardi, S., McDonald, D.W., Lopes, C.V.: Multi-label classification of short text: A study on wikipedia barnstars. In: Analyzing Microtext: Papers from the 2011 AAAI Workshop (2011)
3. Li, T., Ogihara, M.: Content-based music similarity search and emotion detection. In: Proceeding of IEEE International Conference on Acoustic, Speech and Signal Processing, vol. 5, pp. 705–708, Canada (2006)
4. Boutell, M.R., Luo, J., Shen, X., Brown, C.M.: Learning multi-label scene classification. Pattern Recognition **37**(9), 1757–1771 (2004)
5. Clare, A.J., King, R.D.: Knowledge discovery in multi-label phenotype data. In: Siebes, A., De Raedt, L. (eds.) PKDD 2001. LNCS (LNAI), vol. 2168, p. 42. Springer, Heidelberg (2001)
6. Bhattarai, A., Ras, V., Dasgupta, D.: Classification of Clinical Conditions: A Case Study on Prediction of Obesity and Its Co-morbidities. Research in Computing Science **41**, 183–194 (2009)

7. Tsoumakas, G., Katakis, I., Vlahavas, I.: A review of multi-label classification methods. In: Proceedings of the 2nd ADBIS Workshop on Data Mining and Knowledge Discovery (ADMKD 2006), pp. 99–109, Thessaloniki, Greece (2006)
8. Tsoumakas, G., Katakis, I., Vlahavas, I.: Mining multi-label data. In: Maimon, O., Rokach, L. (eds.) Data Mining and Knowledge Discovery Handbook, pp. 667–685. Springer US, Boston (2010)
9. Madjarov, G., Kocev, D., Gjorgjevikj, D., Deroski, S.: An extensive experimental comparison of methods for multi-label learning. Pattern Recognition **45**(9), 3084–3104 (2012)
10. Read, J., Pfahringer, B., Holmes, G., Frank, E.: Classifier chains for multi-label classification. In: Buntine, W., Grobelnik, M., Mladenić, D., Shawe-Taylor, J. (eds.) ECML PKDD 2009, Part II. LNCS, vol. 5782, pp. 254–269. Springer, Heidelberg (2009)
11. Kajdanowicz, T., Kazienko, P.: Multi-label classification using error correcting output codes. Applied Mathematics and Computer Science **22**(4), 829–840 (2012)
12. `http://mulan.sourceforge.net/datasets-mlc.html`
13. `http://www.cs.waikato.ac.nz/ml/weka/index.html`
14. Witten, I.H., Frank, E., Hall, M.A.: Data Mining: Practical Machine Learning Tools and Techniques. Morgan Kaufmann, San Francisco (2011)

Study of the Convergence in Automatic Generation of Instance Level Constraints

Irene Diaz-Valenzuela, Jesús R. Campaña, Sabrina Senatore, Vincenzo Loia, M. Amparo Vila and Maria J. Martin-Bautista

Abstract This work deepens in a methodology to generate Instance Level Constraints for Semi-supervised clustering by the study of the inherent nature of the data. The methodology executes a partitional clustering algorithm repetitively, so we study its behaviour according to the number of iterations of the clustering. In this scenario we propose three different stopping criteria to determine how many times the partitional clustering algorithm should be executed to obtain reliable instance level constraints. These criteria are experimentally tested under the document clustering problem.

Keywords Semi-supervised clustering · Instance Level Constraints

1 Introduction

Recently, the interest in Semi-supervised clustering techniques has grown in the data mining and machine learning communities. Semi-supervised clustering techniques improve unsupervised clustering approaches by including side-information to enhance the clustering processes. This side information, also called supervision, comes from some source of external knowledge, typically human expertise. There are two main ways to provide supervision in semi-supervised clustering: providing a small amount of labelled instances or pairwise instance level constraints.

I. Diaz-Valenzuela · J.R. Campaña(✉) · M.A. Vila · M.J. Martin-Bautista
Department of Computer Science and Artificial Intelligence,
University of Granada, 18071 Granada, Spain
e-mail: {idiazval,jesuscg,vila,mbautis}@decsai.ugr.es

S. Senatore · V. Loia
Dipartimento di Informatica, Universitá degli Studi di Salerno, 84084 Fisciano, SA, Italy
e-mail: {ssenatore,loia}@unisa.it

T. Andreasen et al. (eds.), *Flexible Query Answering Systems 2015*,
Advances in Intelligent Systems and Computing 400,
DOI: 10.1007/978-3-319-26154-6_11

Instance Level Constraints are a kind of external information first described by Wagstaff and Cardie [18]. These constraints are defined between pairs of instances and they indicate whether a pair of instances should be grouped in the same cluster or not. This kind of external information arise naturally in many domains such as: gene classification (two co-occurring proteins), Information Retrieval (documents regarding the same topic), etc.

One of the main drawbacks of these methods is the difficulty of finding such external information. Frequently, the use of expert information is quite costly and, sometimes, it is not even available. For example, finding pairwise relationships in a document corpus containing a few thousand of documents would require to read all of them carefully, which could be a very time consuming and arduous task. To overcome this limitation, in [5] we have introduced a methodology aimed at the automatic generation of pairwise feedback in the shape of instance-level constraints. To accomplish this task, the inherent structure of the data is discovered by reiterated iterations of a partitional clustering algorithm. From the output of this initial clustering, it is possible to establish relationships between pairs of elements leading to instance level constraints that could be used in a semi-supervised clustering process.

There are some aspects of that methodology which were not considered in the initial development. Specifically, the stopping criteria, i.e. the number of times that the partitional clustering algorithm should be executed is not defined, but left to the user to determine. The goal of this paper is to study the influence of the number of iterations of the k-means in the quality of the constraints and to propose an automatic way to determine it.

This paper is organised as follows: Section 2 covers some related work regarding semi-supervised clustering. In Section 3 some background is given, describing the basis of the semi-supervised model that we are using. After that, Section 4 describes in detail the method to generate instance level constraints, including our new proposals for stopping criteria in Section 4.1. These proposals are tested experimentally in Section 5 and finally, some conclusions and future work are presented in Section 6.

2 Related Work

Semi-supervised clustering appeared as an alternative to traditional unsupervised approaches where a small quantity of side information is introduced in the clustering process to improve its performance. Existing methods for semi-supervised clustering fall into three categories: constraint-based, distance-based and hybrid. The first category uses user-provided labels or constraints that are enclosed in the algorithm. They modify the objective function to include the information from the pairwise instance constraints [18] or to generate seed clusters using the labelled data [2].

In distance-based approaches, adaptive distance measures are defined by metric learning techniques, which use them for training the clustering in order to satisfy the labels or constraints in the supervised data [19]. Finally, the hybrid methods propose some combination of these two approaches. For instance, a general probabilistic framework unifiying both ideas is presented in [3] and in [11] the authors provide a complete experimentation, testing the performance of an hybrid proposal in contrast with the use of labels and constraints separately.

These techniques are reviewed in [9], [4]. This paper focuses on a type of semi-supervision that comes from *pairwise instance-level* constraints. These constraints were first introduced by Wagstaff et. al [17] and they have been widely used and reformulated since then [20], [16]. Other approaches use the concept of pairwise external information in combination with fuzzy clustering [12], [14]. Traditionally, *instance-level* constraints are generated by human experts with some knowledge about the specific topic under consideration. Some approaches like [21], [20], [1] basically exploit those assumptions, and automatically select which kind of instances could provide specific important information for the clustering process and ask the expert about them. Under this type of process the expert has a role that cannot be ignored, and the information that provides must always be taken into account. Our proposal differs from others in the sense that the role of the expert is superseded in the process.

The proposal made on this paper is an extension of the methodology introduced in [5]. That methodology is designed to be used with the approaches presented in [6], [8], where semi-supervision is introduced in hierarchical clustering. It is our intention to improve that proposal by adding some stopping criteria that help the user to determine how many iterations of the method should be performed to obtain better results.

3 Background

Instance Level Constraints in the sense of [17] are based on two types of information: *must-link* and *cannot-link*. The first type describes the relationship between two instances in the same cluster, whilst the latter indicates that the documents are in different clusters. Formally, as defined in [5]:

Definition 1. MUST-LINK: Given two instances d_i and $d_j \in D$, there is a must link $ML(d_i, d_j)$, if d_i and d_j are in the same cluster. The set of all *must-link* constraints defined for D is called ML.

Definition 2. CANNOT-LINK: Given d_i and $d_j \in D$, if there is a cannot-link $CL(d_i, d_j)$, then d_i and d_j cannot not be in the same cluster. The set of all *cannot-link* constraints defined for D is called CL.

Moreover, it works with the assumption that there is an underlying class structure that assigns each instance to one of c classes C. This method aims to find a mapping F between the calculated partition and the given classification, such that $F : P \rightarrow C$.

4 Constraints Generation

The intervention of human experts is preferable when some kind of supervision is used. However, as human expertise could be expensive and difficult to use, in some approaches, human intervention can be replaced by automatically generated knowledge. The study of the data and their underlying relationships, that could be found through clustering procedures, could provide a valid support for generation instance level constraints.

The previous idea is approached in [5], where a partitional clustering process is used to obtain a partition of the data according to some distance criteria. The algorithm used in that case is k-means [10]. K-means performs a flat clustering that finds a partition $P_K = \{S_1, \ldots, S_k\}$ for a given k by minimising the within-cluster sum of squares, according to the following objective function:

$$J = \sum_{i=1}^{k} \sum_{j=1}^{n} \|x_{i,j} - \mu_i\|^2 \tag{1}$$

where $\|x_{i,j} - \mu_i\|$ is the distance between a data point $x_{i,j}$ and the centroid μ_i of the cluster S_i.

The partition P_K provides an initial idea of the organisation of the input data. Regardless of the possible mistakes and inaccuracies that P_K could present, it is possible to use this information to generate constraints. If it is considered that all instances from a partition should be in the same group, must-link constraints are pretty straightforward. More formally: for each pair of instances (d_i, d_j) that are in the same cluster in the clustering-driven partitioning, there exists a *must-link* constraint $ML(d_i, d_j), \forall (d_i, d_j) \in S_a | S_a \in P_K$. Under this assertion, the set of must-link constraints, ML, contains all pairs of elements that are in the same cluster (considering all clusters independently): $ML = \cup_{i=1}^{k} ML_i$, where ML_i is a set of must-link constraints from a partition S_i.

Similarly, *cannot-link* constraints are defined between pairs of instancess (d_i, d_j) that are in different clusters, i.e. $\forall (d_i, d_j); d_i \in S_a, d_j \in S_b | S_a, S_b \in P_K; S_a \neq S_b$. Under this definition, the set of cannot-link constraints, CL, contains all pairs of elements that are in the different clusters (considering all clusters independently): $CL = \cup_{\forall a,b:a \neq b} CL_{a,b}$, where $CL_{a,b}$ is a set of cannot-link constraints composed of the pair $(d_i, d_j) \in S_a \times S_b | a \neq b$.

This approach is sensitive to the initial configuration of the *k-means* algorithm. However, this can be overtaken by exploiting the random component of the k-means initialisation. The partition returned by k-means depends of the initial centroids μ_i (with $i = 1 \ldots k$) that are randomly generated on each execution of the algorithm. It means that every execution may provide slightly different partitions. Under that assumption, by executing the algorithm repeatedly, the original set of constraints is refined, defining in that way the constraints. In this sense, if two instances d_i and d_j are placed in the same cluster in all executions of the k-means clustering, then there is a *must-link*, $ML(d_i, d_j)$, constraint between them. In the same way, if two instances

Algorithm 1. Constraints generation

Get P_K an initial partition returned by k-means
for all $K_i \in P_K$ **do**
  if $(d_i, d_j) \in K_i$ **then**
    Add (d_i, d_j) to ML
  else
    Add (d_i, d_j) to CL
  end if
end for
for each k-means execution **do** //
  Get $P_a = \{S_1, \ldots, S_k\}$ the partition returned by k-means
  for all $ML(d_i, d_j) \in ML$ **do**
    if $d_i \in S_a$ and $d_j \in S_b$ with $S_a \neq S_b$ **then**
      Remove (d_i, d_j) from ML
    end if
  end for
  for all $CL(d_i, d_j) \in CL$ **do**
    if $d_i, d_j \in S_a$ **then**
      Remove (d_i, d_j) from CL
    end if
  end for
end for

d_i and d_j are never placed in the same cluster in all executions, then there is a *cannot-link* constraint $CL(d_i, d_j)$. This process has been summarised in Algorithm 1.

Considering that we are keeping as constraints only that information coherent throughout all the executions of the k-means clustering algorithm, there are some pairs of instances without an associated constraint. This occurs because the instancess are placed in the same cluster in some executions and in different clusters in others. Moreover, data structure can affect the performance of k-means clustering, which generally tends to produce clusters of relatively uniform size. In case of bad partitioning, some of the resulting clusters could not be used for the constraint generation, because their data relations are considered not good enough for the constraints (for example if they are too big, compared with the remaining clusters). In that case, the constraints are generated considering only the clusters that fit some criteria, discarding all the remaining information.

This method works by *refinement*, i.e. it generates an initial set of constraints that is improved in further executions of the k-means, smaller sets of constraints usually are more precise than the bigger ones, and they come from a higher number of executions.

4.1 Stopping Criteria

The application of this method arises the question of how many executions of the partitional algorithm are needed to generate a proper set of constraints. To determine this, we propose three different stopping criteria:

- At a determinate percentage of the total number of constraints
- Using an user fixed number of iterations n in the k-means.
- Until there is no further changes between two iterations.

The first criterion executes k-means until a certain pre-fixed number of constraints are generated. This number is chosen by the user and can be fixed according to a study similar to the one performed in [5]. On the other hand, the second criterion executes the k-means a number of times that is fixed by the user, independently of the number of constraints that are generated during the process. It allows the user to have a certain control of the execution time. Finally, the third criterion keeps refining the constraint set until there are no further changes. These three proposals will be compared experimentally on Section 5.

5 Experimental Results

These stopping criteria are tested experimentally under the document clustering problem. This problem takes a set of documents and groups them according to a certain criteria. Formally:

Let $D = \{d_1, \ldots, d_m\}$ be a documents collection, $T = \{t_1, \ldots, t_l\}$ a set of terms from the documents set, then $P = \{C_1, \ldots, C_n\}$ is a partition of the document corpus, where each C_i $(i = 1 \ldots, n)$ is a subset of documents, s.t. $\cup_{i=1}^{n} C_i = D$ and $\cap_{i=1}^{n} C_i = \emptyset$

5.1 Data Description

The document datasets that have been used in these tests are: *Web Snippets* dataset [15] and *Reuters-21578* collection [13]. *Web Snippets* dataset contains 2280 short texts (6 to 20 words) taken from Google, unevenly divided into eight categories and from Reuters-21578 we have taken a subset of the well known Reuters-21578 collection containing 2014 documents and ten independent categories: *trade*, *ship*, *wheat-grain*, *gold*, *sugar*, *money-fx*, *interest*, *crude*, *money-supply* and *coffee*. They have been represented under the Vector Space Model using the data preparation techniques proposed in [5].

The semi-supervised clustering technique used is described in [6]. It requires a hierarchical agglomerative clustering algorithm. The algorithm and weights for the semi-supervised approach have been set following the recommendations made in [5].

5.2 *Validity Measures*

To validate the goodness of the clustering process, two different measures have been used: F-measure and Normalized Mutual Information (briefly, NMI).

F-measure evaluates the quality of the clusters by comparing the relationship between the retrieved documents on each cluster and the relevant documents according to their given class labels. F-measure (4) is defined as the harmonic mean of Precision (2) and Recall (3).

$$Precision = \frac{|\text{relevant documents} \cap \text{retrieved documents}|}{|\text{retrieved documents}|} \tag{2}$$

$$Recall = \frac{|\text{relevant documents} \cap \text{retrieved documents}|}{|\text{relevant documents}|} \tag{3}$$

$$F - measure = \frac{2 \times \text{Precision} \times \text{Recall}}{\text{Precision} + \text{Recall}} \tag{4}$$

The Normalized Mutual Information (5) evaluates the elements on each cluster against class labels. It measures and normalises the mutual information between random variables P_α (the optimal partition of the dendrogram) and C (the ground truth given by class labels).

$$NMI(P_\alpha, C) = \frac{2I(P_\alpha, C)}{H(P_\alpha) + H(C)} \tag{5}$$

where $I(P_\alpha, C)$ is the mutual information between the two random variables and H is the Shannon entropy of the variable.

Considering the random component of the constraints generation process, for each experiment we have generated 3 different sets of constraints. Results are shown averaged over multiple executions.

5.3 *Comparison of the Different Stopping Criteria*

On an initial test, the performance of the first stopping criteria, *stopping at a determinate percentage of the total number of constraints*, has been tested. To do so, the method has been set to stop at a number of constraints equal or lower than a prefixed percentage. During that study we have discovered that the sets of constraints that could be generated with that methodology have a maximum and minimum size. Let us remember that the size of the constraint sets depend on the number of k-means executions. Also, some pairs of instances do not have any associated pairwise information, specifically those placed in clusters too big to be considered in the methodology (as described on Section 4). For that reason, the maximum number of constraints that can be generated correspond with the amount coming from the

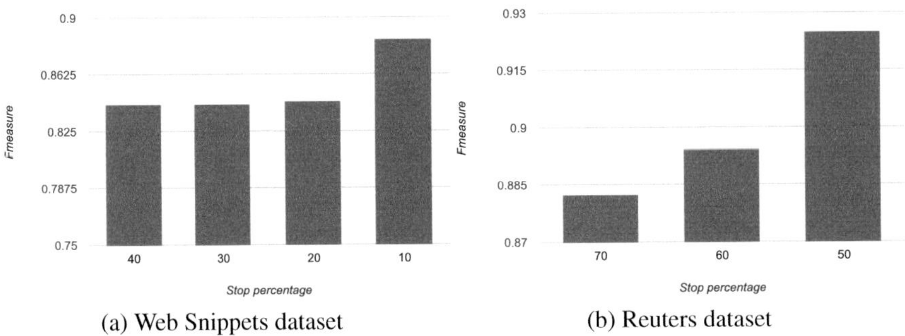

(a) Web Snippets dataset (b) Reuters dataset

Fig. 1 Comparison of the F-measure of different percentages of automatically generated constraints.

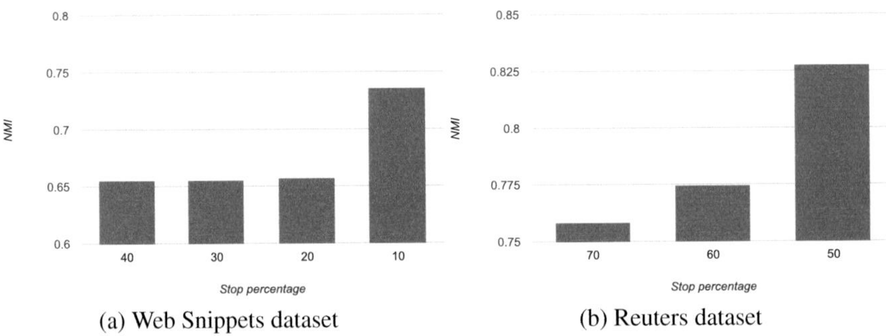

(a) Web Snippets dataset (b) Reuters dataset

Fig. 2 Comparison of the Normalized Mutual Information of different percentages of automatically generated constraints.

initial partition. On the other hand, there is a point of convergence where no further refinement can be achieved, i.e, there are no changes in the constraints set. That point is considered the minimum. Specifically for the data considered in these tests, the maximum of the Web Snippets dataset is located at 40% and the minimum at 10%. For the Reuters dataset the maximum is 70% and the minimum 50%.

To evaluate the performance of the method according to the first criterion, all these ideas are taken into account. For that reason, we have configured the method to generate constraints with sizes from the maximum of each dataset to the minimum, with decrements of 10% of the constraints. The results are shown in Figures 1 and 2.

From the F-measure in Figure 1, it is possible to read that the maximum number of constraints that can be generated (40% for the Web Snippets and 70% for the reuters) has the lower performance. This kind of behaviour is expected as the constraint set is less "refined" and it is also observable with the Normalized Mutual Information in Figure 2. As the constraint sets get smaller, the performance improves, as clearly shown in Figures 1b and 2b for the Reuters dataset. When the constraint sets reach their convergence value corresponding with the minimum percentage of constraints that can be generated, they get their best performance.

To compare the three stopping criteria proposed in this paper, we have taken in consideration the previous results. In Figures 3 and 4 we have represented the F-measure and the NMI of the sets obtained using the three proposed stopping criteria: at a predefined percentage, convergence and at a number of iterations fixed by the user. These values have been chosen to represent the worse performance (maximum) that can be obtained when an user sets a specific percentage value for constraints, the convergence value, and 30 iterations of the k-means algorithm as used in [5].

Figures 3 and 4 show the F-measure and Normalized Mutual Information of the three stopping criteria proposed for this methodology. Let us notice that, in all problems tested, getting the maximum percentage of the constraints is the approach that has the worst performance. However, as the number of iterations of the k-means is lower, it is also the faster method. Using the second criteria, fixing a number of executions of k-means, it is possible to get reasonable results if the number of iterations chosen is long enough. In this example, 30 iterations have provided quite good results for the Web Snippets dataset, but not so good for the Reuters. From that result, it is possible to read that the number of iterations depend of the problem and should be determined experimentally. Nevertheless, this criterion has the advantage of obtaining the constraints in a predictable period of time, i.e. knowing how much time an iteration takes allows the prediction of how long is going to take to execute n. The third criterion, stopping at the convergence, gets us the best constraints but has the drawback of requiring more k-means iterations. From the Figures 3 and 4, it is possible to see that executing k-means until there is a convergence in the number of constraints generated provides higher values of F-measure and Normalized Mutual Information than the other approaches, with differences close to 0.1 in some of the problems. In general, the convergence criterion produces good results with both measures, evidencing that this criterion could be recommended for constraints generation in document clustering problems.

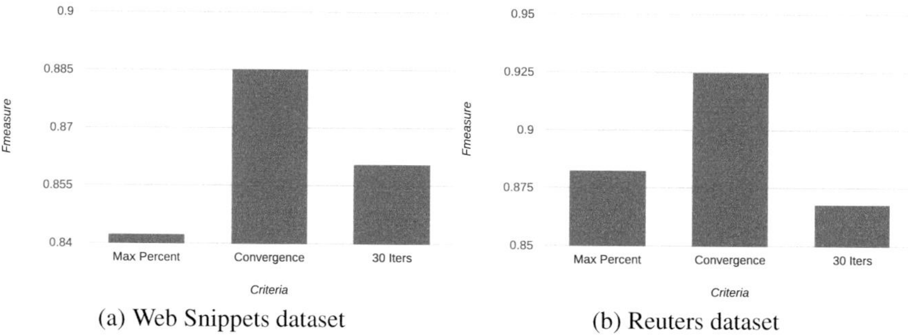

(a) Web Snippets dataset (b) Reuters dataset

Fig. 3 Comparison of the F-measure of different stopping criteria considered.

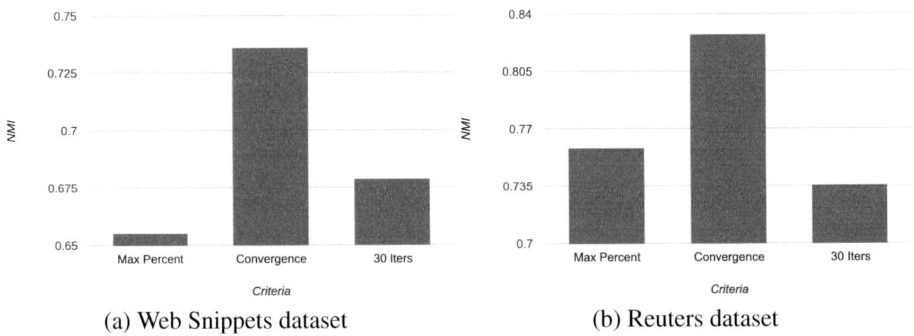

(a) Web Snippets dataset (b) Reuters dataset

Fig. 4 Comparison of the Normalized Mutual Information of different stopping criteria considered.

6 Conclusions and Future Work

In this paper, we have proposed three different stopping criteria for the automatic generation of instance level constraints. This automatic generation is based on the idea of finding the underlying nature of the data by means of repeated executions of a partitional clustering algorithm.

The three different criteria have been tested experimentally in the document clustering problem, showing that all three could be used. Which one is more appropriate for the requirements of each problem should be determined experimentally. However, executing the k-means until there is a convergence in the number of constraints provides higher values of F-measure and Normalized Mutual Information than the other approaches. It means that this criterion is able to obtain a set of constraints with a level of refinement that allows a proper performance of the semi-supervised clustering approach. On the contrary, using a criteria that generates constraints that are not "good enough", like asking for a very high percentage of constraints could lead to a clustering solution with not so good performance.

It remains as a future work the study of the relationship between the performance of the constraints and the execution time required for their generation. It would be also interesting to study the application of this methodology to other semi-supervised clustering problems and algorithms, like the identification of authorities in digital libraries presented in [7].

Acknowledgments This work has been partially funded by the Spanish Ministry of Education under the "Programa de Formación del Profesorado Universitario (FPU)" and grants P10-TIC-6109 and P11-TIC-7460.

References

1. Barr, J., Cament, L., Bowyer, K., Flynn, P.: Active clustering with ensembles for social structure extraction. In: 2014 IEEE Winter Conference on Applications of Computer Vision (WACV), pp. 969–976, March 2014
2. Basu, S., Banerjee, A., Mooney, R.J.: Semi-supervised clustering by seeding. In: Proceedings of the Nineteenth International Conference on Machine Learning, ICML 2002, pp. 27–34. Morgan Kaufmann Publishers Inc., San Francisco (2002)
3. Basu, S., Bilenko, M., Mooney, R.J.: A probabilistic framework for semi-supervised clustering. In: Proceedings of the Tenth ACM SIGKDD International Conference on Knowledge Discovery and Data Mining, KDD 2004, pp. 59–68. ACM, New York (2004)
4. Basu, S., Davidson, I., Wagstaff, K.: Constrained Clustering: Advances in Algorithms, Theory, and Applications, 1st edn. Chapman & Hall/CRC (2008)
5. Diaz-Valenzuela, I., Loia, V., Martin-Bautista, M., Senatore, S., Vila, M.: Automatic constraints generation for semisupervised clustering: experiences with documents classification. Soft Computing, 1–11 (2015). doi:10.1007/s00500-015-1643-3
6. Diaz-Valenzuela, I., Martin-Bautista, M.J., Vila, M.A.: Using a semisupervised fuzzy clustering process for identity identification in digital libraries. In: 2013 Joint IFSA World Congress and NAFIPS Annual Meeting (IFSA/NAFIPS), pp. 831–836 (2013)
7. Diaz-Valenzuela, I., Martin-Bautista, M.J., Vila, M.A., Campaña, J.R.: An automatic system for identifying authorities in digital libraries. Expert Systems with Applications **40**(10), 3994–4002 (2013). `http://www.sciencedirect.com/science/article/pii/S0957417413000134`
8. Diaz-Valenzuela, I., Martin-Bautista, M.J., Vila, M.-A.: A fuzzy semisupervised clustering method: application to the classification of scientific publications. In: Laurent, A., Strauss, O., Bouchon-Meunier, B., Yager, R.R. (eds.) IPMU 2014, Part I. CCIS, vol. 442, pp. 179–188. Springer, Heidelberg (2014)
9. Grira, N., Crucianu, M., Boujemaa, N.: Unsupervised and semi-supervised clustering: a brief survey. In: A Review of Machine Learning Techniques for Processing Multimedia Content, Report of the MUSCLE European Network of Excellence FP6 (2004)
10. Jain, A.K., Dubes, R.C.: Algorithms for clustering data. Prentice-Hall Inc., Upper Saddle River (1988)
11. Li, X., Wang, L., Song, Y., Zhao, X.: A hybrid constrained semi-supervised clustering algorithm. In: 2010 Seventh International Conference on Fuzzy Systems and Knowledge Discovery (FSKD), vol. 4, pp. 1597–1601, August 2010
12. Loia, V., Pedrycz, W., Senatore, S.: P-FCM: a proximity-based fuzzy clustering for user-centered web applications. Int. J. Approx. Reasoning **34**(2–3), 121–144 (2003). doi:10.1016/j.ijar.2003.07.004
13. Ltd., R., Carnegie Group, I.: Reuters-21578 dataset. `http://kdd.ics.uci.edu/databases/reuters21578/reuters21578.html`
14. Pedrycz, W., Loia, V., Senatore, S.: Fuzzy clustering with viewpoints. IEEE Transactions on Fuzzy Systems **18**(2), 274–284 (2010)
15. Phan, X.H., Nguyen, L.M., Horiguchi, S.: Learning to classify short and sparse text & web with hidden topics from large-scale data collections. In: Proceedings of the 17th International Conference on World Wide Web, WWW 2008, pp. 91–100. ACM, New York (2008)
16. Tang, W., Xiong, H., Zhong, S., Wu, J.: Enhancing semi-supervised clustering: A feature projection perspective. In: Proceedings of the 13th ACM SIGKDD International Conference on Knowledge Discovery and Data Mining, KDD 2007, pp. 707–716. ACM, New York (2007)

17. Wagstaff, K., Cardie, C.: Clustering with instance-level constraints. In: Proceedings of the Seventeenth International Conference on Machine Learning, pp. 1103–1110 (2000)
18. Wagstaff, K., Cardie, C., Rogers, S., Schrdl, S.: Constrained k-means clustering with background knowledge. In: Proceedings of the Eighteenth International Conference on Machine Learning, ICML 2001, pp. 577–584. Morgan Kaufmann Publishers Inc., San Francisco (2001)
19. Xing, E.P., Ng, A.Y., Jordan, M.I., Russell, S.: Distance metric learning, with application to clustering with side-information. In: Advances in Neural Information Processing Systems 15, vol. 15, pp. 505–512 (2002)
20. Xiong, S., Azimi, J., Fern, X.: Active learning of constraints for semi-supervised clustering. IEEE Transactions on Knowledge and Data Engineering **26**(1), 43–54 (2014)
21. Zhao, W., He, Q., Ma, H., Shi, Z.: Effective semi-supervised document clustering via active learning with instance-level constraints. Knowledge and Information Systems **30**(3), 569–587 (2012)

Ad-hoc Kalman Filter Based Fusion Algorithm for Real-Time Wireless Sensor Data Integration

Alexander Alexandrov

Abstract The paper describes a new developed software algorithm based on Kalman filtering for multi sensor data fusion designed for Intelligent Wireless Sensor Networks (IWSN). The proposed algorithm is implemented in set of ZigBee 6LowPan based intelligent wireless sensor modules, managed by custom design data integration software platform with SOA architecture. The system consist of ad-hock generated intelligent wireless sensor network for meteorological data collection (include sensors for air temperature, humidity, solar radiation and barometric pressure) and remote control center. Based on the proposed algorithm, the intelligent sensor nodes in the IWSN execute a cluster based decentralized sensor fusion operations of the raw sensor data which reduce the traffic load, minimize the power consumption, decrease the data noise and increase the reliability of the data sent to the control center.

Keywords Sensor fusion · Kalman filter · Software algorithm · Intelligent WSN · Ad-hock · Data integration

1 Introduction

The integration data from several sources is defined as data fusion. The authors Hall and Llinas [1] provided the following well-known definition of data fusion: "data fusion techniques combine data from multiple sensors and related information from associated databases to achieve improved accuracy and more specific inferences than could be achieved by the use of a single sensor alone." Sensor fusion is a process by which data from several different sensors are "fused" to compute something more than could be determined by any one sensor alone. Sensor fusion is a

A. Alexandrov(✉)
Institute of Information and Communication Technology,
Bulgarian Academy of Sciences, Sofia, Bulgaria
e-mail: akalexandrov@iit.bas.bg

T. Andreasen et al. (eds.), *Flexible Query Answering Systems 2015*,
Advances in Intelligent Systems and Computing 400,
DOI: 10.1007/978-3-319-26154-6_12

software process that intelligently combines and integrates data from multiple sensors to improve the measured data reliability and the redundancy.

Combining data from several sensors corrects the deviations and the fluctuations of the measured data and reduces the data flow transferred to the database. This is an important process in many applications, for example, in the management of energy losses in different areas [5], [13]. The key challenge in sensor fusion is effectively separating signal, motion, and noise. In modern sensor systems, the sensor fusion play a significant part in the design of the multiple sensors. Sensor fusion is the combining of sensory data or data derived from sensory data from disparate sources such that the resulting information has less uncertainty than would be possible when these sources were used individually. The term uncertainty reduction in this case can mean more detailed and complete and can refer to the result of an emerging view, such as 3D vision for example [3].

Practically we also can define the data fusion as a process of combination of multiple sources of data to obtain more detailed information. In this context, detailed information means more relevant information.

Data fusion techniques have been extensively employed on multisensory environments with the aim of fusing and aggregating data from different sensors. The goal of using data fusion in multisensory environments is to obtain a lower detection error probability and a higher reliability by using data from multiple distributed sources. Depending on the place of the operations the sensor fusion process can be centralized or decentralized.

Centralized versus decentralized refers to where the fusion of the data occurs. In centralized fusion, the sensor modules simply forward all of the data to a central location, and some entity at the central location is responsible for correlating and fusing the data. In decentralized fusion, the sensor modules or a group of modules take full responsibility for fusing the data. ''In this case, every sensor or platform can be viewed as an intelligent asset having some degree of autonomy in decision-making.''[5] In real sensor networks multiple combinations of centralized and decentralized systems exist.

In the current paper is focused on decentralized sensor data fusion process collecting data from set of homogeneous sensors for temperature, humidity and barometric pressure.

2 Kalman Filter Observation and Transition Models

Basically the equations for the Kalman filter fall into two categories: time update equations and measurement update equations. The time update equations are responsible for predicting (in time) the current state and error estimates to obtain the a priori estimates for the next time step. The measurement update equations are responsible for the feedback i.e. for incorporating a new measurement into the a priori estimate to obtain an improved a posteriori estimate. The time update equations can also be thought of as predictor equations, while the measurement update equations can be thought of as corrector equations. (see Fig.1)

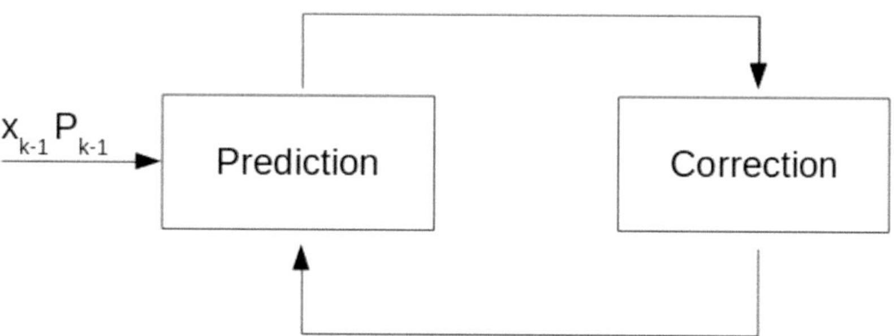

Fig. 1

The prediction phase start with initial estimation of $\hat{x}_{k-1}$ and covariance vector P_{k-1} and proceed with

$$\hat{x}_k = A\hat{x}_{k-1} + \mathrm{B}\mathrm{u}_k + Cw_k \tag{1}$$

$$z_k = \mathrm{H}_k x_k + \mathrm{D}_k v_k \tag{2}$$

where $\hat{x}_k$ is the estimated value, A is the transition state matrix of the process, B is the input matrix, C is the noise state transition matrix, u_k is the known input value, w_k is the noise, z_k is the observation vector, v_k is a variable describing observation noise, and H_k is the matrix of the observed value z_k, and D_k is a matrices describing the contribution of noise to the observation.

The measurement correction adjusts the projected estimate by an actual measurement at that time. In our case we will be focused mainly on the measurement update algorithm.

The current paper doesn't focus in details in the mathematical side of the Kalman filter measured updated equations. This is realized in details in [12, 15, 16].

Based on the papers above we accept that the final extended Kalman filter measurement update equations are formulated as follows:

$$G_k = \frac{P_k H_k}{H_k P_k H_k^t + R_k} \tag{3}$$

$$\hat{x}_k = \hat{x}_k + \mathrm{G}_k(z_k - H_k.\hat{x}_k) \tag{4}$$

$$P_k = (1 - G_k H_k).P_k \tag{5}$$

where G_k is the Kalman gain, P_k is error covariance vector, H_k is as described above the matrix of the observed value vector z_k and R_k is the covariance matrix.

The initial task during the measurement update is to compute the Kalman gain G_k (3). The next step is to actually measure the process to receive, and then to generate an a posteriori state estimate by adding the correction based on the measurement and estimation as in (4). The last step is to obtain an a posteriori error covariance estimate via (5). After each measurement update pair, the process is repeated with the previous a posteriori estimates used to project. This recursive principle is one of the very important features of the Kalman filter – the practical implementation is much easier. The Kalman filter generate recursively a conditions of thc current estimate based on the past measurements.

In the present equation of the filter, each of the measurement error covariance matrix R_k can be measured before the execution of the Kalman filter process. In the case of the measurement error covariance $R_k$especially this makes sense because there is a need to measure the process (while operating the filter). We should be able to take some off-line sample measurements to determine the variance of the measurement error.

In the both cases, we have a good basis for choosing the parameters. Very often superior filter performance (statistically speaking) can be obtained by "modification" of the filter parameter R_k. The modification is usually performed offline, frequently with the help of another Kalman filter. Finally we note that under conditions where R_k is constant, both the estimation error covariance P_k and the Kalman gain $G_k$will stabilize very quickly and then remain constant. If this is the case, the R_k parameter can be pre-computed by running the filter offline.

3 Related Work

Many wireless sensor network (WSN) systems use various data fusion schemes to improve their system performance. Despite of widely using of the fusion processes in practice the analysis of large fusion based WSNs has received little attention. There is a lot of literature on stochastic signal detection based on multi-sensor data fusion. Early works [5, 12] focus on small-scale powerful sensor networks. Recent studies on data fusion have considered the specific properties of WSNs such as sensors' spatial distribution [10, 11, 12] and limited sensing and communication capability [8]. However, these researches focus on analyzing the optimal fusion strategies and algorithms that maximize the system performance of a given network. At the same time, the current paper explores the limits of power and sensing coverage of the battery powered WSNs that are designed based on existing data fusion strategies. Usually the irregular sampling theory has been applied for reconstructing physical fields in WSNs [7, 8].

In contrast from these researches focused on developing discrete schemes to improve the quality of signal reconstruction, we aim to analyze the density of the sensors to achieve the required level of coverage and working life.

Previous works can be separated in two groups, namely, coverage maintenance algorithms and theoretical analysis of coverage performance. These two groups are reviewed in the sources below.

Early work [8, 10, 11] quantifies sensing coverage by the length of target's path where the accumulative observations of sensors are maximum or minimum [8, 9]. These works focus on devising algorithms for finding the target's paths with certain level of coverage. Several algorithms and protocols [10, 12] are designed to maintain sensing coverage using the minimum number of sensors. However, the effectiveness of these schemes largely relies on the assumption that sensors have circular sensing regions and deterministic sensing capability. Some studies [2, 6, 7] on the coverage problem have adopted probabilistic sensing models. The experimental results in [4] show that the coverage of a network can be expanded by the cooperation of sensors using data fusion process. These studies do not quantify the improvement of coverage and energy saving due to data fusion techniques. Out of our focus on analyzing the limits of coverage in WSNs, are the papers describing devise's algorithms and protocols for coverage maintenance.

Theoretical studies of the coverage of large-scale WSNs have been conducted in [1, 2, 3, 5, 7]. Most of the researches [2, 5, 7] focus on deriving the coverage of WSNs. The most important conditions for full coverage are derived for various sensor deployment strategies and algorithms. The coverage of ad-hock deployed networks is studied in [4].

4 Kalman Filter Based Fusion Algorithm for Sensor Data Integration

The proposed algorithms are designed to redundant the data from different sensors that observe the same environment to improve the signal/noise level, and to then to generate more reliable information.

The software algorithms are implemented in the sensor modules of an experimental ZigBee based IWSN with cluster tree topology. The sensor network is ad-hock generated and routed and comprises of sensor nodes and a control center. Each sensor node is battery powered and equipped with integrated sensors for temperature, humidity and barometric pressure. Additionally every sensor node have data processing capabilities and ZigBee based radio communications part.

Sensors are dynamically grouped into clusters called “super-sensors”. Each super sensor consists of a minimum 6 sensor nodes. The energy of a super-sensor is the sum of the energy of all the sensor nodes within it. The distance between two super sensors is the maximum distance between two sensors nodes where, each reside in a different super sensor. Due to their limited power and relatively shorter communication range, sensor nodes perform in-network data fusion.

The data fusion node collects the results from multiple nodes. It fuses the results using Kalman filtering method with its own based on a Fig. 3 decision criterion and sends the fused data to another node or directly to the control center.

The criterion $x_n = \sigma_n^2(\sigma_1^{-2}.\hat{x}_1+\sigma_1^{-2}.\hat{x}_2 + \ldots\ldots\ldots + \sigma_6^{-2}.\hat{x}_6)$ is based on the Central Limit Theorem and the Fraser-Potter fixed interval smoother, where $\sigma_n^2 = (\sigma_1^{-2}+\sigma_1^{-2} + \ldots\ldots\ldots + \sigma_6^{-2})$ is the variance of the combined estimate.

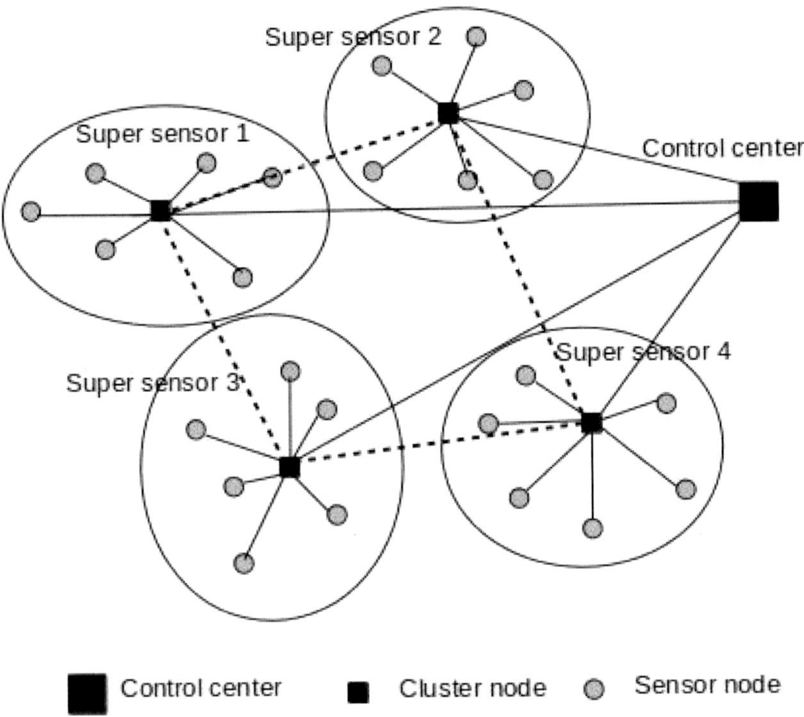

Fig. 2

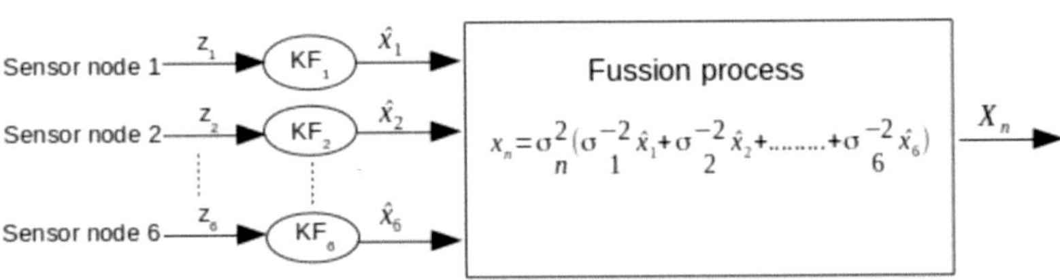

Fig. 3

4.1 Data Fusion Algorithm Solution

The software sensor library provides a concrete implementation of the different type sensors. The library implements sensor fusion, combining inputs from several temperature, humidity and barometric pressure sensors into a composite vector characterizing a specified geographical point. Power-state optimization is included in the library.

Sensors that are not needed for the current measurement are placed in a low-power mode to conserve device battery power. The library's internal state-management module also performs data-rate arbitration.

4.2 *Software Implementations*

The current realization of the fusion algorithm solution consists of a C++ based library and the associated header files. The library includes a working software sensor implementation with support for physical devices - temperature sensor, humidity sensor and barometric pressure sensor.

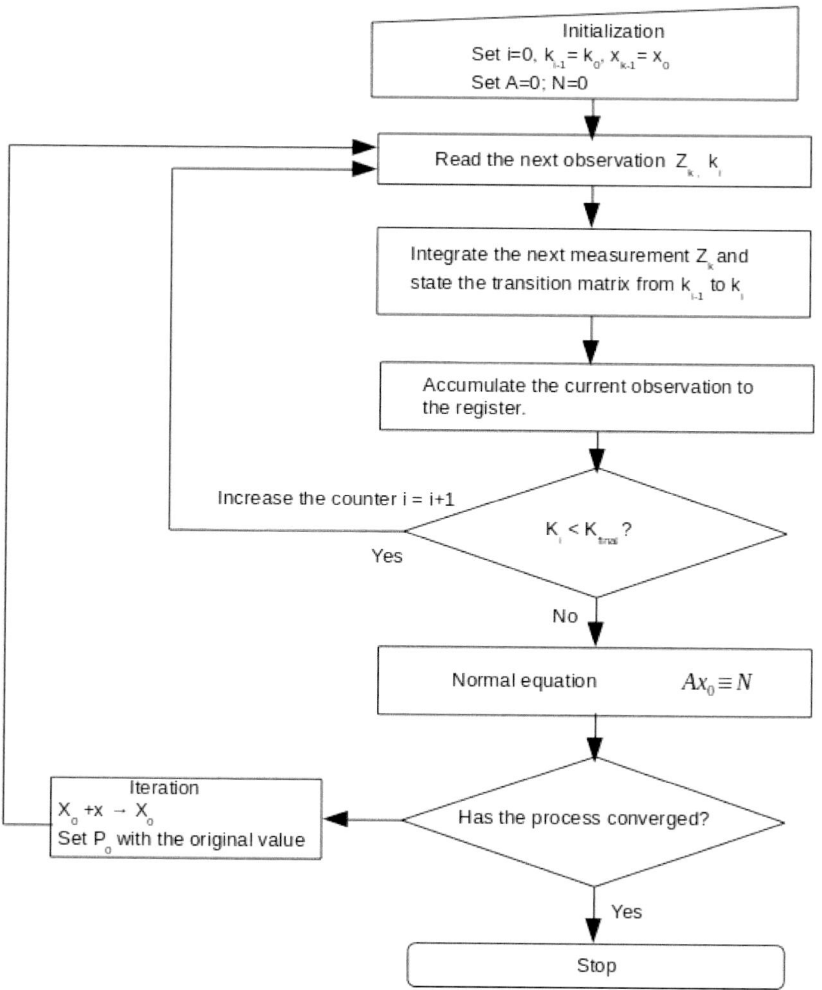

Fig. 4

Additionally, the library interface exports an object-oriented API for accessing both the physical sensors (temperature, humidity and pressure) and the additionally added synthetic sensors generated via the software fusion implementation. A software-generated sensor types with support for temperature, humidity and pressure sensors can be implemented too.

The fusion algorithm solution include the following software library modules:

-SENSOR.TYPE_TEMPERATURE (physical, units of degree Celsius)
-SENSOR.TYPE_PRESSURE (physical, units of mbars)
-SENSOR.TYPE_HUMIDITY (physical, percent %)

The current software solution uses a data-ready interrupt for each physical sensor, enabling a very precise time-stamp to be generated in response to the interrupt.

The software implementation behaves reasonably for all other sensor APIs. Specifically, the implementation will return TRUE or FALSE in response to applications attempting to register sensor listeners. Sensor listeners are not called if the corresponding sensors are not reached in case of malfunction, and power is provided. The implementation have capability to be configured to power down or put to sleep any sensor nodes in the WSN that have no active sensor listeners.

5 Conclusion

The new data fusion algorithm presented in this paper allows the information from different sensors to be combined and integrated in real time. Decentralized Kalman filters (DKF) work as data fusion devices on the local cluster nodes. Such an architecture gives the flexibility and reliability for reconfiguration of a sensor. New clustered ad-hock generated sensor subsystems can easily be added without the need of any redesign of the whole sensor network. The proposed architecture does not require a centralized management and therefore, in the case of failure of some local sensor nodes (each of which includes a CPU with local memory, sensors and wireless module) the overall system will continue to work.

The results from the test of experimental IWSN show that the performance of the overall system is increased because of the reduced communication traffic to the control center and the reliability of the measured data from the sensors stay at high level even if the sensors of some subsystems fail or interconnections are broken.

At the same time the new decentralized Kalman filter algorithm effectively reduces observation and measurement noises.

Based on the proposed algorithm and the ad-hock configured intelligent super-sensors the new developed IWSN execute a cluster based decentralized sensor fusion operations of the raw sensor data which reduce the traffic load, minimize the power consumption, decrease the data noise and increase the reliability of the data sent to the control center.

In a future work, the proposed approach based on Kalman filtering can be used in the development of non-conflict schedules for packets transmission in distributed communication networks [14].

References

1. Hall, D.L., Llinas, J.: An introduction to multisensor data fusion. Proceedings of the IEEE **85**(1), 6–23 (1997)
2. Welch, G., Bishop, G.: An Introduction to the Kalman Filter. Department of Computer Science University of North Carolina, UNC-Chapel Hill, TR 95-041, July 24, 2006
3. Haghighat, M.B.A., Aghagolzadeh, A., Seyedarabi, H.: Multi-focus image fusion for visual sensor networks in DCT domain. Computers & Electrical Engineering **37**(5), 789–797 (2011)
4. Julier, S.J., Uhlmann, J.K.: A new extension of the Kalman filter to nonlinear systems. In: Proceedings of the International Symposium on Aerospace/Defense Sensing, Simulation and Controls, vol. 3 (1997)
5. Luo, R.C., Yih, C.-C., Su, K.L.: Multisensor fusion and integration: approaches, applications, and future research directions. IEEE Sensors Journal **2**(2), 107–119 (2002)
6. Llinas, J., Bowman, C., Rogova, G., Steinberg, A., Waltz, E., White, F.: Revisiting the JDL data fusion model II, Technical Report, DTIC Document (2004)
7. Blasch, E.P., Plano, S.: JDL level 5 fusion model "user refinement" issues and applications in group tracking. In: Proceedings of the Signal Processing, Sensor Fusion, and Target Recognition XI, pp. 270–279, April 2002
8. Durrant-Whyte, H.F., Stevens, M.: Data fusion in decentralized sensing networks. In: Proceedings of the 4th International Conference on Information Fusion, Montreal, Canada, pp. 302–307 (2001)
9. Chen, L., Wainwright, M.J., Cetin, M., Willsky, A.S.: Data association based on optimization in graphical models with application to sensor networks. Mathematical and Computer Modelling **43**(9–10), 1114–1135 (2006)
10. Weiss, Y., Freeman, W.T.: On the optimality of solutions of the max-product belief-propagation algorithm in arbitrary graphs. IEEE Transactions on Information Theory **47**(2), 736–744 (2001)
11. Brown, C., Durrant-Whyte, H., Leonard, J., Rao, B., Steer, B.: Distributed data fusion using Kalman filtering: a robotics application. In: Abidi, M.A., Gonzalez, R.C. (eds.) Data, Fusion in Robotics and Machine Intelligence, pp. 267–309 (1992)
12. Kalman, R.E.: A new approach to linear filtering and prediction problems. Journal of Basic Engineering **82**(1), 35–45 (1960)
13. Atanasova, T., Tashev, T.: Analysis and Evaluation of Energy Losses in Living Environment on the Basis of Cognitive-Expert Classification. Problems of Engineering Cybernetics and Robotics **64**, 11–18 (2011)
14. Kolchakov, K., Monov, V.: An algorithm for non – conflict schedule with diagonal activation of joint sub matrices. In: Proc. of 17-th International Con.
15. Reference on Distributed Computer and Communication Networks (DCCN-2013), Moscow, October 07-10, pp. 180–187 (2013)
16. Bruno, S., Oussama, K. (eds.) Springer Handbook of Robotics. ISBN 978-3-540-23957-4
17. Elmenreich, W.: Sensor Fusion in Time-Triggered Systems, (PDF), p. 173. Vienna University of Technology, Vienna (2002)

On Bilateral Matching between Multisets

Maciej Krawczak and Grażyna Szkatuła

Abstract In the paper we defined a new measure of remoteness between multisets. The development of the measure is based on the definition of sets perturbation originally developed by the authors. The sets perturbation definition is here extended to multisets perturbation, it means perturbation of one multiset by another multiset and/or vice-versa. In general these two measures are different, it means asymmetrical, and therefore can be called the bilateral measure of matching between two multisets. Therefore the measure cannot be considered as a distance between multisets.

Keywords Multisets · Nominal values · Measure of remoteness · Measure of perturbation

1 Introduction

There are various data mining tasks which require comparing objects. In general such objects are described by specified set of attributes, and then each object is described by an appropriate set of attributes' values. Usually the comparison of the objects is done via defining a distance measure between the objects. It is obvious that selection of the distance measure between the objects is of crucial importance, for example, in many classification and grouping algorithms. From the mathematical point of view, distance is defined as a quantitative degree showing how far apart two objects are. Still there are not so many scientific works dedicated to nominal-valued attributes, meanwhile a lot of work has been performed on continuous-valued attributes. It is believed that nominal attributes

M. Krawczak · G. Szkatuła
Systems Research Institute, Polish Academy of Sciences, Warsaw, Poland
e-mail: szkatulg@ibspan.waw.pl

M. Krawczak(✉)
Warsaw School of Information Technology, Warsaw, Poland
e-mail: krawczak@ibspan.waw.pl

T. Andreasen et al. (eds.), *Flexible Query Answering Systems 2015*,
Advances in Intelligent Systems and Computing 400,
DOI: 10.1007/978-3-319-26154-6_13

cases are more difficult to handle. It is understood that an attribute is nominal-valued if it can take one of a finite number of possible nominal values. Nominal-valued data are very characteristic, namely such values do not have neither a natural ordering nor an inherent order, in general.

For comparison of objects there are commonly used different measures of objects' similarity or dissimilarity, and in the case of nominal-valued attributes definitions of the similarity (or dissimilarity) measures become less trivial. Common distance measures cannot be applied for description of objects similarity if the objects are described by nominal-valued attributes. Therefore to compare objects described by nominal attributes we must be equipped with efficient mathematical tools which ensure satisfactory comparisons of two sets of nominal elements.

In classical set theory sets constitute a collection of distinct elements. In a *multiset* (sometimes also shortened to *mset* or *bag*) repeating of any elements is allowed. This way, a multiset is understood as a collection of elements with additional information about the multiplicity of occurring elements. Even if such collection of elements was considered for hundreds years ago but it is assumed that Richard Dedekind introduced multisets to the modern mathematics in 1898. Complete surveys of the multisets theory can be found in numerous papers where appropriate operations and their properties are investigated [1, 2, 9, 10]. There are different notations for expressing multisets. For example a multiset containing a single occurrence of a, three occurrences of b and two occurrences of c can be described in the following ways: $\{a,b,b,b,c,c\}$, $\{a^1,b^3,c^2\}$, $\{a,b,c\}_{1,3,2}$, $\{a1,b3,c2\}$, $\{1/a,3/b,\ 2/c\}$ or $\{(1,a),(3,b),(2,c)\}$ etc. or using square brackets is also acceptable. In this paper we apply the sixth formalism.

This paper is a continuation as well as extension of authors' previous papers related to sets remotrness based on the perturbation of sets idea [3-6]. The term *perturbation* was applied in a general sense as an impact of one object on another object, and this way describes what object can be considered as primary and that as secondary in currying information. Here, instead of crisp sets we examine multisets with nominal elements and then we introduce a new measure of proximity between two multisets. This consideration is based on the multisets theory and its basic operations. Then we introduce *a measure of perturbation of one multiset by another multiset*. The new measure defines changes of one multiset after adding another multiset, and vice versa. In general such a measure is not symmetric, and it means that a value of the measure of perturbation of the first multiset by the second multiset can be different than a value of the measure of perturbation of the second multiset by the first multiset. The measure of perturbation is ranged between 0 and 1, where while 0 is the lowest level of the multisets perturbation, 1 is interpreted as the highest level of the multisets perturbation. The measure described in this way is asymmetrical and therefore cannot be considered as a distance between multisets.

This paper is organized as follows. In Section 2 we present the description of the perturbation methodology and the mathematical properties of the measure of multisets perturbation are studied. In Appendix we recall the basic definitions and notions of functions in multisets context.

2 Matching between Multisets

As it was said there are data described, at least partially, by attributes represented by nominal values. In data mining tasks like classification or grouping the fundamental problem is comparing objects. This problem is equivalent to the problem of comparing sets which just describe the objects. Let us assume, that each object is represented by multisets.

Here, we first describe some necessary preliminaries concerned with the multiset theory [1, 2, 9, 10], and next we introduce the perturbation paradigm of one multiset by another multiset.

Let us consider a non-empty and finite set V of nominal elements, $V = \{v_1, v_2, ..., v_L\}$. First, let us recall the following definition of a multiset.

Definition 1. *A multiset S drawn from an ordinary set* V *can be represented by a set of ordered pairs:*

$$S = \{(k_S(v), v)\},\ \forall v \in V \tag{1}$$

where a function $k_S : V \rightarrow M=\{0,1,2, \ldots\}$ *is called a counting function.*

The value $k_S(v)$ specifies the number of occurrences of the element $v \in V$ in the multiset *S*. The element which is not included in the multiset *S* has a counting index equal zero. Detailed description of the multiset *S* defined by (1), drawn from the finite set V, can be given in the following way

$$S = \{(k_S(v_1), v_1), (k_S(v_2), v_2),\ \ldots, (k_S(v_L), v_L)\} \tag{2}$$

where the element $v_1 \in V$ appears $k_S(v_1)$ times, the element $v_2 \in V$ appears $k_S(v_2)$ times and so on. In general, the value $k_S(v_i)$, $i = 1, 2, ..., L$, specifies the number of occurrences of the element $v_i \in V$ in the multiset *S*, wherein $k_S(v_i) \geq 0$, $\forall i \in \{1,2,...,L\}$. Note that $v_i \in V$ is the ordinary set notation. The multiset *S* drawn from a set *V* is call *empty*, denoted by Ø, if $k_S(v_i) = 0$ $\forall v_i \in V$. For simplicity, the case when $k_S(v_i) = 0$ for an element $v_i \in V$ can be omitted, when it does not lead to confusion in this paper. The above definition can be easy interpreted by considering the following example.

Example 1. Let us consider a set $V = \{a,b,c,d,e,f,g\}$. An exemplary multiset S containing three occurrences of a, two occurrences of d, a single occurrence of e and five occurrences of f can be described in the following way: $S = \{(3,a),(0,b),(0,c),(2,d),(1,e),(5,f),(0,g)\}$. For $k_S(v) = 0$ the element $v \in V$ may be omitted to simplify the notation and then this multiset S can be rewritten in a simplified form as $S = \{(3,a),(2,d),(1,e),(5,f)\}$.

Definition 2. *Let us assume that S is a multiset drawn from a set V. The support or the root of S, denoted by* S^*, *is an ordinary set defined as follows:* $S^* = \{v \in V:\ k_S(v) > 0\}$.

Thus $\forall v \in V$ such that $k_S(v) > 0$ this implies $v \in S^*$, and $\forall v$ such that $k_S(v) = 0$ this implies $v \notin S^*$. Note that the characteristic function of S^* can be described as $\chi_{S^*}(v) = \min\{k_S(v),1\}$. For example, support of the multiset $S=$ $\{(3,a),(2,d),(5,f)\}$ drawn from a set V can be described as $S^* = \{a,d,f\}$.

Definition 3. *Let us assume that S is a multiset drawn from a set V. The cardinality of the multiset S, denoted by* $card(S)$ *or* $|S|$, *is defined as the total number of occurrences of its elements, i.e.,* $card(S) = \sum_{i=1}^{L} k_S(v_i)$.

For example, if $S = \{(3,a),(3,b)\}$ then $card(S) = 6$.

Next, we will recall the definition of a multiset space [2].

Definition 4. *Let us assume that V is a non-empty and finite (crisp) set. The multiset space, denoted by* $[V]^m$, *is a set of all multisets whose elements are in* V *such that no element occurs more than m times.*

Let us assume that there is a collection of multisets $S_n \in [V]^m$ drawn from the set V, where V is a finite set of nominal elements, and $V = \{v_1, v_2, \ldots, v_L\}$, described by

$$S_n = \{(k_{S_n}(v_1),v_1),(k_{S_n}(v_2),v_2),\ \ldots,(k_{S_n}(v_L),v_L)\}, \tag{3}$$

where counting functions $k_{S_n}(v_i)$ specify the number of occurrences of the element $v_i \in V$ in the multiset S_n, $\forall v_i \in V$, $i \in \{1,2,\ldots,L\}$, $k_{S_n} : V \to \{1,2,\ldots,m\}$.

Within the theory of ordinary sets there are several operators, analogous operators are also defined for multisets. Here, we recall the basic definitions and notions related to multisets [9]:

- *The union of multisets,* $S_1 \cup S_2 = \{(k_{S_1 \cup S_2}(v), v) : \forall v \in V, k_{S_1 \cup S_2}(v) = \max\{k_{S_1}(v), k_{S_2}(v)\}\}$.

  For example, if $S_1 = \{(3,a),(3,b)\}$ and $S_2 = \{(5,a),(1,b)\}$ then the union of these multisets is as follows $S_1 \cup S_2 = \{(5,a),(3,b)\}$.
- *The intersection of multisets,* $S_1 \cap S_2 = \{(k_{S_1 \cap S_2}(v), v) : \forall v \in V, k_{S_1 \cap S_2}(v) = \min\{k_{S_1}(v), k_{S_2}(v)\}\}$.

  For example, for the two multisets $S_1 = \{(3,a),(3,b)\}$ and $S_2 = \{(3,a),(1,b)\}$ their intersection is calculated as follows $S_1 \cap S_2 = \{(3,a),(1,b)\}$.
- *The arithmetic addition of multisets,* $S_1 \oplus S_2 = \{(k_{S_1 \oplus S_2}(v), v) : \forall v \in V, k_{S_1 \cap S_2}(v) = \min\{k_{S_1}(v), k_{S_2}(v)\}\}$

  For example, the arithmetic addition of two following multisets $S_1 = \{(3,a),(3,b)\}$ and $S_2 = \{(5,a),(1,b)\}$ is described by the new multiset $S_2 \oplus S_1 = \{(8,a),(4,b)\}$.
- *The arithmetic subtraction of multisets,*

  $$S_1 \Theta S_2 = \{(k_{S_1 \Theta S_2}(v), v) : \forall v \in V, k_{S_1 \Theta S_2}(v) = \max\{k_{S_1}(v) - k_{S_2}(v), 0\}\}.$$

  For example, if $S_1 = \{(3,a),(3,b)\}$ and $S_2 = \{(5,a),(1,b)\}$ then $S_1 \Theta S_2 = \{(0,a),(2,b)\}$ and $S_2 \Theta S_1 = \{(2,a),(0,b)\}$. It is worth noting that the arithmetic subtraction can be asymmetric in the considered case $S_1 \Theta S_2 \neq S_2 \Theta S_1$.

- *The symmetric difference of multisets,* $S_1 \Delta S_2 = \{(k_{S_1 \Delta S_2}(v), v) : \forall v \in V, k_{S_1 \Delta S_2}(v) = |k_{S_1}(v) - k_{S_2}(v)|\}$.

  For example, for two multisets $S_1 = \{(3,a),(3,b)\}$ and $S_2 = \{(5,a),(1,b)\}$ the symmetric difference is the following multiset $S_1 \Delta S_2 = \{(2,a),(2,b)\}$.

Now we will introduce a corollary wherein there are gathered the most important (from this paper point of view) multiset operators.

Corollary 1. Let us assume that we have multisets S_1 and S_2 drawn from the set V. The multisets operations $\oplus$, $\cup$, $\cap$ satisfy the following properties:

1. Commutativity: $S_1 \oplus S_2 = S_2 \oplus S_1$; $S_1 \cup S_2 = S_2 \cup S_1$; $S_1 \cap S_2 = S_2 \cap S_1$.
2. Associativity: $S_1 \oplus (S_2 \oplus S_3) = (S_1 \oplus S_2) \oplus S_3$; $S_1 \cup (S_2 \cup S_3) = (S_1 \cup S_2) \cup S_3$; $S_1 \cap (S_2 \cap S_3) = (S_1 \cap S_2) \cap S_3$.
3. Idempotence: $S_1 \cup S_1 = S_1$; $S_1 \cap S_1 = S_1$; $S_1 \oplus S_1 \neq S_1$ for S_1 being not empty set.
4. Identity laws: $S_1 \cup \emptyset = S_1$; $S_1 \cap \emptyset = \emptyset$; $S_1 \oplus \emptyset = S_1$.
5. Distributivity: $S_1 \oplus (S_2 \cup S_3) = (S_1 \oplus S_2) \cup (S_1 \oplus S_3)$;
   $S_1 \oplus (S_2 \cap S_3) = (S_1 \oplus S_2) \cap (S_1 \oplus S_3)$;
   $S_1 \cup (S_2 \cap S_3) = (S_1 \cup S_2) \cap (S_1 \cup S_3)$;
   $S_1 \cap (S_2 \cup S_3) = (S_1 \cap S_2) \cup (S_1 \cap S_3)$.

In this place, it is worth to give the following comments. Assuming that we have two multisets with finite support drawn from the set V. It can be easily noticed that the operator $\oplus$ is stronger than both operators $\cup$ and $\cap$ in the following sense $(S_1 \cap S_2) \subseteq (S_1 \cup S_2) \subseteq (S_1 \oplus S_2)$. Additionally, the following property is also satisfied: $card(S_1 \cup S_2) + card(S_1 \cap S_2)$ $= card(S_1) + card(S_2)$.

The above described operations performed on the multisets will be presented by the following example.

Example 2. Let us consider two multisets $S_1 = \{(3,a),(3,b)\}$, $S_2 = \{(5,a),(1,b)\}$. Thus, we obtain the following values of the operations performed on this multisets

$$S_1 \cup S_2 = S_2 \cup S_1 = \{(5,a),(3,b)\}$$
$$S_1 \cap S_2 = S_2 \cap S_1 = \{(3,a),(1,b)\}$$
$$S_1 \oplus S_2 = S_2 \oplus S_1 = \{(8,a),(4,b)\}$$
$$S_1 \Theta S_2 = \{(2,b)\},\ S_2 \Theta S_1 = \{(2,a)\}$$
$$S_1 \Delta S_2 = S_2 \Delta S_1 = \{(2,a),(2,b)\}.$$

There are several already existing methods based on computing distances between them for comparing the multisets. In these methods distances between multisets are defined in different ways [7, 8]. Instead of considering distance measure we introduce a new asymmetric measure of remoteness between two multisets. The concept of multisets perturbation will be explained in what follows.

Let us assume, that S_1 and S_2 are multisets, such that both belong to the prescribed multiset space $S_1, S_2 \subseteq [V]^m$. Let us consider attaching of one multiset to another, with attention paid to the direction of attaching the considered multisets. The first case is such that the multiset S_1 is attached to the second multiset S_2 - we say that "the second multiset is perturbed by the first multiset" or, in other words, the first multiset S_1 perturbs the second multiset S_2. It is natural to consider the opposite, the counterpart, case, namely the second multiset S_2 is attached to the first multiset S_1 - and then we say that the second multiset S_2 perturbs the first multiset S_1, or the first multiset S_1 is perturbed by the second multiset S_2. In this way we defined a novel *concept of perturbation* of one multiset by another multiset. One case is denoted by $(S_1 \mapsto S_2)$ and it is said that the first multiset S_1 perturbs the second multiset S_2, the opposite case is denoted by $(S_2 \mapsto S_1)$. The interpretation of the multisets perturbation is following, namely by a multiset representing the arithmetic subtraction of multisets:

$$(S_1 \mapsto S_2) = S_1 \Theta S_2 = \{(k_{S_1 \mapsto S_2}(v), v) : \ \forall v \in V, \ k_{S_1 \mapsto S_2}(v) = \max\{k_{S_1}(v) - k_{S_2}(v), 0\}\} \quad (4)$$

In the counterpart case, the perturbation of the multiset S_1 by the multiset S_2 the definition of a resulting multiset $(S_2 \mapsto S_1)$ is similar and defined as follows:

$$(S_2 \mapsto S_1) = S_2 \Theta S_1 = \{(k_{S_2 \mapsto S_1}(v), v) : \ \forall v \in V, \ k_{S_2 \mapsto S_1}(v) = \max\{k_{S_2}(v) - k_{S_1}(v), 0\}\} \quad (5)$$

An illustrative interpretation of the proposed concept of the perturbation of one multiset by another multiset is presented in the following example.

Example 3. Let us consider the set $V = \{a, b, c, d, e\}$ and an exemplary two multisets $S_1 = \{(1, a), (1, e)\}$ and $S_2 = \{(1, a), (1, d), (3, e)\}$, $S_1, S_2 \subseteq [V]^3$, drawn from the set V. The perturbation of the multiset S_2 by the multiset S_1 is empty multiset because the following condition ($S_1 \mapsto S_2) = S_1 \Theta S_2 = \emptyset$ is satisfied. On the other hand, the perturbation of the multiset S_1 by the multiset S_2 is the following multiset $(S_2 \mapsto S_1) = S_2 \Theta S_1 = \{(1, d), (2, e)\}$.

Now we face a problem of measuring the multisets perturbations, and in order to fix the range of such measurements of multisets perturbation we propose the following normalized way delivered in the following definition.

Definition 5. *Let us assume that* S_1 *and* S_2 *are the multisets drawn from a set V, such that* $S_1, S_2 \subseteq [V]^m$. *The measure of perturbation of the multiset* S_2 *by the multiset* S_1, *denoted by* $Per_{MS}(S_1 \mapsto S_2)$, *is defined in the following manner:*

$$Per_{MS}(S_1 \mapsto S_2) = \frac{card(S_1 \Theta S_2)}{card(S_1 \oplus S_2)}. \tag{6}$$

In other words, the perturbation of one multiset by another multiset is measured as a ratio of two cardinal numbers of two multisets, the arithmetic subtraction of the multisets and the arithmetic addition of the multisets. Thus Eq. (6) can be rewritten in the following way

$$Per_{MS}(S_1 \mapsto S_2) = \frac{\sum_{i=1}^{L}(k_{S_1}(v_i) - k_{S_1 \cap S_2}(v_i))}{\sum_{i=1}^{L}(k_{S_1}(v_i) + k_{S_2}(v_i))} \tag{7}$$

The counterpart case, namely the perturbation of the multiset S_1 by the multiset S_2 is defined in a similar way

$$Per_{MS}(S_2 \mapsto S_1) = \frac{card(S_2 \Theta S_1)}{card(S_2 \oplus S_1)} = \frac{\sum_{i=1}^{L}(k_{S_2}(v_i) - k_{S_2 \cap S_1}(v_i))}{\sum_{i=1}^{L}(k_{S_2}(v_i) + k_{S_1}(v_i))} \tag{8}$$

The illustration of the multisets perturbation idea is presented in the following example.

Example 4. Let us consider the set $V = \{a, b, d, e\}$ and exemplary two multisets $S_1, S_2 \subseteq [V]^3$, where $S_1 = \{(1,a),(1,e)\}$ and $S_2 = \{(1,a),(1,d),(3,e)\}$. Due to Definition 5 we can calculate the following values of the measures of perturbation of one multiset by another, namely:

$$Per_{MS}(S_1 \mapsto S_2) = \frac{\sum_{i=1}^{4}(k_{S_1}(v_i) - k_{S_1 \cap S_2}(v_i))}{\sum_{i=1}^{4}(k_{S_1}(v_i) + k_{S_2}(v_i))} = \frac{0+0+0+0}{2+0+1+4} = 0$$

$$Per_{MS}(S_2 \mapsto S_1) = \frac{\sum_{i=1}^{4}(k_{S_2}(v_i) - k_{S_2 \cap S_1}(v_i))}{\sum_{i=1}^{4}(k_{S_1}(v_i) + k_{S_2}(v_i))} = \frac{0+0+1+2}{2+0+1+4} = \frac{3}{7}.$$

The multisets perturbation is not symmetrical in general, and in the next corollary we deliver a property of the multisets perturbation measure range.

Corollary 2. *Measure of perturbation of the multiset S_2 by the multiset S_1, satisfies the following condition*

$$0 \le Per_{MS}(S_1 \mapsto S_2) \le 1. \tag{9}$$

Proof. 1) We first prove the first inequality $Per_{MS}(S_1 \mapsto S_2) \ge 0$. It should be noticed that the inequality $k_{S_1 \cap S_2}(v_i) \le k_{S_1}(v_i)$, $\forall i \in \{1, 2, ..., L\}$ is satisfied, so $k_{S_1}(v_i) - k_{S_1 \cap S_2}(v_i) \ge 0$. Using Definition 5 we obtain the following inequality

$$Per_{MS}(S_1 \mapsto S_2) = \frac{\sum_{i=1}^{L}(k_{S_1}(v_i) - k_{S_1 \cap S_2}(v_i))}{\sum_{i=1}^{L}(k_{S_1}(v_i) + k_{S_2}(v_i))} \ge 0 .$$

2) Let us prove now the second inequality, $Per_{MS}(S_1 \mapsto S_2) \le 1$. Due to the inequality $k_{S_1}(v_i) - k_{S_1 \cap S_2}(v_i) \le k_{S_1}(v_i) + k_{S_2}(v_i)$, $\forall i \in \{1, 2, ..., L\}$ which is satisfied we obtain the following inequality

$$Per_{MS}(S_1 \mapsto S_2) = \frac{\sum_{i=1}^{L}(k_{S_1}(v_i) - k_{S_1 \cap S_2}(v_i))}{\sum_{i=1}^{L}(k_{S_1}(v_i) + k_{S_2}(v_i))} \le \frac{\sum_{i=1}^{L}(k_{S_1}(v_i) + k_{S_2}(v_i))}{\sum_{i=1}^{L}(k_{S_1}(v_i) + k_{S_2}(v_i))} = 1.$$

In the following corollaries we present several important properties of the measure of the perturbation of one multiset by another multiset.

Corollary 3. *The following condition is fulfilled* $Per_{MS}(S_1 \mapsto S_2) = 0$ *if and only if* $k_{S_1}(v_i) = k_{S_1 \cap S_2}(v_i)$, $\forall i \in \{1, 2, ..., L\}$.

Corollary 4. *If the condition* $k_{S_2}(v_i)=0 \;\; \forall i \in \{1,2,...,L\}$ *is fulfilled and* $\exists v_i \in \{1,2,...,L\}$ that $k_{S_1}(v_i)>0$, *then the condition* $Per_{MS}(S_1 \mapsto S_2)=1$ *is satisfied.*

Proof. Suppose that that condition $k_{S_2}(v_i)=0, \;\; \forall i \in \{1,2,...,L\}$ is satisfied. We obtain

$$Per_{MS}(S_1 \mapsto S_2)=\frac{\sum_{i=1}^{L}(k_{S_1}(v_i)-k_{S_1 \cap S_2}(v_i))}{\sum_{i=1}^{L}(k_{S_1}(v_i)+k_{S_2}(v_i))}=\frac{\sum_{i=1}^{L}(k_{S_1}(v_i)-0)}{\sum_{i=1}^{L}(k_{S_1}(v_i)+0)}=1.$$

The corollaries 3 and 4 describe the necessary as well as sufficient conditions of the lower and upper bound of the measure of the multisets perturbation. The next corollary states a very important property related to well-known idea of similarity of two sets extended to similarity of two multisets.

Corollary 5. *The sum of the measures of the perturbation of an arbitrary multiset* S_2 *by another multiset* S_1 *satisfies the following equality*

$$Per_{MS}(S_1 \mapsto S_2)+Per_{MS}(S_2 \mapsto S_1)=1-\frac{\sum_{i=1}^{L}2 \cdot k_{S_1 \cap S_2}(v_i))}{\sum_{i=1}^{L}(k_{S_1}(v_i)+k_{S_2}(v_i))} \tag{10}$$

Proof. We obtain the following expression $Per_{MS}(S_1 \mapsto S_2)+Per_{MS}(S_2 \mapsto S_1)=$

$$=\frac{\sum_{i=1}^{L}(k_{S_1}(v_i)-k_{S_1 \cap S_2}(v_i))}{\sum_{i=1}^{L}(k_{S_1}(v_i)+k_{S_2}(v_i))}+\frac{\sum_{i=1}^{L}(k_{S_2}(v_i)-k_{S_2 \cap S_1}(v_i))}{\sum_{i=1}^{L}(k_{S_2}(v_i)+k_{S_1}(v_i))}=\frac{\sum_{i=1}^{L}(k_{S_1}(v_i)-k_{S_1 \cap S_2}(v_i))+\sum_{i=1}^{L}(k_{S_2}(v_i)-k_{S_2 \cap S_1}(v_i))}{\sum_{i=1}^{L}(k_{S_1}(v_i)+k_{S_2}(v_i))}=$$

$$=\frac{\sum_{i=1}^{L}(k_{S_1}(v_i)+k_{S_2}(v_i))-\sum_{i=1}^{L}2 \cdot k_{S_2 \cap S_1}(v_i))}{\sum_{i=1}^{L}(k_{S_1}(v_i)+k_{S_2}(v_i))}=1-\frac{2\sum_{i=1}^{L}k_{S_1 \cap S_2}(v_i))}{\sum_{i=1}^{L}(k_{S_1}(v_i)+k_{S_2}(v_i))}.$$

Following [3, 5, 6], the second term on the right side of the equation (10) can be treated as *the measure of similarity of the multisets* S_1 and S_2.

3 Conclusions

Nowadays data representation have various forms such as real, complex, fuzzy, intuitionistic fuzzy, multisets and so on, but from other point of view the data representation can be divided into continuous-valued, ordinal, interval, ratio or nominal-valued. Dealing with data, in principle represented by mixed forms, we must be equipped with different procedures of comparing or matching for information retrieval or tasks of information classification and information clustering. In general data mining tasks of information represented by mixed forms require development of various procedures oriented to separate data form representations. Just in this paper we deliver a novel insight into data comparison or into similarity of information.

Continuing our research related to problems of classification of objects as well as grouping of objects [3, 4], in this paper we extended the idea of perturbation one set by another set [5, 6] into multisets. It means we defined the idea of perturbation of one multiset by another multiset in the case of data represented by nominal values. Some mathematical properties of the measure of perturbation of multisets are explored.

In the authors opinion the methodology presented here is of practical significance. If we represent each multi-attributes object in the form of the multisets and define a description of a group of objects (or the class of objects) as a K-tuple of multisets (i.e., an ordered collection of multisets), we can extend concept of perturbation on all multisets within describing the considered groups.

It seems that the approach presented in this paper can be applied to defining an effective measure of remoteness between objects which are describe by many value of attributes, especially in the case, when several versions of objects may exist with different values of this attributes, e.g. in pattern recognition, text processing, data analysis and in other fields. Of course the presented methodology needs further research.

Acknowledgment This work has been partially supported by the National Science Centre Grant UMO-2012/05/B/ST6/03068.

Appendix. Brief Introduction to Multisets

The multisets are very useful structures arising in many areas of mathematics and computer science. A complete survey of multisets theory can be found in many papers where several operations and their properties are investigated [1, 2, 7-10]. In this section the basic definitions and notions of functions in multisets context are presented.

Definition 6. *Let us assume that S is a multiset drawn from a set V. The dimensionality of the multiset S, denoted by dim(S) or / S /, is defined as the total number of various elements, i.e., dim(S)* $= \sum_{i=1}^{L} \chi_S(v_i)$.

For example, if $S = \{(3,a),(3,b)\}$ then $dim(S) = 2$.

Definition 7. *Let us assume that S is a multiset drawn from a set V. The complement of the multiset S in the multiset space* $[V]^m$ *is the multiset* S^C *such that* $k_{S^C}(v) = m - k_S(v)$, $\forall v \in V$.

For example, if multiset $S = \{(5,a),(1,b)\}$ drawn from the set $V = \{a,b,c\}$ belongs to the multiset space $[V]^8$ then the complement of it is S^C = $\{(3,a),(7,b),(8,c)\}$.

Definition 8. *Let us assume that S is a multiset drawn from a set V. The upper cut of the multiset S, denoted by* $\bar{S}$, *is the ordinary set described as* $\bar{S} = \{(k_{\bar{S}}(v_1), v_1), (k_{\bar{S}}(v_2), v_2), \ldots, (k_{\bar{S}}(v_L), v_L)\}$, *where*

$$k_{\bar{S}}(v_i) = \begin{cases} 1 & \text{if } k_S(v_i) = 0 \\ 0 & \text{if } k_S(v_i) \geq 1 \end{cases}, \quad i \in \{1,2,\ldots,L\}, \ \forall v \in V .$$

The upper cut of the multiset S drawn from the set V is an ordinary set $\bar{S}$ with elements which do not belong to the multiset S, but do belong to the set V.

An ordinary set $\bar{S}$ is treated as a special case of a multiset. For example, for the multiset $S = \{(5,a),(1,b)\}$ drawn from the set $V = \{a,b,c\}$ the upper cut of the multiset S is $\bar{S} = \{c\}$.

Example 5. Let us consider the following set $V = \{a,b,c\}$. In Table 1 there are shown a few exemplary functions on multisets drawn from the set V belonging to the space $[V]^5$.

Table 1 The functions S^C, S^* and $\bar{S}$ of exemplary multisets

Multiset S	The complement S^C in $[V]^5$	The support S^*	The upper cut $\bar{S}$
$\{(2,b)\}$	$\{(5,a),(3,b),(5,c)\}$	$\{b\}$	$\{a,c\}$
$\{(3,a),(5,b)\}$	$\{(2,a),(5,c)\}$	$\{a,b\}$	$\{c\}$
$\{(2,a),(1,b),(3,c)\}$	$\{(3,a),(4,b),(2,c)\}$	$\{a,b,c\}$	Ø

Now, let us consider a finite set V of nominal elements. Assume that S_1 and S_2 are two multisets drawn from a set V, $S_1, S_2 \subseteq [V]^m$, Eq. (2). Below, we present several selected rules of comparison of the multisets.

Definition 9. *The multisets S_1 and S_2 are equal or the same, denoted by $S_1 = S_2$, if $\forall v \in V$ the condition $k_{S_1}(v) = k_{S_2}(v)$ is satisfied.*

The following condition holds: if $S_1 = S_2$ than $S_1^* = S_2^*$, however the converse need not to hold.

Definition 10. *The multisets S_1 and S_2 are similar if $\forall v \in V$, $k_{S_1}(v) > 0$ if and only if $k_{S_2}(v) > 0$.*

The similar multisets have equal support sets but need not be equal themselves. For example, multisets $S_1 = \{(3,a),(2,d),(5,f)\}$ and $S_2 = \{(5,a),(1,d),(3,f)\}$ are similar but not equal.

Definition 11. *The multiset S_1 is a sub-multiset of a multiset S_2, denoted as $S_1 \subseteq S_2$, if $\forall v \in V$ condition $k_{S_1}(v) \leq k_{S_2}(v)$ is satisfied.*

It is easy to see that the empty multiset Ø is a sub-multiset of every multiset.

It should be noted that the sum of the support of the multiset *S* drawn from a set V and the upper cut of the multiset *S* gives the set *V*. Now we will introduce new corollaries which are illustrated by examples.

Corollary 6. *Let us consider a multiset S drawn from the set V. The following property is satisfied: $\bar{S} \oplus S^* = V$, where $\bar{S}$ and S^* are the ordinary sets.*

Example 6. Let us consider the multiset $S = \{(13,a),(12,d)\}$ drawn from the set *V*={*a, b, c, d*}. Its support is $S^* = \{a,d\}$ and the upper cut is $\bar{S} = \{b,c\}$. The addition of its support and the upper cut provide the entire set V and can be written as follows $S^* \oplus \bar{S} = \{a,d\} \oplus \{b,c\} = \{a,b,c,d\}$.

Corollary 7. *Let us consider two multisets* S_1, S_2 *drawn from a set* V. *The following condition* $k_{S_1 \cap S_2}(v_i) + k_{S_1 \cup S_2}(v_i) = k_{S_1}(v_i) + k_{S_2}(v_i)$ *is satisfied.*

References

1. Abo-Tabl, A.E.-S.: Topological approximations of multisets. Journal of the Egyptian Mathematical Society **21**, 123–132 (2013)
2. Girish, K.P., Sunil, J.J.: Multiset topologies induced by multiset relations. Information Sciences **188**, 298–313 (2012)
3. Krawczak, M., Szkatuła, G.: On perturbation measure of clusters: application. In: Rutkowski, L., Korytkowski, M., Scherer, R., Tadeusiewicz, R., Zadeh, L.A., Zurada, J.M. (eds.) ICAISC 2013, Part II. LNCS, vol. 7895, pp. 176–183. Springer, Heidelberg (2013)
4. Krawczak, M., Szkatuła, G.: An approach to dimensionality reduction in time series. Information Sciences **260**, 15–36 (2014)
5. Krawczak, M., Szkatuła, G.: On Perturbation Measure of Sets – Properties. Journal of Automation, Mobile Robotics & Intelligent Systems **8**, 41–44 (2014)
6. Krawczak, M., Szkatuła, G.: On asymmetric matching between sets. Information Sciences **312**, 89–103 (2015)
7. Petrovsky, A.B.: Cluster analysis in multiset spaces. In: Goldevsky, M., Mayr, H. (eds.) Information Systems Technology and its Applications, pp. 199–206. Gesellschaft fur Informatik, Bonn (2003)
8. Petrovsky, A.B.: Methods for the Group Classification of multi-attribute Objects. Scientific and Technical Information Processing **37**(5), 357–368 (2010)
9. Singh, D., Ibrahim, A.M., Yohanna, T., Singh, J.N.: A systematization of fundamentals of multisets. Lecturas Matematicas **29**, 33–48 (2008)
10. Singh, D., Ibrahim, A.M., Yohanna, T., Singh, J.N.: An overview of the applications of multisets. Novi Sad J. Math. **37**(2), 73–92 (2007)

Modular Neural Network Preprocessing Procedure with Intuitionistic Fuzzy InterCriteria Analysis Method

Sotir Sotirov, Evdokia Sotirova, Patricia Melin, Oscar Castilo and Krassimir Atanassov

Abstract Modular neural networks (MNN) are a tool that can be used for object recognition and identification. Usually the inputs of the MNN can be fed with independent data. However, there are certain limits when we may use MNN, and the number of the neurons is one of the major parameters during the implementation of the MNN. On the other hand, the greater number of neurons slows down the learning process. In the paper, we propose a method for removing the number of the inputs and, hence, the neurons, without removing the error between the target value and the real value obtained on the output of the MNN's exit. The method uses the recently proposed approach of InterCriteria Analysis, based on index matrices and intuitionistic fuzzy sets, which aims to detect possible correlations between pairs of criteria. The coefficients of the positive and negative consonance can be combined for obtaining the best results and smaller number of the weight coefficients of the neural network.

Keywords Modular Neural Network · InterCriteria analysis · Intuitionistic fuzziness

---

S. Sotirov(✉) · E. Sotirova · K. Atanassov
Intelligent Systems Laboratory, "Prof. Dr. Asen Zlatarov" University, Burgas, Bulgaria
e-mail: {ssotirov,esotirova}@btu.bg

P. Melin · O. Castilo
Tijuana Institute of Technology, Tijuana, México
e-mail: {pmelin,ocastillo}@tectijuana.mx

K. Atanassov
Bioinformatics and Mathematical Modelling Department,
IBPhBME – Bulgarian Academy of Sciences, Sofia, Bulgaria
e-mail: krat@bas.bg

T. Andreasen et al. (eds.), *Flexible Query Answering Systems 2015*,
Advances in Intelligent Systems and Computing 400,
DOI: 10.1007/978-3-319-26154-6_14

1 Introduction

One of the open questions in biology is the ability to adapt fast to new environments, called evolvability [1]. A typical feature of evolvability is the fact that many biological entities are modular, especially many biological processes and structures that can be modeled as a networks, such as metabolic pathways, gene regulation, protein interactions and brains [1–7].

The networks are modular if they contain highly connected clusters of nodes that are sparsely connected to nodes in other clusters [4, 8, 9].

Despite its importance and continuous research there is no agreement on why modular biological entities can evolve [4, 10, 11]. Modular systems look more adaptable [12] than the monolithic networks [13, 14]. There are many papers dedicated to this problem, for example [14].

In this paper, we introduce a combination between the intuitionistic fuzzy InterCriteria Analysis (ICA) method and modular neural network. Through the intuitionistic fuzzy sets-based InterCriteria method we obtain positive and negative consonance coefficients between criteria for evaluating different objects.

The consonance parameters are based on two functions: $\mu_A(x)$ and $\nu_A(x)$ that define the degrees of membership and of non-membership, respectively, of an element x to the set A, evaluated in [0; 1]-interval.

The membership and non-membership coefficients are presented by the two different index matrices. We propose the usage of these two index matrices to reduce some of the inputs in the module neural network. During the process, we are forming the pairs that can be removed as the inputs of the each neural network. The number of the pairs is the same as a number of the modules of the modular neural network. On the input of the each module, we put the reduced number of the parameters according the forming pairs. On the output of the module neural network, we obtain the best result depending on the average error and the minimum number of the weight coefficients.

2 Short Remarks on the InterCriteria Analysis Method

The ICA-method [16] is based on two concepts: intuitionistic fuzzy sets and index matrices.

The Intuitionistic Fuzzy Sets (IFSs, see [17, 18, 19, 20]) represent an extension of the concept of fuzzy sets, as defined by Zadeh [30], exhibiting function $\mu_A(x)$ defining the membership of an element x to the set A, evaluated in the [0; 1]-interval. The difference between fuzzy sets and intuitionistic fuzzy sets (IFSs) is in the presence of a second function $\nu_A(x)$ defining the non-membership of the element x to the set A, where $\mu_A(x) \in [0; 1]$, $\nu_A(x) \in [0; 1]$, under the condition

$$\mu_A(x) + \nu_A(x) \in [0; 1].$$

The IFS itself is formally denoted by:

$$A = \{\langle x, \mu_A(x), \nu_A(x)\rangle \mid x \in E\}.$$

Comparison between elements of any two IFSs, say A and B, involves pairwise comparisons between their respective elements' degrees of membership and non-membership to both sets.

The second concept on which the proposed method relies is the concept of index matrix, a matrix which features two index sets. The theory behind the index matrices is described in [15]. Here, following the description of the ICA approach, given by [16, 22], we will start with the index matrix M with index sets with m rows $\{O_1, \ldots, O_m\}$ and n columns $\{C_1, \ldots, C_n\}$, where for every p, q $(1 \le p \le m,\ 1 \le q \le n)$, O_p in an evaluated object, C_q is an evaluation criterion, and $e_{O_pC_q}$ is the evaluation of the p-th object against the q-th criterion, defined as a real number or another object that is comparable according to relation R with all the rest elements of the index matrix M.

$$M = \begin{array}{c|ccccccc} & C_1 & \cdots & C_k & \cdots & C_l & \cdots & C_n \\ \hline O_1 & e_{O_1,C_1} & \cdots & e_{O_1,C_k} & \cdots & e_{O_1,C_l} & \cdots & e_{O_1,C_n} \\ \vdots & \vdots & \ddots & \vdots & \ddots & \vdots & \ddots & \vdots \\ O_i & e_{O_i,C_1} & \cdots & e_{O_i,C_k} & \cdots & e_{O_i,C_l} & \cdots & e_{O_i,C_n} \\ \vdots & \vdots & \ddots & \vdots & \ddots & \vdots & \ddots & \vdots \\ O_j & e_{O_j,C_1} & \cdots & e_{O_j,C_k} & \cdots & e_{O_j,C_l} & \cdots & e_{O_j,C_n} \\ \vdots & \vdots & \ddots & \vdots & \ddots & \vdots & \ddots & \vdots \\ O_m & e_{O_m,C_1} & \cdots & e_{O_m,C_j} & \cdots & e_{O_m,C_l} & \cdots & e_{O_m,C_n} \end{array},$$

From the requirement for comparability above, it follows that for each i, j, k it holds the relation $R(a_{O_iC_k}, a_{O_jC_k})$. The relation R has dual relation $\overline{R}$, which is true in the cases when relation R is false, and vice versa.

For the needs of our decision making method, pairwise comparisons between every two different criteria are made along all evaluated objects. During the comparison, it is maintained one counter of the number of times when the relation R holds, and another counter for the dual relation.

Let $S_{k,l}^{\mu}$ be the number of cases in which the relations $R(e_{O_iC_k}, e_{O_jC_k})$ and $R(e_{O_iC_l}, e_{O_jC_l})$ are simultaneously satisfied. Let also $S_{k,l}^{\nu}$ be the number of cases in which the relations $R(e_{O_iC_k}, e_{O_jC_k})$ and its dual $\overline{R}(e_{O_iC_l}, e_{O_jC_l})$ are simultaneously satisfied. As the total number of pairwise comparisons between the object is $m(m-1)/2$, it is seen that there hold the inequalities:

$$0 \le S_{k,l}^{\mu} + S_{k,l}^{\nu} \le \frac{m(m-1)}{2}.$$

For every k, l, such that $1 \le k \le l \le n$, and for $m \ge 2$ two numbers are defined:

$$\mu_{C_k,C_l} = 2\frac{S_{k,l}^{\mu}}{m(m-1)},\quad \nu_{C_k,C_l} = 2\frac{S_{k,l}^{\nu}}{m(m-1)}$$

The pair constructed from these two numbers plays the role of the intuitionistic fuzzy evaluation of the relations that can be established between any two criteria C_k and C_l. In this way the index matrix M that relates evaluated objects with evaluating criteria can be transformed to another index matrix M^* that gives the relations among the criteria:

$$M^* = \begin{array}{c|ccc} & C_1 & \cdots & C_n \\ \hline C_1 & \langle \mu_{C_1,C_1}, \nu_{C_1,C_1} \rangle & \cdots & \langle \mu_{C_1,C_n}, \nu_{C_1,C_n} \rangle \\ \vdots & \vdots & \ddots & \vdots \\ C_n & \langle \mu_{C_n,C_1}, \nu_{C_n,C_1} \rangle & \cdots & \langle \mu_{C_n,C_n}, \nu_{C_n,C_n} \rangle \end{array}.$$

From practical considerations, it has been more flexible to work with two index matrices M^μ and M^ν, rather than with the index matrix M^* of IF pairs.

The final step of the algorithm is to determine the degrees of correlation between the criteria, depending on the user's choice of μ and ν. We call these correlations between the criteria: 'positive consonance', 'negative consonance' or 'dissonance'. Let $\alpha, \beta \in [0; 1]$ be the threshold values, against which we compare the values of μ_{C_k, C_l} and ν_{C_k,C_l}. We call that criteria C_k and C_l are in:

- (α, β)-positive consonance, if $\mu_{C_k,C_l} > \alpha$ and $\nu_{C_k,C_l} < \beta$;
- (α, β)-negative consonance, if $\mu_{C_k,C_l} < \beta$ and $\nu_{C_k,C_l} > \alpha$;
- (α, β)-dissonance, otherwise.

Obviously, the larger α and/or the smaller β, the less number of criteria may be simultaneously connected with the relation of (α, β)-positive consonance. For practical purposes, it carries the most information when either the positive or the negative consonance is as large as possible, while the cases of dissonance are less informative and are skipped.

3 Main Results

The modular neural networks [35, 36, 37] are one of the tools that can be used for object recognition and identification.

The idea is to the inputs of the any of the modules of the MNN to enter few of the inputs distributed using the data from the ICA. Every module is two layers Multilayer Perceptron and the output of the second layer of the ANN is a_i^2 (for $i \in (1, \ldots, n)$, where the n is the maximal number of the modules).

In the [34], it is shown the distribution of the results according the coefficients of membership μ. The number of this zones is 11, starting with strong negative consonance to strong positive consonance according the Table 1.

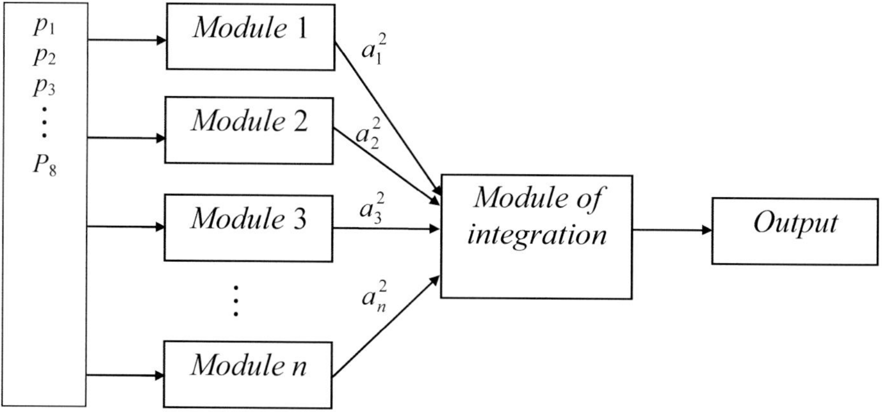

Fig. 1 The structure of the MNN

Table 1 Scale with predefined of intercriteria relations, [34]

N:	From ÷ to	Type of intercriteria relations
1	0÷5%	strong negative consonance
2	5÷15%	negative consonance
3	15÷25%	weak negative consonance
4	25÷33%	weak dissonance
5	33÷43%	Dissonance
6	43÷57%	strong dissonance
7	57÷67%	Dissonance
8	67÷75%	weak dissonance
9	75÷85%	weak positive consonance
10	85÷95%	positive consonance
11	95÷100%	strong positive consonance

On the input of the neural network, we put the experimental data for obtaining cetane number, based on certain correlations with the rest criteria of measurement of crude oil. We work with data for 140 crude oil probes [32], measured against 8 criteria: 1 – Density at 15°C g/cm3; 2 – 10% (v/v) ASTM D86 distillation, °C ; 3 – 50% (v/v) ASTM D86 distillation, °C; 4 – 90% (v/v) ASTM D86 distillation, °C; 5 – Refractive index at 20°C; 6 – H2 content, % (m/m), 7 – Aniline point, °C, 8 – Molecular weight g/mol.

The same data we use as input data of the InterCriteria Analysis method, applied to the whole 140×8 matrix, and a software application that implements the ICA algorithm returns the results in the form of two index matrices in Table 2 and Table 3, containing the membership and the non-membership parts of the IF correlations discovered between each pair of criteria (28 pairs). Table 4 contains a degree of uncertainty that depends on the IF pairs, giving the InterCriteria correlations, but we have not used it for the analysis.

Table 2 Membership parts of the intuitionistic fuzzy pairs, giving the InterCriteria correlations

μ	1	2	3	4	5	6	7	8
1	1,000	0,699	0,770	0,658	0,956	0,176	0,446	0,703
2	0,699	1,000	0,787	0,597	0,676	0,408	0,640	0,775
3	0,770	0,787	1,000	0,777	0,728	0,394	0,665	0,921
4	0,658	0,597	0,777	1,000	0,627	0,468	0,674	0,771
5	0,956	0,676	0,728	0,627	1,000	0,134	0,404	0,661
6	0,176	0,408	0,394	0,468	0,134	1,000	0,730	0,473
7	0,446	0,640	0,665	0,674	0,404	0,730	1,000	0,743
8	0,703	0,775	0,921	0,771	0,661	0,473	0,743	1,000

Table 3 Non-membership parts of the intuitionistic fuzzy pairs, giving the InterCriteria correlations

ν	1	2	3	4	5	6	7	8
1	0,000	0,288	0,217	0,326	0,042	0,822	0,552	0,295
2	0,288	0,000	0,204	0,391	0,312	0,580	0,348	0,213
3	0,217	0,204	0,000	0,212	0,261	0,595	0,325	0,068
4	0,326	0,391	0,212	0,000	0,359	0,518	0,312	0,215
5	0,042	0,312	0,261	0,359	0,000	0,866	0,596	0,339
6	0,822	0,580	0,595	0,518	0,866	0,000	0,270	0,527
7	0,552	0,348	0,325	0,312	0,596	0,270	0,000	0,257
8	0,295	0,213	0,068	0,215	0,339	0,527	0,257	0,000

Table 4 Degrees of uncertainty that depend on the intuitionistic fuzzy pairs, giving the InterCriteria correlations

π	1	2	3	4	5	6	7	8
1	0,000	0,013	0,013	0,016	0,002	0,002	0,002	0,002
2	0,013	0,000	0,009	0,012	0,012	0,012	0,012	0,012
3	0,013	0,009	0,000	0,011	0,011	0,011	0,011	0,011
4	0,016	0,012	0,011	0,000	0,014	0,014	0,014	0,014
5	0,002	0,012	0,011	0,014	0,000	0,000	0,000	0,000
6	0,002	0,012	0,011	0,014	0,000	0,000	0,000	0,000
7	0,002	0,012	0,011	0,014	0,000	0,000	0,000	0,000
8	0,002	0,012	0,011	0,014	0,000	0,000	0,000	0,000

In the case of using the ANN, we choose to reduce the MNN and use only 3 modules from the structure from Fig. 2.

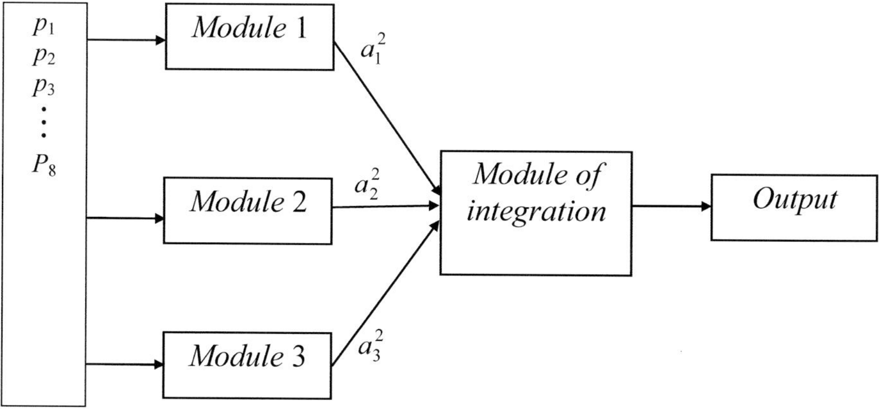

Fig. 2 The structure of the MNN with 3 modules

The output O is calculated according the following formula

$$O = \frac{a_1^2 g_1 + a_2^2 g_2 + a_3^2 g_3}{g_1 + g_2 + g_3}$$

where:

- a_1^2 – output of the first module
- a_2^2 – output of the second module
- a_3^2 – output of the third module
- g_1 – average deviation of the output values of the first module
- g_2 – average deviation of the output values of the second module
- g_3 – average deviation of the output values of the third module

The first module takes the inputs that have strong negative consonance, negative consonance and weak negative consonance.

The last module takes the inputs that have weak positive consonance, positive consonance and strong positive consonance. The middle module takes all others inputs (from 25% to 75%).

In the first module, we can reduce some of the inputs if they have very strong positive consonance (for example the coefficient between the 1 and 5 is 0,956). In this case, we can remove one of the inputs. For the last module, the situation is identical – there is a strong negative consonance. This means that we can also remove some of the inputs. In the middle module, we put the inputs that have independent information (form 25% - 75%). Table 5 shows the distribution of the connections of the inputs according to the values from Table 1.

For the learning process, we set the following parameters: Performance (MSE) = 0.00001; Validation check = 15. The input vector is divided into three different parts:

Training (from 1 to 1000); Validation (from 101 to 120) and Testing (from 121 to 140). For tagret, we use the Cetane number ASTM D613.

For dividing the inputs, we use the intuitionistic fuzzy sets-based approach of InterCriteria Analysis (ICA), which gives dependencies between the criteria, and thus helps us reduce the number of input parameters, yet keeping high enough level of precision.

In the MNN, we put the data for 140 crude oil probes, with 8 criteria (in the inputs of our 3 modules).

In Table 6, we show the values of 10 simulations and we describe a number of parameters that we put on each module. For example – in №1 in module 1 (M1) we put data from parameters 1 and 3.

Table 5 Distribution of the inputs

μ	Inputs	Type of intercriteria dependence
0,956	1-5	strong positive consonance
0,921	3-8	positive consonance
0,787	2-3	weak positive consonance
0,777	3-4	weak positive consonance
0,775	2-7	weak positive consonance
0,771	4-8	weak positive consonance
0,770	1-3	weak positive consonance
0,743	7-8	weak dissonance
0,730	6-7	weak dissonance
0,728	3-5	weak dissonance
0,703	1-8	weak dissonance
0,699	1-2	weak dissonance
0,676	2-5	weak dissonance
0,674	4-7	weak dissonance
0,665	3-7	dissonance
0,661	5-8	dissonance
0,658	1-4	dissonance
0,640	2-7	dissonance
0,627	4-5	dissonance
0,597	2-4	dissonance
0,473	6-8	strong dissonance
0,468	4-6	strong dissonance
0,446	1-7	strong dissonance
0,408	2-6	dissonance
0,404	5-7	dissonance
0,394	3-4	dissonance
0,176	1-6	weak negative consonance
0,134	5-6	negative consonance

Table 6 Result from simulations

№	Description	g_1	g_2	g_3	g_O
1.	M1(1,3,8); M2(2, 4, 7); M3(5,6)	2.1536	2.1608	2.1754	2.1447
2.	M1(1,3); M2(2, 4, 7); M3(5,6)	2.1690	2.1515	2.1591	2.1453
3.	M1(5,8); M2(2, 4, 7); M3(6)	2.2780	2.2336	3.4661	2.5670
4.	M1(3, 5); M2(2, 4, 7); M3(6)	2.1557	2.1550	3.4678	2.5656
5.	M1(1, 8); M2(2, 4, 7); M3(6)	2.1488	2.1587	3.4630	2.4873
6.	M1(3, 8); M2(2, 4, 7); M3(1)	2.2749	2.1999	7.3056	4.9669
7.	M1(3, 8); M2(2, 4, 7); M3(1, 5)	2.2832	2.1907	7.3773	4.9678
8.	M1(3, 8); M2(2, 4, 7); M3(5)	2.2459	2.1771	7.0274	4.7302
9.	M1(3); M2(2, 4, 7); M3(1,6)	6.5784	2.1639	2.3348	4.4438
10.	M1(8); M2(2, 4, 7); M3(1,6)	6.2030	2.2673	2.1628	4.1199
11.	M1(3,5); M2(2,7); M3(4,7)	2.5608	2.7334	2.2069	2.2762

Where g_1 – average deviation of the output values of the whole neural network In row 1, it is described the situation when we put parameters 1, 3 and 8 in module 1. In Module 2 we put parameters 2, 4 and 7. And in Module 3 we put parameters 5 and 6. After simulating the neural network, an average deviation of the MNN O = 2.1447 (g1 = 2.1536, g2 = 2.1608 and g3 = 2.1754) is obtained.

In row 2, is described the situation when we put parameters 1 and 3 in module 1, parameters 2, 4 and 7 in module 2, and parameters with numbers 5 and 6 in module 3. In the study parameter 8 is removed due to the high value of membership coefficient μ = 0.92148 (between parameter 3 and 8). After simulating the neural network an average deviation of the MNN O = 2.1453 (g1 = 2.1690, g2 = 2.1515 and g3 = 2.1591) is obtained. The removed parameter 8 does not have substantial influence on the result, because decreasing the number of the weight coefficients, and along with this the error of the output values also decrease.

The additional information from parameter 8 is not so much because the value of the membership coefficient μ = 0.92148 is high, and parameters 3 and 8 are very dependent.

In row 3 is described the situation when we put parameters 5 and 8 in module 1, parameters 2, 4 and 7 in module 2 and parameter 6 in module 3. In the study, parameters 1 and 3 are removed due to the high value of membership coefficient μ = 0.956 (between parameter 1 and 5) and μ = 0.921 (between parameter 3 and 8). After simulating the neural network an average deviation of the MNN O = 2.5670 (g1 = 2.2780, g2 = 2.2336 and g3 = 3.4661) is obtained. The removed parameters 1and 3 do not have substantial influence on the result, because of the decreasing the number of the weight coefficients, and along with this the error of the output values also decreases.

The additional information from parameters 1 and 3 is not so much because the value of the membership coefficients is high, and the parameters are very dependent.

If we remove some of the parameters with low μ (and high ν) and put them in module 3 we obtain the following values.

In row 11 is described the situation when we put parameters 3 and 5 in module 1, parameters 2 and 7 in module 2, and parameters 4 and 7 in module 3. In the study parameters 6 and 8 are removed due to the high value of non-membership coefficient ν = 0.866 (between parameter 6 and 5) and membership coefficient μ = 0.921 (between parameter 3 and 8). After simulating the neural network an average deviation of the MNN O = 2.2762 (g1 = 2.5608, g2 = 2.7334 and g3 = 2.2069) is obtained. The removed parameters 6 and 8 do not have substantial influence on the result, because of the decreasing the number of the weight coefficients, and along with this the error of the output values also decreases.

The additional information from parameters 6 and 8 is not so much because the value of the membership coefficients is high, and the parameters are very dependent.

It is interesting the case when we compare two rows with similar results. In row 6 is described the situation when we put parameters 3 and 8 in module 1, parameters 2, 4 and 7 in module 2, and parameter 1 in module 3. In the study, parameters 5 and 6 are removed due to the high value of membership coefficient μ = 0.956 (between parameter 1 and 5) and non-membership coefficient ν = 0.866 (between parameter 5 and 6). After simulating the neural network an average deviation of the MNN O = 4.9669 (g1 = 2.2749, g2 = 2.1999 and g3 = 7.3056) is obtained. If we add in module 3 another parameter – 5 the result is similar. The additional information from the parameters 1 and 5 is not so much (the membership coefficient μ = 0.956), and parameter 5 does not add the needed information.

4 Conclusions

The number of the neurons is one of the major parameters during the realization of the MNN. On the other hand, the larger the number of neurons, the slower the learning process. Here, we use the integration of intuitionistic fuzzy InterCriteria Analysis method for reducing the number of input parameters of the modular neural network. This leads to the reduction of the weight matrices, and thus allows implementation of the neural network in limited hardware and saving time and resources in training.

Acknowledgements The authors are grateful for the support provided by the National Science Fund of Bulgaria under grant DFNI-I-02-5/2014.

References

1. Pigliucci, M.: Is evolvability evolvable? Nat. Rev. Genet. **9**, 75–82 (2008)
2. Alon, U.: An introduction to systems biology: design principles of biological circuits. CRC Press, Boca Raton (2006)
3. Carroll, S.: Chance and necessity: the evolution of morphological complexity and diversity. Nature **409**, 1102–1109 (2001)

4. Wagner, G.P., Pavlicev, M., Cheverud, J.M.: The road to modularity. Nat. Rev. Genet. **8**, 921–931 (2007)
5. Hintze, A., Adami, C.: Evolution of complex modular biological networks. PLoS Comput. Biol. **4**, e23 (2008)
6. Mountcastle, V.: The columnar organization of the neocortex. Brain **120**, 701–722 (1997)
7. Guimera, R., Amaral, L.: Functional cartography of complex metabolic networks. Nature **433**, 895–900 (2005)
8. Lipson, H.: Principles of modularity, regularity, and hierarchy for scalable systems. J. Biol. Phys. Chem. **7**, 125–128 (2007)
9. Striedter, G.: Principles of brain evolution. Sinauer Associates, Sunderland (2005)
10. Wagner, G., Mezey, J., Calabretta, R.: Modularity: understanding the development and evolution of complex natural systems. Natural selection and the origin of modules. MIT Press, Cambridge (2001)
11. Espinosa-Soto, C., Wagner A.: Specialization can drive the evolution of modularity. PLoS Comput. Biol. **6** (2010)
12. Suh, N.P.: The principles of design, vol. 226. Oxford University Press, Oxford. Kashtan, N., Alon, U.: Spontaneous evolution of modularity and network motifs. Proc. Natl. Acad. Sci. USA 102, 13 773–13 778 (2005)
13. Kashtan, N., Noor, E., Alon, U.: Varying environments can speed up evolution. Proc. Natl. Acad. Sci. USA 104, 3 711–13 716 (2007)
14. Parter, M., Kashtan, N., Alon, U.: Environmental variability and modularity of bacterial metabolic networks. BMC Evol. Biol. 7 (2007)
15. Atanassov, K.: Index Matrices: Towards an Augmented Matrix Calculus. Springer, Cham (2014)
16. Atanassov, K., Mavrov, D., Atanassova, V.: InterCriteria decision making: A new approach for multicriteria decision making, based on index matrices and intuitionistic fuzzy sets. Issues in Intuitionistic Fuzzy Sets and Generalized Nets **11**, 1–8 (2014)
17. Atanassov, K.: Intuitionistic fuzzy sets. In: Proc. of VII ITKR's Session, Sofia, June 1983 (in Bulgarian)
18. Atanassov, K.: Intuitionistic fuzzy sets. Fuzzy Sets and Systems, Elsevier **20**(1), 87–96 (1986)
19. Atanassov, K.: Intuitionistic Fuzzy Sets: Theory and Applications. Physica, Heidelberg (1999)
20. Atanassov, K.: On Intuitionistic Fuzzy Sets Theory. Springer, Berlin (2012)
21. Atanassova, V., Mavrov, D., Doukovska, L., Atanassov, K.: Discussion on the threshold values in the InterCriteria Decision Making approach. Notes on Intuitionistic Fuzzy Sets **20**(2), 94–99 (2014)
22. Atanassova, V., Doukovska, K., Atanassov, D.: Mavrov – InterCriteria decision making approach to eu member states competitiveness analysis. In: Proc. of the International Symposium on Business Modeling and Software Design, BMSD 2014, June 24–26, Luxembourg, Grand Duchy of Luxembourg, pp. 289–294 (2014)
23. Bellis, S., Razeeb, K.M., Saha, C., Delaney, K., O'Mathuna, C., Pounds-Cornish, A., de Souza, G., Colley, M., Hagras, H., Clarke, G., Callaghan, V., Argyropoulos, C., Karistianos, C., Nikiforidis, G.: FPGA implementation of spiking neural networks - an initial step towards building tangible collaborative autonomous agents. In: Proc. of FPT 2004, Int. Conf. on Field-Programmable Technology, pp. 449–45. The University of Queensland, Brisbane, December 6–8, 2004

24. Hagan, M., Demuth, H., Beale, M.: Neural Network Design. PWS Publishing, Boston (1996)
25. Haykin, S.: Neural Networks: A Comprehensive Foundation. Macmillan, NY (1994)
26. Himavathi, S., Anitha, D., Muthuramalingam, A.: Feedforward Neural Network Implementation in FPGA Using Layer Multiplexing for Effective Resource Utilization. IEEE Transactions on Neural Networks **18**(3), 880–888 (2007)
27. Karantonis, D.M., Narayanan, M.R., Mathie, M., Lovell, N.H., Celler, B.G.: Implementation of a real-time human movement classifier using a triaxial accelerometer for ambulatory monitoring. IEEE Trans. Inform. Technol. Biomed. **10**(1), 156–167 (2006)
28. Meissner, M., Schmuker, M., Schneider, G.: Optimized Particle Swarm Optimization (OPSO) and its application to artificial neural network training. BMC Bioinformatics **7**(1), 125 (2006)
29. Rumelhart, D., Hinton, G., Williams, R.: Training representation by back-propagation errors. Nature **323**, 533–536 (1986)
30. Zadeh, L.A.: Fuzzy Sets. Information and Control **8**, 333–353 (1965)
31. Zwe-Lee, G.: Wavelet-based neural network for power disturbance recognition and classification. IEEE Transactions on Power Delivery **19**(4), 1560–1568 (2004)
32. Stratiev, D., Marinov, I., Dinkov, R., Shishkova, I., Velkov, I., Sharafutdinov, I., Nenov, S., et al.: Opportunity to improve diesel fuel cetane number prediction from easy available physical properties and application of the least squares method and the artificial neural networks. Energy & Fuels (2015)
33. InterCriteria Research Portal. http://intercriteria.net/
34. Atanassov, K., Atanassova, V., Gluhchev, G.: InterCriteria Analysis: Ideas and problems. Notes on Intuitionistic Fuzzy Sets **21**(1), 81–88 (2015). ISSN 1310–4926
35. Nagata, S., Kimoto, T., Asakawa, K.: Control of mobile robots with neural networks. In: INNS, p. 349 (1988)
36. Sawai, H., Waibe, A., et al.: Parallelism, Hierarchy, Scaling in Time-Delay Neural Networks for Spotting Japanese Phonemes/CV-Syllables. IJCNN **11**, 81–88 (1989)
37. Melin, P.: Modular Neural Networks and Type-2 Fuzzy Systems. SCI, vol. 389. Springer, Heidelberg (2012)

Part III
Aspects and Extensions of Flexible Querying and Summarization

Extracting Fuzzy Summaries from NoSQL Graph Databases

Arnaud Castelltort and Anne Laurent

Abstract Linguistic summaries have been studied for many years and allow to sum up large volumes of data in a very intuitive manner. They have been studied over several types of data. However, few works have been led on graph databases. Graph databases are becoming popular tools and have recently gained significant recognition with the emergence of the so-called NoSQL graph databases. These databases allow users to handle huge volumes of data (e.g., scientific data, social networks). There are several ways to consider graph summaries. In this paper, we detail the specificities of NoSQL graph databases and we discuss how to summarize them by introducing several types of linguistic summaries, namely structure summaries, data structure summaries and fuzzy summaries. We present extraction methods that have been tested over synthetic and real database experimentations.

Keywords Linguistic summaries · Graph databases · NoSQL · Fuzzy graph mining

1 Introduction

Representing data with graphs is now a proven practice. Graphs are used in many applications ranging from linguistics to chemistry and social networks. For instance, graphs allow one to intuitively illustrate the relationships between people, as well as those between people and the organizations they belong to.

Graphs are recognized to play an important role within the pattern recognition field [8]; indeed, they are a key technology for retrieving relevant information, such as in fraud detection [18] or social/biological interactions. Relevant information can be retrieved either via data mining methods or predefined queries [1].

A. Castelltort · A. Laurent(✉)
LIRMM, University of Montpellier, Montpellier, France
e-mail: {arnaud.castelltort,anne.laurent}@lirmm.fr
http://www.lirmm.fr

T. Andreasen et al. (eds.), *Flexible Query Answering Systems 2015*,
Advances in Intelligent Systems and Computing 400,
DOI: 10.1007/978-3-319-26154-6_15

Even if graphs are ubiquitous, their representation and use have taken many forms in the literature and in real applications [3, 17]. Theoretical works have attempted to formalize the representations and treatments of graphs More recently, they have been used in the semantic web framework, especially with ontologies. However, with the emergence of big data, the increasing volume and complexity of data and treatments, researchers have realized that robust database management systems are required. Additionally, classical relational database engines are not the solution, as shown in the performance comparisons done on this topic [7]. NoSQL is reputed for its suitability to handle big data [12]. NoSQL graph databases have thus been proposed as the best alternative for managing huge volumes of graph data and complex queries. There are several NoSQL engines; among them, Neo4j is one of the most popular.

In the NoSQL graph database model, the objects considered are nodes and relationships. Complex information on both nodes and relationships are managed as properties with $(key, value)$ pairs. In addition to properties, nodes and relationships may be labeled with types. For instance, Fig. 1 shows the relationships between people and the places they live in. The relationship types depict whether people *own* or *rent* their housing. The type of housing can be an apartment or a house: we define this value as the node type. Node information (e.g., age of people) and relationships (e.g., monthly rental fees) can be provided as node and relationship properties (e.g., $key = age$).

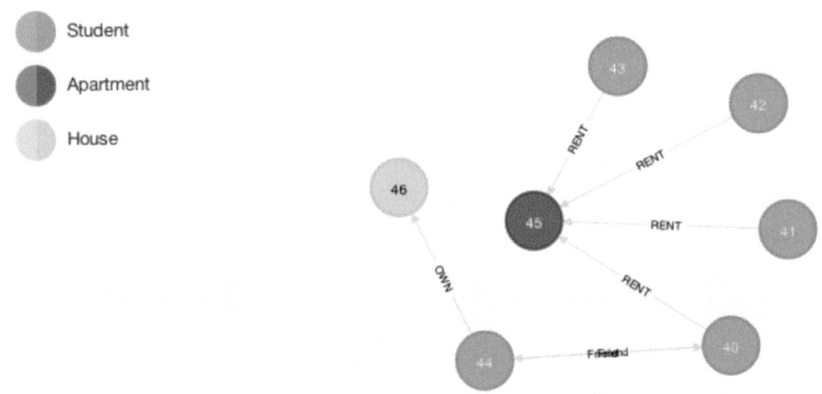

Fig. 1 Example of a Graph Database

As the amount and the complexity of information increase, the need for summaries increases as well. The literature is rich with propositions on linguistic summaries of relational databases [5, 10, 14]. Linguistic summaries are based on protoforms, the first one being $Qy\ are\ P$ where Q stands for a fuzzy quantifier, y are the objects to be summarized and P is a possible value, such as in *Most students are young*. Linguistic summaries have been extended in many works, for instance to handle time series [2, 13]. However, these works cannot be easily applied to NoSQL graph databases. The summaries we aim to discover must indeed be transposable to linguistic summaries. Moreover such databases combine several criteria that have never been considered altogether: node and relationship types, complex information contained in the $(key, value)$ properties.

In this paper, we thus propose several types of linguistic summaries of NoSQL graph databases. The paper is organized as follows. Section 2 reports existing work on NoSQL graph databases and linguistic summarization. Section 3 introduces our definitions for linguistic summaries of NoSQL graph databases. In Section 4, we introduce the queries used to extract the summaries, by highlighting the power of NoSQL graph databases using the Cypher language. Section 5 concludes the paper and discusses perspectives for future work.

2 Background

Our work is strongly related to NoSQL graph databases and data summarization. We therefore review the basics of these two topics below.

2.1 *Graphs*

General Concepts. Graphs have been studied for a long time by mathematicians and computer scientists. A graph can be directed or not, labeled or not. It is defined as follows.

Def. 1 (Graph). *A graph G is given by a pair (V, E) where V stands for a set of vertices and E stands for a set of edges with $E \subseteq \{V \times V\}$.*

Def. 2 (Directed Graph). *A directed graph G is given by a pair (V, E) where V stands for a set of vertices and E stands for a set of edges with $E \subseteq (V \times V)$. That is E is a subset of all ordered pairs of V.*

When used in real world applications, graphs need to be provided with the capacity to label nodes and relations, thus leading to the so-called labeled graphs, or property graphs.

Def. 3 (Labeled Directed Graph). *A labeled oriented graph G, also known as oriented property graph, is given by a quadruplet (V, E, α, β) where V stands for a set of vertices and E stands for a set of edges with $E \subseteq (V \times V)$, α stands for the set of attributes defined over the nodes, and β the set of attributes defined over the relations.*

NoSQL graph databases [17] are based on these concepts, attributes and values over the attributes being stored thanks to the $(key, value)$ paradigm which is very common in NoSQL databases. Fig. 3 shows a graph and its structure in $(key, value)$ pairs.

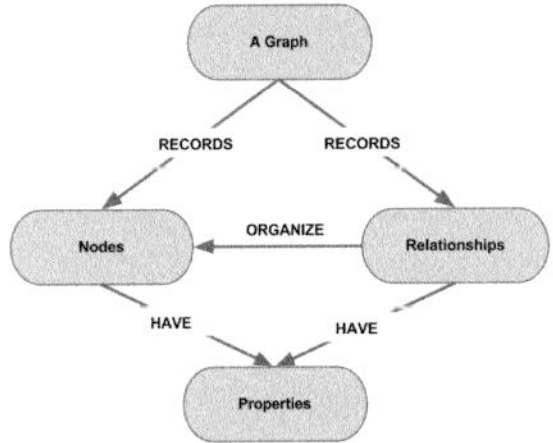

Fig. 2 Labeled Graph

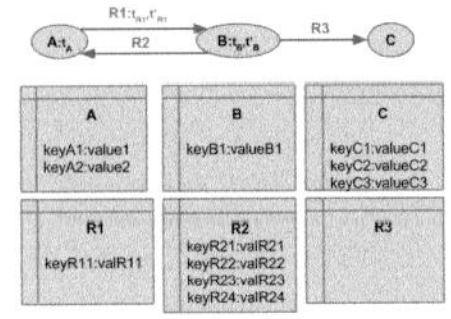

Fig. 3 Node and Relation Properties

Studies have shown that these technologies present good performances, much better than classical relational databases for representing and querying such large graph databases.

Graph Summarization. Data Summarization has been extensively studied in the last decades to produce linguistic sentences, such as *Most of the students are young* [20]. These approaches are based on the so-called *protoforms* (*e.g.*, $Qy\ are\ P$) where Q is a fuzzy quantifier, y are the objects to summarize and P is a (fuzzy) predicate. They focus on relational data where source data are represented in the form of tuples defined over a schema. For instance, the tuples (John, 23, 45000), (Mary, 32, 60000) and (Bill, 38, 55000) are three tuples defined on the schema (Name, Age, Salary). The fuzzy quantifiers are defined over the [0, 1] universe of proportions. We may for instance consider two quantifiers Few and $Most$ which membership functions are displayed by Fig. 4.

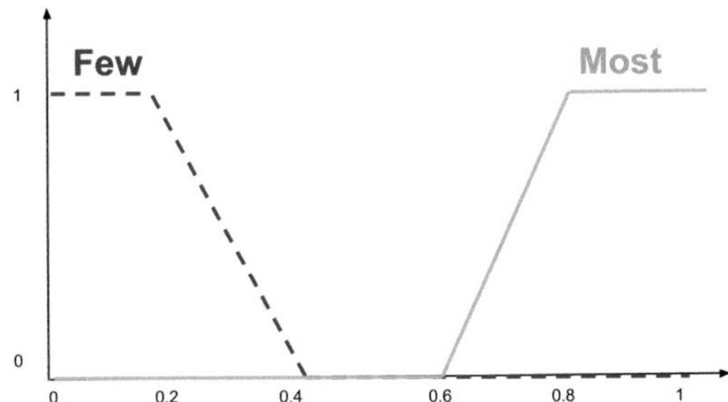

Fig. 4 Example of Fuzzy Quantifier Membership Functions

The quality of linguistic summaries can be assessed by many measures, the seminal one being T, the *degree of truth* that can be simply computed with a σ-count:

$$T(Qy's\ are\ P) = \mu_Q\left(\frac{1}{n}\sum_{i=1}^{n}\mu_P(y_i)\right)$$

where n is the number of objects (y_i) that are summarized, and μ_P , and μ_Q are the membership functions of the summarizer and quantifier, respectively.

There are various ways to examine summaries: Researchers have focused on the design of protoforms, quality measures, efficient algorithms, etc [5].

We will not recall here all the literature on fuzzy linguistic summaries which has been amply presented in previous works. We will focus instead on the subject of graph data. In this framework, two main characteristics have to be highlighted. First, graph databases are not provided with a strict and given schema such as relational data. In fact, they are close to semi-structured data. Second, graph databases focus on relationships.

Summarizing graph data has been considered for many years, aiming for instance at compressing such data with the use of supernodes as shown by Fig. 5 from [16].

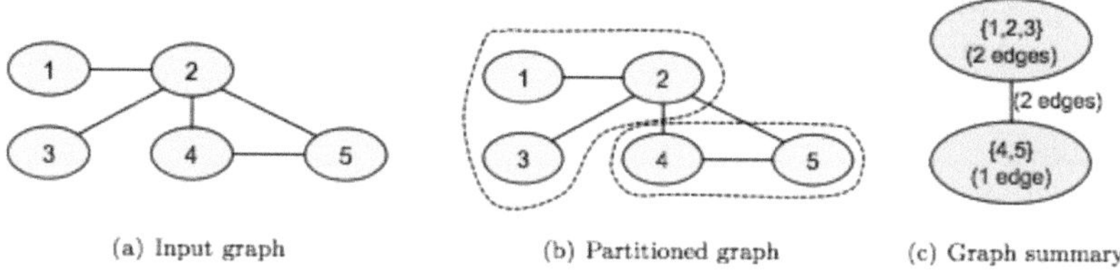

Fig. 5 Graph Summarization using Supernodes [16]

To some extent, graph summarization can be assimilated to graph mining. Graph (and tree) mining is seen as the problem of extracting frequent patterns (subgraphs/subtrees) from a large graph. It is often presented as an extension of the so-called itemset mining methods. Such methods have been successfully applied to large graphs by considering efficient approaches [9, 15, 21].

Several works in the literature have focused on schema extraction in the context of semi-structured graph data, i.e., XML data. The schema extraction problem consists in identifying a schema S from a given set of XML data documents D, such that S captures the structural information of the documents in D in the most minimal way. The schema extraction process is also referred to as schema inference [4]. The underlying structure of a given collection of XML documents can be described using Document Type Definitions (DTD), XML Schema, or via a more general representation such as tree or a graph. The structure extraction techniques in the literature aim to infer three kinds of representations: tree or graph summaries, DTD or XML Schema.

2.2 NoSQL Graph Databases

NoSQL graph databases [17] are based on graph concepts with the following additional points:

– Nodes and relationships are labeled with types;
– Properties are defined over the nodes and relationships stored according to the (key, value) paradigm, which is very common in NoSQL databases.

It should be noted that types are distinguished from properties, as in NoSQL engines such as Neo4j. These types appear in Fig. 1 as colors for nodes (e.g., Student, House) and as labels for relationships (e.g., Owns).

Generalizing the definition of labeled directed graphs, we propose a formal definition of NoSQL graph databases hereafter:

Def. 4 (NoSQL Graph Database). *A NoSQL graph database G is given by a tuple* $(V, E, \theta, \tau, \alpha, \beta)$ *where*

– α *stands for the set of node properties defined by (key:value) pairs;*
– β *stands for the set of edge properties defined by (key:value) pairs;*
– θ *stands for the set of node types;*
– τ *stands for the set of edge types;*
– V *stands for a set of vertices with* $\forall v \in V, v = (id_v, t_v, \kappa_v)$ *s.t.* $t_v \subseteq \theta$ *stands for the types of v,* $\kappa_v \subseteq \alpha$ *stands for the properties of v and* id_v *is the vertice identifier;*
– E *stands for a set of edges with* $\forall e \in E, e = (id_e, (v_e^1, v_e^2), t_e, \lambda_e)$ *s.t.* $(v_e^1, v_e^2) \in \{V \times V\}$, $t_e \subseteq \tau$ *stands for the types of e,* $\lambda_e \subseteq \beta$ *stands for the properties of e and* id_e *is the edge identifier.*

The set of properties of a node v is denoted by α_v, the set of types is denoted by θ_v. The set of properties of a relation e is denoted by β_e, the set of types is denoted by τ_e.

Fig. 3 shows a graph and its structure in (*key*, *value*) pairs. In this figure, we have:

– $\alpha = \{(keyA1, valueA1), \ldots\}$
– $\beta = \{(keyR11, valueR11), \ldots\}$
– $V = \{(A, \{t_A\}, \{(keyA1, valueA1), (keyA2, valueA2)\}), (B, \{t_B, t'_B\}, \{(keyB1, value B1)\}), (C, \emptyset, \{(keyC1, valueC1), (keyC2, valueC2), (keyC3, valueC3)\})$
– $E = \{(R1, (A, B), \{t_{R1}, t'_{R1}\}, \{(keyR11, valueR11)\}), \ldots\}$

There exist several NoSQL graph database engines (OrientDB, Neo4J, HyperGraphDB, etc.) [3]. Neo4J is considered the best perfomer [19]. All NoSQL graph databases require developers and users to use graph concepts to query data. Queries are called traversals, refering to the action of visiting elements, i.e. nodes and relationships. There are three main ways to traverse a graph:

– Programmatically: using an API;
– By functional traversal: using a traversal based on a sequence of functions applied to a graph;
– By declarative traversal: explicitly expressing what we want to do and not how we want to do it. The database engine then defines the best way to achieve this goal.

In this paper, we focus on declarative queries over a NoSQL graph database. The Neo4j language is called Cypher.

For instance, in Fig. 6, a query to return the customers who have visited the "Ritz hotel is displayed. Those customers are both displayed in the list and circled in red in the graph.

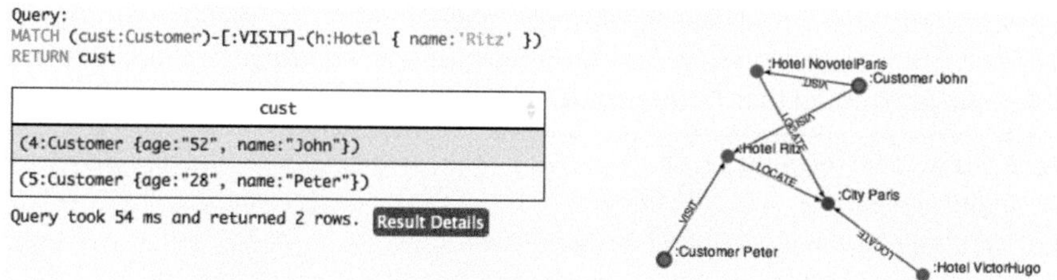

Fig. 6 Displaying the Result of a Cypher Query

Cypher clauses are similar to SQL ones. It is based on a "ASCII art" way of writing graph elements. For example, directed relations are written using the -[]->symbol. Types and labels are written after a semi-column (:).

More specifically, queries in Cypher have the following syntax[1]:

Listing 1.1 Query example on a Graph

```
[START]
[MATCH]
[OPTIONAL MATCH WHERE]
[WITH [ORDER BY] [SKIP] [LIMIT]]
RETURN [ORDER BY] [SKIP] [LIMIT]
```

As shown above, Cypher is comprised of several distinct clauses which are listed below in Listing 1.2.

Listing 1.2 Cypher Clauses

```
START: Starting points in the graph, obtained via index lookups or by element IDs.
MATCH: The graph pattern to match, bound to the starting points in START.
WHERE: Filtering criteria.
RETURN: What to return.
CREATE: Creates nodes and relationships.
DELETE: Removes nodes, relationships and properties.
SET: Set values to properties.
FOREACH: Performs updating actions once per element in a list.
WITH: Divides a query into multiple, distinct parts.
```

These operations can even be extended to fuzzy queries [6]. In this work, they are used for computing the linguistic summaries introduced below.

[1] http://docs.neo4j.org/refcard/2.0/
http://docs.neo4j.org/chunked/milestone/cypher-query-lang.html

3 Building Fuzzy NoSQL Graph Summaries

Below, we define the protoforms for summing up NoSQL graph databases.

3.1 Structure Summaries

Structure summaries are meant to retrieve the structure of the graph embedded in element types, which could somehow be associated with relational database schema. Such summaries could thus be associated with with schema mining in the literature.

Def. 5 (Structure Summary). *Let* $G = (V, E, \theta, \tau, \alpha, \beta)$ *be a NoSQL graph database. A structure summary S of G is defined as* $S = (a \text{ --}[r]\text{->} b, Q)$ *where* $a, b \in \theta$ *(node types),* $r \in \tau$ *(relation type) and Q is a fuzzy quantifier.*

The structure summary can be expressed in a linguistic form as follows: In G, Q of the a r b

Example 1. *In the toy example,* $(Student \text{ --}[rent]\text{->} apartment, Most)$*, expressed as* "Most of the students rent an apartment", *is a structure summary.*

3.2 DataStructure Summaries

Data structure summaries are meant to refine structure summaries. They allow one to differentiate cases when schema depend on the value of properties. For instance, depending on their salary, employees may rent apartments instead of houses.

Def. 6 (Data Structure Summary). *Let* $G = (V, E, \theta, \tau, \alpha, \beta)$ *be a NoSQL graph database. A Data Structure Summary S is defined as* $S = (a.X \text{ --}[r.Z]\text{->} b.Y, Q)$ *with* $a, b \in \theta$ *(node types),* $r \in \tau$ *(relation type),* $X, Y \subseteq \alpha$ *(node properties),* $Z \subseteq \beta$ *(relation properties) and Q a fuzzy quantifier.*

Example 2. *In the toy example,* $(Student(Age : 28) \text{ --}[rent(fees : 1200)]\text{->} apartment, Few)$ *is a data structure summary.*

Such summaries are extended in order to allow fuzzy linguistic labels in the refinement. Indeed, it would be both difficult and useless to define summaries on single values such as "the age is 28, as in fuzzy data mining for fuzzy association rule mining. Using fuzzy linguistic labels makes it possible to retrieve fuzzy linguistic summaries where young students and low rental fees are considered.

Def. 7 (Fuzzy Data Structure Summary). *Let* $G = (V, E, \theta, \tau, \alpha, \beta)$ *be a NoSQL graph database. Let* F_α *and* F_β *be sets of fuzzy properties. A Fuzzy Data Structure Summary S is defined as* $S = (a.X \text{ --}[r.Z]\text{->} b.Y, Q)$ *with* $a, b \in \theta$ *(node types),* $r \in \tau$ *(relation type),* $X, Y \subseteq \alpha \bigcup \mathcal{F}_\alpha$ *(node properties and node fuzzy properties),* $Z \subseteq \beta \bigcup F_\beta$ *(relation properties and relation fuzzy properties) and Q a fuzzy quantifier.*

Example 3. *In the toy example,* $(Student(Age : young) \text{ --}[rent(fees : low)]\text{->} apartment, Most)$ *is a fuzzy structure summary.*

For all these summaries, it is important to discuss the way to assess them. For this purpose, we propose to rely on the extension of the degree of truth.

3.3 *Degree of Truth*

In our framework, the degree of truth determines the extent to which the relationship appearing in the summary is truthful regarding the fuzzy quantifier. For instance, if the summary mentions that *most of the students rent an apartment,* then the degree of truth describes to which extent a high proportion of students rent an apartment.

There are two ways of calculating this degree. Indeed, it may consist in computing the proportion of students who rent an apartment over the whole student population, or the proportion of students who rent an apartment over the number of students who rent their housing. These two definitions are provided below.

Def. 8 (Degree of Truth of NoSQL Graph Summaries). *Given a graph database G and a summary* $S = a \dashv\!\![r]\!\!\vdash\!> b, Q$*, the degrees of Truth of S in G are defined as:*

$$Truth_1(S) = \mu_Q\left(\frac{count(distinct(S))}{count(distinct(a))}\right)$$

$$Truth_2(S) = \mu_Q\left(\frac{count(distinct(S))}{count(distinct(a \dashv\!\![r]\!\!\vdash\!> (?)))}\right)$$

The second type of degree of truth is also called the *diversity of target source* denoted by D_T later on in this paper.

Example 4. *In the toy example from Fig. 1, the degree of truth of the summary "*$S = Student \dashv\!\![rent]\!\!\vdash\!> apartment, Most$*" is given by the membership degree to the fuzzy quantifier*[2] *of the ratio between the number of times a relationship appears between a Student and an Apartment (s)he rents over the number of relations of type* $Rents$ *starting from a Student node. In this example, we thus have* $Truth_2(S) = \mu_{Most}\left(\frac{4}{5}\right) = \mu_{Most}(0.8)$

The ratio appearing in the definition of the degree of truth can be compared to the $confidence$ in association rule mining which acts as a conditional probability.

4 Extracting the Summaries

Below we detail how to extract the above-defined summaries from NoSQL graph databases and some first results.

[2] We do not mention here the detailed membership function of the $Most$ quantifier which can be defined in a very classical manner as done in the literature of fuzzy quantifiers and fuzzy summaries.

4.1 Queries

Most of the treatments we propose can be performed effectively by defining queries over the NoSQL graph database, so as to obtain a more declarative than procedural way to extract summaries. This property is based on the fact that NoSQL graph databases provide powerful pattern-matching features, as intented in inductive relational databases [11].

The structure summaries can be retrieved by considering Cypher queries such as:

```
MATCH (a)-[r]->(m)
RETURN DISTINCT labels(a), type(r), labels(m), count(r)
```

In the above query, all the structures are retrieved together, along with the number of times they appear.

The query from Listing 1.3 extracts the summaries from a graph and calculates the degree of truth for every summary.

Listing 1.3 Retrieving Structure Summaries

```
MATCH (a)-[r]->(b)
WITH DISTINCT labels(a) AS labelsA, type(r) AS typeR, labels(b) AS labelsC, toFloat(count(*)↩
      ) AS countS
MATCH (a1)-[r2]->(m)
WHERE labels(a1)= labelsA AND type(r2)= typeR
WITH DISTINCT labelsA, typeR, labelsC, countS, labels(a1) AS labelsA1, type(r2) AS typeR2, ↩
     count(*) AS count2
RETURN labelsA, typeR, labelsC,
      tofloat(countS)/ count2 AS Truth,
      MuMost(countS/count2) as TruthMost,
      MuFew(countS/count2) as TruthFew,
      MuVeryFew(countS/count2) as TruthVeryFew
```

Where μ_{Most}, μ_{Few} and $\mu_{VeryFew}$ respectively stand for the membership function of the fuzzy subsets $Most$, Few and $VeryFew$ defined on the [0, 1] universe which are implemented as the $MuMost$, $MuFew$ and $MuVeryFew$ functions in Cypher.

4.2 Experimental Results

Experiments have been run on synthetic and real databases with a Java implementation run on an Intel Core i5 2.4 Ghz. The Movie real database deals with Directors, Movies and Actors[3]. It contains about 12,000 movies and 50,000 actors. Below is an example of the results from the synthetic database:

Summary	$TruthMost$	$TruthFew$	$TruthVeryFew$
Student-Rents-Apartment	$\mu_{Most}(4/5)$	$\mu_{Few}(4/5)$	$\mu_{VeryFew}(4/5)$
Student-Friend-Student	$\mu_{Most}(2/5)$	$\mu_{Few}(2/5)$	$\mu_{VeryFew}(2/5)$
Student-Owns-House	$\mu_{Most}(1/5)$	$\mu_{Few}(1/5)$	$\mu_{VeryFew}(1/5)$

[3] http://neo4j.com/developer/example-data

The Listing 1.4 displays some examples of fuzzy data structure summaries.

Listing 1.4 Fuzzy Data Structure Summaries Extracted from the Movie Database

```
DS12:(Person(old)-[DIRECTED]->Movie,Most); Truth(DS12)=1.0
DS16:(Person(young)-[DIRECTED]->Movie,VeryFew); Truth(DS16)=0.21
DS17:(Person(middleAge)-[DIRECTED]->Movie,VeryFew); Truth(DS17)=0.47
DS18:(Person(old)-[DIRECTED]->Movie,VeryFew); Truth(DS18)=0.0
```

5 Conclusion and Perspectives

In this paper, we have presented an approach to summarize NoSQL graph databases in the form of linguistic summaries. These databases are quite specific with respect to existing work, as they combine several difficulties: a focus on relationships, node and relationship types management, management of complex (key:value) pair information management, etc.

Several types of linguistic summaries are proposed. They can be easily extracted using existing declarative query languages, making it possible to deploy them on every commercial tool. Experiments have been made on Neo4j using the Cypher query language, and have demonstrated the interest of our proposal.

Our work opens several perspectives. First, we plan to run more experiments to test scalability. The propositions could also be extended to several graphs. Moreover, we might try and integrate other types of summaries for managing time and space information. Another perspective is to consider extended linguistic summaries containing several relationships.

References

1. Aggarwal, C.C., Wang, H. (eds.): Managing and Mining Graph Data. Advances in Database Systems, vol. 40. Springer (2010)
2. Almeida, R.J., Lesot, M., Bouchon-Meunier, B., Kaymak, U., Moyse, G.: Linguistic summaries of categorical series for septic shock patient data. In: Proceedings of the IEEE International Conference on Fuzzy Systems, FUZZ-IEEE 2013, Hyderabad, India, July 7–10, pp. 1–8. IEEE (2013). http://dx.doi.org/10.1109/FUZZ-IEEE.2013.6622581
3. Angles, R., Gutiérrez, C.: Survey of graph database models. ACM Comput. Surv. 40(1) (2008)
4. Bex, G.J., Neven, F., Vansummeren, S.: Inferring XML schema definitions from XML data. In: Koch, C., Gehrke, J., Garofalakis, M.N., Srivastava, D., Aberer, K., Deshpande, A., Florescu, D., Chan, C.Y., Ganti, V., Kanne, C., Klas, W., Neuhold, E.J. (eds.) Proceedings of the 33rd International Conference on Very Large Data Bases, University of Vienna, Austria, September 23–27, pp. 998–1009. ACM (2007)
5. Bouchon-Meunier, B., Moyse, G.: Fuzzy linguistic summaries: where are we, where can we go ? In: CIFEr 2012 IEEE Conf. on Computational Intelligence for Financial Engineering & Economics (CIFEr), pp. 317–324. IEEE (2012)

6. Castelltort, A., Laurent, A.: Fuzzy queries over NoSQL graph databases: perspectives for extending the cypher language. In: Laurent, A., Strauss, O., Bouchon-Meunier, B., Yager, R.R. (eds.) IPMU 2014, Part III. CCIS, vol. 444, pp. 384–395. Springer, Heidelberg (2014)
7. Cattell, R.: Scalable SQL and NoSQL data stores. SIGMOD Record **39**(4), 12–27 (2010)
8. Conte, D., Foggia, P., Sansone, C., Vento, M.: Thirty Years Of Graph Matching In Pattern Recognition. International Journal of Pattern Recognition and Artificial Intelligence (2004)
9. Cook, D.J., Holder, L.B.: Mining Graph Data. John Wiley & Sons (2006)
10. Rasmussen, D., Yager, R.R.: Finding fuzzy and gradual functional dependencies with summary SQL. Fuzzy Sets and Systems **106**, 131–142 (1999)
11. De Raedt, L.: A perspective on inductive databases. SIGKDD Explor. Newsl. **4**(2), 69–77 (2002)
12. Han, J., Haihong, E., Le, G., Du, J.: Survey on noSQL database. In: Proc. of the 6th International Conference on Pervasive Computing and Applications (ICPCA), pp. 363–366 (2011)
13. Kacprzyk, J., Wilbik, A., Zadrozny, S.: An approach to the linguistic summarization of time series using a fuzzy quantifier driven aggregation. Int. J. Intell. Syst. **25**(5), 411–439 (2010). doi:10.1002/int.20405
14. Kacprzyk, J., Zadrozny, S.: Linguistic database summaries and their protoforms: towards natural language based knowledge discovery tools. Inf. Sci. **173**(4), 281–304 (2005). doi:10.1016/j.ins.2005.03.002
15. Kuramochi, M., Karypis, G.: Frequent subgraph discovery. In: Proceedings of the 2001 IEEE International Conference on Data Mining, ICDM 2001, pp. 313–320. IEEE Computer Society, Washington, DC (2001). `http://dl.acm.org/citation.cfm?id=645496.658027`
16. LeFevre, K., Terzi, E.: Grass: Graph structure summarization. In: Proceedings of the SIAM International Conference on Data Mining, SDM 2010, April 29 - May 1, Columbus, Ohio, USA, pp. 454–465. SIAM (2010)
17. Robinson, I., Webber, J., Eifrem, E.: Graph Databases. O'Reilly (2013)
18. Sadowski, G., Rathle, P.: Fraud detection: Discovering connections with graph databases. In: White Paper - Neo Technology - Graphs are Everywhere (2014)
19. ThoughtWorks: Technology advisory board, May 2013. `http://thoughtworks.fileburst.com/assets/technology-radar-may-2013.pdf`
20. Yager, R.R.: A new approach to the summarization of data. Information Sciences **28**(1), 69–86 (1982)
21. Yan, X., Han, J.: gSpan: Graph-based substructure pattern mining. In: Proceedings of the 2002 IEEE International Conference on Data Mining, ICDM 2002, pp. 721–724. IEEE Computer Society, Washington, DC (2002). `http://dl.acm.org/citation.cfm?id=844380.844811`

Analogical Database Queries

William Correa Beltran , Hélène Jaudoin and Olivier Pivert

Abstract In this paper, we introduce a new type of database query inspired from some works in AI about the concept of analogical proportion. The general idea is to retrieve the tuples that participate in a relation of the form "*a* is to *b* as *c* is to *d*". We provide a typology of analogical queries in a relational database context, devise different processing strategies and assess them experimentally.

1 Introduction

Analogical proportions bind together four objects A, B, C, D of the same type in an assertion of the form "A is to B as C is to D". They express the equality or the proximity of two relationships $(A : B)$ and $(C : D)$, connecting respectively A to B and C to D. Two typical illustrations of this notion, expressed in natural language, are: "calf is to cow as foal is to mare", and "aurochs is to ox as mammoth is to elephant". The semantics of the connector "is to" occurring in an analogical proportion is manifold and potentially complex. In the first example, it denotes a filiation relationship whereas in the second one, it expresses some sort of temporal evolution.

Analogical proportions have been investigated in the domains of Artificial Intelligence (see, e.g., [3, 4, 6, 9, 10]) and Cognitive Sciences (see, e.g., [2, 12]) for already several decades (a much more complete bibliography may be found in [11]). On the other hand, there is almost no work in the domain of databases that exploits analogical proportions, if we except an approach aimed to estimate missing values [1]. In the present paper, our objective is to use analogical proportions as the concept underlying a new type of database queries, called analogical queries, that can be used, for instance, to search for combinations of four n-uples bound by an analogical

W.C. Beltran · H. Jaudoin · O. Pivert(✉)
Université de Rennes 1 – Irisa, Lannion, France
e-mail: {William.Correa_Beltran,Helene.Jaudoin,Olivier.Pivert}@irisa.fr

T. Andreasen et al. (eds.), *Flexible Query Answering Systems 2015*,
Advances in Intelligent Systems and Computing 400,
DOI: 10.1007/978-3-319-26154-6_16

relationship. We focus on the case of relations including numerical values. In such a context, one may consider the case of mathematical proportions (a particular case of analogical proportion) that bind four variables A, B, C and D according to the equality $A/B = C/D$ (e.g., $1/3 = 2/6$) in the case of a geometrical proportion, or according to the equality $A - B = C - D$ (e.g., $5 - 3 = 9 - 7$) in the case of an arithmetic one.

More generally, analogical proportions capture the notion of parallels between four entities that can be objects, observations, events, etc. These parallels are of a major importance as they model reproducible transformations from one entity to another. In the particular case where a temporal dimension comes into play, they make it possible to model for instance societal changes or parallels between trajectories of moving objects. In this paper, we focus on the problem of discovering *parallels* between pairs of tuples occurring in a relation. In the case of a Boolean relation, they will serve to identify pairs of tuples (t_a, t_b) and (t_c, t_d) such that t_a and t_b (resp. t_c and t_d) have common properties and such that the differences between t_a and t_b on the one hand, and between t_c and t_d on the other hand, are the same. For example, one may infer from descriptions of animals that *calf* is to *foal* as *cow* is to *mare*. In the case of a relation including numerical values, they will highlight pairs of tuples that differ in the same way (for instance, similar sales trends between two regions observed at distinct times).

The remainder of the paper is organized as follows. In Section 2, we define the notion of an analogical proportion in the setting of relational databases, following a vectorial approach that extends the definition of an arithmetical proportion to the n-dimensional case. This approach is flexible enough to allow for the retrieval of *approximate* analogical proportions. Section 3 describes the different types of analogical queries that may be thought of in a database context and proposes a syntax *à la SQL* for expressing such queries. Section 4 deals with analogical query optimization and proposes three processing strategies. Section 5 describes an experimentation carried out on a real-world dataset and a synthetic one. Section 6 recalls the main contributions and outlines some perspectives.

2 Refresher on Analogical Proportions

2.1 General Notions

The following brief presentation is mainly drawn from [7]. An analogical proportion is an assertion of the form "A is to B as C is to D", denoted by $A : B :: C : D$ where : denotes a relationship (called a *ratio*) and ::, called *conformity*, expresses the parallelism between pairs of entities. An analogical proportion then expresses the conformity between two pairs of entities.

Generally A, B, C, and D represent entities that can be described as sets, multisets, vectors, strings or trees. A possible interpretation of an analogical relationship follows: how A differs from B is (almost) the same as how C differs from D.

Moreover, it is assumed that analogical proportions must satisfy the following properties [7, 10]:

- reflexivity: $(A : B :: A : B)$
- symmetry: $(A : B :: C : D) \Leftrightarrow (C : D :: A : B)$
- exchange of the means: $(A : B :: C : D) \Leftrightarrow (A : C :: B : D)$.

2.2 Geometric Interpretation

Let us first give a definition of analogical proportion based on a geometric point of view, already considered by several authors, see, e.g., [7, 9].

Definition 1 (Analogical proportion). *Let $\mathcal{D} = \{ D_1, \ldots, D_n\}$ be a set of n dimensions and A, B, C and D be four points in $D_1 \times \ldots \times D_n$. The analogical proportion $A : B :: C : D$ is valid if and only if*

$$\overrightarrow{AB} = \overrightarrow{CD}$$

or equivalently,

$$||\overrightarrow{AB} - \overrightarrow{CD}|| = 0.$$

The analogical relationship binding A, B, C, and D can then be represented by the vector $\overrightarrow{AB}$ (or equivalently $\overrightarrow{CD}$).

Definition 1 straightforwardly satisfies the basic properties of analogical proportions and the transitivity property. Indeed, when conformity is the equality relationship, the properties of symmetry, reflexivity and transitivity trivially hold. Moreover, if we assume that $A = \langle a_1, \ldots, a_n\rangle$, $B = \langle b_1, \ldots, b_n\rangle$, $C = \langle c_1, \ldots, c_n\rangle$ and $D = \langle d_1, \ldots, d_n\rangle$, one has $(b_i - a_i = d_i - c_i) \equiv (c_i - a_i = d_i - b_i)$ and the exchange of the means also holds.

The transitivity property allows us to define the notion of equivalence class for analogical proportions. Such an equivalence class groups together pairs of points representing the same vector, i.e., ratios of the same value.

Definition 2 (Analogical equivalence class). *Let $\mathcal{D} = \{ D_1, \ldots, D_n\}$ be a set of n dimensions and $\mathcal{P} = \{ (x, y) \mid x, y \in D_1 \times \ldots \times D_n \}$.*
One denotes by $[x, y]$, the analogical equivalence class of (x, y), i.e., the subset of elements (x', y') of $\mathcal{P}$ such that $\overrightarrow{xy} = \overrightarrow{x'y'}$.

Example 1. Let A, B, C, D, E, F, G, H and I be points defined in a n-dimensional space.
Assume that $\overrightarrow{AB} = \overrightarrow{CD}$, and that $\overrightarrow{GH} = \overrightarrow{FD} = \overrightarrow{EI}$.
One has two analogical equivalence classes $[A, B] = \{(A, B), (C, D)\}$ and $[G, H] = \{(G, H), (F, D), (E, I)\}$ that represent the following analogical proportions:

- in $[A, B]$, $A : B :: C : D$ and then $A : C :: B : D$
- in $[G, H]$, $G : H :: F : D$ and then $G : F :: H : D$,
- in $[G, H]$, $G : H :: E : I$ and then $G : E :: H : I$,
- in $[G, H]$, $F : D :: E : I$ and then $F : E :: D : I$.

Equivalence classes provide a more compact view of the analogical proportions that exist in a n-dimensional space by highlighting only the corresponding ratios.

A More Flexible Definition. The modelling of analogical proportions in the setting of relational databases must comply with the properties of the relational model. Each tuple of a relation is an element of the Cartesian product of the active domains of a set of attributes $\{A_1, \ldots, A_m\}$. One assumes here that the active domains are subsets of $\mathbb{R}$. Each tuple t may be represented as an n-dimensional point, denoted by $(t_1, \ldots, t_n)$. Consequently, Definition 1 is well suited to the setting of relational databases: four tuples t_A, t_B, t_C, t_D, of a relation are bound by an analogical relationship if and only if $\overrightarrow{t_A t_B} = \overrightarrow{t_C t_D}$. However, strict equality of vectors may be difficult to obtain when dealing with real-world datasets. It is then necessary to make Definition 1 more flexible by relaxing the equality relationship between the two vectors involved in an analogical relationship. One needs to assess the "distortion" between two vectors, i.e., the extent to which $||\overrightarrow{t_A t_B} - \overrightarrow{t_C t_D}||$ is close to 0.

In order to make the dimensions commensurable when attributes are defined on different domains, we assume that the coordinates of the vectors are normalized and belong to the interval [0, 1]. To this aim, each value v of the active domain of an attribute is replaced by:

$$\frac{v - min_{att}}{max_{att} - min_{att}} \tag{1}$$

where min_{att} and max_{att} denote respectively the minimal value and the maximal value of the attribute domain.

Several strategies may be used to assess the extent to which $||\overrightarrow{t_A t_B} - \overrightarrow{t_C t_D}||$ is close to 0. Different norms may be used, such as the Minkowsky norm (p-norm) that gives the length of the "correction vector" that allows to switch from $\overrightarrow{t_A t_B}$ to $\overrightarrow{t_C t_D}$ [9], or, the infinity norm that gives the maximal coordinate value of this correction vector.

Definition 3 (Analogical dissimilarity). *Let t_A, t_B, t_C, and t_D four n-uples and $\overrightarrow{u} = \overrightarrow{t_A t_B}$, $\overrightarrow{v} = \overrightarrow{t_C t_D}$. We define:*

$$ad_p(\overrightarrow{t_A t_B}, \overrightarrow{t_C t_D}) = ad_p(\overrightarrow{u}, \overrightarrow{v}) = (\sum_{i \in \{1,\ldots,n\}} |u_i - v_i|^p)^{1/p}. \tag{2}$$

With the infinity norm, which is a special case, we have:

$$ad_\infty(\overrightarrow{t_A t_B}, \overrightarrow{t_C t_D}) = ad_\infty(\overrightarrow{u}, \overrightarrow{v}) = max_{i \in \{1,\ldots,n\}} |u_i - v_i|. \tag{3}$$

In all cases, the closer the analogical dissimilarity is to 0, the closer the two vectors are, and the more the analogical proportion is valid.

It is shown in [9] that Definition 3 satisfies the basic properties of analogical proportions.

With such a gradual view of analogical proportion, it is not possible to get a proper definition of equivalence classes due to the lack of a transitivity property. However, the following property makes it possible to define clusters based on analogical proportions (which will be the basis of a query optimization technique described in Section 4).

Theorem 1 (Distance property). *The dissimilarity measure from Definition 3 is a metric.*

The proof is straightforward. Recall, in particular, that the triangular inequality is true for every p-norm.

3 Analogical Queries

The general idea underlying what we call "analogical queries" is to retrieve from a relation those tuples that are involved in an analogical proportion. Five kinds of analogical queries may be thought of:

1. find the tuples in analogical proportion on a given set X of attributes, i.e., find the quadruples (t_a, t_b, t_c, t_d) such that $t_a.X : t_b.X :: t_c.X : t_d.X$ holds with a validity degree at least equal to a specified threshold λ;
2. find the tuples that are in analogical proportion with a given tuple t_a on a given set X of attributes, i.e., find the triples (t_b, t_c, t_d) such that $t_a.X : t_b.X :: t_c.X : t_d.X$ holds with a validity degree at least equal to a specified threshold λ;
3. find the pairs of tuples that are in analogical proportion with two given tuples t_a and t_b on a given set X of attributes: in other words, find the pairs (t_c, t_d) such that $t_a.X : t_b.X :: t_c.X : t_d.X$ holds with a validity degree at least equal to a specified threshold λ;
4. find the tuples that form an analogical proportion with three given tuples t_a, t_b, and t_c on a given set X of attributes: in other words, find the t_d's such that $t_a.X : t_b.X :: t_c.X : t_d.X$ holds with a validity degree at least equal to a specified threshold λ;
5. find the extent to which an analogical proportion between four given tuples (t_a, t_b, t_c, t_d) on a given set X of attributes is true, i.e., compute $1 - dist(\overrightarrow{t_a t_b}, \overrightarrow{t_c t_d})$.

From a syntactic point of view, an analogical query must specify i) the relation concerned and the attributes to be returned; ii) the attributes on which the analogical proportion must hold; ii) the threshold considered. Hereafter, we use a syntax *à la SQL* for expressing the five types of queries listed above.

Type 1: `find` x, y, z, t `projected on` A_π
`from` r

where (x is to y) as (z is to t) according to A_α
with threshold λ

where the set of attributes A_π is assumed to include a key of r (in order to identify the objects that are involved in the analogical proportion) and A_α is the set of attributes on which the analogical proportion must hold.

Type 2: find x, y, z projected on A_π
from r
where (x is to y) as (z is to $K = k_1$) according to A_α
with threshold λ

where K is assumed to be the key of relation r.

Type 3: find x, y projected on A_π
from r
where (x is to y) as ($K = k_1$ is to $K = k_2$) according to A_α
with threshold λ.

Type 4: find x projected on A_π
from r
where (x is to $K = k_1$) as ($K = k_2$ is to $K = k_3$)
according to A_α
with threshold λ.

Type 5: find validity in r
of ($K = k_1$ is to $K = k_2$) as ($K = k_3$ is to $K = k_4$)
according to A_α.

In the following section, we describe three query evaluation strategies suited to analogical queries. Due to space limitation, we only illustrate these methods on type 3 queries, but they can be straightforwardly adapted to the four other kinds of analogical queries.

4 Query Processing

Three strategies are presented hereafter: i) a "naive" one based on nested loops; ii) a method exploiting classical indexes on some attributes involved in the analogical proportion targeted; iii) a strategy exploiting an index structure referencing clusters of tuples in analogical proportion. In the following, it is assumed that the attribute values are normalized (cf. Equation 1) and the definition of analogical dissimilarity *ad* is based on the infinity norm (cf. Equation 3). Then, the validity of the analogical proportion $(x : y :: z : t)$ is defined as $1 - ad(\overrightarrow{xy}, \overrightarrow{zt})$ and it belongs to the interval $[0, 1]$.

4.1 Naive Strategy

The naive evaluation strategy relies on sequential scans of the relation, using nested loops. For instance, a type 3 query leads to a data complexity in $\theta(n^2)$ where n is the cardinality of the relation concerned, see Algorithm 1.

```
Data: t_a, t_b, λ
Result: Set S of pairs of tuples (t_c, t_d) in analogical proportion with t_a : t_b
S ← ∅;
for each tuple t_c of r do
    for each tuple t_d of r do
        if 1 − ad(→t_a t_b, →t_c t_d) ≥ λ then
            S := S ∪ {(t_c, t_d)}
        end
    end
end
```

Algorithm 1. Naive algorithm for type 3 queries

4.2 Classical-Index-Based Strategy

The idea is to exploit a property of the infinite norm in order to limit the number of disk accesses. Let us consider a type 3 query based on the condition:

$$ad_{\infty}(\overrightarrow{xy},\ \overrightarrow{t_a t_b}) \leq 1 - \lambda \tag{4}$$

where t_a (resp. t_b) is the tuple whose key is equal to k_1 (resp. k_2). Let us denote: $\overrightarrow{t_a t_b} = \langle u_1,\ u_2,\ \ldots,\ u_p \rangle$. Tuple x (resp. y) is represented by $\langle x_1,\ x_2,\ \ldots,\ x_p \rangle$ (resp. $\langle y_1,\ y_2,\ \ldots,\ y_p \rangle$). Thus, $\overrightarrow{xy} = \langle y_1 - x_1,\ y_2 - x_2,\ \ldots,\ y_p - x_p \rangle$. According to Equation 3, we have:

$$\begin{aligned}(4) &\Leftrightarrow max_{i \in \{1,\ldots,p\}} |u_i^* - (y_i^* - x_i^*)| \leq 1 - \lambda \\ &\Leftrightarrow \forall i \in \{1, \ldots, p\},\ |u_i^* - (y_i^* - x_i^*)| \leq 1 - \lambda.\end{aligned} \tag{5}$$

where x_i^* (resp. u_i^*) represents the normalized value of attribute A_i in the tuple x (resp. u), cf. Formula 1. Now, let us consider an attribute A_k of domain $[min_k,\ max_k]$. We have:

$$(4) \Rightarrow \frac{|u_k - (y_k - x_k)|}{max_k - min_k} \leq 1 - \lambda \Rightarrow |u_k - (y_k - x_k)| \leq 1 - \lambda' \tag{6}$$

where $\lambda' = \lambda * (max_k - min_k)$. Then:

$$(4) \Rightarrow u_k - 1 + \lambda' + x_k \leq y_k \leq u_k + 1 - \lambda' + x_k. \quad (7)$$

Now, if one has available an index I_k on attribute A_k, one may avoid a data complexity in $\theta(n^2)$ by using two nested loops, as described in Algorithm 2. The idea is to first filter (using the index) the pairs of tuples so as to retain those which satisfy the analogical proportion on attribute A_k, then to scan these pairs and check the analogical proportion on the other attributes from A_α.

Data: t_a, t_b, λ
Result: Set $\mathcal{S}$ of pairs of tuples (t_c, t_d) in analogical proportion with $t_a : t_b$
$\mathcal{S} \leftarrow \emptyset$;
for *each entry* v *of* I_k **do**
  H_{1k} := set of tuple addresses associated with v;
  for *each entry* v' *of* I_k *s.t.* $u_k - 1 + \lambda' + v \leq v' \leq u_k + 1 - \lambda' + v$ **do**
    H_{2k} := set of tuple addresses associated with v';
    $H_{3k} := H_{3k} \cup (H_{1k} \times H_{2k})$;
  end
end
$A_\beta \leftarrow A_\alpha - A_k$;
$\mathcal{S} \leftarrow \{(t_c, t_d) \in H_{3k} \mid ad(\overrightarrow{t_a.A_\beta\, t_b.A_\beta}, \overrightarrow{t_c.A_\beta\, t_d.A_\beta}) \leq 1 - \lambda\}$;

Algorithm 2. Index-based algorithm for type 3 queries

If several attributes are indexed, the potential gain is even more important. Let us denote by *Ind* the set of attributes for which an index is available. One builds H_{3i} for all i such that $A_i \in Ind$. Then, one computes the intersection of these sets, one accesses the corresponding pairs of tuples and one checks whether condition 7 holds for the remaining attributes. If so, the pair of tuples is added to the result.

4.3 Cluster-Based Strategy

The idea is to create clusters grouping together the pairs of tuples that are in analogical proportion. The metric underlying the clustering process is ad_∞ (cf. Formula 3). The criteria related to the choice of the clustering algorithm are the following:

- it must be incremental (since the database may evolve and one must avoid performing the clustering process each type an analogical query has to be evaluated);
- the number of clusters does not have to be known a priori.

A good candidate is the crisp (nonfuzzy) version of the algorithm named *l-fcdmed-select* proposed in [8], itself based on the algorithm introduced in [5]. The clusters obtained may then be used as an index structure, as illustrated hereafter. First, we propose to have an access to the clusters through a relation *ClusterTable* (see Table 1) whose key is denoted by cid. Each tuple of this relation represents an element in the cluster identified by cid.

Table 1 Cluster table

cid	*x*	*y*
1	rowid1	rowid2
1	rowid5	rowid6
1	rowid4	rowid9
2	rowid1	rowid3
2	rowid7	rowid10
3	rowid6	rowid7

We also assume the presence of two tables giving respectively the maximal distortion (ad) value inside a cluster (Table *MaxDist* of schema (*cid*, *maxdist*)) and the minimal distortion value between the elements of every pair of clusters (Table *MinDist* of schema (cid_1, cid_2, *mindist*)).

Let us consider again the query "find the pairs of tuples that are in analogical proportions with (t_a, t_b), with a validity degree at least equal to $\lambda = 0.3$". Tuples t_a and t_b may belong to several clusters. However, by definition of the clusters, the pair (t_a, t_b) belongs to only one cluster. To find it, one just has to intersect the set of clusters associated with t_a and that associated with t_b. Let c_{ab} be the cluster containing (t_a, t_b) (if any). Then, if $(1 - \lambda)$ — which corresponds to the maximal dissimilarity value accepted by the user — is greater than the maximal dissimilarity value associated with the cluster, every pair of tuples present in the cluster belongs to the answer. In the opposite case, the pairs of tuples in the cluster must be filtered so as to retain only the satisfactory ones. In both cases, one has also to check the clusters that may be close enough to (t_a, t_b), i.e., those clusters whose minimal dissimilarity value with respect to c_{ab} is less than $1 - \lambda$. The different steps of the processing are described in Algorithm 3.

5 Experimentation

The main objective of the preliminary experimentation described hereafter was to assess the respective performances of the evaluation methods described in the previous section. The experimentation was carried out using a laptop with an Intel Core i5-2520M CPU @ 2.50 GHz, and 4 Gb of RAM.

Concerning the cluster-based method, the data was organized as shown in Table 2, where *vector* corresponds to the vector id, *cluster* indicates the cluster that contains the vector considered, t_a and t_b are the extremities of the vector, and $x_1, \ldots, x_n$ represent its dimensions (Table 2 corresponds to a 2-dimensional case). For instance, the third line of Table 2 expresses that vector 2143, formed from the elements 24 and 67, belongs to cluster 1 and its value for the x_1 and x_2 dimensions are -4.2 and -10.69 respectively.

Data: t_a, t_b, λ
Result: Set $\mathcal{S}$ of pairs of tuples $(t_c,\ t_d)$ in analogical proportion with $t_a : t_b$
c := select cid from *ClusterTable*
where $(x = rowid(t_a)$ and $y = rowid(t_b))$
or $(x = rowid(t_b)$ and $y = rowid(t_a))$;
max := select $maxdist$ from *Max_ad_Table* where $cid = c$;
if $max \leq 1 - \lambda$ **then**
| $\mathcal{S}$:= select x, y from *ClusterTable* where $cid = c$;
else
| $\mathcal{S}$:= select x, y from *ClusterTable*
| where $cid = c$ and $ad(x,\ y,\ rowid(t_a),\ rowid(t_b)) \leq 1 - \lambda$;
end
C := (select cid_2 from *Min_ad_Table* where $cid_1 = c$ and $min\_ad \leq 1 - \lambda) \cup$
(select cid_1 from *Min_ad_Table* where $cid_2 = c$ and $min\_ad \leq 1 - \lambda)$;
for *each cluster c' in C* **do**
| $\mathcal{S} := \mathcal{S} \cup$ (select x, y from *ClusterTable*
| where $cid = c'$ and $ad(x,\ y,\ rowid(t_a),\ rowid(t_b)) \leq 1 - \lambda)$;
end

Algorithm 3. Clustering-based algorithm for type 3 queries

Table 2 Structure used by the cluster-based method

vector	*cluster*	t_a	t_b	x_1	x_2
105	0	1	201	−2.55	−0.65
4702	0	64	86	−4.26	−0.88
2143	1	24	67	−4.2	−10.69
883	1	9	74	0.23	−15.34
2736	2	31	100	12.54	−25.89
5452	2	87	100	13.67	−28.9

In order to optimize query processing, we also defined: i) two hash indexes on attributes t_a and t_b respectively; ii) a hash index on attribute *vector*; iii) a b-tree index on the attributes *cluster*, x_1, and x_2.

In the following, Algorithm 1 is denoted by *naive*, and Algorithm 2 by *CIB* (for classical-index-based). As to the method based on clustering (Algorithm 3), we use five variants: *cl* when no index is used; *cl-v* when a hash index on attribute *vector* is used; *cl-el* when two hash indexes on t_a and t_b respectively are used; *cl-b* when a b-tree on the attributes *vector*, x_1, and x_2 is exploited; and *cl-t* when all of the previously mentioned indexes are used together.

5.1 *Real-World Dataset*

The first dataset we used shows the votes during the first round of the French Presidentials elections in 2012[1], aggregated by département. We only kept the 6 most voted political parties. An extract of this dataset is shown in Table 3, where *vote 1* refers to *Les Verts*; *vote 2* to *UMP*; *vote 3* to the *Front de Gauche*; *vote 4* to *Union pour la Démocratie Française*; *vote 5* to *Debout la République*; and *vote 6* to the *Parti Socialiste*.

Table 3 Votes for some departments

Id	*name*	*vote 1*	*vote 2*	*vote 3*	*vote 4*	*vote 5*	*vote 6*
8	Ardennes	24.5	24.43	9.28	7.52	1.81	28.93
22	Côtes D'Armor	13.58	23.86	12.2	10.6	1.71	33.02
32	Gers	15.9	24.14	12.06	9.95	1.82	31.86
42	Loire	21.55	25.07	11.18	9.75	2.11	26.46
73	Savoie	18.92	28.61	11.47	9.89	2.11	23.64

Table 4 results obtained with a real-world dataset

λ	*naive*	*CIB*	*cl*	*cl-v*	*cl-el*	*cl-b*	*cl-t*
0.9	16.4	13.2	221	209	206	200	211
0.8	18.5	14.6	254	255	238	217	241
0.7	24.1	22	322	288	296	250	260
0.5	29.2	33.5	358	342	327	283	269
0.2	51.2	45	367	369.9	360	269	284

Table 4 shows the results obtained. Each time value corresponds to an average computed over 10 different queries. We see that *cl-b*/*cl-t* clearly outperform *cl*, but both *cl-b* and *cl-t*'s results are much worse than those obtained with the *naive* and *C-I-B* methods. The *C-I-B* strategy is somewhat faster than the *naive* one. The bad performance of the cluster-based processing technique is due to the fact that the data cannot be clustered into well-separated groups, so most of the clusters have to be accessed for each query.

5.2 *Synthetic Dataset*

We then generated a set of 1,124,400 tuples, making sure they could be separated into 10 clearly distinct clusters. Doing this, we were able to set the minimal distance between each pair of clusters, which is used by Algorithm 3.

---

[1] `https://www.data.gouv.fr/fr/datasets/`

Table 5 Results obtained with synthetic data

λ	*nb tuples*	*naive*	*C-I-B*	*cl-t*
0.95	400	1323	234	108
0.9	111,147	2113	592	1065
0.8	119,409	2126	1175	1235
0.5	805,649	5607	2056	4448
0.2	1,122,888	8317	3507	7760

The results appear in Table 5 which shows, for each value of λ, the number of returned tuples, and the times in milliseconds related to the *naive*, the *C-I-B*, and the *cl-t* methods. Again, each of these numbers corresponds to an average computed over 10 queries.

When λ is high, only one cluster is accessed, and the cluster-based strategy is much faster than the *naive* one. However, the processing time increases dramatically as the number of accessed clusters grows. One may also notice that the *cl-t* method is always better than the *naive* one. On the other hand, *cl-t* is better than *C-I-B* only when a single cluster has to be accessed (this is the case of the queries with a λ value equal to 0.95 (which corresponds to a maximal analogical dissimilarity value equal to 0.05), knowing that the minimal inter-cluster distance is 0.1).

From these experimental results, it appears that the clustering-based method is interesting only in the presence of well-separated clusters, with large enough distances between the groups, and a high value of λ specified in the query. When the underlying structure of the dataset is unknown, the *C-I-B* strategy is the best choice.

6 Conclusion

In this paper, we have introduced a new type of database query inspired from some works in AI about the concept of analogical proportion. The general idea is to retrieve the tuples that participate in a relation of the form "*a* is to *b* as *c* is to *d*". We have provided a typology of analogical queries in a relational database context, devised different processing strategies and assessed them experimentally. The most important result is that analogical queries are tractable and may be optimized thanks to simple classical indexes.

In terms of perspectives, the most important one is to keep investigating query processing techniques suitable for analogical queries as the experimental results reported here are still limited. In particular, it would be worth studying whether some more sophisticated cluster-based indexes (taking into account the typicality of the elements in a cluster) could improve the performances.

Aknowledgement This work was partially funded by a grant from the Region Brittany and the Conseil Général des Côtes-d'Armor.

References

1. Beltran, W.C., Jaudoin, H., Pivert, O.: Analogical prediction of null values: the numerical attribute case. In: Manolopoulos, Y., Trajcevski, G., Kon-Popovska, M. (eds.) ADBIS 2014. LNCS, vol. 8716, pp. 323–336. Springer, Heidelberg (2014)
2. Gentner, D.: Bootstrapping the mind: Analogical processes and symbol systems. Cognitive Science **34**(5), 752–775 (2010)
3. Hofstadter, D.: A review of mental leaps: Analogy in creative thought. AI Magazine **16**(3), 75–80 (1995)
4. Kling, R.: A paradigm for reasoning by analogy. Artif. Intell. **2**(2), 147–178 (1971)
5. Krishnapuram, R., Joshi, A., Nasraoui, O., Yi, L.: Low-complexity fuzzy relational clustering algorithms for web mining. IEEE T. Fuzzy Systems **9**(4), 595–607 (2001)
6. Lepage, Y.: Solving analogies on words: An algorithm. In: Proc. of COLING-ACL 1998, pp. 728–735 (1998)
7. Lepage, Y.: (Re-)discovering the graphical structure of Chinese characters. In: SAMAI (workshop colacated with ECAI), pp. 57–64 (2012)
8. Lesot, M.-J., Revault d'Allonnes, A.: Credit-card fraud profiling using a hybrid incremental clustering methodology. In: Hüllermeier, E., Link, S., Fober, T., Seeger, B. (eds.) SUM 2012. LNCS, vol. 7520, pp. 325–336. Springer, Heidelberg (2012)
9. Miclet, L., Bayoudh, S., Delhay, A.: Analogical dissimilarity: Definition, algorithms and two experiments in machine learning. J. Artif. Intell. Res. (JAIR) **32**, 793–824 (2008)
10. Miclet, L., Prade, H.: Handling analogical proportions in classical logic and fuzzy logics settings. In: Sossai, C., Chemello, G. (eds.) ECSQARU 2009. LNCS, vol. 5590, pp. 638–650. Springer, Heidelberg (2009)
11. Prade, H., Richard, G. (eds.): Computational Approaches to Analogical Reasoning: Current Trends. Studies in Computational Intelligence, vol. 548. Springer (2014)
12. Ramscar, M., Yarlett, D.: Semantic grounding in models of analogy: an environmental approach. Cognitive Science **27**(1), 41–71 (2003)

Improving Hadoop Hive Query Response Times Through Efficient Virtual Resource Allocation

Tansel Dokeroglu, Muhammet Serkan Cınar, Seyyit Alper Sert, Ahmet Cosar and Adnan Yazıcı

Abstract The performance of the MapReduce-based Cloud data warehouses mainly depends on the virtual hardware resources allocated. Most of the time, the resources are values selected/given by the Cloud service providers. However, setting the right virtual resources in accordance with the workload demands of a query, such as the number of CPUs, the size of RAM, and the network bandwidth, will improve the response time when querying large data on an optimized system. In this study, we carried out a set of experiments with a well-known Mapreduce SQL-translator, Hadoop Hive, on benchmark decision support the TPC benchmark (TPC-H) database in order to analyze the performance sensitivity of the queries under different virtual resource settings. Our results provide valuable hints for the decision makers who design efficient MapReduce-based data warehouses on the Cloud.

Keywords Hadoop · Hive · Virtual resource allocation · Multi-objective query optimization

1 Introduction

Cloud computing is a new computation paradigm that builds elastic and scalable software systems. Amazon, Google, Microsoft, Salesforce, provide several options for computing infrastructures, platforms, and software systems [1][2][3]. Users pay all costs associated with hosting and querying their data. Recently, extensive academic and commercial research is undertaken to construct self-tuning, efficient, and resource-economic Cloud database services that serve to the benefits of both the

T. Dokeroglu(✉) · M.S. Cınar · S.A. Sert · A. Cosar · A. Yazıcı
Computer Engineering Department of Middle East Technical University,
Universities Street, 6800 Cankaya, Ankara, Turkey
e-mail: {tansel,alper,cosar,yazici}@ceng.metu.edu.tr, mscinar@hacettepe.edu.tr
http://www.ceng.metu.edu.tr

T. Andreasen et al. (eds.), *Flexible Query Answering Systems 2015*,
Advances in Intelligent Systems and Computing 400,
DOI: 10.1007/978-3-319-26154-6_17

customers and the vendors [5][6]. Virtualization that provides the illusion of infinite resources in many respects is the main enabling technology of Cloud computing [7]. This technique is being used to simplify the management of physical machines and provide efficient systems. The perception of hardware and software resources is separated from the actual implementation and the virtual resources are mapped by Cloud to real physical resources. Through mapping virtual resources to physical ones as needed, the virtualization can be used by several databases that are located on physical servers to share and change the allocation of resources according to query workloads [9]. This capability of virtualization provides efficient Cloud Data Warehouses (DW) where each Virtual Machine (VM) has its own operating system and resources (CPU, main memory, network bandwidth, etc.) that are controlled by using a VM Monitor (VMM) [10].

Conventional Data Warehouse (DW) design techniques seek to assign data tables/fragments to a given static database hardware setting optimally. However; it is now possible to use elastic virtual resources provided by the Cloud environment, thus achieve reductions in both the execution time and the monetary cost of a DW system within predefined budget and response time constraints. Finding an optimal virtual resource assignment plan for database tables to VMs for this design problem and providing cost-efficient Cloud DWs in terms of query workload response time and the total ownership price of virtual resources (CPU and/or cores, RAM, hard disk storage, network bandwidth, and disk I/O bandwidth) is known as an NP-Hard problem. In this study, we analyze the performance of a distributed and MapReduce-based DW Hadoop Hive [11, 12, 14] with various virtual resource settings and present our results to design a multi-objective (cost-efficient and high-throughput) DW on the Cloud. The contribution of this work is being the first study that concerns the performance improvement issues of a MapReduce-based DW from the perspectives of both the virtual resource assignment and monetary issues. This is a multi-objective query optimization problem and our approach employs the concept of pareto optimality for alternative solutions of the problem. To the best of our knowledge, this is the first study that concerns the performance issues of a MapReduce-based DW from the perspectives of both the virtual resource assignment and monetary issues. With our proposed method, one can decide on the priority of the virtual resources to be invested for a higher performance multi-objective Cloud DW.

The organization of the paper is as follows: In the second section we give detailed information about the related studies and the pricing scheme of Cloud service providers is explained, in section 3, Hadoop Hive main framework is explained. Section 4 presents our experimental setup and pareto curve for the solution of the problem. Section 5 gives the concluding remarks.

2 Virtual Resource Assignment for Databases

In this section, we give information about some of the studies related to our work. To the best of our knowledge, there is no approach like ours that consider both the optimization of the budgetary concerns and the performance of the queries by taking into account alternative virtual resource allocation and the performance of the MapReduce-based DW queries. Cloud computing has changed the conventional ways of database services accessed by the users. Cloud service providers now provide infrastructure resources and computing capabilities as a service to the users. Allocating resources intelligently to database clients is an important problem while satisfying the client Service Level Agreements (SLAs). Cloud data management services is one of the most significant components of the new model. With this infrastructure, the clients need to optimize both the response time of their databases and the monetary profits. A Virtual Machine (VM) consists of computing resources such as CPU(s), memory, disk(s), and network resources. This is the typical way of usage for most of the Clouds providers. The containers constitute the virtual infrastructure of the application. The Cloud offers data storage. VMs transfer data from these storage resources and use it for processing.

Mariposa was the earliest distributed database system that implements an economic paradigm to solve many drawbacks of wide-area network cost-based optimizers [16]. Clients and servers own an account and users allocate a budget to each of their queries in Mariposa. The processes aim to finish the query in a limited budget by executing portions of it on various sites. A set of similar works are proposed for the problem [17][18]. In a study by Soror et al., an advisor automatically configures a set of VMs for database workloads where the advisor requests domain knowledge [9]. This study concerns with the relational database virtual resource allocations not a MapReduce SQL paradigm like ours and uses the underlying query optimizer to evaluate the alternatives. It does not optimize the monetary issues. Recently, several efficient cost models are proposed for scheduling with regard to monetary cost and/or completion time of the queries on the Cloud [6][5][19][24]. In [20], a VM usage policy is proposed. There is a function of the execution time of the tasks that are submitted. The function permits decisions regarding the amount of virtual resources allocated.

In [25], the cost performance tradeoffs of different execution and resource provisioning plans are simulated, showing that by provisioning the right amount of storage and computing resources, the cost can be reduced significantly. The performance of three workflow applications with different I/O, memory, and CPU requirements has also been compared on Amazon EC2 and a typical high-performance cluster (HPC) to identify what applications achieve the best performance in the Cloud at the lowest cost [19]. Recent research takes interest in various aspects of database and decision support technologies in the Cloud. Different studies investigate the storage and processing of structured data [28], the optimization of join queries, and how to support analysis operations such as aggregation [14]. Cloud data warehousing and OLAP systems also raise various problems related to storage and query performance [29]. Adaptations of these technologies to the Cloud are addressed in [31], or the calculation of OLAP cuboids using the MapReduce runtime environment [32]. There has been a great amount of work for tuning database systems for specific workloads or execution environments [22] [23] [33]. In this study we tune the virtual resources, rather than tuning the data files for given resource settings.

2.1 Pricing Scheme of Cloud Service Providers

In this section, we describe the pricing scheme of a Cloud Service Provider. The Cloud infrastructure provides unlimited amount of storage space, CPU nodes, RAM, and very high speed intra-cloud networking. In our study, all the resources of the Cloud are assumed to be on a network. The CPU nodes, RAM, and I/O bandwidth of each VM are different from each other and can be deployed by using VM Monitors in milliseconds [7]. There are several Cloud Service Providers in the market and they offer different pricing schemes for the services they provide. Every different pricing schema of Cloud server providers can be an opportunity for customers in accordance with the tasks they want to complete. Our scheme is similar to Window's Azure [3]. Configurations such as Extra Small, Small, and Medium VMs are provided by the Cloud service providers. The cost for a small VM (1GHz CPU, 768MB RAM) is $0.02/hr, whereas an A7 (8 x 1.6GHz CPU, 56GB RAM) costs 80 time more, $1.64/hr. Our results show that using few small VMs it is possible to match the performance of large & expensive configurations at a small fraction of their cost.

In our study, we constitute various virtual resource settings, observe their performances, and present some results that will help decision makers design more cost-efficient DW systems.

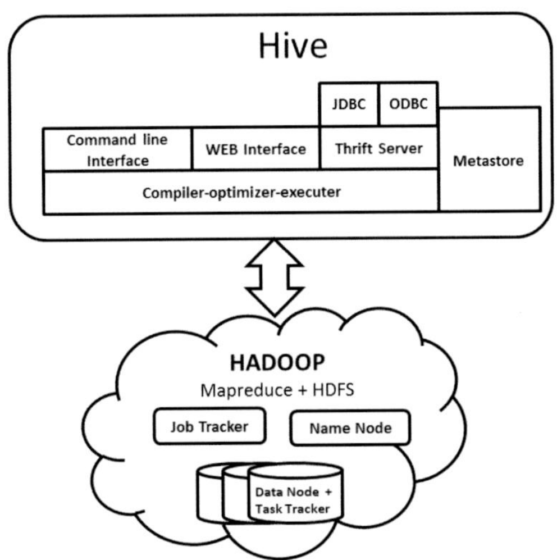

Fig. 1 Main framework of Hadoop Hive.

3 Hadoop Hive and the Querying Language HiveQL

Hadoop is an open source framework that can process big data by using its distributed file infrastructure [4, 11, 13–15]. Hadoop Distributed File System (HDFS) is the underlying file system of Hadoop. It uses MapReduce paradigm to divide the tasks into several parts to provide horizontal scalability (see Figure 1). The input is retrieved from HDFS and assigned to one of the *mappers* that will process data in parallel and produce key-value pairs for the reduce step. Due to its simplicity and efficiency Hadoop has gained significant support from both industry and academia [21]. Hadoop has a MapReduce engine that consists of a JobTracker where MapReduce jobs are submitted. The JobTracker assigns jobs to TaskTracker nodes in the cluster. The JobTracker knows where the related data is located. First, the work is assigned to the actual node where the data resides otherwise it is assigned to the node nearby. This strategy is used to reduce the communication traffic in the network. If a part of the job is not completed due to any reason then it is rescheduled.

Hive is an open source distributed DW providing an SQL-like abstraction on top of Hadoop framework [26]. Hive can execute SQL-based queries on MapReduce that uses (*key*,*value*) pairs. By using Hive, users can benefit from both MapReduce and SQL [8, 30]. The attributes of Hive make it a suitable tool for DW applications where large scale data is analyzed and fast response times are not required. HiveQL is the SQL-like querying language used by Hive that strictly follows the SQL-92 standard. HiveQL gives basic support for indexes and does not support for transactions and materialized views.

4 Evaluation of Performance Results

We carry out our experiments on a private Cloud server, 4U DELL PowerEdge R910 having 32 (64 with Hyper Threading) cores. Each core is Intel Xeon E7-4820 with 2.0GHz processing power. The server has 128GB DDR3 1600MHz virtualized memory and Broadcom Nextreme II 5709 1Gbps NICs. Operating system of the physical server is Windows Server 2012 Standard Edition. 20 Linux CentOS 6.4 virtual machines are installed on this server as guest operating systems. Each virtual machine has 2.0GHz processors and 250GB disk storage. An additional master node is used as NameNode/JobTracker. The latest stabilized versions of Hadoop, release 1.2.1 and Hive version 0.12.0 are used [26]. The splitsize of the files (HDFS block size) is 64MB, replication number is 2, maximum number of map tasks is 2, maximum number of reduce tasks is 2 and map output compression is disabled during the experiments. The Cloud server is dedicated to only our experiments. Therefore, the performance interference from external factors such as network congestion or OS-level contention on shared resources are minimized as much as possible.

TPC-H decision support benchmark 1/10GB databases are chosen as our DW environments [34]. Single standard queries of TPC-H Q1, Q3, Q5, Q6, Q9, Q10, Q14, Q19 are used. The queries are demanding different virtual resources for their executions. Sample HiveQL queries can be seen at the Appendix [34]. Various virtual resource settings and the execution times of the queries that we performed on them are presented in Tables 1 and 2. Figure 2 presents the execution times of the queries with different settings. Figures 3 and 4 present the performance increase of the queries in accordance with the capacity improvements.

Table 1 Experiments with various virtual resource settings for 1GB TPC-H database

#	# of CPUs	RAM	Net.band.	Q1	Q3	Q5	Q6	Q9	Q10	Q14	Q16	Q19
1	1	2GB	10Mbit	232	480	591	234	1851	557	322	359	270
2	1	2GB	100Mbit	98	216	253	62	471	249	100	196	114
3	1	2GB	200Mbit	90	202	253	49	381	225	100	196	104
4	2	1GB	100Mbit	124	183	239	52	401	249	167	171	85
5	2	2GB	200Mbit	73	163	199	41	368	177	71	136	77
6	2	4GB	200Mbit	72	156	193	40	352	170	70	134	81

Table 2 Experiments with various virtual resource settings for 10GB TPC-H database

#	# of CPUs	RAM	Net.band.	Q1	Q3	Q5	Q6	Q9	Q10	Q14	Q16	Q19
7	1	2GB	10Mbit	694	1220	1653	691	4871	1506	651	810	1082
8	1	2GB	100Mbit	162	319	479	113	929	380	162	268	236
9	1	2GB	200Mbit	138	239	443	94	834	318	140	245	217
10	2	1GB	100Mbit	226	495	750	116	1047	537	274	288	356
11	2	2GB	200Mbit	95	215	300	69	641	254	95	201	153
12	2	4GB	200Mbit	94	211	292	67	616	237	94	189	152

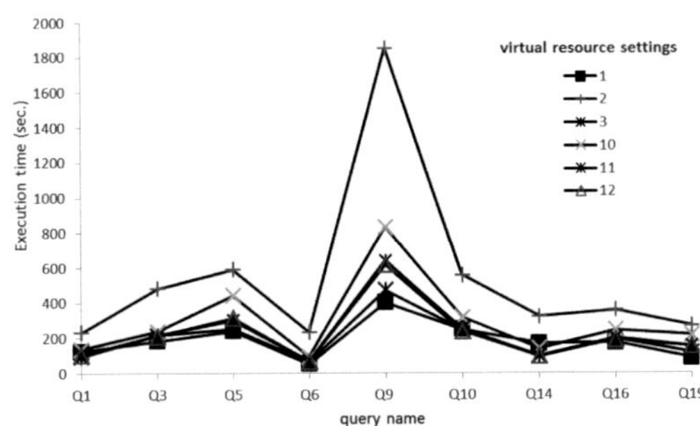

Fig. 2 Execution performance of the queries with several virtual resource settings).

Increasing the number of CPUs (from 1 to 2) improves the performance of the queries by 21.2% and 24.7% for 1GB and 10GB databases respectively. Increasing the network bandwidth (from 100Mbit to 200Mbit) improves the performance by 8.3% 13.5% and increasing the RAM capacity (from 1GB to 2GB) improves the response times of the queries 21.8% and 25.3% on the average. The largest improvements are

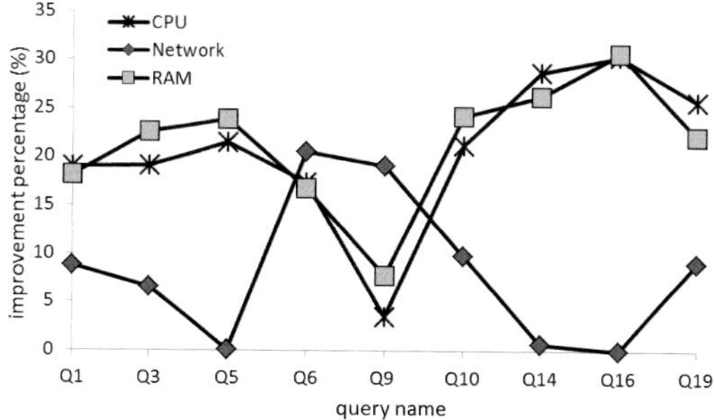

Fig. 3 1GB tpc-h database performance query improvements (increased from 1 to 2 CPUs, 2GB to 4GB RAM, and 100Mbit to 200Mbit network bandwidth).

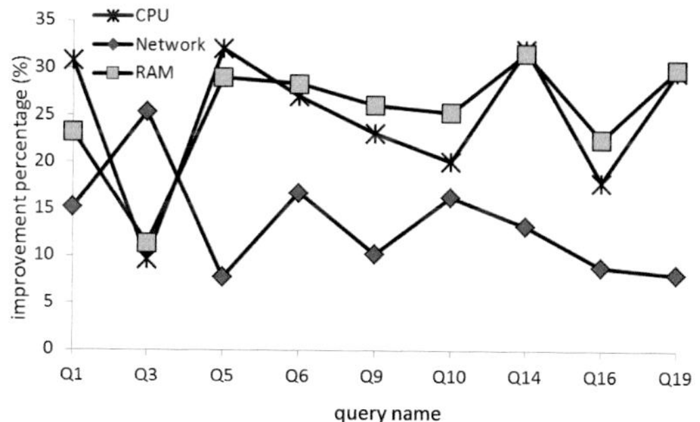

Fig. 4 10GB TPC-H database performance query improvements (increased from 1 to 2 CPUs, 2GB to 4GB RAM, and 100Mbit to 200Mbit network bandwidth).

observed to be within the number of CPUs and the RAM capacity in accordance with the degree of capacity increase ratios. However, DW with larger sizes tend to consume larger network bandwidth with respect to the smaller DWs. Therefore, for a Cloud virtual resource allocation scheme such as our experimental environment, investing the budget to the CPU and RAM seems to be a wiser solution. Of course, the nature of the query affects the resources to be consumed by the virtual environment. As it can be seen from the Figures 3 and 4, some of the queries gain less performance increases although they have higher configurations for the virtual resource settings (for 1GB database query Q9 and for 10GB query Q3 does not gain much performance increase from the new settings). This is due to the changing virtual resource demand of the queries. Some queries demand for CPU and RAM power whereas others are more I/O intensive. In some virtual environment settings, it is possible to allocate I/O bandwidth to the queries. It can also be an interesting area of research.

In Figure 5, the query execution and the monetary cost of the executions are presented on a pareto curve. Although the most expensive virtual resource setting gives the fastest query response time, other solutions giving priority to CPU and RAM

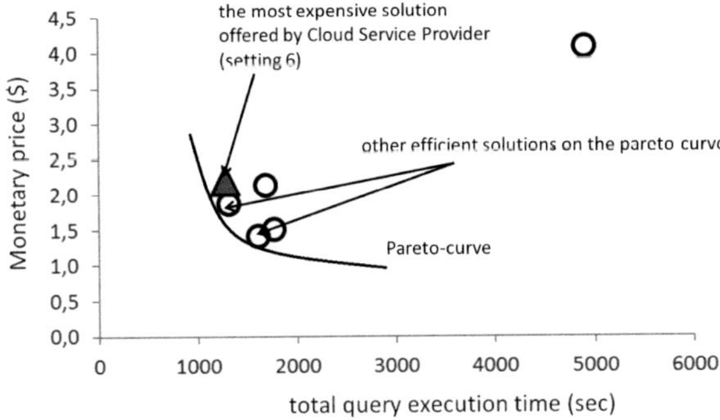

Fig. 5 Pareto curve for the multi-objective query optimization of the query workload 1GB tpc-h.

resources are observed with small response time reductions but lower monetary costs than the default settings of the Cloud Service Provider.

Adding a simple weighted averaging scheme on the non-dominated solution would be a nice way for the set of pareto optimal solutions. However, these α and β values need to be well adjusted for each different computation environment.

5 Conclusions and Future Work

In this study, we analyze the performance of MapReduce-based query execution on Hadoop Hive with various virtual resource settings and observed the performance results under each setting. The distributed environment of Hadoop Hive benefits from more CPU power and RAM capacity than network bandwidth for higher performances. By increasing network bandwidth we can improve the performance of the queries between 8.3%-13.5%, whereas CPU and RAM improve query response time by 21.2%-24.7% and 21.8%-25.3% on the average respectively when the capacities are doubled. However, DWs with larger sizes tend to consume larger network bandwidth with respect to the smaller DWs. On a pareto curve that we obtained from the query execution times and the monetary costs of the query workloads, we observe that there are solutions with execution times that have slight reductions in time but larger gains from the monetary costs.

For future work, we plan to develop an algorithm that predicts the future workload of Hadoop Hive queries and dynamically allocate (near-) optimal virtual resources by using machine learning techniques.

Acknowledgement This study is supported in part by a research grant from TUBITAK with Grant No. 114R082.

References

1. Amazon Web Services (AWS). aws.amazon.com (last accessed September 5, 2014)
2. Google App Engine. `http://code.google.com/appengine/` (last accessed September 5, 2014)
3. Windows Azure Platform. microsoft.com/windowsazure/ (last accessed September 5)
4. Apache Hadoop. `http://hadoop.apache.org/` (last accessed May 1, 2015)
5. Kantere, V., Dash, D., Francois, G., Kyriakopoulou, S., Ailamaki, A.: Optimal service pricing for a cloud cache. IEEE Transactions on Knowledge and Data Engineering **23**(9), 1345–1358 (2011)
6. Kllapi, H., Sitaridi, E., Tsangaris, M.M., Ioannidis, Y.E.: Schedule optimization for data processing ows on the cloud. In: Proceedings of the ACM SIGMOD International Conference on Management of Data, pp. 289–300 (2011)
7. Barham, P., Dragovic, B., Fraser, K., Hand, S., Harris, T., Ho, A., Warfield, A.: Xen and the art of virtualization. ACM SIGOPS Operating Systems Review **37**(5), 164–177 (2003)
8. Thusoo, A., Sarma, J.S., Jain, N., Shao, Z., Chakka, P., Zhang, N., Murthy, R.: Hive-a petabyte scale data warehouse using hadoop. In: ICDE, pp. 996–1005 (2010)
9. Soror, A.A., Minhas, U.F., Aboulnaga, A., Salem, K., Kokosielis, P., Kamath, S.: Automatic virtual machine configuration for database workloads. ACM Transactions on Database Systems (TODS) **35**(1), 7 (2010)
10. Aboulnaga, A., Amza, C., Salem, K.: Virtualization and databases: state of the art and research challenges. In: Proceedings of the 11th International Conference on Extending Database Technology: Advances in Database Technology, pp. 746–747 (2008)
11. Dokeroglu, T., Ozal, S., Bayir, M.A., Cinar, M.S., Cosar, A.: Improving the performance of Hadoop Hive by sharing scan and computation tasks. Journal of Cloud Computing **3**(1), 1–11 (2014)
12. Dokeroglu, T., Sert, S.A., Cinar, M.S.: Evolutionary multiobjective query workload optimization of Cloud data warehouses. The Scientific World Journal (2014)
13. Dean, J., Ghemawat, S.: MapReduce: simplified data processing on large clusters. Communications of the ACM **51**(1), 107–113 (2008)
14. Condie, T., Conway, N., Alvaro, P., Hellerstein, J.M., Elmeleegy, K., Sears, R.: Mapreduce online. In: Proc. of the 7th USENIX Conf. on Networked Systems Design and Implementation (2010)
15. Stonebraker, M., et al.: MapReduce and parallel DBMSs: friends or foes. Communications of the ACM **53**(1), 64–71 (2010)
16. Stonebraker, M., Aoki, P.M., Litwin, W., Pfeffer, A., Sah, A., Sidell, J., Sidell, J.: Mariposa: a wide-area distributed database system. The VLDB Journal **5**(1), 48–63 (1996)
17. Marbukh, V., Mills, K.: Demand pricing and resource allocation in market-based compute grids: a model and initial results. In: ICN 2008, pp. 752–757 (2008)
18. Moreno, R., Alonso-Conde, A.B.: Job scheduling and resource management techniques in economic grid environments. In: Fernández Rivera, F., Bubak, M., Gómez Tato, A., Doallo, R. (eds.) Across Grids 2003. LNCS, vol. 2970, pp. 25–32. Springer, Heidelberg (2004)
19. Berriman, G.B., Juve, G., Deelman, E., Regelson, M., Plavchan, P.: The application of cloud computing to astronomy: a study of cost and performance. In: Sixth IEEE International Conference e-Science Workshops, pp. 1–7 (2010)
20. Tsakalozos, K., Kllapi, H., Sitaridi, E., Roussopoulos, M., Paparas, D., Delis, A.: Flexible use of cloud resources through profit maximization and price discrimination. In: 2011 IEEE 27th International Conference on Data Engineering (ICDE), pp. 75–86 (2011)

21. Abouzeid, A., Bajda-Pawlikowski, K., Abadi, D., Silberschatz, A., Rasin, A.: HadoopDB: an architectural hybrid of MapReduce and DBMS technologies for analytical workloads. Proc. of the VLDB **2**(1), 922–933 (2009)
22. Weikum, G., Moenkeberg, A., Hasse, C., Zabback, P.: Self-tuning database technology and information services: from wishful thinking to viable engineering. In: Proceedings of VLDB, pp. 20–31 (2002)
23. Agrawal, S., Chaudhuri, S., Das, A., Narasayya, V.: Automating layout of relational databases. In: ICDE, pp. 607–618 (2003)
24. Dash, D., Kantere, V., Ailamaki, A.: An economic model for self-tuned cloud caching. In: IEEE 25th International Conference on Data Engineering, ICDE 2009, pp. 1687–1693 (2009)
25. Deelman, E., Singh, G., Livny, M., Berriman, B., Good, J.: The cost of doing science on the cloud: the montage example. In: Proceedings of the 2008 ACM/IEEE Conference on Supercomputing, p. 50 (2008)
26. Hadoop Hive project. `http://hadoop.apache.org/hive/` (last accessed May 1, 2015)
27. Dai, W., Bassiouni, M.: An improved task assignment scheme for Hadoop running in the clouds. Journal of Cloud Computing: Advances, Systems and Applications **2**(1), 1–16 (2013)
28. Chatziantoniou, D., Tzortzakakis, E.: Asset queries: a declarative alternative to mapreduce. ACM SIGMOD Record **38**(2), 35–41 (2009)
29. Mahboubi, H., Darmont, J.: Enhancing XML data warehouse query performance by fragmentation. In: Proceedings of ACM Symposium on Applied Computing, pp. 1555–1562 (2009)
30. Ordonez, C., Song, I.Y., Garcia-Alvarado, C.: Relational versus non-relational database systems for data warehousing. In: Proc. of the ACM 13th Int. Workshop on Data warehousing and OLAP, pp. 67–68 (2010)
31. Armbrust, M., Fox, A., Griffith, R., Joseph, A.D., Katz, R., Konwinski, A., Zaharia, M.: A view of cloud computing. Communications of the ACM **53**(4), 50–58 (2010)
32. Zhou, J., Larson, P.A., Elmongui, H.G.: Lazy maintenance of materialized views. In: Proceedings of the 33rd International Conference on Very Large Data Bases, pp. 231–242 (2007)
33. Storm, A.J., Garcia-Arellano, C., Lightstone, S.S., Diao, Y., Surendra, M.: Adaptive self-tuning memory in DB2. In: Proceedings of VLDB, pp. 1081–1092 (2006)
34. Running TPC-H queries on Hive. `http://issues.apache.org/jira/browse/HIVE-600` (last accessed May 1, 2015)

Appendix: Sample Hadoop Hive Queries

Query Q1 [34]

```
DROP TABLE lineitem;
DROP TABLE q1_pricing_summary_report;

    – create tables and load data
Create external table lineitem (L_ORDERKEY INT, L_PARTKEY INT, L_SUPPKEY INT,
L_LINENUMBER INT, L_QUANTITY DOUBLE, L_EXTENDEDPRICE DOUBLE, L_DISCOUNT DOUBLE,
L_TAX DOUBLE, L_RETURNFLAG STRING, L_LINESTATUS STRING, L_SHIPDATE STRING, L_COMMIT-
DATE STRING, L_RECEIPTDATE STRING, L_SHIPINSTRUCT STRING, L_SHIPMODE STRING, L_COMMENT
STRING) ROW FORMAT DELIMITED FIELDS TERMINATED BY '–' STORED AS TEXTFILE LOCATION
'/tpch/lineitem';
```

```
    – create the target table
CREATE TABLE q1_pricing_summary_report ( L_RETURNFLAG STRING, L_LINESTATUS STRING, SUM_QTY
DOUBLE, SUM_BASE_PRICE DOUBLE, SUM_DISC_PRICE DOUBLE, SUM_CHARGE DOUBLE, AVE_QTY
DOUBLE, AVE_PRICE DOUBLE, AVE_DISC DOUBLE, COUNT_ORDER INT);

    – the query
INSERT OVERWRITE TABLE q1_pricing_summary_report SELECT L_RETURNFLAG, L_LINESTATUS, SUM
(L_QUANTITY), SUM(L_EXTENDEDPRICE), SUM(L_EXTENDEDPRICE*(1-L_DISCOUNT)), SUM(L_EXTEN-
DEDPRICE*(1-L_DISCOUNT)*(1+L_TAX)), AVG(L_QUANTITY),
AVG(L_EXTENDEDPRICE), AVG(L_DISCOUNT), COUNT(1)
FROM
    LINETITEM
WHERE
    L_SHIPDATE<='1998-09-02'
GROUP BY L_RETURNFLAG, L_LINESTATUS
ORDER BY L_RETURNFLAG, L_LINESTATUS;
```

Query Q6 [34]

```
DROP TABLE lineitem;
DROP TABLE q6_forecast_revenue_change;

– create tables and load data
create external table lineitem (L_ORDERKEY INT, L_PARTKEY INT, L_SUPPKEY INT,
L_LINENUMBER INT, L_QUANTITY DOUBLE, L_EXTENDEDPRICE DOUBLE,
L_DISCOUNT DOUBLE, L_TAX DOUBLE,L_RETURNFLAG STRING, L_LINESTATUS STRING,
L_SHIPDATE STRING, L_COMMITDATE STRING, L_RECEIPTDATE STRING,
L_SHIPINSTRUCT STRING, L_SHIPMODE STRING,L_COMMENT STRING)
ROW FORMAT DELIMITED FIELDS TERMINATED BY '–'
STORED AS TEXTFILE LOCATION '/tpch/lineitem';

    – create the target table
CREATE TABLE q6_forecast_revenue_change (REVENEU DOUBLE);

    – the query
INSERT OVERWRITE TABLE q6_forecast_revenue_change
SELECT
    SUM(L_EXTENDEDPRICE*L_DISCOUNT) AS REVENEU
FROM
    LINETITEM
WHERE
    L_SHIPDATE >= '1994-01-01'
    AND L_SHIPDATE < '1995-01-01'
    AND L_DISCOUNT >= 0.05 AND L_DISCOUNT <= 0.07
    AND L_QUANTITY < 24;
```

$\mathcal{MP2R}$: A Human-Centric Skyline Relaxation Approach

Djamal Belkasmi and Allel Hadjali

Abstract Skyline queries have gained much attention in the last decade and are proved to be valuable for multi-criteria decision making. They are based on the concept of Pareto dominance. When computing the skyline, two scenarios may occur: either (i) a huge number of skyline which is less informative for the user or (ii) a small number of returned objects which could be insufficient for the user needs. In this paper, we tackle the second problem and propose an approach to deal with it. The idea consists in making the skyline more permissive by adding points that strictly speaking do not belong to it, but are close to belonging to it. A new fuzzy variant of dominance relationship is then introduced. Furthermore, an efficient algorithm to compute the relaxed skyline is proposed. Extensive experiments are conducted to demonstrate the effectiveness of our approach and the performance of the proposed algorithm.

1 Introduction

Skyline queries [2] are specific and popular example of preference queries. They are based on Pareto dominance relationship. This means that, given a set D of d-dimensional points, a skyline query returns, the skyline S, set of points of D that are not dominated by any other point of D. A point p dominates another point q iff p is better than or equal to q in all dimensions and strictly better than q in at least one dimension. One can see that skyline points are incomparable. This kind of queries provide an adequate tool that can help users to make intelligent decisions in the presence of multidimensional data where different and often conflicting criteria must

D. Belkasmi
DIF-FS/UMBB, Boumerdes, Algeria
e-mail: belkasmi.djamel@gmail.com

A. Hadjali(✉)
LIAS/ENSMA, Poitiers, France
e-mail: allel.hadjali@ensma.fr

T. Andreasen et al. (eds.), *Flexible Query Answering Systems 2015*,
Advances in Intelligent Systems and Computing 400,
DOI: 10.1007/978-3-319-26154-6_18

be taken. Several research studies have been conducted to develop efficient algorithms and introduce multiple variants of skyline queries [9, 12, 16, 18]. However, querying a d-dimensional data sets using a skyline operator may lead to two possible cases: (i) a large number of skyline points returned, which could be less informative for users, (ii) a small number of skyline points returned, which could be insufficient for users. To solve the two above problems, various approaches have been proposed to refine the skyline, therefore reducing its size [1, 3, 4, 6, 10, 11, 14, 15], but only very few works exist to relax the skyline in order to increase the number of skyline results [8, 10]. Goncalves and Tineo [8] propose a flexible dominance relationship using fuzzy comparison operators. This increases the skyline with points that are only weakly dominated by any other point. In [10], some ideas of relaxing the skyline have also been proposed.

In this paper, and taking as starting point the study in [10], we address in deep the problem of skyline relaxation to return more interesting results to the user. In particular, we develop an efficient approach, called $\mathcal{MP2R}$ ($\mathcal{M}$uch $\mathcal{P}$referred $\mathcal{R}$elation for $\mathcal{R}$elaxation), for skyline relaxation. The approach relies on a novel fuzzy dominance relationship *Much Preferred (MP)* which makes more demanding the dominance between the points of D. In this context, a point still belonging to the skyline unless it is much dominated, in the spirit of MP relation, by another skyline point. By this way, much points would be considered as incomparable and then as elements of the new relaxed skyline, denoted S_{relax}. Note that such points are ruled out from the skyline when applying classical Pareto dominance. Furthermore, an algorithm for computing the skyline S_{relax} efficiently is provided. In summary, our main contributions cover the following points:

- We investigate a new variant of fuzzy dominance relation based on the MP relation and provide the semantic basis for a relaxed variant of skyline S_{relax}.
- We develop and implement an algorithm to compute S_{relax} efficiently.
- We conduct a set of experiments to study and analyze the relevance and effectiveness of S_{relax}.

The paper is structured as follows: Section 2 provides some necessary background on fuzzy set theory and skyline queries and a survey on existing approaches. In Section 3, we introduce a new approach for skyline relaxation based on MP dominance relationship. In Section 4, the algorithm to efficiently compute S_{relax} is presented, while Section 5 is devoted to the experimental study. Finally, Section 6 concludes the paper and draws some lines for future works.

2 Background

In this section, we recall some notions on fuzzy set theory and skyline queries. Then, we review some related works.

2.1 Fuzzy Set Theory

The concept of fuzzy sets has been developed by Zadeh [19] in 1965 to represent classes or sets whose limits are imprecise. They can describe gradual transitions between total belonging and rejection. Typical examples of these fuzzy classes are those described with adjectives or adverbs in natural language, as *not expensive*, *fast* and *very close*. Formally, a fuzzy set F on the universe X is described by a membership function $\mu_F : X \longrightarrow [0, 1]$, where $\mu_F(x)$ represents the **degree of membership** of x in F. By definition, if $\mu_F(x) = 0$ then the element x **does not belong to** F, if $\mu_F(x) = 1$ then x **completely belongs to** F, these elements form the **core** of F denoted by $Cor(F) = \{x \in F | \mu_F(x) = 1\}$. When $0 < \mu_F(x) < 1$, we talk about a **partial membership**, these elements form the **support** of F denoted by $Supp(F) = \{x \in F | \mu_F(x) > 0\}$. Moreover, more the value of $\mu_F(x)$ is close to 1, more x belongs to F. Let $x, y \in F$, we say that x is preferred to y iff $\mu_F(x) > \mu_F(y)$. If $\mu_F(x) = \mu_F(y)$, then x and y have the same preference. In practice, F is represented by a trapezoid membership function (t.m.f) $(\alpha, \beta, \varphi, \psi)$, where (α, ψ) is the support and (β, φ) is its core.

2.2 Skyline Queries

Skyline queries [2] are a specific, yet relevant, example of preference queries. They rely on Pareto dominance principle which can be defined as follows:

Definition 1. Let D be a set of d-dimensional data points and u_i and u_j two points of D. u_i is said to dominate in Pareto sense u_j (denoted $u_i \succ u_j$) iff u_i is better than or equal to u_j in all dimensions and better than u_j in at least one dimension. Formally, we write

$$u_i \succ u_j \Leftrightarrow (\forall k \in \{1, .., d\}, u_i[k] \geq u_j[k]) \wedge (\exists l \in \{1, .., d\}, u_i[l] > u_j[l]) \quad (1)$$

where each tuple $u_i = (u_i[1], u_i[2], u_i[3], ..., u_i[d])$ with $u_i[k]$ stands for the value of the tuple u_i for the attribute A_k.

In (1), without loss of generality, we assume that the largest value, the better.

Definition 2. The skyline of D, denoted by S, is the set of points which are not dominated by any other point.

$$u \in S \Leftrightarrow \nexists u' \in D, u' \succ u \quad (2)$$

Example 1. To illustrate the Skyline, let us consider a database containing information on candidates as shown in Table 1. The list of candidates includes the following information: Code, Age, Management experience (man_exp in years), Technical experience (tec_exp in years) and distance work to Home (dist_wh in Km).

Ideally, personnel manager is looking for a candidate with the largest management and technical experience (Max man_exp and Max tec_exp), ignoring the other pieces

Table 1 List of candidates.

code	age	man_exp	tec_exp	dist_wh
M1	32	5	10	35
M2	41	7	5	19
M3	37	5	12	45
M4	36	4	11	39
M5	40	8	10	18
M6	30	4	6	27
M7	31	3	4	56
M8	36	6	13	12
M9	33	6	6	95
M10	40	7	9	20

of information. Applying the traditional skyline on the candidate list shown in Table 1 returns the following candidates: M_5, M_8. As can be seen, such results are the most interesting candidates (see Fig. 1).

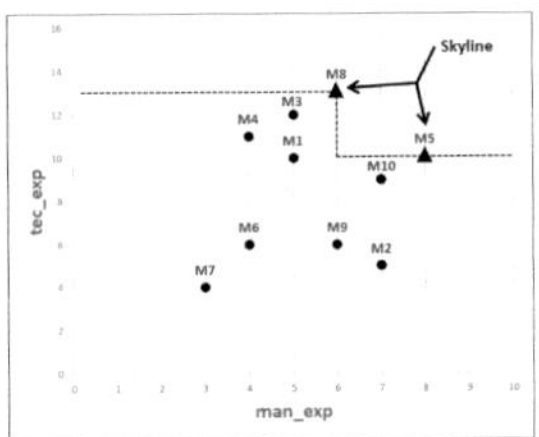

Fig. 1 Skyline of candidates

2.3 *Related Work*

Since its proposal, the skyline queries have been recognized as a useful and practical technique to capture user preferences and integrate them in the querying process. They have been widely used in various types of applications: (decision support, spatial data management, data mining, navigation systems, ...). In the years that followed the emergence of the concept of skyline queries, computing the skyline was the major concern, most of the works were about designing efficient evaluation algorithms under different conditions and in different contexts. In general, the existing algorithms can be classified into two categories: sequential algorithms and index based algorithms. Sequential algorithms scan the entire dataset to compute the skyline and don't require a pre-computed structure (index, hash table, ...). They include: the Block Nested Loops (BNL) [2], Divide and Conquer (D&C) [2], Sort First Skyline (SFS) [5] and Linear Elimination Sort for Skyline (LESS) [7]. Index-based algorithms compute the skyline by accessing just a part of the dataset through the use of indexes (R-tree, B-Tree,

Bitmap Index, ...), let us mention: Bitmap algorithm [17], Index algorithm [17], Nearest Neighbor algorithm [13] and Branch and Bound algorithm [15].

Some research efforts have been made to develop efficient algorithms and to introduce different variants of skyline queries [3, 4, 11, 12, 14, 16]. But only few works have been proposed to address the skyline relaxation issue. In Goncalves and Tineo [8], the problem of skyline rigidity is addressed by introducing a weak dominance relationship based on fuzzy comparison operators. This relationship allows enlarging the skyline with points that are not much dominated by any other point (even if strictly speaking they are dominated).

In [10], Hadjali et al. have introduced some ideas to define some novel variants of Skyline. First, one idea consists in refining the skyline by introducing some ordering between its points in order to single out the most interesting ones. The second idea aims at making the Skyline more flexible by adding some points that strictly speaking do not belong to it, but are not much dominated by any other point in all Skyline dimensions. The third one tries to simplify the skyline either by granulating the scales of the criteria which may enable us to cluster points that are somewhat similar. The last idea addresses the skyline semantics in the context of uncertain data.

Taking inspiration from that work, this papers tackles the problem of skyline relaxation. It develops a complete approach to make the Skyline more permissive where both the semantic basis of the relaxed skyline and its computation are addressed in a deep way.

3 $\mathcal{MP2R}$: An Approach for the Skyline Relaxation

Let a relation $R(A_1, A_2, ..., A_d)$ be defined in a d-dimensional space $\mathbb{D} = (\mathbb{D}_1, \mathbb{D}_2, ..., \mathbb{D}_d)$, where $\mathbb{D}_i$ is the domain attribute of A_i. We assume the existence of a total order relationship on each domain $\mathbb{D}_i$. U is a set of n tuples belonging to the relationship R, $U = (u_1, u_2, ..., u_n)$. Let S be the skyline of U and S_{relax} is the relaxed skyline of U returned by our approach $\mathcal{MP2R}$.

$\mathcal{MP2R}$ relies on a new dominance relationship that allows enlarging the skyline with the most interesting points among those ruled out when computing the initial skyline S. This new dominance relationship uses the relation *"Much Preferred (MP)"* to compare two tuples u and u'. So, u is an element of S_{relax} if there is no tuple $u' \in U$ such that u' is *much preferred* to u (denoted $MP(u', u)$) in all skyline attributes. Formally, we write:

$$u \in S_{relax} \Leftrightarrow \nexists u' \in U, \forall i \in \{1, ..., d\}, MP_i(u'_i, u_i) \tag{3}$$

where, MP_i is a fuzzy preference relation defined on the domain $\mathbb{D}_i$ of the attribute A_i and $MP_i(u'_i, u_i)$ expresses the extent to which the value u'_i is *much preferred* to the value u_i. Since *MP* relation is of a gradual nature, each element u of S_{relax} is associated with a degree ($\in [0, 1]$) expressing the extent to which u belongs to S_{relax}.

In fuzzy set terms, one can write:

$$\mu_{S_{relax}}(u) = 1 \quad \max_{u' \in U} \min_{i} \mu_{MP_i}(u'_i, u_i) = \min_{u' \in U} \max_{i}(1 - \mu_{MP_i}(u'_i, u_i)) \tag{4}$$

As for MP_i relation on $\mathbb{D}_i$, its semantics can be provided by the formulas (5) (see also Fig. 2). In terms of t.m.f., MP_i writes $(\gamma_{i1}, \gamma_{i2}, \infty, \infty)$, and denoted $MP_i^{(\gamma_{i1}, \gamma_{i2})}$. It is easy to check that $MP_i^{(0,0)}$ corresponds to the regular preference relation expressed by means of the crisp relation *"greater than"*.

$$\mu_{MP_i^{(\gamma_{i1}, \gamma_{i2})}}(u'_i, u_i) = \begin{cases} 0 & \text{if } u'_i - u_i \leq \gamma_{i1} \\ 1 & \text{if } u'_i - u_i \geq \gamma_{i2} \\ \frac{(u'_i - u_i) - \gamma_{i1}}{\gamma_{i2} - \gamma_{i1}} & \text{else} \end{cases} \tag{5}$$

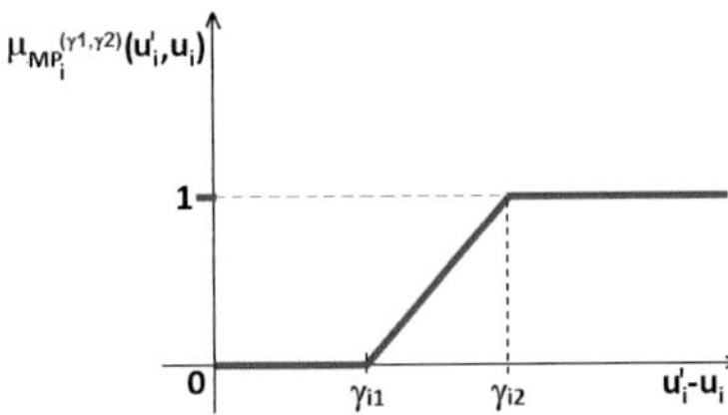

Fig. 2 The membership function $\mu_{MP_i^{(\gamma_{i1}, \gamma_{i2})}}$

Let $\gamma = ((\gamma_{11}, \gamma_{12}), \cdots, (\gamma_{d1}, \gamma_{d2}))$ be a vector of pairs of parameters where $MP_i^{(\gamma_{i1}, \gamma_{i2})}$ denotes the MP_i relation defined on the attribute A_i and $S_{relax}^{(\gamma)}$ denotes the relaxed skyline computed on the basis of the vector γ. One can easily check that the classical Skyline S is equal to $S_{relax}^{(\mathbf{0})}$, where $\mathbf{0} = ((0, 0), \cdots, (0, 0))$.

We say that $MP_i^{(\gamma_{i1}, \gamma_{i2})}$ is more constrained than $MP_i^{(\gamma'_{i1}, \gamma'_{i2})}$ if and only if $(\gamma_{i1}, \gamma_{i2}) \geq (\gamma'_{i1}, \gamma'_{i2})$ (i.e., $\gamma_{i1} \geq \gamma'_{i1} \wedge \gamma_{i2} \geq \gamma'_{i2}$).

Definition 3. Let γ and γ' be two vectors of parameters. We say that $\gamma \geq \gamma'$ if and only if $\forall i \in \{1, \cdots, d\}$, $(\gamma_{i1}, \gamma_{i2}) \geq (\gamma'_{i1}, \gamma'_{i2})$.

Proposition 1. *Let γ and γ' be two vectors of parameters. The following property holds: $\gamma' \leq \gamma \Rightarrow S_{relax}^{(\gamma')} \subseteq S_{relax}^{(\gamma)}$.*

Proof. Let $\gamma' \leq \gamma$ and let $u \in S_{relax}^{(\gamma')} \Rightarrow \nexists u' \in U, \forall i \in \{1, \cdots, d\}, MP_i^{(\gamma'_{i1}, \gamma'_{i2})}(u'_i, u_i)$
$\Rightarrow \nexists u' \in U, \forall i \in \{1, \cdots, d\}, \mu_{MP_i^{(\gamma'_{i1}, \gamma'_{i2})}}(u'_i, u_i) > 0$
$\Rightarrow \nexists u' \in U, \forall i \in \{1, \cdots, d\}, u'_i - u_i > \gamma'_{i1} \Rightarrow \forall u' \in U, \forall i \in \{1, \cdots, d\}, u'_i - u_i \leq \gamma'_{i1}$
$\Rightarrow \forall u' \in U, \forall i \in \{1, \cdots, d\}, u'_i - u_i \leq \gamma_{i1} \Rightarrow \nexists u' \in U, \forall i \in \{1, \cdots, d\}, u'_i - u_i > \gamma_{i1}$

$\Rightarrow \nexists u' \in U, \forall i \in \{1, \cdots, d\}, \mu_{MP_i^{(\gamma_{i1}, \gamma_{i2})}}(u'_i, u_i) > 0$

$\Rightarrow \nexists u' \in U, \forall i \in \{1, \cdots, d\}, MP_i^{(\gamma_{i1}, \gamma_{i2})}(u'_i, u_i) \Rightarrow u \in S_{relax}^{(\gamma)} \Rightarrow S_{relax}^{(\gamma')} \subseteq S_{relax}^{(\gamma)}$

□

Lemma 1. *Let* $\gamma = ((0, \gamma_{12}), \cdots, (0, \gamma_{d2}))$ *and* $\gamma' = ((\gamma'_{11}, \gamma'_{12}), \cdots, (\gamma'_{d1}, \gamma'_{d2}))$, *the following holds:* $S_{relax}^{(0)} \subseteq S_{relax}^{(\gamma)} \subseteq S_{relax}^{(\gamma')}$

Example 2. Let us come back to the skyline calculated in Example 1. Assume that the *"much preferred"* relations corresponding to the skyline attributes (man_exp and tec_exp) are respectively given by:

$$\mu_{MP_{man\_exp}^{(1/2,2)}}(u', u) = \begin{cases} 0 & \text{if } u' - u \leq 1/2 \\ 1 & \text{if } u' - u \geq 2 \\ 2/3(u' - u) - 1/3 & \text{else} \end{cases} \tag{6}$$

$$\mu_{MP_{tec\_exp}^{(1/2,4)}}(u', u) = \begin{cases} 0 & \text{if } u' - u \leq 1/2 \\ 1 & \text{if } u' - u \geq 4 \\ 2/7(u' - u) - 1/8 & \text{else} \end{cases} \tag{7}$$

Now, applying the $\mathcal{MP2R}$ approach, to relax the skyline $S = \{M_5, M_8\}$ found in example 1, leads to the following relaxed skyline $S_{relax} = \{(M_5, 1), (M_8, 1), (M_3, 0.85), (M_{10}, 0.85), (M_1, 0.66), (M_2, 0.66), (M_4, 0.57)\}$, see Table 2.

Table 2 Degrees of the elements of S_{relax}

Mat	M5	M8	M3	M10	M1	M2	M4	M6	M7	M9
$\mu_{S_{relax}}$	1	1	0.85	0.85	0.66	0.66	0.57	0	0	0

One can note that some candidates that were not in S are now elements of S_{relax} (such M_{10} and M_4) see Fig. 3 . As can be seen, S_{relax} is larger than S. Let us now take a glance at the content of S_{relax}, one can observe that (i) the skyline elements of S are still elements of S_{relax} with a degree equal to 1 ; (ii) Appearance of new elements recovered by our approach whose degrees are less than 1 (such as M_3). Interestingly, the user can select from S_{relax}:

- the Top-k elements (k is a user-defined parameter) : elements of S_{relax} with highest degrees, or
- the subset of elements , denoted $(S_{relax})_\sigma$, with a degrees higher than a threshold σ provided by the user.

In the context of example 2, it is easy to check that $Top - 6 = \{(M_5, 1), (M_8, 1), (M_3, 0.85), (M_{10}, 0.85), (M_1, 0.66), (M_2, 0.66)\}$ and $(S_{relax})_{0.7} = \{(M_5, 1), (M_8, 1), (M_3, 0.85), (M_{10}, 0.85)\}$.

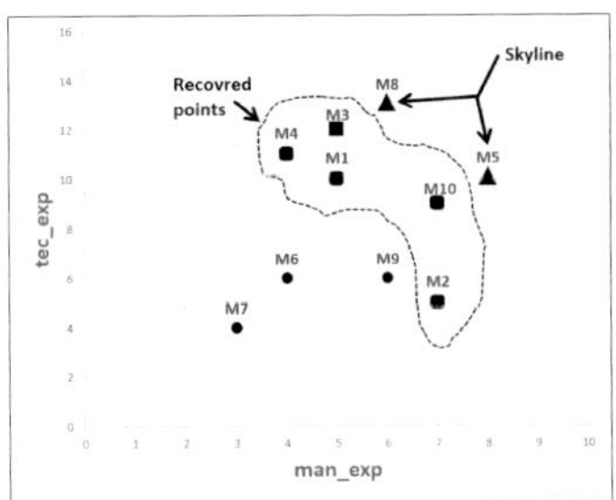

Fig. 3 Skyline relaxation

4 S_{relax} Computation

According to the definition of S_{relax}, a tuple u belongs to S_{relax} if: it does not exist a tuple u' which is *much preferred* to u w.r.t. all skyline attributes. So, to compute S_{relax}, we proceed in two steps (see Fig. 4):

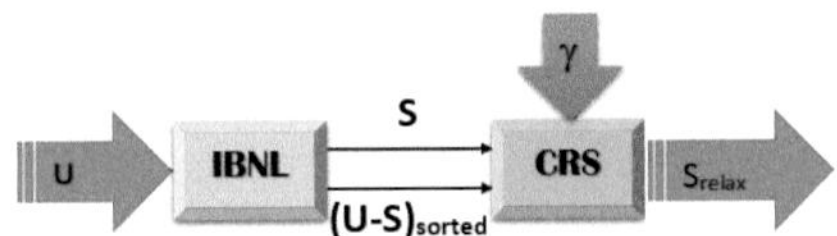

Fig. 4 The process of skyline relaxation

First, the classical skyline (S) is computed using an Improved Basic Nested Loop algorithm ($IBNL$), see algorithm 1. The dataset U is sorted in ascending order by using a monotonic function Mf (e.g., the sum of attributes skyline multiplying by -1 attribute values whose criterion is MAX). Sorting allows the following property:

$$\forall u, v \in U \mid Mf(u) \leq Mf(v) \Longrightarrow \neg(v \succ u) \tag{8}$$

SkylineCompare function evaluates the dominance , in the sense of Pareto, between u_i and u_j on all skyline dimensions and returns the result in status. It may be equal to: 0 if $u_i = u_j$, 1 if $u_i \succ u_j$, 2 if $u_i \prec u_j$, 3 if they are incomparable.
Secondly, we introduce an efficient algorithm called CRS (Computing Relaxed Skyline) to relax S using a vector of parameters γ (see algorithm 2).

Algorithm 1. IBNL

```
Input: A set of tuples U
Output: A skyline S
Sort(U);
for i := 1 to n − 1 do
    if ¬u_i.dominated then
        for j := i + 1 to n do
            status = 0;
            if ¬u_j.dominated then
                evaluate SkylineCompare(u_i,u_j,status);
                switch status do
                    case 1
                        u_i.dominated= true;
                    case 2
                        u_j.dominated= true;
        if ¬u_i.dominated then
            S = S ∪ {u_i};
return S;
```

Algorithm 2. CRS

```
Input: A set of tuples U; Skyline S; γ a vector of parameters;
Output: A relaxed skyline S_relax;
begin
    for i = 1 to n do
        if u_i ∉ S then
            for j = 1 to n do
                for k = 1 to d do
                    evaluate μ_MP_k(u_i, u_j);
                compute min_k(μ_MP_k);
            compute max_j(min_k(μ_MP_k)); μ_S_relax(u_i) = 1 − max_j(min_k(μ_MP_k));
        if μ_S_relax(u_i) > 0 then
            S_relax = S_relax ∪ {u_i};
    rank u_i in decreasing order w.r.t. μ_S_relax(u_i);
    return S_relax;
```

5 Experimental Study

In this section, we present the experimental study that we have conducted. The goal of this study is to prove the effectiveness of $\mathcal{MP2R}$ and its ability to relax small skylines with the most interesting tuples.

5.1 Experimental Environment

All experiments were performed under Linux OS, on a machine with an Intel core i7 2,90 GHz processor, a main memory of 8 GB and a 250 GB of disk. All algorithms were implemented with Java. Dataset benchmark is generated using method described in [2] following three conventional distribution schema (correlated, anti-correlated and independent). For each dataset, we consider three different sizes (10K, 50K and 100K). Each tuple contains an integer identifier (4 bytes), 8 decimal fields (8 bytes) with values belonging to the interval [0,1], and a string field with length of 10 characters. Therefore, the size of one tuple is 78 bytes.

5.2 Experimental Results

We vary a collection of parameters that impact the result. This collection includes the dataset size [D] (1OK, 50K, 100K), dataset distribution schema [DIS] (independent, correlated, anti-correlated), the number of dimensions [d] (2, 4, 6, 8) and the relaxation thresholds [$\gamma = (\gamma_{i1}, \gamma_{i2})$ for i$\in \{1, \ldots, d\}$] where ($\gamma_{i1}, \gamma_{i2} \in$[0,1] and $\gamma_{i1} \leq \gamma_{i2}$). These parameters are set as follows: D=10K; DIS="Correlated"; d=2; γ=((0.25,0.5),(0.25,0.5)). In our experiment, we consider that the less the value, the better is. Thus, this study addresses the following points:

- Impact of [DIS] on the size of relaxed skyline and on the computation time,
- Impact of [d] on the size of relaxed skyline and on the computation time,
- Impact of [D] on the size of relaxed skyline and on the computation time,
- Impact of ($MP_i^{(\gamma_{i1},\gamma_{i2})}$) on the size of relaxed skyline.

Impact of DIS. Fig. 5 shows that for anti-correlated and independent Distribution, our approach provides more tuples than the correlated data context. This is due to the type of data. We observe that the efficiency of CRS to relax the skyline is very high for all types of data. We note that the execution time of the CRS, for the three distributions, is almost similar.

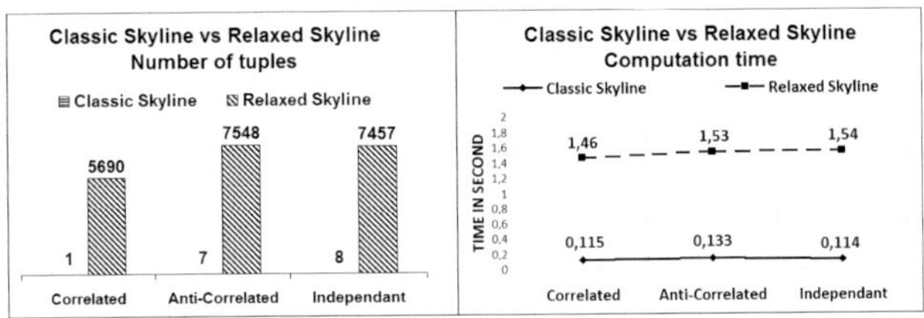

Fig. 5 Impact of [DIS].

Impact of the Number of Dimension. It is well-known that the number of dimensions increases the classic skyline; this phenomenon is known as "the problem of dimensionality". The CRS leads to the same behavior where data are highly correlated (correlation coefficient = 0.9952) see Fig. 6. From 2 to 8 dimensions, the relaxed skyline size changes from 5690 to 7519 tuples and the execution time increases from 1.46 to 2.45 second.

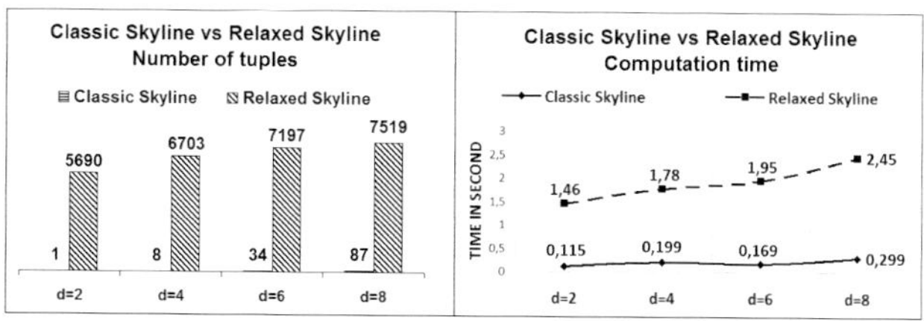

Fig. 6 Impact of [d]

Impact of the Dataset Size. Fig. 7 shows that the size of relaxed skyline is proportional to the data size, which confirms CRS's ability to relax the skyline with respect to the size of Dataset but it is time consuming.

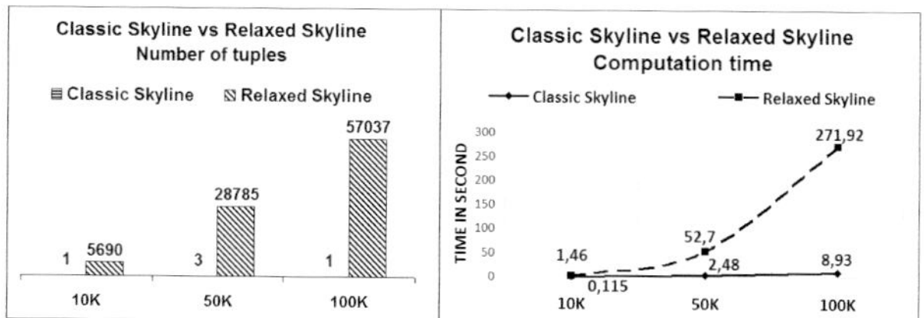

Fig. 7 Impact of [D]

Impact of ($MP_i^{(\gamma_1,\gamma_2)}$) Dominance Relationship. Now, we show the influence of the "Much Preferred" dominance relationship ($MP_i^{(\gamma_1,\gamma_2)}$) on the size of relaxed skyline.Tthe idea is to vary both thresholds (γ_{i1} and γ_{i2}) of the relationship. For the sake of simplicity, and since the data are normalized, we will apply the same function $MP_i^{(\gamma_1,\gamma_2)}$ for all skyline dimensions. Note that the size of the skyline is equal to 1 and we will analyze the variation of the number of tuples whose degree $\mu_{S_{relax}}(u) > 0$. The following scenarios are worth to be discussed :

Scenario 1: In this scenario, we fix γ_{i1} and vary γ_{i2} to increase the relaxation zone. We observe the following cases:

Case 1: $\gamma_{i1} = 0$ and $\gamma_{i2} \in \{0.25; 0.5; 0.75; 1\}$. Fig. 8 shows the results obtained. The analysis of these curves shows that the size of relaxed skyline increases when the value of γ_{i2} increases. We also note that there are no tuples whose degrees of relaxation is equal to 1 (this is due to the value of $\gamma_{i1} = 0$)

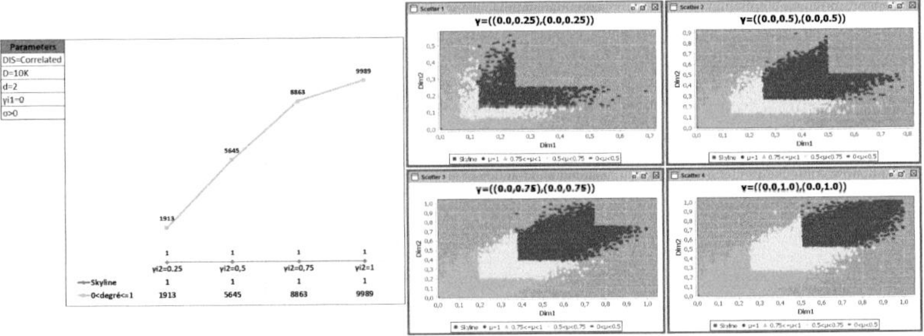

Fig. 8 Fix γ_{i1} and vary γ_{i2} **(case 1)**

Case 2: $\gamma_{i1} = 0.25$ and $\gamma_{i2} \in \{0.25; 0.5; 0.75; 1\}$. In this case, the size of relaxed skyline increases also when the value of γ_{i2} increases. We note the appearance of tuples whose degree is equal to 1 (see Fig. 9).

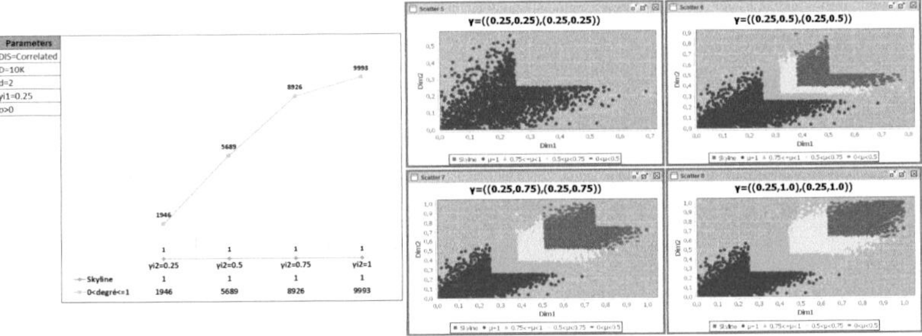

Fig. 9 Fix γ_{i1} and vary γ_{i2} **(case 2)**

Case 3: $\gamma_{i1} = 0.5$ and $\gamma_{i2} \in \{0.5; 0.75; 1\}$.. The same results are obtained in this case but the number of tuples whose degrees are equal to 1 is more important than in the case 2 (see Fig. 10).

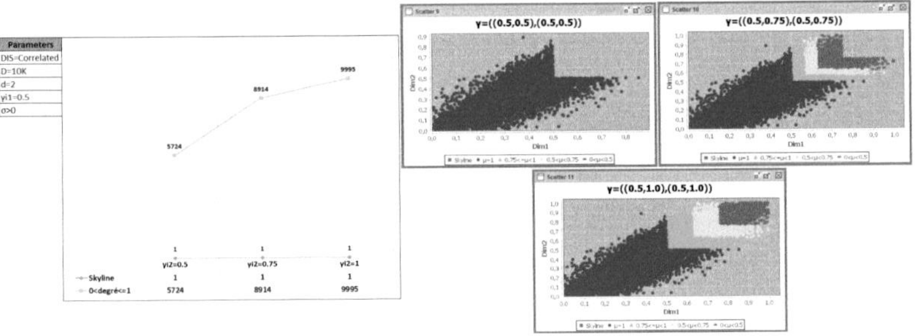

Fig. 10 Fix γ_{i1} and vary γ_{i2} **(case 3)**

Case 4: $\gamma_{i1} = 0.75$ and $\gamma_{i2} \in \{0.75; 1\}$. The same results are obtained in this case but the number of tuples whose degrees are equals to 1 is more important than in the cases 2 and 3 (see Fig. 11). This means that more the values of γ_{i1} and γ_{i2} are close to 1, the larger the number of tuples whose degrees are equal to 1.

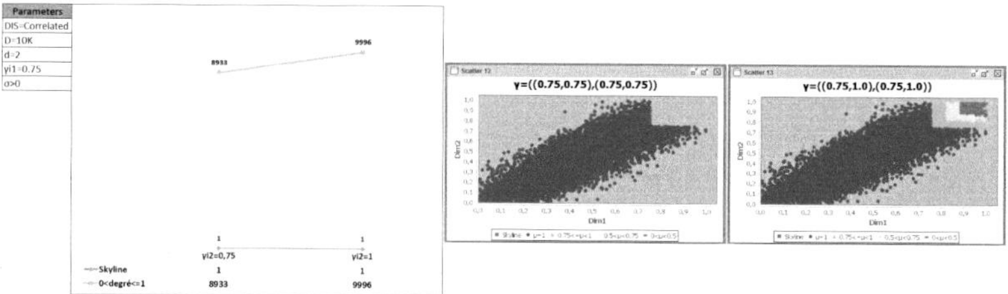

Fig. 11 Fix γ_{i1} and vary γ_{i2} **(case 4)**

Scenario 2: In this scenario, we vary both thresholds. The obtained results are shown in Fig. 12. The analysis of these curves shows that the relaxation function becomes more permissive when thresholds move away from the origin.

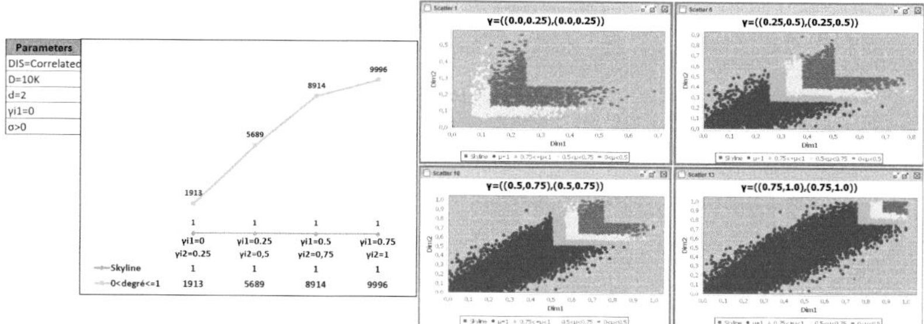

Fig. 12 Varying γ_{i1} and γ_{i2}

Finally, the choice of values for $\gamma = (\gamma_{i1}, \gamma_{i2})$ is extremely important in the relaxation process. This choice also depends on the domains values of the different dimensions used in the skyline.

6 Conclusion

In this paper, we addressed the problem of skyline relaxation, especially less skylines. An approach for relaxing the skyline, called $MP2R$, is discussed. The key concept of this approach is a particular relation named *much preferred* whose semantics is user-defined. In addition, a new algorithm called CRS to compute the relaxed skyline is proposed. The experimental study we done has shown that, on the one hand, the $MP2R$ approach is a good alternative to tackle the relaxation issue of classic skyline and, on the other hand, the computation cost of S_{relax} is quite reasonable. $MP2R$ involves various parameters, which can be used to control the size of the relaxed skyline.

As for future works, we will explore the possibility of using multidimensional index (R-Tree and its variants) to accelerate the computation of S_{relax}.

References

1. Abbaci, K., Hadjali, A., Lietard, L., Rocacher, D.: A linguistic quantifier-based approach for skyline refinement. In: IFSA/NAFIPS, pp. 321–326 (2013)
2. Börzsönyi, S., Kossmann, D., Stocker., K.: The skyline operator. In: ICDE, pp. 421–430 (2001)
3. Chan, C.Y., Jagadish, H.V., Tan, K., Tung, A.K.H., Zhang, Z.: Finding k-dominant skylines in high dimensional space. In: ACM SIGMOD, pp. 503–514 (2006)
4. Chan, C.-Y., Jagadish, H.V., Tan, K.-L., Tung, A.K.H., Zhang, Z.: On high dimensional skylines. In: Ioannidis, Y., Scholl, M.H., Schmidt, J.W., Matthes, F., Hatzopoulos, M., Böhm, K., Kemper, A., Grust, T., Böhm, C. (eds.) EDBT 2006. LNCS, vol. 3896, pp. 478–495. Springer, Heidelberg (2006)
5. Chomicki, J., Ciaccia, P., Meneghetti, N.: Skyline queries, front and back. SIGMOD Record (2013)
6. Endres, M., Kießling, W.: Skyline snippets. In: Christiansen, H., De Tré, G., Yazici, A., Zadrozny, S., Andreasen, T., Larsen, H.L. (eds.) FQAS 2011. LNCS, vol. 7022, pp. 246–257. Springer, Heidelberg (2011)
7. Godfrey, P., Shipley, R., Gryz, J.: Maximal vector computation in large data sets. In: VLDB (2005)
8. Goncalves, M., Tineo, L.J.: Fuzzy dominance skyline queries. In: Wagner, R., Revell, N., Pernul, G. (eds.) DEXA 2007. LNCS, vol. 4653, pp. 469–478. Springer, Heidelberg (2007)
9. HadjAli, A., Pivert, O., Prade, H.: Possibilistic contextual skylineswith incomplete preferences. In: SoCPaR, pp. 57–62
10. Hadjali, A., Pivert, O., Prade, H.: On different types of fuzzy skylines. In: Kryszkiewicz, M., Rybinski, H., Skowron, A., Raś, Z.W. (eds.) ISMIS 2011. LNCS, vol. 6804, pp. 581–591. Springer, Heidelberg (2011)
11. Hüllermeier, E., Vladimirskiy, I., Prados Suárez, B., Stauch, E.: Supporting case-based retrieval by similarity skylines: basic concepts and extensions. In: Althoff, K.-D., Bergmann, R., Minor, M., Hanft, A. (eds.) ECCBR 2008. LNCS (LNAI), vol. 5239, pp. 240–254. Springer, Heidelberg (2008)
12. Khalefa, M.E., Mokbel, M.F., Levandoski, J.J.: Skyline query processing for incomplete data. In: IEEE ICDE, pp. 556–565 (2008)
13. Kossmann, D., Ramsak, F., Rost, S.: Shooting stars in the sky: an online algorithm for skyline queries. In: VLDB, pp. 275–286 (2002)
14. Lin, X., Yuan, Y., Zhang, Q., Zhang, Y.: Selecting stars: the k most representative skyline operator. In: ICDE (2007)

15. Papadias, D., Tao, Y., Fu, G., Seeger, B.: An optimal and progressive algorithm for skyline queries. In: ACM SIGMOD (2003)
16. Pei, J., Jiang, B., Lin, X., Yuan, Y.: Probabilistic skylines on uncertain data. In: VLDB, pp. 15–26 (2007)
17. Tan, K., Eng, P., Ooi, B.C.: Efficient progressive skyline computation. In: VLDB, pp. 301–310 (2001)
18. Yiu, M.L., Mamoulis, N.: Efficient processing of top-k dominating queries on multi-dimensional data. In: VLDB, pp. 483–494 (2007)
19. Zadeh, L.A.: Fuzzy sets. Information and Control, 338–353 (1965)

Qualifying Ontology-Based Visual Query Formulation

Ahmet Soylu and Martin Giese

Abstract This paper elaborates on ontology-based end-user visual query formulation, particularly for users who otherwise cannot/do not desire to use formal textual query languages to retrieve data due to the lack of technical knowledge and skills. Then, it provides a set of quality attributes and features, primarily elicited via a series of industrial end-user workshops and user studies carried out in the course of an industrial EU project, to guide the design and development of successor visual query systems.

Keywords Visual query formulation · Ontologies · Data retrieval · End-user programming · Usability

1 Introduction

Today the storage and retrieval of *structured data* is fundamental to organisations. Particularly, *value creation* processes rely on *domain experts'* ability to reach the data of interest in a timely fashion. However, usually an army of *IT experts* mediates between domain experts and databases in an inherently time-consuming way, since *end users* often lack necessary technical *skills* and *knowledge* and have low tolerance to formal textual query languages, such as SQL and SPARQL. A substantial time could be freed-up by providing domain experts with the flexibility to pose relatively complex ad hoc queries in an easy and intuitive way.

A. Soylu(✉) · M. Giese
Department of Informatics, University of Oslo, Oslo, Norway
e-mail: {ahmets,martingi}@ifi.uio.no

A. Soylu
Faculty of Informatics and Media Technology, GjøVik University College, GjøVik, Norway
e-mail: ahmet.soylu@hig.no

T. Andreasen et al. (eds.), *Flexible Query Answering Systems 2015*,
Advances in Intelligent Systems and Computing 400,
DOI: 10.1007/978-3-319-26154-6_19

A *visual query system* (VQS) (cf. [8]) is a *visual programming* paradigm (cf. [7]) that uses visual representations to depict the domain of interest and data requests. A VQS allows its users to formalise their *information needs* into queries by *directly manipulating* (cf. [28]) domain elements. VQSs are oriented towards a wide spectrum of users; however, people with no technical background are expected to benefit the most. In this respect, a VQS is also an *end-user development/programming* paradigm (cf. [22]).

Early VQSs typically employ vocabulary extracted from database schemas; however, it is known that conceptual models are more natural than logical models for end users [29] (i.e., *semantic gap*). This is inherently evident, since ontologies are *problem domain* artefacts while models are *solution domain* artefacts (cf. [26]). Therefore, ontology-based VQSs emerged (e.g., [9, 30]) to address this *conceptual mismatch*. Yet, today a considerable amount of enterprise data resides in relational databases, rather than triple stores. Nevertheless, recent advances on *ontology-based data access* (OBDA) technologies (cf. [21, 33]) enable in-place access to relational data over ontologies (e.g., [25]), and hence raise ontology-based visual query formulation as a viable and promising approach for querying a variety of structured data sources.

A VQS, called *OptiqueVQS* [30, 32], has been developed for non-technical users as a part of an OBDA platform for *Big Data* sources in the course of an industrial EU project, namely *Optique*[1] [13, 14]. And, in this paper, based on the results of requirement elicitation and evaluation efforts for OptiqueVQS, particularly a series of industrial end-user workshops and usability studies, notable concepts and aspects of ontology-based VQSs are reviewed and a set of quality attributes and features, backed with the relevant literature, are suggested.

2 Data Retrieval

A VQS is a *data retrieval* (DR) paradigm, which differs from *information retrieval* (IR) (cf. [2]). In DR, an information need has to be exact and complete, and is defined over a deterministic model with the aim of retrieving all and only those objects that exactly match the criteria. However, in IR, an information need is typically incomplete and loosely defined over a probabilistic model with the aim of retrieving relevant objects. In other words, DR systems have no tolerance for missing or irrelevant results and a single erroneous object implies a total failure; while IR systems are variably insensitive to inaccuracies and errors, since they often interpret the original user query and the matching is assumed to indicate the likelihood of the relevance.

In ontology-based DR, ontologies, apart from being a natural communication medium, also act as *superstructures* to seamlessly *federate* distributed data sources and allow extracting implicit information from data with *reasoning*. And, data does not necessarily need to reside in *triple stores* anymore. OBDA technologies built on *data virtualisation* make it possible to query legacy relational data sources over ontologies, while enjoying the benefits of well-established query optimisation and

[1] `http://optique-project.eu`

evaluation support available for traditional database systems, without ever needing to migrate or transform (i.e., *materialise*) data into triples. A set of *mappings* relate the elements of the ontology to data sources (e.g., [25]).

3 Visual Query Systems

A VQS should drive the capabilities of the output medium and human visual system optimally to increase the magnitude of *preconscious* processing (e.g., recognition vs. recall) and to foster *innate* user reactions (e.g., clicking vs. typing).

A VQS needs to support two complementary but adverse activities, that is *exploration* and *construction*. The former relates to the activities for understanding and finding domain concepts and relationships relevant to the information need at hand (aka understanding the reality of interest), while the latter concerns the compilation of relevant concepts and constraints into formal information needs (i.e., queries) (cf. [8]). A primary concern in the design of a VQS is the selection of appropriate visual *representation* and *interaction paradigms* that support these two activities. The use of real life metaphors, and analogies with familiar situations or concepts are important. Interested readers are referred to Catarci et. al [8] and Katifori et. al [18] for an overview of representation paradigms, such as *forms* and *diagrams*, and interaction paradigms, such as *navigation* and *filtering*.

Finally, one should note that query formulation is an *iterative* process, that is a user explores the conceptual space, formulates a query, inspects the results, and reformulates the query until she/he reaches to the desired query, and each iteration could be considered an *attempt*.

3.1 On User Types

VQS users could be diverse, such as *casual users*, *domain experts*, and *IT experts*. Casual users use computers in their daily life/work for basic tasks (e.g., typing documents, sending e-mails, and web browsing); they have low tolerance on formal languages and are unfamiliar with the technical details of an information system. Domain experts have an in-depth knowledge and understanding of the semantics of their expertise domain, while IT experts have technical knowledge and skills on a wide spectrum of topics, such as programming languages, databases etc. Note that this categorisation is neither complete nor mutually exclusive – e.g., a domain expert could be also an IT expert.

Casual users and domain experts with no technical background are the primary group of users that could benefit the most from a VQS. The ultimate benefit of a VQS is questionable for IT experts, since in many cases they might find working on a textual language more efficient and non-limiting (cf. [8, 28]). Noteworthy, the selection of representation and interaction paradigms should not only consider the *user characteristics*, but also the *frequency of interaction*, the *variance of query tasks*, and *query complexity* (cf. [8]).

3.2 Usability and Expressiveness

VQS are built on two competing notions, namely *usability* and *expressiveness*.

Usability. The usability of a VQS is measured in terms of the level of *effectiveness*, *efficiency* and *user satisfaction*. Effectiveness is measured in terms of *accuracy* and *completeness* that users can achieve while constructing queries. Note that, typically, in IR systems, effectiveness is measured in terms of *precision*, *recall*, and *f-measure* (harmonic mean of precision and recall) over the result set; however, as stated earlier, for a DR system, a single missing or irrelevant object implies failure. Therefore, for a VQS, effectiveness is rather measured in terms of a binary measure of success (i.e., correct/incorrect query), which could be accompanied with a fuzzy measure (i.e., rate of correctness) weighting and combining different type of query errors (cf. [19, 36]). As far as SPARQL is concerned, the measure of correctness could be built on the semantic similarity between the user query and correct query (cf. [12]). Since query formulation is an iterative process, allowing and incorporating multiple attempts into effectiveness measure is also a sensible approach.

Efficiency is typically measured in terms of the time spent to complete a query, while user satisfaction measured through surveys, interviews etc. (e.g., examining the attitude of users after experiencing with the subject system or language). The use of a standard questionnaire, such as *system usability scale* (SUS) [4], is beneficial since it becomes possible to rank and compare SUS scores of similar tools.

Usability evaluation is usually done in terms of *query writing* and *query reading* tasks with respect to variable user and query types, such as the structure and lengths of queries. Query reading tasks could be replaced with query correction and completing tasks, since users are generally poor in articulating complex structures in human language, but better in doing, and a user needs to understand a query first in order to correct/complete it. For an end-user query tool, the representative target group should not include people with substantial technical background regardless of their expertise in the subject textual language and ontology language, since they possess semantic knowledge on programming languages, systems, frameworks, and tools, which is non-volatile and easily transferable while using a new tool/language (cf. [28]).

Expressiveness. The expressiveness of a VQS refers to its ability and breadth to characterise the domain knowledge and information needs and is bounded by the underlying formal language. A VQS is typically less expressive than the underlying formality, due to the *usability-expressiveness trade-off*. The design of a VQS should be informed by the type of query tasks and domain knowledge needed by the end users. Then, the *perceived complexity* plays a determining role. If a visually comprehensible solution for a query task or domain knowledge is not found, it is better left out and delegated to the IT experts. This is because, generally, the benefits gained by incorporating rarely used complex functionality does not make up for the loss in usability. In this respect, the categorisation of domain knowledge and query tasks are important for defining a scope and evolution path for a VQS.

As far as the domain knowledge is considered, for instance *multiple inheritance*, *disjointness*, *subproperties*, *inverse properties*, and *multiple ranges* are comparatively harder to communicate compared to *classes/subclasses*, *instances*, *relationships*, and *attributes*. One should also take into account the propagative effect of reasoning, that is, in an ontology, explicit restrictions attached to a concept are inherited by its subconcepts and the interpretation of a concept also includes the interpretations of all its subconcepts (cf. [10]). This effect drastically increases the number of possible properties for and between concepts.

One could categorise queries into *disjunctive* and *conjunctive* queries with respect to logical connectives used; from a topological perspective into *linear queries*, *queries with branching*, and *cyclic queries*; and from a non-topological perspective into queries that involve *aggregation*, *negation*, and *quantification* (cf. [30]). While the use of existential quantification remains implicit, queries including universal quantification, negation, and aggregation are quite esoteric for end users; this even applies to skilled users, particularly for universal quantifiers (cf. [20]). In this regard, queries could be categorised into three levels with respect to need and perceived complexity. The first level refers to linear and tree-shaped conjunctive queries, while the second level refers to queries with disjunctions, aggregation and cycles. The third level refers to queries with universal quantifiers and negation.

4 Research Landscape

Formal textual languages, *keyword search* (e.g., [3]), *natural language interfaces* (e.g., [11]), and *visual query languages* (VQL) (e.g., [17]) are other means for querying structured data. Formal textual languages are inaccessible to end users, since they demand a sound technical background. Keyword search and natural language interfaces remain insufficient for querying structured data, due to low expressiveness and ambiguities, respectively. VQLs provide a formal visual syntax and notation, and are comparable to formal textual languages in terms of their accessibility to end users. A VQS differs from a VQL, since it is built on an arbitrary set of user actions that effectively capture a set of syntactic rules specifying a (query) language, and hence offers a good usability-expressiveness balance. Existing approaches for ontology-based visual query formulation are either meant for semantic search and browsing of linked data (e.g., [5, 6]) or purely for visual query formulation (e.g., [15, 30]).

An example of the former paradigm is *Rhizomer* [5], as shown in Figure 1. Rhizomer combines faceted search with navigation to enable joining and filtering information from multiple objects. In the present context, a primary issue is that *browsing* is not the same as *querying*, although some commonalities exist. The fact that databases include large data sets makes browsing an inefficient mechanism and necessitates interaction at a conceptual rather than data level.

An example of the latter paradigm is *OptiqueVQS* [32], as shown in Figure 2. OptiqueVQS combines different representation and interaction paradigms through a *widget-based* architecture. OptiqueVQS employs OWL 2 ontologies and graph navigation forms the backbone of query formulation. Since OWL 2 axioms per se are not well-suited for a graph-based navigation, OptiqueVQS employs an approach projecting ontologies into navigation graphs [1].

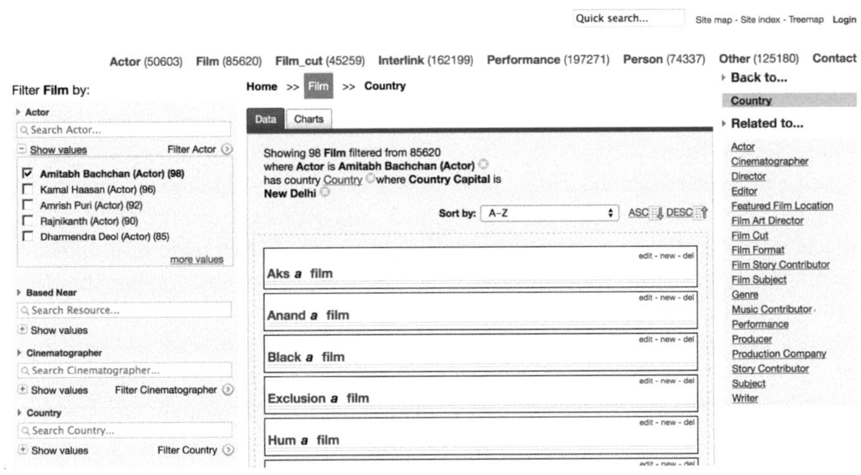

Fig. 1 Rhizomer with an example query.

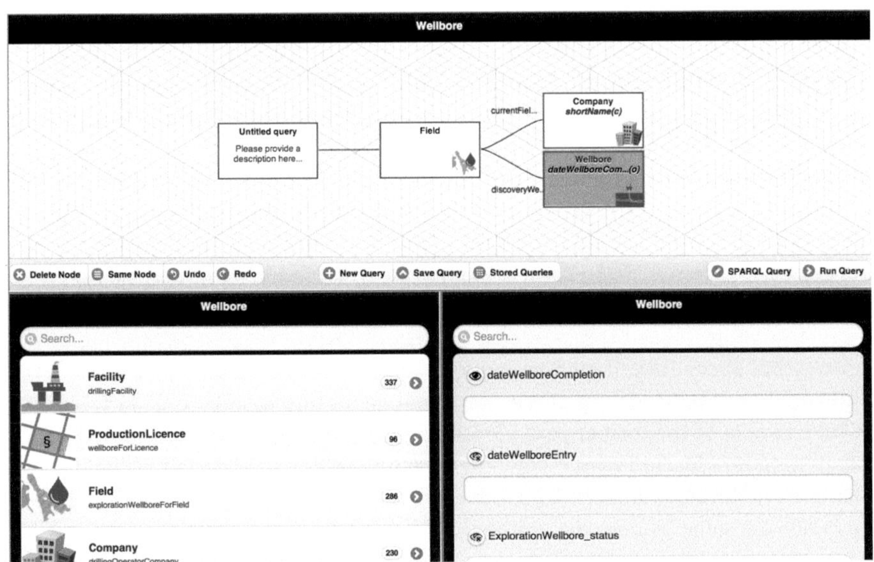

Fig. 2 OptiqueVQS with an example query.

Table 1 Quality attributes mapped to a cost-benefit model.

Cost	**Attribute**	**Benefit**	**Attribute**
Technology	A7, A8	Functionality	A2, A6, A7
Learning	A1, A10	Flexibility	A2, A5, A6
Developing	A1, A3, A4	Usability	A1, A3, A10
Debugging	A3, A9	Quality	A1, A3, A10

5 Quality Attributes and Features

Quality attributes are non-functional requirements that effect run-time behaviour, design, and user experience. From an end-user development perspective, the goal is to derive a set of quality attributes for VQSs, which effectively increases the benefits gained and decreases the cost of adoption for end users (cf. [34]). The design and development process of OptiqueVQS and industrial partners involved provide a real-life basis for this purpose.

Optique strives to offer an *end-to-end* ontology-based data access platform that brings together different related components, such as for query formulation, transformation, and execution (see [14]). In the course of the Optique project, a total of four industrial workshops were conducted with the use-case partners, namely Siemens AG[2] and Statoil ASA[3]. In the first set of workshops, unstructured interviews were conducted with domain experts and they were observed in their daily work routines. Shortly after the first set of workshops, a paper mock-up was demonstrated to the domain experts and further discussions were held. A running prototype was developed iteratively with representative users in the loop. At the second round of workshops, domain experts experimented with the prototype in a formal think aloud session and provided further feedback. A usability study with casual users was also conducted (see [32]).

Along with encouraging results, a set of interrelated and non-exhaustive quality attributes and *features* realising them were collected, which are reported in the following. Note that these quality attributes are strongly interrelated, for instance, usability of a VQS is generally positively affected by every other quality attribute, or adaptivity inherently supports visual scalability as it provides means to eliminate less important concepts and properties.

In Table 1, the proposed quality attributes are mapped into the cost-benefit model suggested by Sutcliffe [34] for end-user development. The cost for adopting a specific end-user development tool includes: the actual cost of the software plus effort necessary to install it (*technology*), the time taken to understand the language or tool (*learning*), the effort necessary to develop applications using the adopted technology

[2] `http://www.siemens.com`

[3] `http://www.statoil.com`

(*development*), and the time taken to test and debug the designed system (*debugging*); while the benefits set against include: the extent of functionality which using the technology can deliver (*functionality*), the flexibility to respond to new requirements (*flexibility*), the usability of applications produced (*usability*), and the overall quality of the applications produced (*quality*).

5.1 Quality attributes

Usability (A1). Usability is measured in terms of effectiveness, efficiency, and satisfaction. The effort required to learn and formulate queries is typically proportional to the usability of a VQS, which also transitively effects the usability and quality of the formulated queries, as it is the VQS that visualises them and provides necessary affordances.

Modularity (A2). Modularity is the degree to which a system's components are independent and interlocking. A highly modular VQS ensures flexibility and extensibility, so that new components could be easily introduced to adapt to changing requirements and to extend and enrich the functionality provided.

Scalability (A3). In the present context, scalability refers to the ability of a VQS to visualise and deal with large ontologies, data sets, and queries. A scalable VQS increases the usability and quality of formulated queries by increasing their comprehensibility and makes formulation and debugging easier against large ontologies, queries and result sets.

Adaptivity (A4). Adaptivity refers to the ability of a system to alter its behaviour, presentation, and content with respect to context. A VQS could reduce the effort required for query formulation by adaptively offering concepts and properties for instance with respect to previously executed queries.

Adaptability (A5). In contrary to adaptivity, adaptability is a manual process, where users customise application to their own needs and contexts. An adaptable VQS could provide flexibility against changing requirements, e.g., by activating a new presentation module.

Extensibility (A6). Extensibility refers to the ability and the degree of effort required to extend a system. An extensible VQS provides flexibility to extend and enrich functionality to meet changing requirements or to improve the system.

Interoperability (A7). Interoperability for a VQS is its ability to communicate and exchange data with other applications in an organisation or a digital ecosystem.

Interoperability contributes to the functionality of a VQS by allowing it to utilise or feed other applications.

Portability (A8). Portability refers to the ability of a VQS to query arbitrary domains, rather than only a specific domain, without high manual tuning and re-coding. A portable system reduces the effort required for installation and configuration.

Data Quality (A9). Data is often imperfect, for instance, records might be missing/incomplete, new facts might derived as a result of reasoning process, and multiple data sources might be involved. The debugging of formulated queries becomes easier, if means to detect such situations are provided.

Reusability (A10). Reusability refers to the ability of a VQS to utilise queries as consumable resources. Reusability could decrease the learning effort by utilising stored queries as learning resources and increase the quality and usability by allowing more complex queries to be built by modifying the existing ones.

5.2 Features

A set of prominent *features* were identified to support the realisation of proposed quality attributes, and are presented in the following.

View and Overview (F1). A VQS needs to provide a overview of the constructed query continuously and at the same time should focus the user to the active phase of the task at hand. This balance is to ensure maximum end-user *awareness* and *control* (cf. [8]) (A1).

Exploration and Construction (F2). Means for exploration and construction need to be intertwined adequately. Exploration does not necessarily need to be at the conceptual level in terms of concepts and properties, but also at the instance level, in terms of cues (i.e., sample results) and instance level browsing (cf. [8, 27]) (A1).

Non-local Navigation (F3). The navigation in an ontology might be a tedious process, particularly when the source concept and the target concept(s) of an intended path are considerably distant from each other. Suggesting paths for given two distant concepts could enhance the user experience (cf. [24]) (A1).

Collaborative Query Formulation (F4). Users can formulate more complex queries and/or improve effectiveness and efficiency by collaborating with each other actively or passively. Such collaboration could be between end users and IT experts or between end users (cf. [23]) (A1).

Query-Reuse (F5). An end user should be able to reuse an existing query stored in the system either as it is or could modify it to construct more complex queries and/or to improve the effectiveness and efficiency (A1 and A10). Query reuse could indeed be considered a passive form of collaboration (cf. [23]) (F4).

Spiral/Layered Design (F6). Complex functionalities could hinder the usability for less competent users. Therefore, system functionality should be distributed into different layers (cf. [28]) (A1). With such an approach, users can also view ontology at different levels of detail (A3); the system or users can tailor available functionality with respect to their needs (A4 and A5), and new functionalities could be added without overloading the interface (A6).

Gradual Access (F7). An ontology might represent a large conceptual space with many concepts and relationships, and the amount of information that can be communicated on a finite display is limited. Therefore, gradual and on-demand access to the relevant parts of an ontology is necessary to cope with large ontologies (cf. [18]) (A1 and A3).

Ranked Suggestions (F8). Even with a small ontology, users can be confronted with many options to choose from. Ranking ontology elements with respect to context, e.g., query log, improves the user efficiency and filters down the amount of knowledge presented (cf. [31]) (A1, A3, and A4). Ranking is a form of passive collaboration as it utilises previously formulated queries to provide adaptive suggestions and gradual access (F4 and F7).

Domain-Specific Representations (F9). Variety in data necessities specific visualisation paradigms that suit well to the data at hand, such as maps for geospatial data. This ensures contextual delivery of data leading to immediate grasping (cf. [35]) (A1). The availability of domain-specific representations provides users and system with the opportunity to select paradigm(s) that fits best to context (A4 and A5).

Multi-paradigm and Multi-perspective (F10). A multi-paradigm and perspective approach combines multiple representation and interaction paradigms and query formulation approaches, such as visual query formulation and textual query editing, to meet diverse type of users, tasks, etc. (cf. [8, 18]) (A1). Moreover, the system and users can adapt presentation (A4 and A5) and can select among various paradigms depending on their user role (F4), task (F2), and data (F9).

Modular Architecture (F11). A modular architecture allows new components to be easily introduced and combined in order to adapt to changing requirements and to support diverse user experiences (cf. [30]) (A1, A2 and A6). This could include alternative/complementary components for query formulation, exploration, visualisation, etc. with respect to context (A3, A4, A5, F9, and F10).

Data Exporting (F12). Domain experts usually feed analytics tools with the data extracted for sense-making processes, however they are not expected to have skills to transform data from one format to another. Therefore affordances to export data in different formats are required to ensure that the system fits into organisational context, workflows (A7) and broader user experience (A1).

Domain-Agnostic Backend (F13). Domain independence is the key for a VQS to access different ontologies and datasets without any extensive manual customisation and code change (cf. [19]) (A8).

Provenance (F14). Provenance support particularly should enable users to inspect the derived facts and the source of data if multiple data sources are involved (A9). Such a feature increases the user trust for the system and the returned results sets (A1).

Auxiliary Support (F15). While formulating a query, users might benefit greatly from auxiliary support, such as cues (i.e., example results), numeric indicators (e.g., whether the query will return something or not), and autocompletion for text fields etc. (A1).

Iterative Formulation (F16). A query is often not formulated in one iteration, therefore a VQS should enable users to engage in a formulate, inspect, and reformulate cycle (cf. [16]) (A1).

6 Conclusion

In this paper, a set of key concepts and aspects for ontology-based visual query formulation was reviewed and a set of quality attributes and features was suggested largely from an end-user perspective.

Acknowledgment This research is funded by the Seventh Framework Program (FP7) of the European Commission under Grant Agreement 318338, "Optique".

References

1. Arenas, M., et al.: Faceted search over ontology-enhanced RDF data. In: CIKM 2014 (2014)
2. Baeza-Yates, R., Ribeiro-Neto, B.: Modern Information Retrieval. Addison Wesley (1999)
3. Bobed, C., et al.: Enabling keyword search on Linked Data repositories: An ontology-based approach. International Journal of Knowledge-Based and Intelligent Engineering Systems **17**(1) (2013)
4. Brooke, J.: Usability evaluation in industry, chap. SUS - A quick and dirty usability scale. Taylor and Francis (1996)
5. Brunetti, J.M., et al.: From overview to facets and pivoting for interactive exploration of semantic web data. International Journal on Semantic Web and Information Systems **9**(1) (2013)
6. Brunk, S., Heim, P.: tFacet: hierarchical faceted exploration of semantic data using well-known interaction concepts. In: DCI 2011 (2011)
7. Burnett, M.M.: Visual programming. In: Webster, J.G. (ed.) Wiley Encyclopedia of Electrical and Electronics Engineering. John Wiley & Sons (1999)
8. Catarci, T., et al.: Visual query systems for databases: A survey. Journal of Visual Languages and Computing **8**(2) (1997)
9. Catarci, T., et al.: An ontology based visual tool for query formulation support. In: ECAI 2004 (2004)
10. Grau, B.C., et al.: Towards query formulation and query-driven ontology extensions in OBDA systems. In: OWLED 2013 (2013)

11. Damljanovic, D., et al.: Improving habitability of natural language interfaces for querying ontologies with feedback and clarification dialogues. Web Semantics: Science, Services and Agents on the World Wide Web **19**, (2013)
12. Dividino, R., Groner, G.: Which of the following SPARQL queries are similar? why? In: LD4IE 2013 (2013)
13. Giese, M., et al.: Scalable end-user access to big data. In: Rajendra, A. (ed.) Big Data Computing. CRC (2013)
14. Giese, M., et al.: Optique - Zooming In on Big Data Access. IEEE Computer **48**(3) (2015)
15. Haag, F., et al.: Visual SPARQL querying based on extended filter/flow graphs. In: AVI 2014 (2014)
16. Harth, A.: VisiNav: A system for visual search and navigation on web data. Web Semantics: Science, Services and Agents on the World Wide Web **8**(4) (2010)
17. Harth, A., et al.: Graphical representation of RDF queries. In: WWW 2006 (2006)
18. Katifori, A., et al.: Ontology visualization methods - A survey. ACM Computing Surveys **39**(4) (2007)
19. Kaufmann, E., Bernstein, A.: Evaluating the usability of natural language query languages and interfaces to Semantic Web knowledge bases. Web Semantics: Science, Services and Agents on the World Wide Web **8**(4) (2010)
20. Kawash, J.: Complex Quantification in Structured Query Language (SQL): A Tutorial Using Relational Calculus. Journal of Computers in Mathematics and Science Teaching **23**(2) (2004)
21. Kogalovsky, M.R.: Ontology-Based Data Access Systems. Programming and Computer Software **38**(4) (2012)
22. Lieberman, H., et al.: End-user development: an emerging paradigm. In: Lieberman, H., Paternó, F., Wulf, V. (eds.) End-User Development. Springer, Netherlands (2006)
23. Marchionini, G., White, R.: Find what you need, understand what you find. International Journal of Human-Computer Interaction **23**(3) (2007)
24. Popov, I.O., Schraefel, M.C., Hall, W., Shadbolt, N.: Connecting the dots: a multi-pivot approach to data exploration. In: Aroyo, L., Welty, C., Alani, H., Taylor, J., Bernstein, A., Kagal, L., Noy, N., Blomqvist, E. (eds.) ISWC 2011, Part I. LNCS, vol. 7031, pp. 553–568. Springer, Heidelberg (2011)
25. Rodriguez-Muro, M., Calvanese, D.: Quest, a system for ontology based data access. In: OWLED 2012 (2012)
26. Ruiz, F., Hilera, J.R.: Using ontologies in software engineering and technology. In: Calero, C., Ruiz, F., Piattini, M. (eds.) Ontologies for Software Engineering and Software Technology. Springer-Verlag (2006)
27. Schraefel, M.C., et al.: mSpace: improving information access to multimedia domains with multimodal exploratory search. Communications of the ACM **49**(4) (2006)
28. Shneiderman, B.: Direct Manipulation: A Step Beyond Programming Languages. Computer **16**(8) (1983)
29. Siau, K.L., et al.: Effects of query complexity and learning on novice user query performance with conceptual and logical database interfaces. IEEE Transactions on Systems, Man and Cybernetics - Part A: Systems and Humans **34**(2) (2004)
30. Soylu, A., et al.: OptiqueVQS - towards an ontology-based visual query system for big data. In: MEDES 2013 (2013)
31. Soylu, A., Giese, M., Jimenez-Ruiz, E., Kharlamov, E., Zheleznyakov, D., Horrocks, I.: Towards exploiting query history for adaptive ontology-based visual query formulation. In: Closs, S., Studer, R., Garoufallou, E., Sicilia, M.-A. (eds.) MTSR 2014. CCIS, vol. 478, pp. 107–119. Springer, Heidelberg (2014)

32. Soylu, A., et al.: Experiencing OptiqueVQS: A Multi-paradigm and Ontology-based Visual Query System for End Users. Universal Access in the Information Society (2015) (in press)
33. Spanos, D.E., et al.: Bringing relational databases into the Semantic Web: A survey. Semantic Web **3**(2) (2012)
34. Sutcliffe, A.: Evaluating the Costs and Benefits of End-user Development. ACM SIGSOFT Software Engineering Notes **30**(4), 1–4 (2005)
35. Tran, T., et al.: SemSearchPro - Using semantics throughout the search process. Web Semantics: Science, Services and Agents on the World Wide Web **9**(4) (2011)
36. Yen, M.Y.M., Scamell, R.W.: A Human Factors Experimental Comparison of SQL and QBE. IEEE Transactions on Software Engineering **19**(4) (1993)

A Scalable Algorithm for Answering Top-K Queries Using Cached Views

Wissem Labbadi and Jalel Akaichi

Abstract Recently, various algorithms were proposed to speed up top-k query answering by using multiple materialized query results. Nevertheless, for most of the proposed algorithms, a potentially costly view selection operation is required. In fact, the processing cost has been shown to be linear with respect to the number of views and can be exorbitant given the large number of views to be considered. In this paper, we address the problem of identifying the top-N promising views to use for top-k query answering in the presence of a collection of views. We propose a novel algorithm, called *Top-N rewritings algorithm*, for handling this problem, which aims to achieve significant reduction in query execution time. Indeed, it considers minimal amount of rewritings that are likely necessary to return the top-k tuples for a top-k query. We consider, also, the problem of how, efficiently, exploit the output of the *Top-N rewritings algorithm* to retrieve the top-k tuples through two possible solutions. The results of a thorough experimental study indicate that the proposed algorithm offers a robust solution to the problem of efficient top-k query answering using views since it discards non-promising query rewritings from the view selection process.

Keywords Top-k query processing · Top-k query answering using materialized views · Top-N views

1 Introduction

In front of the increasing number of web applications, users are often overwhelmed by the variety of relevant data. Therefore, the need for ranking queries that allow users to retrieve only a limited set of the most interesting information is becoming indispensable. Top-k queries appear as an effective solution to this problem.

W. Labbadi(✉) · J. Akaichi
BESTMOD Lab-ISG of Tunis, 41 City Bouchoucha 2000, Bardo, Tunisia
e-mail: {wissem.labbadi,jalel.akaichi}@isg.rnu.tn

T. Andreasen et al. (eds.), *Flexible Query Answering Systems 2015*,
Advances in Intelligent Systems and Computing 400,
DOI: 10.1007/978-3-319-26154-6_20

Nowadays, top-*k* query processing is becoming widely applied in many domains such as information retrieval, multimedia search and recommendation generation [2], due to the capacity of such queries to retrieve a limited number (*k*) of highest ranking tuples according to some monotone combining function defined on a subset of the attributes of a relation [3].

Different techniques for flexible top-*k* query processing in a distributed databases context have been surveyed in [2]. In this paper, we deal with the distinguished approach consisting of answering top-*k* queries using views. A view in this context is a materialized version of a previously posed query. In the literature, the problem of using materialized views for speeding up top-*k* query answering in relational databases is known as the top-*k* query answering using views (top-*k* QAV) problem. One of the reasons for encouraging top-*k* query processing through materialized views is the promise of increased performance. For example, views may be of small size, so if an answer can be obtained by processing data from one or more views the answer could be obtained much faster [4]. That is a new top-*k* query may be answered through these materialized views resulting in better performance than making use only of the base relation from the database. Moreover, in [5], authors have provided theoretical guarantees for the adequacy of a view to answer a top-*k* query, along with algorithmic techniques to compute the query via a view when this is possible.

Several algorithms have been proposed for processing such queries using views. Hristidis et al. [6] first considered the problem of using materialized query results for speeding up top-k query processing in relational databases. In the PREFER system, Hristidis et al. focused on selecting one best view which can be used for answering a top-*k* query. However, as mentioned in [4], their setting is highly restrictive as it assumes that only one view could be utilized to obtain an answer to a new query, and it also makes a strong assumption that all attributes of the underlying base table are always utilized for all top-*k* queries. Overcoming the limitations of PREFER, Das et al. [4] propose a new algorithm, called LPTA, capable of combining the information from multiple views to answer a top-*k* query. Ryeng et al. [8] extend these techniques to answer top-*k* queries in a distributed environment.

Though LPTA overcomes many of the limitations of PREFER, unfortunately it still suffers from lack of scalability. To overcome these limitations, Xie et al. [7] propose the LPTA^{+} algorithm which provides significantly improved efficiency compared to the state-of-the-art LPTA algorithm [4].

For general flexible accessing to information, the most general approach is the one based on the fuzzy set theory, which allows for a larger variety of flexible terms and connectors [10]. Top-*k* queries, in this context, are considered only as a special case of fuzzy ones [11]. Various efficient algorithms have been proposed for processing queries involving either preferences or fuzzy predicates [12, 13].

The main objective in this paper is to determine only the top-N rewritings based on views which are able to return the top-*k* tuples to a user query. To do that, we propose the *Top-N rewritings algorithm* to generate, without computing all possible ones, the N best rewritings necessary to return the top-*k* tuples as answers to the user ranking query. It tries to achieve significant reduction in query execution time as it considers minimal amount of rewritings to answer a top-*k* query. It generates the best rewriting fitting user preference first, then the next best rewriting second until having the N required rewritings. Its originality is for searching the N best rewritings, it considers only N rewritings and it is certain that are the best ones. *Top-N rewritings algorithm* is inspired from the reference MiniCon algorithm [1] generating the maximally contained rewriting of a conjunctive query Q.

The rest of this paper is organized as follows. We discuss related work in section 2. In section 3 we present formal definitions of the problem considered in this paper and we propose the *Top-N rewritings algorithm*. In section 4 we present a detailed set of experiments showing that the performance of the *Top-N rewritings algorithm* can be order of magnitude better than the state-of-the-art algorithms and section 5 concludes the paper.

2 Literature Survey

In this section, we first survey the relevant works that have dealt with the problem of answering top-*k* query using materialized views. Second, we present similar works having addressed the issue of using only the best views to answer top-*k* queries.

The problem of using materialized views for answering top-*k* queries in relational databases has attracted many research works in the domain of data integration, query optimization, and data warehouse design due to its relevance for speeding up top-*k* query processing. Several algorithms have been proposed for processing such queries using views. Hristidis et al. [6] first considered the problem of using materialized views to speed up top-*k* query processing in a relational database. Each view stores all tuples of a relational table ranked according to different ranking functions. In the PREFER system, Hristidis et al. focused on finding among these materialized views the best one which can be used for answering a top-*k* query. But, as mentioned in [4], their setting is highly restrictive, as it cannot exploit multiple materialized views. Due to this assumption, the selected view should be defined as a materialized version of a previously posed query selecting all the tuples in the relation not only some of them like the case of a top-*k* query requesting all the values ranked according to some monotone combining function defined on a subset of the attributes of the relation. An additional strong assumption is that all attributes of the underlying base table are always utilized for all top-*k* queries.

In order to achieve improved performance, researchers propose to restrict the size of materialized views from a copy of the entire relation to a fixed number of tuples such as the result of a previous top-*k* query. Such views are called the top-*k* views [3], [4], [6]. Then, a new top-*k* query may be answered through these materialized top-*k* views resulting in better performance than making use only of the base relation from the database. Das et al. [4] propose a novel algorithm, called LPTA, which overcomes the limitations of [6] by combining the information from multiple views to answer a top-*k* query as long as they contain enough tuples to satisfy the new query. This algorithm is a nontrivial adaptation of the Threshold Algorithm TA proposed by Fagin et al.[15] and independently by Guntzer et al. [18] and Nepal et al. [19], requiring the solution of linear programming optimizations in each iteration. Ryeng et al. [8] extend the techniques proposed in [4] to answer top-*k* queries in a distributed environment. They assume that access to the original database is available through the network interface, thus exact top-*k* answers can always been found by forming a "remainder" query which can be utilized to fetch tuples not available in the views. Although LPTA overcomes many of the limitations of PREFER, however it still suffers from several significant limitations. Firstly, it bases one's argument on the assumption that either each top-*k* view is a complete ranking of all tuples in the database, or that the base views, which are complete rankings of all tuples in the database according to the values of each attribute, are available. These assumptions may often be unrealistic in practice. The second limitation of the LPTA algorithm proposed in [4] is that it calls iteratively a sub-procedure to calculate the upper bound on the maximum value achievable by a candidate result tuple. It has been shown in [4] that for scenarios with higher dimensionality, the cost of the iterative invocation of the sub-procedure may be very high. Finally, LPTA [4] requires a potentially costly view selection operation. Thus, the view selection algorithm in [4] requires the simulation of the top-*k* query process over the histograms of each attribute, and the processing cost is linear with respect to the number of views. This cost can be prohibitive given a large pool of views.

To overcome the limitations of LPTA [4], Xie et al. [7] propose the LPTA$^+$ algorithm, an improved version of LPTA [4] which operates in an environment in which base views are not available and the cached views include only the top-*k* tuples which need not cover the whole relation. By taking advantage of the fact that the contents of all views are cached in memory, LPTA$^+$ can usually avoid the iterative calling of the sub-procedure invoked by LPTA [4] and can be leveraged to answer a new top-*k* query efficiently without any need for view selection, thus greatly improving the efficiency over the LPTA algorithm.

In [8], Ryeng et al. approve the ideas proposed by Das et al. [4] and show that is more beneficial to execute a remainder query, rather than restarting a cached query and retrieving additional tuples. Authors in [8] present a framework called ARTO an algorithm with remainder top-*k* queries which is different from existing approaches [16] that require the servers to recompute the query from scratch. Eventually, ARTO does not evaluate the entire query, but it retrieves as many tuples k' ($k' < k$) as possible from the local cache, and in order to retrieve the missing tuples, it defines the remainder query that provides the remaining $k - k'$ tuples that are not stored in the cache. More detailed, Ryeng et al. split the top-*k* query into a top-k' query ($k' < k$) that is answerable from cache, and a remainder ($k - k'$) query that provides the remaining tuples that were not retrieved from the top-k' query.

Finally, Hristidis et al. [3] use and extend the core algorithm described in the PREFER system [6] to answer ranked queries in an efficient pipelined manner using materialized ranked views. Thus the PREFER system precomputes a set of materialized views that provide guaranteed query performance. Then, to cope with space limitation reserved for storing materialized views, Hristidis et al. in [3] present an algorithm that selects a near optimal set of views necessary to answer a top-*k* query. In more details, they present an algorithm that computes the top-*k* results of a query by using the minimal prefix of a ranked view and show that the performance can be improved by storing only prefixes of views, as opposed to whole views, and utilizing the space to precompute and save more views.

A more general approach of flexible accessing information is that based on the fuzzy set theory, which allows for a larger variety of flexible terms and connectors [10]. In this context, the top-*k* queries are considered only as a special case of fuzzy ones [11].

In [12], authors deal with fuzzy query processing in a distributed database context where user queries and view descriptions may include fuzzy predicates. This work considers the case where the query involves only one fuzzy predicate. The main objective was to match a fuzzy query against the descriptions of fuzzy views in order to determine the set of views which only provide satisfactory answers, those whose satisfaction degree with respect to the fuzzy query is over a threshold specified by the user. This matching mechanism is based on a graded inclusion, itself relying on a fuzzy implication.

In addition, authors in [12] propose to apply the cutting techniques on views according to a fixed β level of cuts in order to determine subsets of the views such that these subsets only contain answers whose satisfaction degree relatively to the query is above a user defined threshold α.

Jaudoin et al. [13] deal with answering a query Q involving preferences in the presence of imprecise views. Each query Q is associated with a threshold $\alpha \in [0, 1]$ that specifies the minimal expected satisfaction degree attached with the answers, namely that of α-certain answers. The idea was to approximate this set of answers by authorizing exceptions that concern only data whose plausibility degree with respect to the description of their original source is less than $1-\alpha$. Thus, according to Jaudoin et al. [13], a data from a source is considered an answer to a given query Q, either if it has a low plausibility degree with respect to that source, or it has a high enough satisfaction degree with respect to Q.

To the best of our knowledge, there were three research works that have addressed the issue of top-*k* query answering using only a set of best views for an efficient processing of user queries. In [12], the authors propose to determine, based on a matching mechanism, the set of satisfactory views to answer a fuzzy query. This paper dealt with fuzzy query processing in a context of large-scale distributed relational databases. The data sources were assumed to be accessible through fuzzy views where each view contains a set of tuples (μ/t) where μ is a satisfaction degree expressing the extent to which the tuples satisfy the fuzzy constraint that defines the view. The idea was to match a fuzzy query against the fuzzy views in order to determine the set of views which only provide answers whose satisfaction degree is over the threshold specified by the user. The matching mechanism measures the subsumption degree between the description of a fuzzy view and the description of the fuzzy query submitted by the user. This matching mechanism is based on a graded inclusion, itself relying on a fuzzy implication.

In [3], an attempt to rank approximate rewritings based on views to answer a query by means of a tolerant method was proposed. In this work, the authors studied the problem of ranking query rewritings based on views in the setting of conjunctive queries with simple interval constraints X ∈ [*a*, *b*], where X is an attribute and *a* and *b* are constants. They proposed to attach a degree ∈ [0, 1] to each rewriting Q' of a query Q, which corresponds to the probability for a tuple returned by Q' to be an answer to Q. The idea was, instead of computing all the approximate rewritings of a query and then ranking them, to adapt a well-known regular query rewriting algorithm, namely Minicon [1], in order to directly generate the k best rewritings ranked in decreasing order of their degree, reducing thereby the combinatorics relative to the potentially high number of MCDs formed by the MiniCon.

Finally, in [14], authors suggested to answer ranked queries using ranked views. They proposed an algorithm that computes the top-N results of a query by using the minimal prefix of a ranked view. Globally, the idea was an extension of the core algorithm in the described PREFER [6] and MERGE systems. The proposed algorithm selects a near optimal set of views under space constraints and then merges the ranked results from these views by retrieving the minimum prefix from each of them. In this paper, the authors finished by concluding that the performance can be improved by storing only prefixes of views, as opposed to whole views, and utilize the space to precompute and save more views.

3 Answering Top-K Query Using Top-N Rewritings

In this work, we propose an improvement of the previous attempts proposed to solve the problem of TOP-*K* QUERY ANSWERING USING VIEWS [4, 6, 7]. To do that, we present in this section an algorithm called the *Top-N rewritings algorithm* inspired from the MiniCon one [1]. Then, we briefly propose two techniques to explore the output of the *Top-N rewritings algorithm* in order to generate the top k ranked tuples for the query.

We formulate in the following the problem considered in this paper.

PROBLEM 1. (TOP-*K* QUERY ANSWER USING TOP-N VIEWS). Given a collection of views $V = \{V_1, \ldots, V_T\}$ over a relational data set schema R, and a Top-*k* query Q requesting a set of ranked tuples from R such that $|Q(R)| = k$ according to a preference function, determine efficiently the subset $U \subseteq V$ of N best views ($N < T$) which can provide the top *k* ranked tuples for Q combining all the information conveyed by the views in U.

The desirability or preference of a given view is measured according to a value or score assigned to this view using a score function.

3.1 Top-N Rewritings Solution

The essence of the *Top-N rewritings algorithm* is to generate the N best rewritings for a top-*k* query Q without computing its all possible rewritings. It tries to achieve significant reduction in query execution time as it considers minimal amount of rewritings that are likely necessary to return the Top-*K* tuples for a submitted Top-*k* ranking query. It generates the best rewriting fitting user preference first, then the next best rewriting second until having the N required rewritings. We mention that the numbers (N) of the rewritings to find and (*K*) of the desired tuples are different and N is an internal variable fixed by the system depending on the *K* value. Our algorithm differs from similar research works on top-N rewritings retrieval process. Actually it generates the top-N ordered rewritings without computing them or searching all possible ones for a given ranking query. Its originality is for searching the N best rewritings, it considers only N rewritings and it is certain that these are the best ones. *Top-N rewritings algorithm* is inspired from the MiniCon [1] one generating the maximally contained rewriting of a conjunctive query Q. MiniCon provides a maximally contained rewriting for Q and not an equivalent one because in virtual information integration systems, it is assumed that the sources are incomplete including only a subset of answers and the views defined over the global schema may not refer to all databases stored information (Open World Assumption). MiniCon can be used as a part of our algorithm after modifying it. The first modification concerns the input and the output of the algorithm. The input of *Top-N rewritings algorithm* is a top-*k* query, a set of crisp views ordered according to their scores and the number N. The score of a view, noted S (V), is between 0 and 1 and describes the average of the preference scores attached to the different values composing V. Each value *v* of V is assigned with a value or score calculated using a score function *f*(*v*), which indicates the desirability or preference of *v*. The simplest and most intuitive function is simply a linear scoring function [4, 6, 20, 21]. The output is the N best rewritings according to the submitted query. The second modification is in the MCD formation step. In fact, this step is not done for all views at the algorithm beginning. Nevertheless, it is performed for the most important view in the views' list then for the second most important view, etc. The higher view score is, the most important the view is for a rewriting. The second step of the MiniCon algorithm consisting in combining the MCDs is also modified. In fact, *Top-N rewritings algorithm* searches the best rewritings using the MCDs of the most important view, then the second most important rewritings by combining the MCDs of the most important view and the MCDs of the second most important view, etc. A rewriting importance is evaluated by a monotonic scoring function aggregating the elementary scores of the different views composing the rewriting. Our algorithm stops once it reaches the N needed rewritings.

Top-N Rewritings Algorithm. The pseudocode for *Top-N rewritings* is given in Algorithm 1. The input of the algorithm is a sorted views list according to their associated scores S (V_i), the number of rewritings N and the top-*k* query Q. The sorted views (V_1,..., V_n) are examined according to their orders in the list (see the following running example): the best view is used first then the second until obtaining the N needed rewritings. When a view V_i is chosen from the ordered list of views, the algorithm looks for the MCDs that can be formed to cover the query sub-goals. Next, the algorithm searches the rewritings that can be formed using only the MCDs of V_i, then the rewritings that can be formed using the MCDs of V_i and the MCDs of the previous examined views from V_1 to V_{i-1} (if these MCDs exist). The rewriting research is facilitated by the monotonic propriety of the scoring function. In fact, the research begins with the rewritings formed by one MCD then by two MCDs then by three MCDs, etc. The limit number of MCDs in a rewriting depends on the number of the previous observed views. Our algorithm abandons the search ones it reaches the number N of rewritings. The found rewritings are the best that can be searched.

Algorithm 1. *Top-N rewritings* ($V = \{V_1, \ldots, V_T\}, Q = (f, k)$)

```
Input
  Q: a top-k query
  N: the number of query rewritings
  V: a list of n sorted views
     V= {v1, v2,...,vn |S(v1) < S(v2) < ... < S(vn)}
Output
  Result: top-N rewritings
Begin
 Result ←— Ø
 NbRewriting ←— 0
 NbViews ←— 1
 While (NbRewriting < N) and (NbViews <= n)
Begin
   MCD [NbViews] ←— createMCD (V[NbViews], Q)
   NbVR ←— 1
  While (NbRewriting < N) and (NbVR <= nbsubgoal) and
     (NbVR <= NbViews)
```

```
  Begin
   rewritings ← findRewritings (MCD[NbViews],
                  NbVR, Q)
   If (card (rewritings) ≠ 0) Then
    x ← 1
    While (x <= card (rewritings) and (NbRewriting < N)
    Begin
     Result ← add (rewritings[x], Result)
     x ← x+1
     NbRewriting ← NbRewriting+1
    End
  End if
  NbVR ← NbVR+1
 End
 NbViews ← NbViews+1
 End
  Display (Result)
End.
```

Running Example. Assume we have four views (from V_1 to V_4) to rewrite a top-*k* ranking query Q. Each view has a score (S (V_i)) such that S (V_i) is higher than S (V_{i+1}).

Table 1 Ranked rewritings generation.

Query rewritings	View score
R_1:-V_1	S(V_1) >
R_3:-V_1, V_2	S(V_2) >
R_2:-V_2	
R_7:-V_1,V_2, V_3	S(V_3) >
R_6:-V_1, V_3	
R_5:-V_2, V_3	
R_4:-V_3	
R_{15}:-V_1,V_2, V_3, V_4	S(V_4)
R_{14}:-V_1, V_3, V_4	
R_{13}:-V_2, V_3, V_4	
R_{12}:-V_1, V_2, V_4	
R_{11}:-V_3, V_4	
R_{10}:-V_1, V_4	
R_9:-V_2, V_4	
R_8:-V_4	

Assuming also that each combination of views is a rewriting for Q. Using *Top-N rewritings algorithm*, the rewritings are generated, in Table 1, from the most to the least important (from R_1 to R_{14}): R_1 is the best rewriting as it is formed by the view V_1 having the highest score for Q. Then R_2 is generated having the score of V_2 then R_3, etc. The generation of R_2 is done before R_3 because a rewriting having the least number of views is the most important in the case of equal scores. *Top-N rewritings algorithm* stops once it reaches the number N of rewritings. If N is 1, it stops once R_1 is found without exploring the whole search space as R_1 is the best rewriting. When it finds R_2, it returns R_1 and R_2, etc.

4 Empirical Results

In this section, we investigate the effectiveness of the *Top-N rewritings algorithm* for efficiently answering top-*k* queries using multiple materialized views by comparing its performance to the LPTA$^+$ [7]. The experiments were conducted based on two real datasets of ALBB and NLBB statistics respectively and two synthetic datasets.

Both datasets of ALBB and NLBB are collected from the Baseball Statistics website [22], which contains the career statistics information of the American League Baseball (ALBB) and National League Baseball (NLBB) players until the year 2009. The two datasets have more than 3500 tuples and the different used queries were requiring the top-*k* tuples of the attribute 'Win%'.

The synthetic datasets contain each one 100 k tuples and number of attributes between 8 and 10. In the uniform (UNIFORM) dataset, attribute values are sampled from a uniform distribution in which values have equal spreads. The Zipf dataset (ZIPF) contains attributes where values are sampled from a Zipf-ran distribution in which value spreads follow a Zipf distribution and are randomly assigned to attribute values.

Weights for the score functions in all views are generated randomly, and all views are cached in memory.

The goal of these experiments is to study the performance of the *Top-N rewritings algorithm* compared to the LPTA$^+$-based algorithms. We compare the query processing time of a set of queries provided by real users which ask for top ranked tuples in the presence of a set of views. The experiments were done in three steps. First, we compare the performance of the *Top-N rewritings algorithm* and LPTA$^+$ for queries which ask for the top-100 tuples using a set of 100 views. Second, we vary the number of views stored in the cache and we compare the processing times of queries where the parameter *k* is fixed to 100. Finally, we fix the number of views in the cache pool to 100 and we vary *k* from 10 to 100.

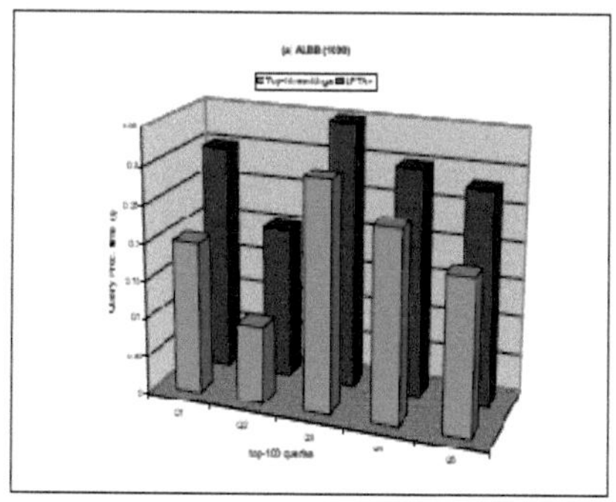
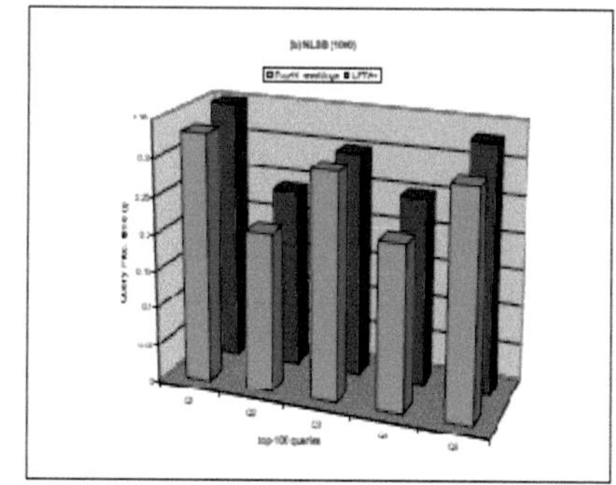
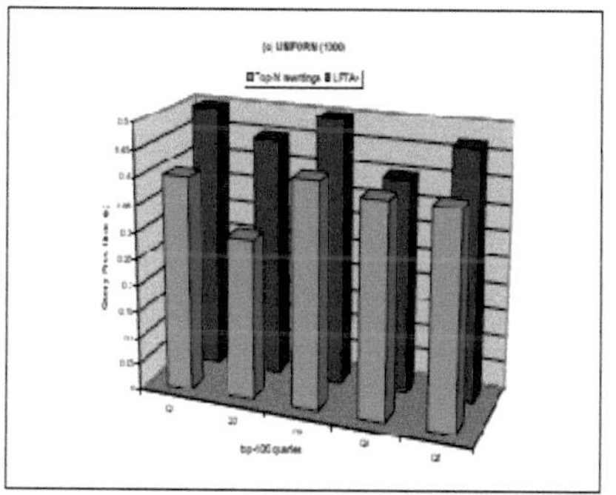
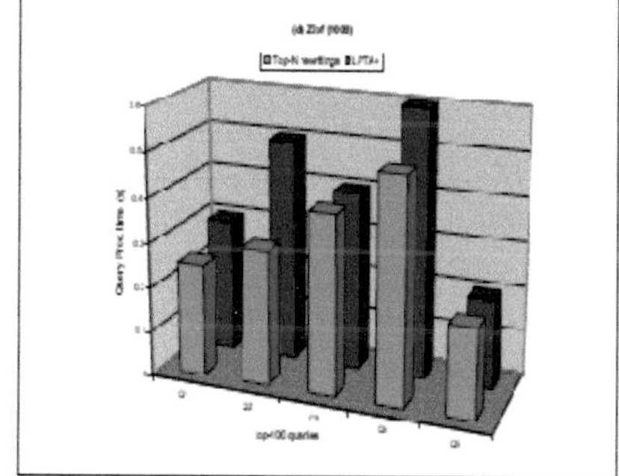
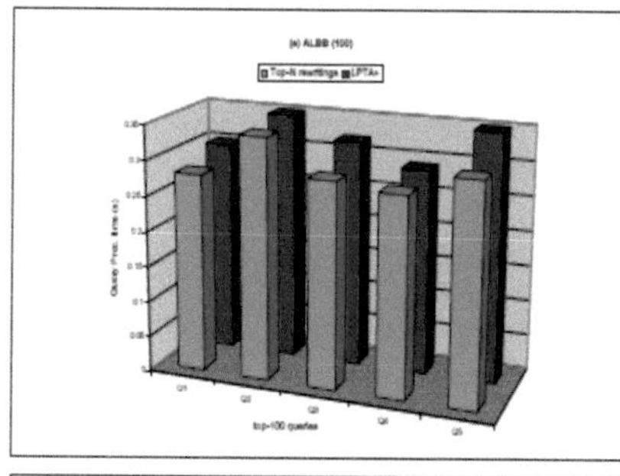
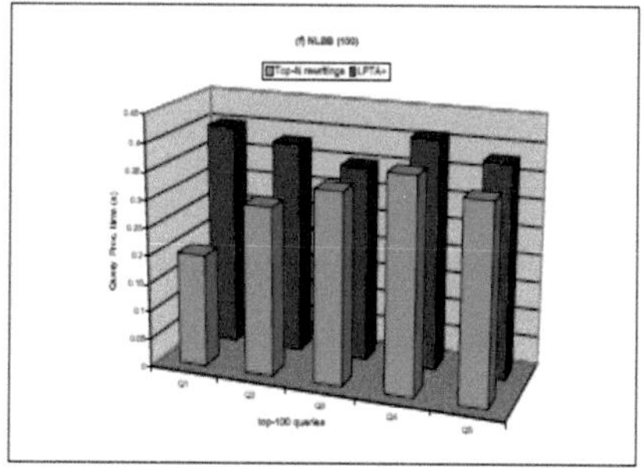
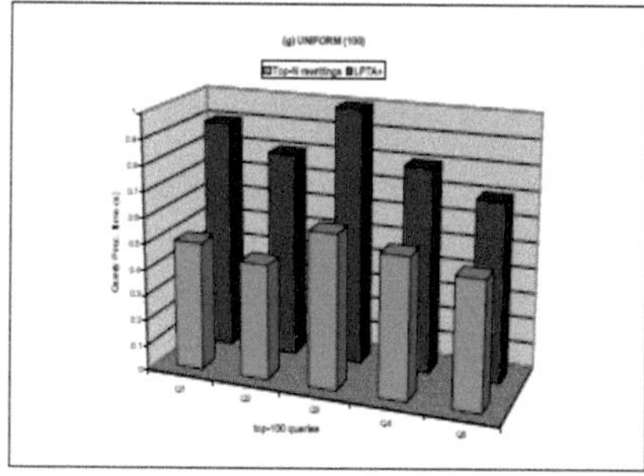
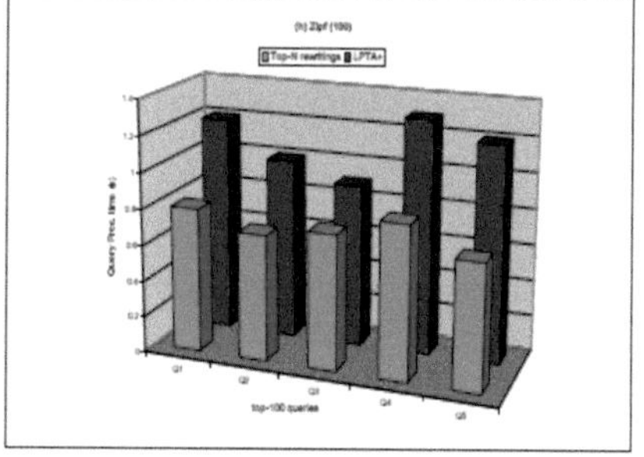

Fig. 1 LPTA+ vs. Top-*N* rewritings: (a-d) results on 4 datasets with each view containing 1000 tuples; (e-h) results on 4 datasets with each view containing 100 tuples.

The two algorithms *Top-N rewritings* and LPTA$^+$ were implemented using Java SE 5, and all experiments were run on a Linux machine with a Core i7 CPU, OpenSUSE 11.

In Fig. 1, we compare the performance of the *Top-N rewritings algorithm* and LPTA$^+$ for queries which ask for the top-100 tuples in the presence of a set of 100 views. Fig. 1 (a–d) considers the setting in which each view contains 1000 tuples. We can see that, for all four datasets, the *Top-N rewritings algorithm* is faster than LPTA$^+$ in most cases. Similar results are obtained for the setting in which each view contains 100 tuples (Fig. 1 (e–h)).

In Fig. 2, we compare the performance of the *Top-N rewritings algorithm* and LPTA$^+$ when varying the number of views in the cache pool and using the dataset UNIFORM. Here we consider queries where *k* is fixed to 100. We can see clearly from this figure that the performance of both algorithms decreases when the number of views increases. Also, the *Top-N rewritings* is faster then LPTA$^+$ in most settings. Results obtained using the other datasets were omitted for lack of space.

In Fig. 3, we compare the performance of *Top-N rewritings algorithm* and LPTA$^+$ when fixing the number of cached views to 100 and varying the value *k* in each query from 10 to 100. We can see clearly from this figure that the performance of the two algorithms degrades as *k* increases, and *Top-N rewritings* is faster than LPTA$^+$. This result was obtained using the UNIFORM dataset.

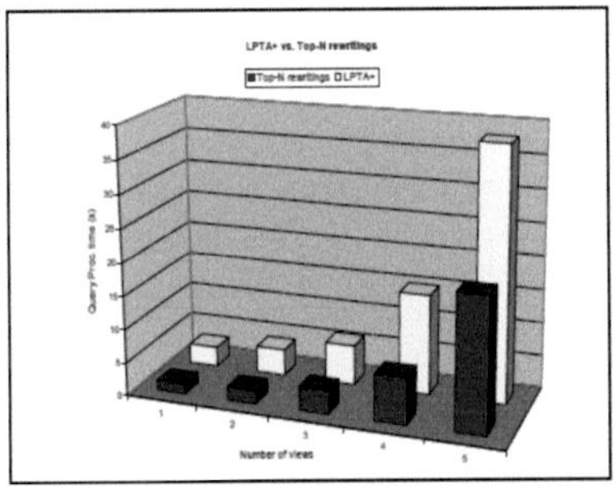

Fig. 2 Performance comparison between LPTA+ and Top-*N* rewritings when varying the number of cached views on the dataset UNIFORM.

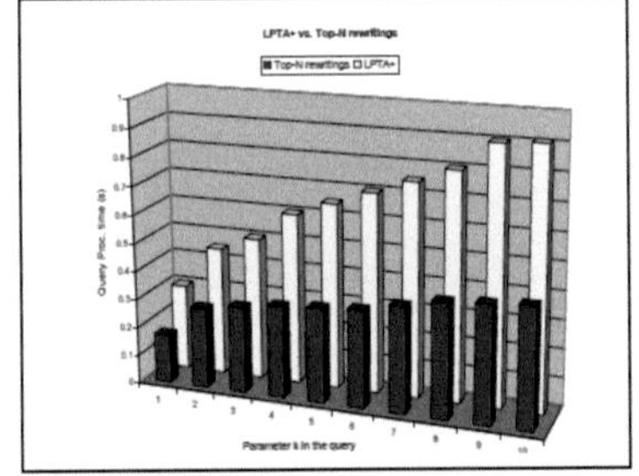

Fig. 3 Performance comparison between LPTA+ and Top-*N* rewritings when varying the value k of a query on the dataset UNIFORM.

5 Conclusion

In this paper, we considered the problem of how to efficiently answer a conjunctive top-*k* query by using top-N views among a collection of previously cached ones.

The proposed approach attaches a degree (score) ∈ [0, 1] to each rewriting Q' of a query Q, which corresponds to the probability for a tuple returned by Q' to be an answer to Q. The *Top-N rewritings algorithm*, presented in this paper, is a novel algorithm for generating ordered top-N rewritings for a top-*k* query without computing or searching all possible ones.

References

1. Pottinger, R., Levy, A.Y.: A scalable algorithm for answering queries using views. In: VLDB, pp. 484–495, San Francisco, CA, USA (2000)
2. Ilyas, I.F., Beskales, G., Soliman, M.A.: A survey of top-k query processing techniques in relational database systems. ACM Comput. Surv. **40**(4) (2008)
3. Hristidis, V., Papakonstantinou, Y.: Algorithms and applications for answering ranked queries using ranked views. VLDB Journal **13**(1), 49–70 (2004)
4. Das, G., Gunopulos, D., Koudas, N., Tsirogiannis, D.: Answering top-k queries using views. In: VLDB, pp. 451–462 (2006)
5. Baikousi, E., Vassiliadis, P.: View usability and safety for the answering of top-k queries via materialized views. In: DOLAP, pp. 97–104 (2009)
6. Hristidis, V., Koudas, N., Papakonstantinou, Y.: PREFER: a system for the efficient execution of multi-parametric ranked queries. In: SIGMOD, pp. 259–270 (2001)
7. Xie, M., Lakshmanan, L.V.S., Wood, P.T.: Efficient top-k query answering using cached views. In: EDBT/ICDT, pp. 18–22 (2013)
8. Ryeng, N.H., Vlachou, A., Doulkeridis, C., Nørvåg, K.: Efficient distributed top-*k*query processing with caching. In: DASFAA, pp. 280–295 (2011)
9. Baikousi, E., Vassiliadis, P.: View usability and safety for the answering of top-k queries via materialized views. In: DOLAP, pp. 97–104 (2009)
10. Bosc, P., Prade, H.: An Introduction to the Treatment of Flexible Queries and Uncertain or Imprecise Databases. In: Motro, A., Smets, P. (eds.) Uncertainty Management in Information Systems, pp. 285–324. Kluwer Academic Publishers, Dordrecht (1997)
11. Bosc, P., Hadjali, A., Jaudoin, H., Pivert, O.: Flexible querying of multiple data sources through fuzzy summaries. In: DEXA Workshop, pp.350–354 (2007)
12. HadjAli, A., Pivert, O.: Towards fuzzy query answering using fuzzy views – a graded-subsumption-based approach. In: An, A., Matwin, S., Raś, Z.W., Ślęzak, D. (eds.) Foundations of Intelligent Systems. LNCS (LNAI), vol. 4994, pp. 268–277. Springer, Heidelberg (2008)
13. Jaudoin, H., Pivert, O.: Rewriting fuzzy queries using imprecise views. In: Eder, J., Bielikova, M., Tjoa, A.M. (eds.) ADBIS 2011. LNCS, vol. 6909, pp. 257–270. Springer, Heidelberg (2011)
14. Jaudoin, H., Colomb, P., Pivert, O.: Ranking approximate query rewritings based on views. In: Andreasen, T., Yager, R.R., Bulskov, H., Christiansen, H., Larsen, H.L. (eds.) FQAS 2009. LNCS, vol. 5822, pp. 13–24. Springer, Heidelberg (2009)
15. Fagin, R., Lotem, A., Naor, M.: Optimal aggregation algorithms for middleware. In: PODS (2001)
16. Zhao, K., Tao, Y., Zhou, S.: Efficient top-k processing in large-scaled distributed environments. Data and Knowledge Engineering **63**(2), 315–335 (2007)

17. Halevy, A.Y.: Answering queries using views: A survey. VLDB Journal **10**(4), 270–294 (2001)
18. Guntzer, U.: Optimizing Multifeature Queries in Image Databases. VLDB Journal (2003)
19. Ramakrishna, M.V., Nepal, S.: Query processing issues in image (multimedia) databases. In: ICDE (1996)
20. Das, G., Gunopulos, D., Koudas, N., Sarkas, N.: Ad-hoc top-k query answering for data streams. In: VLDB, pp. 183–194 (2007)
21. Yu, A., Agarwal, P.K., Yang, J.: Processing a large number of continuous preference top-k queries. In: SIGMOD, pp. 397–408 (2012)
22. Baseball statistics. http://www.databaseBaseball.com

The Browsing Issue in Multimodal Information Retrieval: A Navigation Tool Over a Multiple Media Search Result Space

Umer Rashid, Marco Viviani, Gabriella Pasi and Muhammad Afzal Bhatti

Abstract In the field of Multimodal Information Retrieval, one of the issue to tackle is how to effectively browse the search result space. In addressing this issue, it is particularly important to take into consideration that, especially nowadays, data is highly semantically interlinked. In this scenario, we present a tool to navigate and visualize the results produced by the evaluation of a query over a set of multiple media objects. The search result space can be represented via a graph-based data model where (i) multiple media objects are represented as nodes with multiple modalities of information associated with them, and (ii) media objects can be connected via different kinds of relationships. Our idea is to give to the user the possibility to navigate the space of the results of a query, constituted by multiple media objects, as s/he was exploring a graph of connected entities. As a preliminary work, in this paper we only deal with textual information for building similarity relationships among media objects and part-of relationships in the case of media objects belonging to a same (multimedia) document. This way, we show how a user can navigate and visualize the result space following different links connecting media objects. We illustrate our navigation and visualization tool with different examples.

1 Introduction

Information exists in multiple media formats; the most popular ones consist of text, audio, image and video objects. Data belonging to these different formats are referred in the literature as *multiple media objects* [1]. Media containers having a collection of

U. Rashid(✉) · M.A. Bhatti
Quaid-i-Azam University, Islamabad, Pakistan
e-mail: {umerrashid,mabhatti}@qau.edu.pk

M. Viviani · G. Pasi
University of Milano-Bicocca, Milan, Italy
e-mail: {marco.viviani,pasi}@disco.unimib.it

T. Andreasen et al. (eds.), *Flexible Query Answering Systems 2015*,
Advances in Intelligent Systems and Computing 400,
DOI: 10.1007/978-3-319-26154-6_21

multiple media objects are referred as *multimedia documents* [2]. A Web page is the most common example of a multimedia document and nowadays a significant number of queries submitted to Web search engines usually aims at retrieving multiple media objects [3].

Multimodal Information Retrieval (MMIR) [4] is the discipline at the basis of multiple media objects retrieval. With respect to 'traditional' IR, MMIR amplifies many of the research problems at the base of search over textual data. According to [4], the main challenges in retrieving multiple media objects include their acquisition, normalization, indexing, querying and browsing. Proposing innovative solutions to all of these challenges is out of the scope of this paper, and in the literature several proposals exist that address each of these specific problems. Instead, in this paper we focus in particular on the issue of browsing the search results produced by a query evaluation. On the one hand, most of the existing search engines available over the Web enable separate searches within different media types. They restrict users to specify media types in which they are interested to perform a search. They usually provide a structured (non-blended) integration of search results presented on separate Web pages or panels [5]. Alternatively, aggregated-search interfaces are usually designed to integrate the search results from multiple sources into a single Web page, providing this way blended integrations of the search results [6]. The aggregated presentation of the search results is considered an easier way to find the relevant information as compared to conventional web search interfaces [5]. Interactive multimedia search systems provide the exploration of multimedia contents via navigation without putting extra effort in query reformulation [7].

Our approach arises in this line of research. Exploiting the fact that nowadays data (and, in particular, multiple media objects) are highly interconnected, we represent the search result space via a high-level graph-based data model that can be used to browse and visualize in a blended (integrated) way the retrieved results. Multiple media objects are represented as nodes in the graph, and each node can be characterized by different modalities of information [2, 8] associated with it. For example, an audio object may be accessed via its textual and/or acoustic modalities, while an image object may characterized by textual and visual modalities of information. Nodes are connected through different kinds of links, which represent similarity relationships among media objects. Different similarity relationships are computed for each modality of information characterizing the media objects. This way, the navigation and visualization tool gives to the user the possibility to browse the space of the results of a query as s/he was exploring a graph of connected entities. The set of retrieved results are both presented as an expandable linear list – which can be further navigated in a Web-like manner following the relationships among multiple media objects – and visualized in a graph-based way.

The paper is organized as follows. In Section 2 we discuss the background and the motivations of our research. In Section 3, we provide a detailed description of our proposed model as well as its instantiation on a concrete scenario. In Section 4 we illustrate our browsing tool with some examples of navigation in the search result space. Finally, in Section 5, we draw the conclusions and we discuss future research directions.

2 Background and Motivations

Traditional or general purpose search engines provide search within almost all types of media objects. Users have a choice to specify a single media type in which they are interested in when formulating a query, and the search results are presented in separate media tabs or panels. In most cases, the main focus of this kind of search engines is to produce a simple ranked list of the retrieved results [9]. General purpose search engines are generally not focused towards a blended integration of the search results. Some of the general purpose search engines like Google and Bing nowadays also provide content-based search for only image objects, and they are also exploring cross-media retrieval at a limited scale (Google for example provides few search results from image objects in Web search). The full presentation of search results via a full integration of multiple media objects still remains a challenging task because it requires to "determine the relevance of a source to a search task" and to "organize the search results with multiple sources" [5].

Although some researchers have recently explored aggregated search, blended integration, navigation and visualization of the search result space [5, 10, 11] – by proposing either data models or concrete navigation tools (or both) – these challenging aspects have not been specifically discussed in the context of multiple media search.

Concerning the proposed data models, several approaches are based on a graph-based representation of the search result space, but most of them elaborate collective relationships among multimedia document instead of representing relationship among different media contents that may exist in multimedia documents. For example, the data model proposed by Wiesener et al. [12] defines node descriptors to represent the multimedia content within multimedia documents and the links identify semantic relationship among nodes by considering their descriptors. Szegõ in [13] proposes a simple directed graph-based tree model for multimedia documents. In this model, nodes are labeled by atomic predicates. The nodes represent document models, and labels of the edges demonstrate atomic transformations. A pair of document models are semantically interconnected by an edge if and only if a set of transformations can convert one document model into the other. Rigamonti et al. [14] represent multimedia documents as a collection of raw data, annotations, and links. Raw data include media sources as media objects; annotations contain keywords utilized in indexing documents, and a link identifies the relationship of a multimedia document with the rest of documents in the collection. Different kinds of semantic links can be established (e.g., thematic, temporal, reference relationships). Lazaridis et al. [15] proposed I-Search as a multimodal search engine. The I-Search establishes a *have-a* relationship by putting similar media objects with distinct modalities of information in a multimedia document or container called content object (for example I-Search encapsulates text, audio, image, and video objects regarding a *Fighter Jet Plane* in a content object). The I-Search project can identify semantic relationships among media objects having distinct modalities of information and belonging to different content objects. Recently, the work of Sabetghadam et al. [16] has addressed the issue of reachability of relevant objects in a graph modeled collection by taking into account multiple media objects and their relationships. This paper shows how adding semantic links among media objects and considering their different associated modalities boosts the potential recall of the retrieval. In this work the authors do not implement a concrete browsing tool.

Among the approaches which implement visual tools for navigating the retrieved multimedia documents space (approaches that provide the exploration of archives through document to document navigation), FaericWorld [14] provides combined search, browsing and visualization of the multimedia documents containing audio-visual contents through a graph-based data model. The navigation is possible by following links that directly connect multimedia documents that exploit thematic, temporal and reference relationships. Visual Islands [11] provides text-based retrieval of a broadcast-news corpus. The multimedia documents contain news related images and textual descriptions. Images are retrieved against textual query terms and clustered in the form of image islands, i.e., thumbnails of similar images in consecutive locations. The Media Finder [17] provides image and video search over multiple social networks and related events. The Media Finder utilizes extracted textual information in the establishment of links among multimedia documents. At the basis of the tool there is a common document model/schema to align the results retrieved over different social network sites.

According to Marti Hearst [18], "Navigation structures lend themselves more successfully to books, information collections, personal information, Web sites, and retrieval results than to vast collections such as the Web". In her work, the author has discussed the role of browsing, navigation, and visualization of search results and graph-based structures. She has also elaborated the concept of faceted navigation within some of the existing search tools (e.g., the Aduna Autofocus enterprise search system; the relationBrowser faceted visualization; the Fathumb faceted navigation interface). Although in her work Hearst has widely addressed the issues of search, navigation and visualization of text, audio, image and video based information, her research did not primarily focus on the issue of multimodal retrieval in a blended integrated way.

According to best of our knowledge, a browsing tool able to exploit the semantic relationships across a set of retrieved multiple media objects to offer a blended integration has not been discussed yet in the literature. In this first implementation of our tool, we illustrate the advantages for the user to have the possibility to navigate the search result space as s/he was exploring a graph of connected entities, taking advantage of the different modalities of information connected to media objects.

3 A Graph-Based Browsing Model

In this section we describe a graph-based data model that is used to represent the search result space over multiple media objects. Based on this model, in Section 3.2 we describe the tool that we have developed to provide a user with the possibility of visualizing and navigating in a search result space composed by multiple media objects.

3.1 The Data Model

The search result space is represented as a graph $G = \langle V, E \rangle$ where (i) multiple media objects and multimedia documents containing them (if they exist) are represented as nodes belonging to V and (ii) both the relationships among media objects, and between media objects and multimedia documents are represented by the edges in E.

The set V is composed of two kinds of nodes: the nodes representing *multimedia documents*, and the nodes representing *media objects*[1]. We assume that a multimedia document is composed of a variable number of media objects (text, audio, image, video). Here we denote by n_d a multimedia document. A text object is denoted by n_t; such a node is characterized by only a textual dimension denoted by d_t. An audio object is denoted by n_a and it may be multi-modal; this means that it can be characterized by both a textual dimension denoted by d_t and an acoustic dimension denoted by d_a. The text associated with an audio object may be, for example, either the transcription of the audio recording or some annotations provided by listeners. An image object, denoted by n_i may have a textual dimension denoted by d_t and a visual dimension denoted by d_v. A video object, which we denote by n_v, is the most complex, since it may have a textual dimension denoted by d_t, an acoustic dimension denoted by d_a and a visual dimension denoted by d_v.

The set of edges E is composed of two kind of links: *part-of* links and *similarity* links. The relationship among a media object node and the (possibly) multimedia document node it belongs to is represented via a part-of link.

A similarity link represents a similarity relationship between a pair of media objects. In graph-based data model we have three kinds of similarity links, because they are established among media objects by exploiting textual, acoustic, and visual modalities of information separately. For example, a link between a video object and an audio object can be established by considering the similarity between their associated acoustic dimensions; in the same way, a link between a text object and an image object can be established by considering the similarity between their associated textual dimensions. A *t-link* represents textual similarity between a pair of media objects. Text, audio, image, and video objects can be potentially interlinked with each other through *t-links*. In the same way, *a-links* and *v-links* represent acoustic and visual similarity between a pair of media objects. Audio and video objects can be interlinked with each other through *a-links*, while image and video objects through *v-links*.

3.2 Instantiation of the Data Model to a Real Scenario

In this paper, we do not instantiate our data model by exploiting all the modalities of information connected to media objects, but we only consider the textual modality.

---

[1] For the sake of clarity and conciseness, we will simply refer in the rest of the paper to multimedia documents and media objects, without specifying that they are nodes.

This means, with respect to the presented data model, that we generate only *t-links* between media objects.

We make use of the multimedia document collection generated and used by the I-Search Project[2], the aim of which is to provide a unified framework for multimodal content indexing, sharing, search and retrieval. To our purposes, we used in particular the Rich Unified Content Description (RUCoD) format and a dataset containing 10,305 Content Objects (COs) as XML documents, both provided by the project. A CO, according to RUCoD, is a multimedia document which can result from the combination of one or more media objects. It is like a container that encapsulates media objects defining a same concept. Each CO may contain free text in the form of textual descriptions or user tags/annotations, the URI of the media preview, the URI of the actual media source, and low-level descriptors of multiple media objects associated with the content object. In our model, each content object is represented as a multimedia document (node) n_d, and each media object (text, audio, image and video object) connected to a content object is represented by the appropriate node (i.e., n_t, n_a, n_i, n_v).

Within the search result space, media objects are connected via 'part-of' links to the multimedia documents (content objects) they belong to. Concerning 'similarity links', as illustrated before we exploited only the textual information (in the form of either free text or annotations)associated with the retrieved media objects (belonging to a given CO). In order to compute the textual similarity between media objects, several similarity measures can be chosen [19, 20]. We consider the simple Jaccard's similarity measure to build *t-links*. Jaccard's similarity is based on the comparison of terms (keywords) found in documents. In the chosen dataset, although free text connected to media objects may be available, in most of the cases it is very short (i.e., keywords or short annotations are provided). For this reason, in order to avoid sparsity problems and the complexity connected to the population of a vector space, we have implemented the Jaccard's similarity measure. To approximate a threshold to establish t-links among media objects, we do not have implemented machine learning techniques or some statistical method, since it is not the primary objective of this research. Thus, the threshold has been approximated by considering the mean Jaccard's similarity of a selected node with respect to all of the other nodes in the search result space.

4 The Browsing Tool

In this section we describe our proposed browsing tool, which is composed of the following panels: (i) a *query formulation* panel (although the purpose of this paper is not to define an interface for query formulation, we have provided it to allow a user to easily specify both textual query terms and media object types in searching); (ii) a *result list* panel, to present to the user the retrieved multimedia documents (in our case COs) retrieved in answer to the specified query; (iii) a *media view* panel, to

[2] `http://iti.gr/iti/projects/I-SEARCH.html`

present the media objects contained in a retrieved multimedia document (in a CO); and (*iv*) a *navigation panel* which allows the user the navigation within the search result space. The main interface of our tool is shown in Fig. 1. We now describe each of the panels in detail.

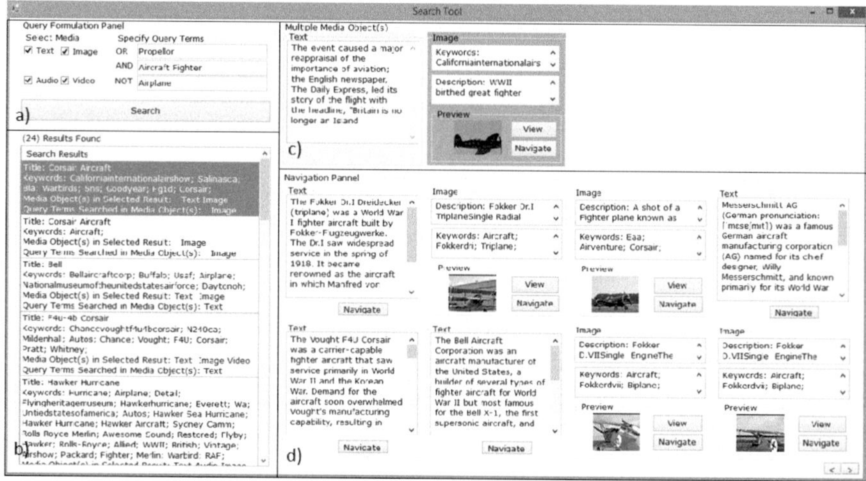

Fig. 1 The interface of the browsing tool with its four panels: the query formulation (*a*), the result list (*b*), the media view (*c*) and the navigation (*d*).

In our tool the search is based on classical indexing techniques, and on the Vector Space Model. The search tool has been implemented via Lucene and C#. In our applicative scenario, the postings in the inverted index file include both COs and media objects in which the searched keywords exist. The query formulation panel allows a user to specify textual query terms and Boolean operators (AND, OR, NOT) among query terms. The user can also specify the media objects s/he is interested to retrieve. This way, the query formulation panel allows aggregated search within any combination of media objects.

The result list panel presents the initial set of the (ranked) retrieved COs. In the result list a user can see the title and the keywords connected to the retrieved COs. Each result in the COs list can be clicked to visualize the media objects contained in the selected CO.

The media view panel provides the presentation of media objects contained in the selected CO from the ranked list. The media view panel not only presents media object contained in the selected CO from the rank list, it also highlights media objects in which keywords have been effectively found. In the media view panel, the user can view textual details, keywords, descriptions, media previews and links to actual media objects depending on the type of the retrieved media objects. The user can also directly open audio, image, or video objects from the panel and, by simply clicking on specific media objects, s/he can view their connected media objects in the navigation panel.

This last panel provides to the user the possibility to expand her/his browsing experience following the similarity relationships among the multiple media objects that have been retrieved. The navigation panel represents media objects in a grid. These media objects are the adjacent ones to the selected media object. A media object can be selected either from the media view panel or from the navigation panel to navigate in the search result space.

To give to the user the possibility of better understanding the relationships among her/his results, we have implemented the possibility to export a graph-based visualization of the results. We will illustrate this visualization in the next section, which describes a concrete example of navigation using our tool.

4.1 A Navigation Example

We have described the main components and features of the tool we proposed, and we have provided a single screenshot (Figure 3) illustrating altogether the navigation

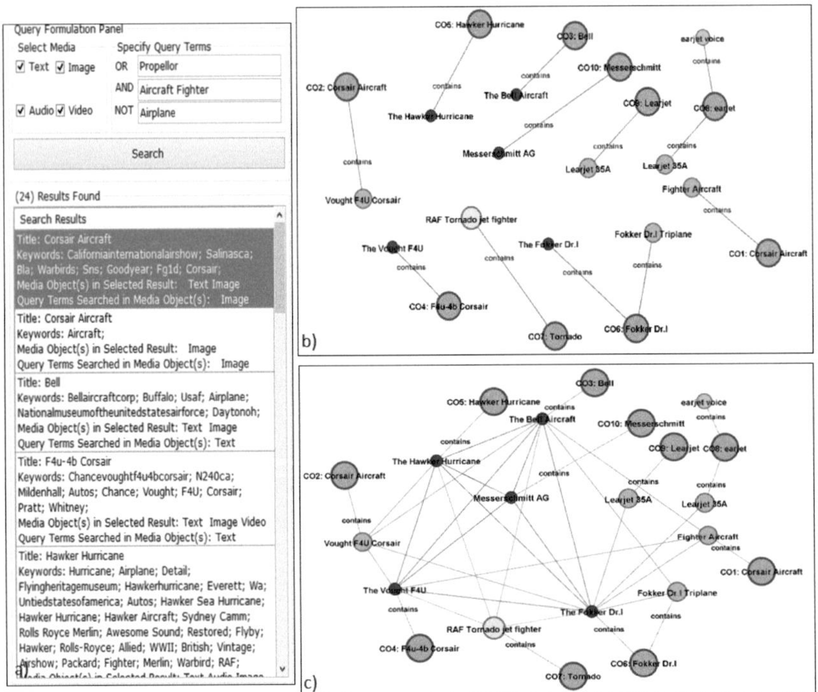

Fig. 2 The first step of navigation (*a*) and the graph-based visualization of the retrieved results in the traditional ranked list (*b*) and in the proposed graph-based representation of the search result space (*c*).

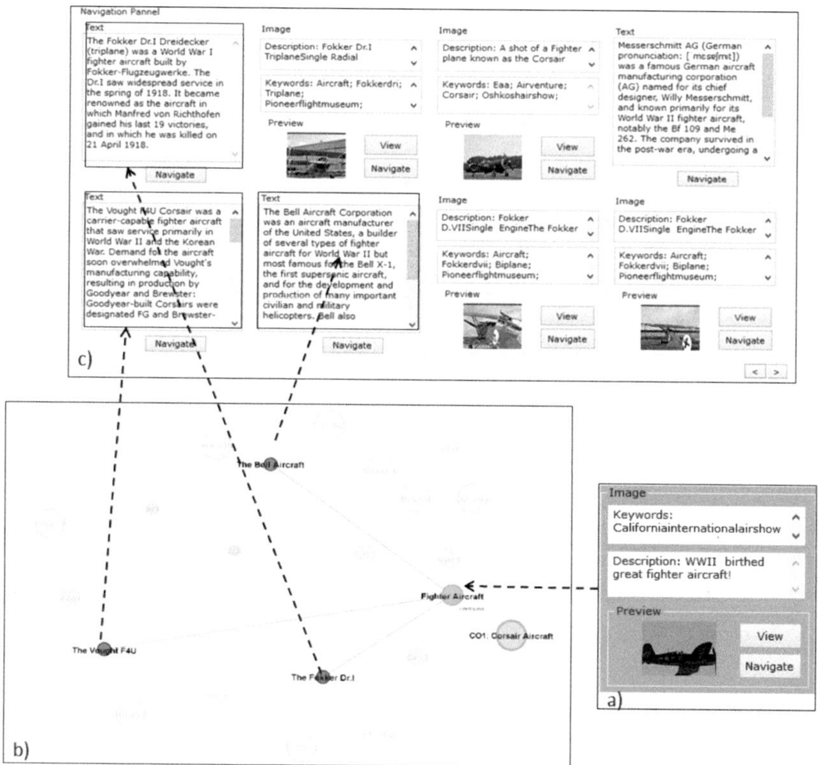

Fig. 3 An image object selected from the media view panel (*a*), the graph-based visualization of the 'adjacent' text objects (*b*) and the text objects presented to the user in the navigation panel (*c*).

steps a user can make across the different panels. In this section, we describe step-by-step the possible interactions of the user with the application, and we present a graph-based representation of the result space connected to each navigation step.

Figure 2 (*a*) illustrates the first step of navigation: a user has introduced in the query formulation panel some query terms (i.e., `Propellor`, `Fighter AND Aircraft`, `NOT Airplane`) and s/he is intended to retrieve all kinds of media objects (i.e., `text`, `image`, `audio`, `video`) relevant to the specified query. As a result, the user obtains in the result list panel all the multimedia documents (content objects in the specific example) which satisfy the query terms. The COs contain one or more media objects in which textual query terms have been matched in the retrieval phase (content objects and media objects connected via *part-of* links). Media objects belonging to the different COs in the results panel are connected via *t-links* by exploiting their textual similarity (Figure 2 (*c*)) unlike in traditional lists of ranked results (Figure 2 (*b*)). This clearly emerges from the figures, where bigger circles represent multimedia documents, and smaller circles with different colors

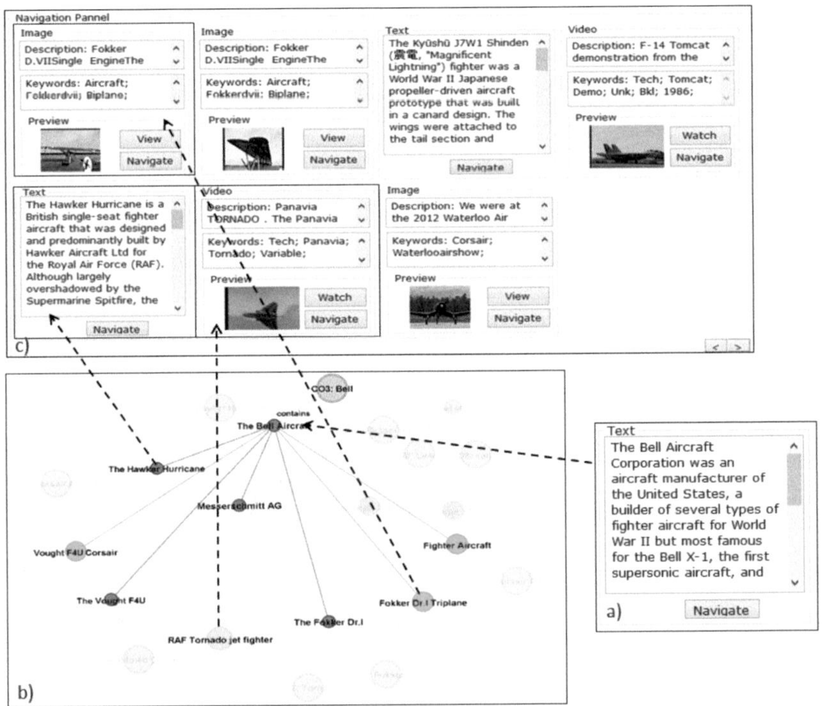

Fig. 4 An text object selected from the navigation panel (a), the graph-based visualization of the 'adjacent' media objects (b) and some of the media objects presented to the user in the navigation panel (c).

represent different media objects. The connections among media objects provide a better exploration of the search results in the search result space as compared to the traditional presentation of search results via ranked lists.

As a second step of the navigation, a user can select any of the COs from the ranked list in the results panel. All of the media objects which belong to the selected CO are presented in the media view panel, but only the media objects in which query terms have been effectively found are highlighted (see Figure 1). As a third step of navigation, the user can further explore any of the highlighted media objects in the media view panel to view its 'adjacent' media objects in the search result space. In Figure 3 (a), for example, a user has selected, from the media view panel, the retrieved image object `Fighter Aircraft` connected to `CO1: Corsair Aircraft`. As illustrated by the graph-based representation (Figure 3 (b)), this image object is semantically connected to other three text objects (i.e., `The Fokker Dr. I`, `The Vought F4U` and `The Bell Aircraft`), which are presented to the user in the navigation panel (Figure 3 (c)) and which belong, in their turn, to other three content objects (i.e., `CO6: Fokker Dr. I`, `CO4: F4u-4b Corsair` and `CO3: Bell`).

The navigation panel presents media objects 'adjacent' to the selected media objects either from the media view panel or from the navigation panel itself. So, as a fourth step of navigation, a user can continue the exploration of the search result space by selecting media objects from the navigation panel. As illustrated in Figure 4 (*a*), a user has decided to further explore the text object `The Bell Aircraft`, which is further semantically connected to other media objects (Figure 4 (*b*)). These media objects are presented to the user in the navigation panel (Figure 4 (*c*)). For the sake of simplicity and conciseness, in the figure we do not have illustrated all the correspondences between the graph-based visualization (*b*) and the media objects in the navigation panel (*c*), since some of the media objects would be displayed on further pages of the navigation panel.

5 Conclusion and Further Research

An effective browsing of search results in a blended integration remains one of the open issue in Multimodal Information Retrieval, and it requires further investigation. In this paper an approach for navigating a search result space over multiple media objects has been proposed. The approach exploits a graph-based data model that connects, within the result space, multiple media objects based on some possible semantic relationships built across the modalities of information associated with them. Each media object is, in fact – being it a text, image, audio or video object – characterized by multiple modalities of information. With an image object, for example, certainly characterized by a visual modality, can also be associated a textual modality (e.g., its caption, the name of the file). In the same way, a video object can be characterized by both a visual modality and an acoustic modality. As a preliminary work, in this paper we have only generated similarity relationships exploiting the textual modality associated with different media objects.

In the future, we aim at investigating other possible modality connections (via *t-links*, *a-links* and *v-links* together). Our main objective is to develop a complete tool providing aggregated search within any combination of media objects by exploiting at the same time textual, acoustic, and visual modalities. We will also investigate aggregated query formulation through textual, acoustic, and visual features, besides continuing to improve the user navigation experience within the result space via visualization of the search results.

References

1. Lauer, C.: Contending with terms: Multimodal and Multimedia in the Academic and Public Spheres. J. Computers and Composition **26**, 225–239 (2009)
2. Rafailidis, D., Manolopoulou, S., Daras, P.: A unified framework for multimodal retrieval. J. Pattern Recognition **4**, 358–3370 (2013)
3. Tjondronegoro, D., Spink, A., Jansen, B.J.: A study and comparison of multimedia Web searching: 1997–2006. J. American Society for Information Science and Technology **60**, 1756–1768 (2009)

4. Bozzon, A., Fraternali, P.: Chapter 8: Multimedia and multimodal information retrieval. In: Ceri, S., Brambilla, M. (eds.) Search Computing. LNCS, vol. 5950, pp. 135–155. Springer, Heidelberg (2010)
5. Sushmita, S., Joho, H., Lalmas, M., Villa, R.: Factors affecting click-through behavior in aggregated search interfaces. In: 19th ACM International Conference on Information and Knowledge Management, pp. 519–528. ACM (2010)
6. Bron, M., Van Gorp, J., Nack, F., Baltussen, L.B., de Rijke, M.: Aggregated search interface preferences in multi-session search tasks. In: Proceedings of the 36th International ACM SIGIR Conference on Research and Development in Information Retrieval, pp. 15–20. ACM, Japan (2013)
7. Mei, T., Rui, Y., Li, S., Tian, Q.: Multimedia search Re-ranking: A literature survey. ACM Computing Surveys 46(38) (2014)
8. Kalamaras, I., Malassiotis, S., Tzovaras, D., Mademlis, S.: Novel framework for retrieval and interactive visualization of multimodal data. J. Electronic Letters on Computer Vision and Image Analysis **12**, 28–29 (2013)
9. Lauer, C.: Precision-recall is wrong for multimedia. J. IEEE MultiMedia **18**, 04–07 (2009)
10. Kopliku, A., Pinel-Sauvagnat, K., Boughanem, M.: Aggregated search: A new information retrieval paradigm. ACM Computing Surveys **46**(41) (2014)
11. Zavesky, E., Chang, S.F., Yang, C.C.: Visual islands: intuitive browsing of visual search results. In: Proceedings of International Conference on Content-Based Image and Video Retrieval, pp. 617–626. ACM (2008)
12. Wiesener, S., Kowarschick, W., Bayer, R.: Semalink: an approach for semantic browsing through large distributed document spaces. In: Proceedings of the Third Forum on Research and Technology Advances in Digital Libraries, pp. 86–94. IEEE (1996)
13. Szegõ, D.: A logical framework for analyzing properties of multimedia web documents. In: Workshop on Multimedia Discovery and Mining, ECML/PKDD, pp. 19–30. (2003)
14. Rigamonti, M., Lalanne, D., Ingold, R.: Faericworld: browsing multimedia events through static documents and links. In: Baranauskas, C., Abascal, J., Barbosa, S.D.J. (eds.) INTERACT 2007. LNCS, vol. 4662, pp. 102–115. Springer, Heidelberg (2007)
15. Lazaridis, M., Axenopoulos, A., Rafailidis, D., Daras, P.: Multimedia search and retrieval using multimodal annotation propagation and indexing techniques. Signal Processing: Image Communication **28**, 351–367 (2013)
16. Sabetghadam, S., Lupu, M., Bierig, R., Rauber, A.: Reachability analysis of graph modelled collections. In: Hanbury, A., Kazai, G., Rauber, A., Fuhr, N. (eds.) ECIR 2015. LNCS, vol. 9022, pp. 370–381. Springer, Heidelberg (2015)
17. Rizzo, G., Steiner, T., Troncy, R., Verborgh, R., Redondo Garcia, J.L.: What fresh media are you looking for?: retrieving media items from multiple social networks. In: Proceedings of International Workshop on Socially-Aware Multimedia, pp. 15–20. ACM, Japan (2012)
18. Hearst, M.: Search User Interfaces. Cambridge University Press, UK (2009)
19. Huang, A.: Similarity measures for text document clustering. In: Proceedings of the Sixth New Zealand Computer Science Research Student Conference, pp. 49–56 (2008)
20. Chim, H., Deng, X.: Efficient phrase-based document similarity for clustering & Retrieval. IEEE Transactions on Knowledge and Data Engineering **29**, 1217–1229 (2009). New Zealand

A Unified Framework for Flexible Query Answering over Heterogeneous Data Sources

Roberto De Virgilio, Antonio Maccioni and Riccardo Torlone

Abstract The lack of familiarity that most users have with information systems has led to a variety of methods to access data in a flexible way (such as keyword search, faceted search, and similarity search). However, capabilities of flexible query answering are hard to integrate in one system since they are based on different data representations and relies on different techniques for query answering. The problem becomes more involved if we need to query heterogeneous data sources. To address such variety in one fell swoop, we propose FleQSy, a framework that relies on a "meta" approach for accessing heterogeneous data with different methods for flexible query answering. In FleQSy structured and semi-structured data sources are modeled as graphs and query answering consists of a multi-phase process that leverages the commonalities of the various search techniques. We show the effectiveness of our approach in different application scenarios that require easy-to-use and elastic methods for data access.

1 Introduction

One of the main purposes of information systems is to satisfy the information need of different kinds of users. This achievement becomes more challenging when users are unaware of the content and organization of the underlying data and ignore query languages. This scenario is quite common, as many data access points are directly exposed on the Web to random users. In these cases, flexible query answering capabilities are usually provided to take away the barriers that non-expert users encounter in information search. In the evaluation of flexible queries, we want the "best" answers (usually called top-k) that might match the query only approximately, or we want to consider queries expressed without syntactical or structural constraints. An example of relaxed query answering is the common method of searching data over

R. De Virgilio · A. Maccioni(✉) · R. Torlone
Università Roma Tre, Rome, Italy
e-mail: {dvr,maccioni,torlone}@dia.uniroma3.it

T. Andreasen et al. (eds.), *Flexible Query Answering Systems 2015*,
Advances in Intelligent Systems and Computing 400,
DOI: 10.1007/978-3-319-26154-6_22

the Web, where users just input their keywords of interest and the engine attempts to find the more relevant web pages accordingly.

Since flexible query answering methods differ considerably from one another their co-existence in the same system is difficult to achive. Flexible query capabilities usually require, in fact, ad-hoc representations of data and replicas of the databases used for exact querying. Moreover, since in a typical scenario the organizations's information is scattered among several sources that are managed by different systems, the problem becomes harder. This situation is common in the scenario of Enterprise Search [13, 19], where users would like to express queries over very diverse data of an enterprise by means of a simple search interface. Taking apart the problem of data integration, a relevant problem here is that these kind of queries are usually imprecise and vague, which introduces an uncertainty and an overhead in the search process. For instance, in relational keyword search, this uncertainty leads to an enormous number of join operations to find candidate answers. Therefore, it is hard for an Enterprise Search tool to provide different flexible capabilities of query answering over heterogeneous data. Actually, in current systems (such as, Microsoft Azure Search and Google Search Appliance) these capabilities are provided in isolation, each one relying on an ad-hoc data representation.

To address such heterogeneity, we propose a framework, called FleQSy, that provides different capabilities of flexible query answering over an unified representation of semi-structured and structured data. The basic idea is to identify the common operations required by different techniques for flexible query answering and combine them in a unified and high-level representation of the search process. This "meta" approach allows us to avoid data transformations that involve unnecessary replication of data and computational overhead. Structured and semi-structured data are represented in FleQSy in a uniform way using a graph-based data model, which is general enough for representing a variety of data sources and strongly simplifies the integration of heterogeneous data. The three phases of the meta-approach involve: (i) the analysis of the query and the identification of the portions in the data graph that can be used for the computation of the answer(s), (ii) the assessment of each portion in terms of relevance for the query and of similarity among the other portions, and (iii) the generation of the answers by combining the most relevant and diverse portions.

FleQSy has been designed with practical scenarios in mind: it aims at facilitating the development and the reuse of tools and applications including, for example, vertical search engines equipped with plug-and-play capabilities of flexible query answering. To validate our approach, we have tested the framework over different application scenarios, namely: keyword search over both relational and semantic data, and approximate pattern matching over graph databases.

The rest of the paper is organized as follows. In Section 2 we introduce the problem and in Section 3 we illustrate our unified approach. Section 5 discusses related works, whereas Section 4 discusses the current implementation of FleQSy. Finally, in Section 6, we sketch conclusions and future works.

2 Flexible Query Answering

We consider a scenario in which user specifies a query Q over a data set D of semi-structured data and expects as results a set of answers $a_1, a_2, \ldots a_n$ that "best fit" Q. We assume that the database D and the query Q are both composed by a set of tuples $< A_1 : v_1, \ldots, A_k : v_k >$, where in D the v_i's are constants, whereas in Q they can be both variables and constants. Exact answers to Q over D are obtained by finding a function s, called a *substitution*, that is the identity on constants and maps variables to constants such that $s(Q)$ is contained in D, with the risk of finding no result since the query should conform to the way in which data is stored in D. Unlike exact query answering, answers to flexible queries are not found only by instantiating the variables of Q, but also introducing a mechanism for finding a "matching" between Q and D that tries to capture the intention of the user. In this respect, we consider two general types of flexible query answering (and combination thereof): *approximate answering* and *query relaxation*. Note that often in literature the terms approximate, relaxed and flexible are used interchangeably.

In order to define the notion of approximate query answering in a more precise way, we consider three simple *edit operations* over a set of tuples T consisting in the addition, the deletion or the substitution of a tuple.

Definition 1. Approximate Query Answering] An approximate query answer to a query Q over a data set D is a set of tuples $a = \{t_1, t_2, \ldots t_n\}$ such that there exists a substitution s for Q and a set of edit operation E over a such that $s(Q) = E(a)$.

Another way to query the database D in a flexible way is by relaxing the query. Usually, relaxed queries use a simplified syntax. We therefore assume that the query is expressed in some formalism and simply call *search* the expression of the query in such formalism.

Definition 2. Relaxed Query Answering] A relaxed query answer to a search S over a data set D is a set of tuples $a = \{t_1, t_2, \ldots t_n\}$ of D such that there is a query Q capturing S and a substitution s for Q such that $s(Q) = a$.

Basically, relaxed query answering needs an interpretation of the user need. The result of the interpretation is the query Q, formed by a set of tuples, to be evaluated over the database, possibly in an approximate way. Thus, Q can be evaluated with either an approximate or exact answering process.

FleQSy ranks the answers in order of relevance with respect to the query. The relevance is measured with a scoring function σ that takes into account the "vicinity" between the query Q and the answer a in a given application domain. In this respect, FleQSy is independent of the specific scoring function chosen, so that different implementation of σ can be used for different problems.

The framework we envisage should guarantee a number of important properties: they represent our main goals in the design of FleQSy.

- *Monotonicity:* a ranking is monotonic if a_i is more relevant than a_{i+1} according to the scoring function σ. Consequently, a query answering process is monotonic if it generates the answers following a monotonic ranking. From a practical point of view, this means to return the best (top-k) answers in the first generated instead of processing blocks of n candidates, with $n > k$, out of which selecting the best k. We believe that monotonicity is a relevant feature for flexible query answering because, since relaxations are less selective on the database, flexible query answering intrinsically tends to generate (much) more answers than required (i.e. the exact and the relaxed/approximated answers). Differently than the state-of-the-art, where searching and ranking are computed separately, FleQSy combines the generation with a relevance assessment of partial answers. It follows that, if query answering is monotonic, the ranking task is needless and the *time-to-result* is improved. For example, asynchronous applications such as data visualization apps can start the rendering process before the query answering has terminated. To guarantee the monotonicity preserving the correctness, in some problems, we rely on a theoretical result inspired by the Threshold Algorithm [14].

- *Scalability:* this property concerns the ability of the algorithms to handle an increase of the size of the database with a gradual increase of computational time (i.e., the time spent to evaluate a query is sub-proportional to the size of the database). More specifically, since we deal with flexible query answering problems that are mostly NP-hard, our goal is to relax those problems in order to reach a polynomial (possibly linear) time complexity of the algorithms with respect to the size of the database. At the same time, we do not want to decrease the practical effectiveness of the process.
- *Distributed Implementation:* this property is a practical (and today common) way to achieve *scalability*. More precisely, it is intended as the possibility to scale a problem through the execution of parallel algorithms over a distributed network of data sources that are located on different computer machines.
- *System-independence*: this property means that the models and the algorithms employed in this work have to be independent of the technology used for their implementation. This will be achieved by always considering an abstract model of the data that can be implemented on a multitude of systems. This enables interoperability at data level of different database systems.

3 FleQSy: A Unified Approach

3.1 Graph Data Modeling

FleQSy manages structured and semi-structured data using a single model, called a Data Graphs, as follows.

Definition 3. Data Graph] Given a set of labels Σ, a directed data graph G is denoted by $G = \{N, E, l\}$ where N is the set of vertices, $E \subseteq N \times N$ is the set of edges and l is a labeling function, $l : \{N \cup E\} \rightarrow \Sigma$. For each node $n \in N$ and for each edge $e \in E$, $l(n)$ and $l(e)$ denote the label of n and e, respectively.

In FleQSy, paths are the basic unit of information. Therefore, we consider our graph composed by a set of data paths, as follows.

Definition 4. Data Path] Given a data graph $G = \{N, E, l\}$, a simple data path is a sequence $l(n_1) - l(e_1) - l(n_2) -- \ldots - l(e_{z-1}) - l(n_z)$ where $n_i \in N$, $e_i \in E$.

In some cases, we consider a more specialized version of data path, called *Full Data Path*, that goes from a *source* node and end into a *sink* node. In a data graph, we call *sources* the nodes with no in-coming edges and *sinks* the nodes with no out-going edges.

Definition 5. Full Data Path] Given a data graph $G = \{N, E, l\}$, a data path is a sequence $l(n_1) - l(e_1) - l(n_2) - \ldots - l(e_{z-1}) - l(n_z)$ where $n_i \in N$, $e_i \in E$, n_1 is a source and n_z is a sink.

For simplicity, a (full) data path is hereafter abbreviated to path.

So far we tackled problems over different kind of data sources: relational, RDF and graph databases. We need to map these databases over a data structure conforming Definition 3. RDF is the standard language for the representation of semantic information. With RDF databases the graph modeling is straightforward because RDF is already a logical data graph model. In a RDF graph, nodes are classes, entities and literal values, while edges represent semantic relationships between nodes. Classes, entities and edges are labeled with URIs in our data graph.

Unlike RDF, relational databases are not natively conceived as graphs. In FleQSy, we model a relational database in terms of a pair of graphs, one representing the schema (*schema graph*) and another representing the instance (*data graph*). The schema graph is materialized in terms of schema paths that track primary and foreign key constraints among attributes, while the data graph is just a conceptual notion (i.e. we do not duplicate the instance). The data graph conforms to the Definition 3, where nodes represent table names, attribute names, tuple ids and values.

3.2 *Logical Architecture*

The high-level architecture of FleQSy is in Fig. 1. Many different kind of flexible queries Q can be submitted to FleQSy through the User Interface. Then, the Query Analyser extracts from Q the information needed by the Query Engine for computing the answers $a_1, a_2, \ldots, a_k$ to output.

We access the sources taking advantage of a Path-Based View of the graph, where *paths* are first class citizens. The Path-Based View is a logical component that is sometimes implemented through a virtual view (i.e. data paths are computed at run time); while other times it is implemented with path-based indexes [6].

3.3 *The Approach for Query Answering*

The query answering process in FleQSy takes as input a flexible query Q and the database G. It produces the top-k answers for Q in G according to the relevance metric.

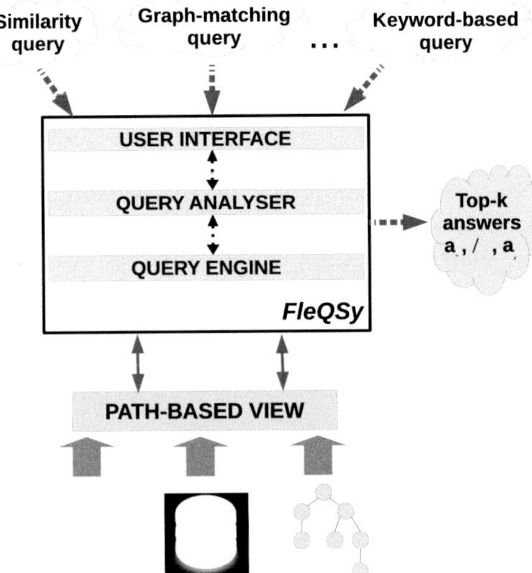

Fig. 1 FleQSy: The Unified Framework

It is based on a meta-approach composed of three main phases. Basically, a specific problem is tackled in FleQSy by instantiating an algorithm for each of the phases.

- PRE- PROCESSING: the query Q is analysed by the QUERY ANALYSER to individuate the criteria that the final answers should conform to. Such criteria are constraints that answers have to satisfy. They regard both the content and the structure of the answers. Then, we use the path-based view to search for the data paths in G that match the criteria. At the end of these lookups, we have all the paths P of G that are "relevant" for Q. Let us suppose, for instance, a keyword-based query where the constraint is the inclusions of the input keywords in the answers. Therefore, during the pre-processing, we find the paths matching (e.g., containing) such keywords.
- CLUSTERING: we group together the paths of P that are similar with respect to the criteria individuated during the pre-processing. Each group is also called a *cluster* and CL is the set of all the clusters. FleQSy assesses the relevance of the paths in P by using the same scoring function used for the answers. The paths are ordered inside the clusters according to their relevance.
- BUILDING: we generate the final answers by combining, at each iteration, the most relevant paths in every cluster. The algorithms used for the building decides how to combine the paths. In order to generate them in a monotonic order, the algorithms rely on theoretical results inspired by [14]. They are expressed in terms of theorems that we do not discuss here since they involve metrics used in the specific problems [9, 10].

Correctness and completeness of the process are less meaningful in the context of flexible query answering and therefore, we analyse the accuracy of the approach

Table 1 Summary of results obtained with FleQSy

Query Answering Problem	Theoretical Time Complexity	Experimental Time Complexity	Monotonicity	Distributed Implementation
Keyword search over relational databases	$O(\|P\|)$	$O(\|P\|)$	•	○
Keyword search over RDF databases	$O(\|P\|)$	$O(\|P\|)$	•	•
Approximate Graph Matching over graph databases	$O(\|P\|^2)$	$O(\|P\|^2)$	•	○

• Fulfilled ○ Not fulfilled yet P = measure of database size

in terms of *precision* and *recall*. Intuitively, we are able to find relevant answers to the query since the clusters are built on top of the criteria individuated by the pre-processing and every cluster tries to contribute to the generation of the current answer.

4 FleQSy at Work

This section describes how FleQSy can be implemented to solve specific problems of flexible query answering. We consider three scenarios, achieving the results that are summarized in Table 1. The table shows theoretical and experimental results as well as the fulfillment of the properties identified in Section 2. We did not include in the table a column for the system-independence property since it has been always satisfied. In the following we show how each phase of the meta-approach have been implemented for solving these problems.

Keyword Search over Relational Databases. We aim at overcoming with FleQSy the barriers that prevent traditional RDBMSs to implement existing keyword search solutions. In this context, the answers are joining tuple trees (JTT) whose values match the input keywords. Thus, the core problem is the construction of the JTTs. There are various approaches proposed in the literature that are classified in two different categories: *schema-based* and *schema-free*. Schema-based approaches usually implement a middleware layer in which: first, the portion of the database that is relevant for the query is identified and then, using the database schema and the constraints, a (possibly large) number of SQL statements is generated to retrieve the tuples matching the keywords of the query. As the number of SQL queries resulting from the interpretation of Q is high, their execution is expensive. Moreover, they generally waste computational time producing many empty results. Conversely, schema-free approaches first build an in-memory, graph-based, representation of the database, and then use graph exploration techniques to select the subgraphs that connect nodes matching the keywords of the query. Clearly, this is not scalable and for certain huge data-sets is even unfeasible.

Our technique builds top-k JTTs by combining incrementally the shortest (join) paths (i.e. data paths) that involve the tuples relevant to the query. The PATH- BASED VIEW has all the paths in the relational schema (schema paths) involving attributes linked by primary and foreign keys. During the *pre-processing* phase we identify all the schema paths ending in an attribute that contains a value matching one of the keywords in Q. From these schema paths, we generate a set of data paths P, each one containing (at least) one of the keywords in the sink. The rest of the nodes in these data paths are variables. We instantiate these variables with tuples in the last phase. In the *clustering* phase we groups together the paths of P ending in a node matching the same input keyword (i.e. one cluster for each keyword). As our scoring function is the number of tuples in the answer, the clusters are kept ordered according to the length of the paths, with the shortest paths coming first. The *building* phase generates iteratively the most relevant answers. This is done iteratively by picking, in each step, the shortest paths from each cluster and then combining them in order to find an answer. We find an answer with the tuples of a combination of paths having a node (i.e. tuple) in common. The search proceeds in this way with longer data paths in the clusters until we have not found k answers yet. More in detail, such paths are enriched with data by traversing them backward. This step only requires simple selection and projection operations over the database. If the backward navigation is not able to generate an answer with the current combination of most-relevant paths, the paths are navigated forward. The forward navigation uses the information retrieved in the backward phase, without further accessing the database. We show that, in this way, answers are retrieved in order of relevance. This eliminates the need to compare different answers and allows us to return the results to the user as soon as they are built.

Inspired by [23], we have developed the first technique exploiting only an RDBMS. This meets the system-independent goal of our framework. For more details about this implementation, we suggest to read the paper in [11].

Keyword Search over RDF Data. Answers to this problem are sub-graphs of the source dataset containing query keywords. A generic approach would: (i) identify the vertices of the RDF graph holding the data matching the input keywords, (ii) traverse the edges to discover the connections (i.e. trees or sub-graphs) between them to build n candidate answers (with $n > k$), and (iii) rank answers according to a relevance criteria in order to return the top relevant k. This kind of approach is hard to scale over distributed data because the unawareness about the location of the relevant data does not allow to optimize the distributed join processing. In FleQSy we solve the problem more efficiently since we avoid the step in (iii) generating directly the best k answers, both in a sequential and parallel fashion. For the distributed version we exploit the MapReduce paradigm in order to benefit from the features (i.e. load balancing, fault tolerance, job scheduling, etc.) that current implementations (e.g., Hadoop or variants thereof) give to developers. By exploiting a path-store for RDF, we reverse the distributed keyword search over RDF from a graph-parallel problem to a data-parallel problem.

The three phases of FleQSy work as follows. In the *pre-processing* phase all the paths P, whose sink matches one of the input keywords, are generated via the PATH-BASED VIEW. This task uses an index to speed up the retrieval of the paths [6]. Note that we assume that users enter keywords corresponding to attribute values, that are necessarily within the sink's labels.

To some extent, the sequence of edge labels in an RDF path characterizes the path. While we cannot advocate the presence of a schema, we can say that such a sequence is a template for the path. For instance, let `work1-author-aut1-name-Dante` be an RDF data path, the corresponding template is `author-name`. In the *clustering* phase, we group the paths of P according to their template. Paths within a cluster are considered homogeneous. The *building* phase is done by picking and combining the paths with greatest relevance from each cluster. We devised two different strategies to combine the paths in the *building*, explained as follows.

1. *Linear Strategy:* guarantees a linear time complexity with respect to the size of the input. Basically, the final answers are the connected components of the most relevant paths of the clusters. The whole process in the distributed version requires only one round of MapReduce.
2. *Monotonic Strategy:* generates the answers in order according to their relevance. As the linear strategy, it computes the connected components from the most relevant paths in the clusters. Unlike the previous strategy, the path interconnection is not the only criterion to form an answer. At this point every connected components is locally analysed to check if it fulfills the monotonicity, i.e. we check if the answer that we are generating is the optimum by relying on some theoretical results [9]. The overall complexity of the query answering is, in this case, quadratic with respect to the size of the input. The distributed version of this strategy completes in $2 \times k$ rounds of MapReduce, at most.

More details about the instantiation of the problem for both sequential and distributed version can be found in [8, 9].

Approximate Graph Matching over Graph Databases. Graphs establish a general way to represent information. Social networks, Semantic Web and bioinformatics are examples of domains in which it is natural to represent data in form of graphs. Pattern matching is a very common (perhaps even the most common) graph query operation. In fact, some graph query languages, such as SPARQL, exclusively support pattern matching queries. They require to search for the subgraphs in the database that strictly conform the query. However, there are use cases where the user would benefit from a more flexible way to answer these queries. For instance, in the context of linked open data the users are random citizens that want to use data provided by governments and organizations. A user should know the OWL language to understand the organization of the data. Moreover, data do not always conform strictly to the ontology of reference, so that users write their queries by trial and error in order to avoid empty results. Instead, with FleQSy we get approximate answers, that is to retrieve the best sub-graphs of the database that best match (not necessarily in an exact way) the input query. In literature, the problem is solved by relaxing the graph isomorphism problem with heuristics or by indexing sub-components of the dataset (e.g., sub-graphs, sub-trees, paths). These methods are able to reach a polynomial time complexity but they are still not practicable for systems exposed on the Web.

In FleQSy, the *pre-processing* phase decomposes Q in query paths and retrieves the paths P from the PATH- BASED VIEW based on a matching between the final constant node (if any) of the query paths and the last node of the data paths. The paths of P retrieved through the same query path are grouped together (*clustering* phase). Again, the paths within a cluster are ordered according to their relevance. The relevance in this case measures the similarity between the data path and the corresponding query path. Note that the same path can be inserted in different clusters, possibly with a different relevance. The most relevant paths are those that best approximate the query paths among all the paths in our data-set. In the *building* phase, we combine the paths coming from different clusters (i.e. picking the most relevant ones) and we check if their intersections conform those in Q. If yes, the combination is an answer for Q. More details about this problem in FleQSy can be found in [10], where the graph database is an RDF data-set.

5 Related Work

FleQSy was inspired by frameworks in the context of data modeling, data retrieval and data analysis. Apache MetaModel [2] is a framework for modeling heterogeneous data sources, including, among the others, relational databases, XML, JSON and CSV files, NoSQL document stores, NoSQL extensible column stores. MetaModel relies on a unified view of the underlying sources, implemented by connectors and API for CRUD and SQL-like operations. Similarly, SOS [3] provides a common programming interface to NoSQL systems and hides the specific details of the systems. UnQL addresses the problem of data heterogeneity through a unified language based on a SQL extension [24]. MetaModel and UnQL try to avoid duplications due to the query rewriting mechanism of the connectors, but unfortunately, they do not provide features for flexible query answering. In the context of graph processing systems, PowerGraph [17] proposes the GAS paradigm that subsumes the models of other graph-based systems and provides very general primitives for graph analytics. In a similar way to GAS paradigm, FleQSy provides a high-level abstraction for flexible query answering through one round of the *pre-processing*, *clustering* and *building* functions.

To the best of our knowledge there is no approach integrating flexible query evaluation over heterogeneous data, since existing works focus on single problems separately. In the context of relaxed query answering, there are works that propose to search over different databases by keywords [18, 23]. Unfortunately, these approaches heavily suffer from performance drawbacks because of high computational complexity [7]. We addressed the problem of structured keyword search in FleQSy inspired by a recent trend that proposes to solve the problem beyond the traditional dichotomy of *schema-based* and *schema-free* approaches [4].

For what concerns the monotonicity, there exist works focusing on the searching of top-k answers [16, 20]. Klee [20] addresses top-k algorithms on a distributed environment of peers and proposes variations of the Threshold Algorithm (TA) [14], where however each answer is computed on the same peer. The authors in [16] proposes a unified solution for different top-k computations, where a set of scoring metrics are given. Our monotonic query answering differs from TA because we can use non-aggregative and non-monotonic relevance metrics and, unlike [16], we use one metric at a time.

The problem of information interoperability is today fairly common since organizations use different kinds of databases. These issues were usually solved at data level with data/schema integration methodologies [12, 21]. More recently, the setting of the problem has changed considerably since enterprises store their data leveraging on different kind of database systems [5, 22], which clearly makes the traditional techniques obsolete. At the moment, the most adopted methodology to integrate different systems is to act at software level, as for instance, with Enterprise Architecture Integration methodologies. Few solutions proposing data infrastructures with heterogeneous database systems are arising under the name of multi-modal databases [1, 15]. ArangoDB [1] embeds documents, graphs, and key-values in the same database system. FoundationDB [15] implements several data model layers on top of a key-value store.

6 Conclusion and Future Work

In this paper we presented FleQSy, a novel meta-approach that can be used to provide a variety of flexible query-answering capabilities over structured and semi-structured data. In FleQSy data sources are represented in a uniform way using a simple but powerful graph-based data model. The approach leverages the commonalities of various search techniques and is structured in three abstract phases, which are instantiated according to the specific application domain. We validated the framework by applying FleQSy to different application scenarios showing how they can benefit from flexible querying capabilities provided by FleQSy. The experiments demonstrated that approximate and relaxed query answering can be efficiently embedded into a unified system that operates over heterogeneous data sources.

There are several directions for future work. We are currently working on the aspect of distributability and we are exploring a querying process where the answers are built by combining data coming from different sources. In the future, we would like to consider more flexible querying problems, more features of the framework as well as more types of data sources.

References

1. ArangoDB: ArangoDB (2015). https://www.arangodb.com/
2. ASF: Apache MetaModel (2015). http://metamodel.apache.org/
3. Atzeni, P., Bugiotti, F., Rossi, L.: Uniform access to non-relational database systems: the SOS platform. In: Ralyté, J., Franch, X., Brinkkemper, S., Wrycza, S. (eds.) CAiSE 2012. LNCS, vol. 7328, pp. 160–174. Springer, Heidelberg (2012)

4. Baid, A., Rae, I., Li, J., Doan, A., Naughton, J.F.: Toward scalable keyword search over relational data. PVLDB **3**(1), 140–149 (2010)
5. Borthakur, D.: Petabyte scale databases and storage systems at facebook. In: SIGMOD, pp. 1267–1268 (2013)
6. Cappellari, P., De Virgilio, R., Maccioni, A., Roantree, M.: A path-oriented RDF index for keyword search query processing. In: Hameurlain, A., Liddle, S.W., Schewe, K.-D., Zhou, X. (eds.) DEXA 2011, Part II. LNCS, vol. 6861, pp. 366–380. Springer, Heidelberg (2011)
7. Coffman, J., Weaver, A.C.: An empirical performance evaluation of relational keyword search techniques. TKDE **26**(1), 30–42 (2014)
8. De Virgilio, R., Maccioni, A.: Distributed keyword search over RDF via MapReduce. In: Presutti, V., d'Amato, C., Gandon, F., d'Aquin, M., Staab, S., Tordai, A. (eds.) ESWC 2014. LNCS, vol. 8465, pp. 208–223. Springer, Heidelberg (2014)
9. De Virgilio, R., Maccioni, A., Cappellari, P.: A linear and monotonic strategy to keyword search over RDF data. In: Daniel, F., Dolog, P., Li, Q. (eds.) ICWE 2013. LNCS, vol. 7977, pp. 338–353. Springer, Heidelberg (2013)
10. De Virgilio, R., Maccioni, A., Torlone, R.: Approximate querying of RDF graphs via path alignment. Distributed and Parallel Databases, 1–27 (2014)
11. De Virgilio, R., Maccioni, A., Torlone, R.: Graph-driven exploration of relational databases for efficient keyword search. In: Third International Workshop on Querying Graph Structured Data (EDBT/ICDT Workshops), pp. 208–215 (2014)
12. De Virgilio, R., Orsi, G., Tanca, L., Torlone, R.: Semantic data markets: a flexible environment for knowledge management. In: 20th ACM Conf. on Information and Knowledge Management, CIKM 2011, pp. 1559–1564 (2011)
13. Dmitriev, P., Serdyukov, P., Chernov, S.: Enterprise and desktop search. In: WWW, pp. 1345–1346 (2010)
14. Fagin, R., Lotem, A., Naor, M.: Optimal aggregation algorithms for middleware. In: PODS, pp. 102–113 (2001)
15. FoundationDB: FoundationDB (2015). `https://foundationdb.com/`
16. Ge, S., Hou, U.L., Mamoulis, N., Cheung, D.W.: Efficient all top-k computation - A unified solution for all top-k, reverse top-k and top-m influential queries. TKDE **25**(5), 1015–1027 (2013)
17. Gonzalez, J.E., Low, Y., Gu, H., Bickson, D., Guestrin, C.: Powergraph: distributed graph-parallel computation on natural graphs. In: OSDI, pp. 17–30 (2012)
18. Li, G., Ooi, B.C., Feng, J., Wang, J., Zhou, L.: Ease: an effective 3-in-1 keyword search method for unstructured, semi-structured and structured data. In: SIGMOD (2008)
19. Li, Y., Liu, Z., Zhu, H.: Enterprise search in the big data era: Recent developments and open challenges. PVLDB **7**(13), 1717–1718 (2014)
20. Michel, S., Triantafillou, P., Weikum, G.: Klee: a framework for distributed top-k query algorithms. In: VLDB, pp. 637–648 (2005)
21. Papotti, P., Torlone, R.: Schema exchange: a template-based approach to data and metadata translation. In: Parent, C., Schewe, K.-D., Storey, V.C., Thalheim, B. (eds.) ER 2007. LNCS, vol. 4801, pp. 323–337. Springer, Heidelberg (2007)
22. Qiao, L., et al.: On brewing fresh espresso: linkedin's distributed data serving platform. In: SIGMOD, pp. 1135–1146 (2013)
23. Qin, L., Yu, J.X., Chang, L.: Keyword search in databases: the power of RDBMS. In: SIGMOD, pp. 681–694 (2009)
24. UnQL: Unql (2015). `http://unql.sqlite.org/`

Queries with Fuzzy Linguistic Quantifiers for Data of Variable Quality Using Some Extended OWA Operators

Janusz Kacprzyk and Sławomir Zadrożny

Abstract We present a new type of a query with a fuzzy linguistic quantifier that makes it possible to account for a variable quality of pieces of data involved. We assume the so called horizontal quantification query introduced in our works (cf. Kacprzyk and Ziółkowski [10] and Kacprzyk, Ziólkowski and Zadrożny [9]), an additional degree of the quality of a piece of data (corresponding to a particular query condition), and then use some extended OWA (ordered weighted aggregation) operator with importance (weight) qualification for the implementation of a linguistic quantifier driven aggregation of satisfaction of the particular query conditions. Finally, we show how the new approach can be accommodated in our FQUERY for Access database querying system.

1 Introduction

This paper concerns a very relevant and useful, both from a theoretical and practical points of view, concept of a *database query with a fuzzy linguistic quantifier* introduced by Kacprzyk and Ziółkowski [10], and Kacprzyk, Ziółkowski and Zadrożny [9] in the late 1980s, and then considerably developed over the next decades, which has inspired many other works in fuzzy querying and related areas like linguistic data summarization (cf. Kacprzyk and Zadrożny [8]).

To best present the very essence of queries with linguistic quantifiers in the setting employed in this paper, we will start with a basic concept of a fuzzy query which was first suggested by Tahani [12] which may be exemplified as "find all employees in a personnel database who are *young* and *well qualified*" with the linguistic terms "young" and "well qualified" represented by suitable fuzzy sets.

J. Kacprzyk(✉) · S. Zadrożny
Systems Research Institute Polish Academy of Sciences, ul. Newelska 6, 01-447 Warsaw, Poland

WIT – Warsaw School of Information Technology, ul. Newelska 6, 01–447 Warsaw, Poland
e-mail: {kacprzyk,zadrozny}@ibspan.waw.pl

T. Andreasen et al. (eds.), *Flexible Query Answering Systems 2015*,
Advances in Intelligent Systems and Computing 400,
DOI: 10.1007/978-3-319-26154-6_23

For our purposes it is more convenient to consider fuzzy querying in the context of a flexible querying interface supporting an extended version of SQL as proposed by Kacprzyk and Zadrożny [6, 7] and Bosc et al. [2]. Basically, a traditional querying language is there extended to support *linguistic terms* in queries exemplified by fuzzy values like "young" and fuzzy relations (fuzzy comparison operators) like "much greater than" in the following SQL query:

```
SELECT *
FROM   employees
WHERE  (age IS "young") AND                                   (1)
       (salary IS "MUCH GREATER THAN USD 50,000")
```

Such a simple and straightforward extension of SQL to cover linguistic terms, represented in our context by fuzzy sets, has been later further extended in various directions, and one of the most relevant is certainly to *queries with linguistic quantifiers* such as "most", "almost all", "much more than a half", etc. For our purposes, it is most appropriate and convenient to consider the linguistic quantifiers to play the role of flexible aggregation operators, that is, aggregation of satisfaction degrees of not all query conditions but, for instance, of most conditions. In our approach, we will equate linguistic quantifiers with appropriate fuzzy sets which is very effective and efficient.

The concept of a (fuzzy) query with linguistic quantifiers was proposed in the mid-1980s in the seminal papers by Kacprzyk and Ziółkowski [10], and then Kacprzyk, Ziółkowski and Zadrożny [9]. Basically, they proposed to use the fuzzy linguistic quantifiers to aggregate conditions in the `WHERE` clause of the SQL `SELECT` statement as, e.g. in

"*Most* of conditions among: age IS *young*, salary IS *high*,...' are to be satisfied"

which can be written as:

```
SELECT *
FROM   employees
WHERE  most_of {(age IS "young") AND ...                      (2)
       ...(salary IS "MUCH GREATER THAN USD 50,000")}
```

This solution was simple and efficient, and intuitively appealing, so that it has been extensively further developed, notably by Kacprzyk and Zadrożny [8], and some commercial implementation have also been reported.

Later, Bosc et al. (cf. e.g. [3]) proposed another approach to use linguistic quantifiers with subqueries or against groups of rows as, e.g., in

```
SELECT    deptno
FROM      employees
GROUP BY  deptno                                              (3)
HAVING    most_of (young are well-paid)
```

We will be concerned here with the authors' approach depicted in (2).

Let us now proceed to the presentation of the motivation for this paper. Basically, this work has roots in our participation at the Large International River Project at the International Institute for Applied Systems Analysis (IIASA) in Laxenburg, Austria, in which we have been commissioned to develop a database querying system for a specialized database containing large sets of data related to river water quality. The system was meant to be employed in a decision support system (DSS) to be used by regional authorities from some administrative units along a large river. Those people had a limited computer proficiency, and wanted to ask questions and formulate requests to their databases in a natural language, very far from the strict SQL type queries.

After a close interaction with those regional authorities, the domain experts in this case, it was found that while they ask questions like "at which point along the river in our administrative unit there is a *serious water pollution*", which was obvious to them, but not natural to the traditional SQL based querying system, they did meant by a "serious pollution" that, for instance, "*most* of the *relevant* water quality indicators considerably exceed some *limits* set by some specialists or authorities". Notice that, obviously, not all limits should have been exceeded to yield a perception of a serious pollution so that a conventional fuzzy query should have been inappropriate! This has inspired us to explicitly include the fuzzy linguistic quantifier (e.g., most) in the query, specifically in the WHERE clause.

What can readily be seen is that in any practical case, like the one mentioned above, i.e. related to querying water quality databases, there is a serious problem. Namely, virtually all the data (practically the values of water quality indicators) have to be either downloaded from some public sources that are often maintained by public administration, universities, etc. or – which more often happens, notably for more specialized and sophisticated data – purchased from some specialized agencies and companies. Not always, however, data purchased for a high price are better.

Unfortunately, in such a case of different data sources and providers, there immediately emerges a serious problem of *data quality*. That is, in practice we have to deal with handling queries formulated by users against data that are of a highly variable quality. Obviously, for low quality data the results of querying would be questionable. That is why the term "relevant" was included in the example of a query given above.

To be brief, the goodness of data is often considered from the point of view of *validity* and *reliability* of the methods and instruments used in data collection. Basically, by the *validity* we mean the extent to which a measurement fulfills what it supposed to do while by the *reliability* we mean the consistence, stability, or dependability of the data. Whenever we employ some measurement, we want to be sure that the data obtained are true and accurate, when we measure a variable (here the value of a water quality indicator), our measurements should provide dependable and consistent results. Clearly, if a measurement is valid, it is also reliable. But if is reliable, it may or may not be valid, etc.

For the purpose of this paper we will not deal in much detail with the validity, reliability, and other aspects of goodness of data, because this is out of scope, and will somehow combine all of them under the term of *data quality*, for simplicity and operability. A more detailed analysis will be included in a next paper.

So, in our case we will include another aspect, related to the term "relevant", that will be included in the query, that is *data quality* which will be assumed to be from the unit interval, from 1 for the highest quality data to 0 for the poorest quality data, through all intermediate values. It can be done by experts who could evaluate the data themselves and their providers, or using some automated analysis of, for instance, some blogs, comments, etc. that can be found from the Internet. The determination of these values of data quality will not be considered in this paper. Just a value from [0, 1] will be assumed. Obviously, the higher the data quality, the higher the degree of relevance as mentioned.

The new queries with fuzzy linguistic quantifiers, with an account of data quality with respect to particular pieces of data, proposed in this paper can be therefore written in a general form by using and then extending the prototypical example (2) as:

```
SELECT *
FROM   river_points
WHERE  most_of {(ind_1 EXCEEDS_LIMIT_TO_DEGREE
       0.7 | qual_1 IS 0.9)                              (4)
       AND ...
       AND(ind_n EXCEEDS_LIMIT_TO_DEGREE 0.5
       | qual_n  IS 0.4)}
```

We will now discuss in more detail first the main tools and techniques to handle such queries with linguistic quantifiers, starting with the basic version without an account for data quality, i.e. (2), and then proceeding to the one with an account for data quality, i.e. (4).

2 Imprecise (Fuzzy) Queries and Queries with Fuzzy Linguistic Quantifiers

In this paper we deal with a particular extensions of fuzzy queries, the queries with linguistic quantifiers initiated by Kacprzyk and Ziółkowski [10], and then Kacprzyk, Ziółkowski and Zadrożny [9], written schematically as (2). Basically, a *non-standard aggregation scheme* of the fulfillment degrees of partial conditions, as for example in "*most* of the conditions among those specified have to be satisfied" is allowed. This brings in an essential extension of the classical schemes with the use of the *conjunction* and *disjunction*, and makes it possible to express often complex dependencies among the partial query conditions by using aggregation driven by *linguistic quantifiers* such as "most" or "almost all" which is very intuitively appealing to the human user. We will employ Kacprzyk and Ziółkowski's [10], and then Kacprzyk, Ziółkowski and Zadrożny's [9] so called "horizontal quantification" (cf. e.g. [1]), and not the so called "vertical quantification" due to Bosc and Lietard [1]).

It is easy to notice that the basic structure of the query with a fuzzy linguistic quantifiers (2) boils down to Zadeh's [20] calculus of linguistically quantified propositions which is basically meant to model natural language expressions like

$$\text{“}\textit{Most}\text{ Swedes are }\textit{tall}\text{”} \tag{5}$$

where "most" is an example of a linguistic quantifier; other examples of linguistic quantifiers include "almost all", "much more than 50%" etc.

For our purposes, we are only concerned with the *relative* quantifiers such that:

- their semantics refers to the proportion of elements possessing a certain property (e.g., the set of tall Swedes) among all the elements of the universe of discourse (e.g., the set of all Swedes);
  and *nondecreasing* such that:
- the larger such a proportion the higher the truth value of a proposition containing such a linguistic quantifier.

A linguistically quantified proposition exemplified by (5) might be formally written in a general form, playing the role of a *protoform*, as

$$QxP(x) \tag{6}$$

where Q denotes a linguistic quantifier (e.g., *most*), $X = \{x\}$ is a universe of discourse (e.g., a set of Swedes), and $P(x)$ is a predicate corresponding to a certain property (e.g., *tall*).

The truth value of (6) may be computed as follows. The relative quantifier Q is equated with a fuzzy set defined in [0, 1]. In particular, for a regular nondecreasing quantifier its membership function μ_Q is assumed to be nondecreasing and normal, i.e.,

$$x \le y \Rightarrow \mu_Q(x) \le \mu_Q(y); \quad \mu_Q(0) = 0; \quad \mu_Q(1) = 1 \tag{7}$$

The particular $y \in [0, 1]$ correspond to the proportions of elements possessing property P, and $\mu_Q(y)$ assesses the degree to which a given proportion matches the semantics of Q. For example, Q = "most" might be given as:

$$\mu_Q(y) = \begin{cases} 1 & \text{for } y > 0.8 \\ 2y - 0.6 & \text{for } 0.3 \le y \le 0.8 \\ 0 & \text{for } y < 0.3 \end{cases} \tag{8}$$

The predicate P is modeled by a fuzzy set $P \in \mathcal{F}(X)$, where $\mathcal{F}(X)$ is a family of fuzzy sets defined in X, with its membership function μ_P.

Then, the truth degree of (6) is:

$$\text{Truth}(QxP(x)) = \mu_Q(\frac{1}{n}\sum_{i=1}^{n}\mu_P(x_i)) \tag{9}$$

where n is the cardinality of X.

Now, we will show how the above calculus of linguistically quantified proposition can be implemented using Yager's [17] operators.

3 OWA Operators and Extended, Importance Quantified OWA Operators

The *ordered weighted averaging operators*, or OWA operators, for short, were introduced by Yager [17] (cf. also Yager and Kacprzyk [18] or Yager, Kacprzyk and Beliakov [19] for comprehensive expositions). They are a special class of aggregation operators which can provide a wide array of aggregation modes, in particular – for our purposes – being able to simply and uniformly model a large class of fuzzy linguistic quantifiers.

An OWA operator of dimension p is a mapping $F_W : [0, 1]^p \rightarrow [0, 1]$ if associated with F_W is a weighting vector $W = [w_1, \ldots, w_p]^T$ such that: $w_i \in [0, 1]$, $w_1 + \cdots + w_p = 1$, and

$$F_W(x_1, \ldots, x_p) = w_1 b_1 + \cdots + w_p b_p \tag{10}$$

where b_i is the i-th largest element among $\{x_1, \ldots, x_p\}$. $B = [b_1, \ldots, b_p]^T$ is called an ordered argument vector if each $b_i \in [0, 1]$, and $j > i$ implies $b_i \geq b_j$, $i = 1, \ldots, p$.

Then (10) may be also written as

$$F_W(x_1, \ldots, x_p) = W^T B \tag{11}$$

Example 1. Let $W^T = [0.2, 0.3, 0.1, 0.4]$, and calculate $F_W(0.6, 1.0, 0.3, 0.5)$. Thus, $B^T = [1.0, 0.6, 0.5, 0.3]$, and $F_W(0.6, 1.0, 0.3, 0.5) = W^T B = 0.55$.

An important problem that now occurs is how to determine the weighting vector of an OWA operator. It may be determined using, for instance (cf., e.g., [4, 21]:

1. Experimental data: let a set of data $\{(a_1^j, \ldots, a_p^j, y^j)\}_{j=1,\ldots,n}$ be; then assuming that $y_j = F_W(a_1^j, \ldots, a_p^j)$ $\forall j$ the weights vector W best matching this assumption is sought for. Filev i Yager [4] formalized that problem as the search for the weights vector minimizing the sum of squared errors $\sum_{j=1}^n (y_j - \sum_{i=1}^p w_i b_i^j)^2$. Assuming a specific parametrized form of the weight vector $w_i = e^{\lambda_i} / \sum_{j=1}^p e^{\lambda_j}$, this optimization problem may be reduced to unconstrained optimization.
2. A fixed value of certain characteristic features of the OWA operator such as, e.g., ORness [17]: a weights vector W is sought so that to obtain this fixed value, possibly optimizing the value of another characteristic feature.
3. A linguistic quantifier in the sense of Zadeh: the OWA operator is meant to represent a given linguistic quantifier.

For our purposes, in the context of queries with fuzzy linguistic quantifiers, the last option is most relevant, i.e., using a linguistic quantifier Q as a starting point. An early approach given in Yager [17] serves well the purpose and may be used here:

$$w_k = \mu_Q(k/p) - \mu_Q((k-1)/p) \quad \text{for} \quad k = 1, \ldots, p \tag{12}$$

Some examples of the w_i's associated with the classical quantifiers are:

- If $w_p = 1$, and $w_i = 0$, for each $i \neq p$, then this corresponds to $Q =$ "all";
- If $w_1 = 1$, and $w_i = 0$, for each $i \neq 1$, then this corresponds to $Q =$ "at least one",

and the intermediate cases as, e.g., a half, most, much more than 75%, a few, almost all, etc., which correspond to linguistic quantifiers, may be obtained using (12).

Thus, we will generally write the formulas for the calculation of the truth value of the linguistically quantified proposition that corresponds to the traditional query with a fuzzy linguistic quantifier (2):

$$\text{truth}(Qy'\text{s are } F) = \text{OWA}_Q(\text{truth } (y_i \text{ is } F)) = W^T B \tag{13}$$

It is easy to notice that the OWA weights employed in the above formulas, which are related to the particular rank order positions of the particular pieces of evidence (degrees of satisfaction of the particular query conditions rank ordered from the highest to the lowest) can only be used for the basic query with a fuzzy linguistic quantifier (2) but not for the new type of a query with a fuzzy linguistic quantifiers with an account for data quality introduced in this paper and generally written as (4). In this case, we associate an evaluation degree from the unit interval of the quality of data source which, in the water quality example, corresponds to the quality of data on some specific water quality indicator. This can result from the quality or functioning of a sensor, quality of a data set fetched from a public source or purchased from a specialized agency, etc. In our case, for the purpose of this study that aims at a proposal of a new type of query with a fuzzy linguistic quantifier, we will not be concerned with how it is calculated just assuming, for the time being, that its is evaluated by experts and given as a number from [0, 1].

Therefore, we should use some extension of the OWA operators in which additional weights may be taken into account which are coupled with the particular query conditions and "stay" with them even after the OWA specific rank ordering. There are known many approaches to such extensions of the OWA operators, starting from the early work by Yager [17], to newer and more advanced approaches exemplified by Torra's [13] weighted OWA operators (WOWA) or Xu and Da's [16] XWA operators, to name a few (cf. also Torra and Lv [14], Xu [15], etc).

We will employ the source Yager's [17] work that will best show the very essence of our extended approach and make further extensions possible. It is termed an OWA operator with importance qualification and is meant for fuzzy linguistic quantifier based OWA operators.

We have therefore a vector of data (pieces of evidence, here values of water quality indicators) $A = [a_1, \ldots, a_n]$, a linguistic quantifier Q which defines the general aggregation mode of the OWA operator, and a vector of importances $V = [v_1, \ldots, v_n]$ such that $v_i \in [0, 1]$ is the importance (here degree of quality) of a_i, $i = 1, \ldots, n$, $(v_1 + \cdots + v_n \neq 1$, in general). The OWA operator with importance qualification works as previously, i.e.,

$$F_V(a_1, \ldots, a_n) = \overline{W}^T \cdot B = \sum_{j=1}^{n} \overline{w}_j b_j \tag{14}$$

but the weights are computed in a different way than in (12), and thus are here denoted as $\overline{W} = [\overline{w}_1, \ldots, \overline{w}_n]$. We order first the pieces of evidence a_i, $i = 1, \ldots, n$, in descending order to obtain B such that b_j is the j-th largest element of $\{a_1, \ldots, a_n\}$. Next, we denote by u_j the importance of b_j, i.e. of the a_i which is the j-th largest; $i, j = 1, \ldots, n$. Finally, the weights $\overline{W}$ are defined as

$$\overline{w}_j = \mu_Q\left(\frac{\sum_{k=1}^{j} u_k}{\sum_{k=1}^{n} u_k}\right) - \mu_Q\left(\frac{\sum_{k=1}^{j-1} u_k}{\sum_{k=1}^{n} u_k}\right) \tag{15}$$

Example 2. Suppose that $A = [a_1, a_2, a_3, a_4] = [0.7, 1, 0.5, 0.6]$, and $V = [u_1, u_2, u_3, u_4] = [1, 0.6, 0.5, 0.9]$. Q = "most" is given by (8).

Then, $B = [b_1, b_2, b_3, b_4] = [1, 0.7, 0.6, 0.5]$, and $\overline{W} = [0.0, 0.47, 0.53, 0.0]$, and $F_V(A) = \sum_{j=1}^{4} \overline{w}_j b_j = 0.0 \cdot 1 + 0.47 \cdot 0.7 + 0.53 \cdot 0.6 + 0.0 \cdot 0.5 = 0.64$.

Such a definition of the weights that are then employed in the OWA operator calculations is in fact characteristic to virtually all newer approaches to the weighted OWA operators, with the meaning of "weighted" being more general than in Torra's WOWA or Xu and Da's XWA operators. For more information, cf. Torra [13], Xu and Da [16], Torra and Lv [14], Xu [15], etc.

An important problem is how to determine importance degrees V in the above expressions. As we have already mentioned, in this paper a new query with a fuzzy linguistic quantifier is proposed with an account of data quality. We assume that the "importance" degree employed in the formulas mentioned is a degree from the unit interval that is provided by an expert in this first version. It is clearly related to a comprehensive aggregation of evaluations of the validity and reliability of the particular pieces of evidence (data). A detailed analysis will be given in a next paper and, for the purpose of this paper, for simplicity and operability, we just rely on a domain expert providing a data quality degree from [0, 1].

4 Linguistic Quantifiers in Queries

As mentioned in the Introduction, the use of the fuzzy linguistic quantifiers in various clauses of the SQL query might be conceived. For the purposes of our new proposal, the best way of presentation is to use the general schemes of queries with fuzzy linguistic quantifiers: the traditional one given by (2), and the new – query with a fuzzy linguistic quantifier with an account of data quality given by (4). The latter can be implemented using tools and techniques, and user interfaces used in our FQUERY for Access package (cf. Kacprzyk and Zadrożny [5, 6, 7, 11]), i.e., as operators aggregating conditions in the `WHERE` clause. Basically, the linguistic quantifiers are here defined according to Zadeh's [20] calculus of linguistically quantified propositions, briefly recalled in the previous section. Such a representation has some advantages and an easy scalability to a varying number of conditions to be aggregated. For example, having the linguistic quantifier "most" defined by (8) one may use it to interpret such a condition, corresponding to (2), as:

$$\text{“\textit{Most} of the predicates } \{P_i\}_{i=1,\ldots,n} \text{ are satisfied”} \tag{16}$$

for any number, n, of predicates.

And in the case of (4), i.e.

$$\text{“\textit{Most} of the \textit{important} predicates } \{P_i\}_{i=1,\ldots,n} \text{ are satisfied”} \tag{17}$$

for any number, n, of predicates, the situation is analogous.

Thus in FQUERY for Access the linguistic quantifiers are first defined in the sense of Zadeh and later, during the query execution, they are interpreted and manipulated in terms of OWA operators.

FQUERY for Access maintains a dictionary of various linguistic terms defined by the user and available for the use in the queries. Among them are linguistic quantifiers in the sense of Zadeh. Each quantifier is identified with a piecewise-linear membership function and assigned a name. When inserting a quantifier into a query the user just picks up its name from the list of quantifiers available in the dictionary. The degrees of importance (data qualities) are just numbers from [0, 1] provided by human experts.

In this section we have therefore presented tools and techniques, and some remarks on implementation within our FQUERY for Access system of a new type of a query with a fuzzy linguistic quantifier with an account for data quality.

5 Concluding Remarks

We presented a new type of a query with a fuzzy linguistic quantifier with an account for a variable quality of pieces of data. We assumed the so called horizontal quantification query introduced in our works (cf. Kacprzyk and Ziółkowski [10], and Kacprzyk, Ziółkowski and Zadrożny [9]), an additional degree of the quality of pieces of data (corresponding to particular query conditions), and then use some OWA (ordered weighted aggregation) operator with importance for the implementation of a linguistic quantifier driven aggregation of satisfaction of the particular query conditions with an additional importance. Moreover, though we employed the standard Yager's [17] approach to importance qualified OWA operators, an extension to the use of other weighted OWA operator based aggregation schemes is natural. We showed how the new approach can be accommodated in our FQUERY for Access database querying system. We also suggested as a further research topic a more detailed analysis of how to derive a comprehensive data quality index by combining, for instance, validity and reliability, and also other aspects.

Acknowledgments This work has been partially supported by the National Science Centre under Grant DEC-2012/05/B/ST6/03068.

References

1. Bosc, P., Lietard, L.: Quantified statements and some interpretations for the OWA operators. In: Yager, Kacprzyk [18], pp. 241–257
2. Bosc, P., Pivert, O.: SQLf: A relational database language for fuzzy querying. IEEE Transactions on Fuzzy Systems **3**(1), 1–17 (1995)
3. Bosc, P., Prade, H.: An introduction to the fuzzy set and possibility theory-based treatment of flexible queries and uncertain or imprecise databases. In: Uncertainty Management in Information Systems, pp. 285–324. Kluwer Academic Publishers, Boston (1996)
4. Filev, D., Yager, R.: On the issue of obtaining OWA operator weights. Fuzzy Sets and Systems **94**, 157–169 (1998)
5. Kacprzyk, J., Zadrożny, S.: FQUERY for Access: fuzzy querying for a windows-based DBMS. In: Bosc, P., Kacprzyk, J. (eds.) Fuzziness in Database Management Systems, pp. 415–433. Physica-Verlag, Heidelberg (1995)
6. Kacprzyk, J., Zadrożny, S.: The paradigm of computing with words in intelligent database querying. In: Zadeh, L., Kacprzyk, J. (eds.) Computing with Words in Information/Intelligent Systems. Part 1. Foundations. Part 2. Applications, vol. 34, pp. 383–398. Springer-Verlag, Heidelberg and New York (1999)
7. Kacprzyk, J., Zadrożny, S.: Computing with words in intelligent database querying: standalone and internet-based applications. Information Sciences **134**(1–4), 71–109 (2001)
8. Kacprzyk, J., Zadrożny, S.: Linguistic database summaries and their protoforms: towards natural language based knowledge discovery tools. Inf. Sci. **173**(4), 281–304 (2005)
9. Kacprzyk, J., Zadrożny, S., Ziółkowski, A.: FQUERY III+: a "human consistent" database querying system based on fuzzy logic with linguistic quantifiers. Information Systems **14**(6), 443–453 (1989)
10. Kacprzyk, J., Ziółkowski, A.: Database queries with fuzzy linguistic quantifiers. IEEE Transactions on System, Man and Cybernetics **16**(3), 474–479 (1986)
11. Kacprzyk, J., Zadrożny, S.: Fuzzy queries in microsoft access v. 2. In: Dubois, D., Prade, H., Yager, R. (eds.) Fuzzy Information Engineering - A Guided Tour of Applications, pp. 223–232. John Wiley & Son, New York (1997)
12. Tahani, V.: A conceptual framework for fuzzy query processing: a step toward very intelligent database systems. Information Processing & Management **13**(5), 289–303 (1977)
13. Torra, V.: The weighted owa operator. International Journal of Intelligent Systems **12**, 153–166 (1997)
14. Torra, V., Lv, Z.: On the wowa operator and its interpolation function. International Journal of Intelligent Systems **24**, 1039–1056 (2009)

15. Xu, Z.: An overview of methods for determining owa weights. International Journal of Intelligent Systems **20**, 843–865 (2005)
16. Xu, Z., Da, Q.: An overview of operators for aggregating information. International Journal of Intelligent Systems **18**, 953–969 (2003)
17. Yager, R.: On ordered weighted averaging aggregation operators in multi-criteria decision making. IEEE Transactions on Systems, Man and Cybernetics **18**, 183–190 (1988)
18. Yager, R., Kacprzyk, J. (eds.): The Ordered Weighted Averaging Operators: Theory and Applications. Kluwer, Boston (1997)
19. Yager, R.R., Kacprzyk, J., Beliakov, G. (eds.): Recent Developments in the Ordered Weighted Averaging Operators: Theory and Practice. Studies in Fuzziness and Soft Computing, vol. 265. Springer (2011)
20. Zadeh, L.: A computational approach to fuzzy quantifiers in natural languages. Computers and Mathematics with Applications **9**, 149–184 (1983)
21. Zadrożny, S., Kacprzyk, J.: Issues in the practical use of the OWA operators in fuzzy querying. J. Intell. Inf. Syst. **33**(3), 307–325 (2009)

A Clustering-Based Approach to the Explanation of Database Query Answers

Aurélien Moreau, Olivier Pivert and Grégory Smits

Abstract This paper describes an approach providing end-users with more insight to better understand the results of their queries. Using a clustering algorithm, the idea is to form subgroups of answers sharing some properties and to discover explanations for each subgroup. The originality of this work is that the data considered for characterizing each cluster of answers is not limited to the attributes used in the query. The objective is to enable the users to comprehend the structure of the results of their queries, using linguistic labels taken from their own vocabulary.

1 Introduction

The general issue of providing answers with additional information is one of the aspects of the domain known as *cooperative query answering* [3], and is certainly a challenging research direction in the database domain. Several types of approaches have recently been proposed that share that general objective. In the approach presented in [8], for instance, the suspect nature of some answers (involved in the violation of one or several functional dependencies) to a request is identified through auxiliary queries. This may be viewed as a form of cooperative answers where additional information (here, the suspect nature of an answer, possibly with a degree) is provided to the user. In [7], the authors take advantage of the lineage of answers for finding causes for a query result and computing a degree of responsibility of a tuple with respect to an answer, as a basis for explaining unexpected answers to a query. The idea there is that "tuples with high responsibility tend to be interesting explanations to query answers".

Another example of explanation needs is when the set of answers obtained can be clustered in clearly distinct subsets of similar or close answers. Then, it may be interesting for the user to know what meaningful differences exist between the

A. Moreau · O. Pivert(✉) · G. Smits
Irisa, University of Rennes 1, Technopole Anticipa, 22305 Lannion Cedex, France
e-mail: {aurelien.moreau,olivier.pivert,gregory.smits}@irisa.fr

T. Andreasen et al. (eds.), *Flexible Query Answering Systems 2015*,
Advances in Intelligent Systems and Computing 400,
DOI: 10.1007/978-3-319-26154-6_24

tuples leading to the answers that may explain the discrepancy in the result. For instance, if one looks for possible prices for houses to let obeying some (possibly fuzzy) specifications, and that two clusters of prices are found, one may discover, e.g., that this is due to two categories of houses having, or not, some additional valuable equipment. This latter issue, which constitutes the topic of the present paper, has been previously dealt with in [2]. Here, we propose an alternative approach that first uses a clustering algorithm to detect groups of answers (a group corresponds to elements that have similar values on the attributes from the projection clause of the query) – this is the description step, that makes use of a fuzzy vocabulary. Then we look for common properties between the elements of each cluster (that are not possessed by elements from other clusters) for the other attributes – this is the characterization step.

The remainder of the paper is structured as follows. Section 2 provides a refresher on fuzzy partitions. In Section 3, we describe the principle of the approach. Experimental results are presented in Section 4. A flexible extension is proposed in Section 5 while related work is discussed in Section 6. Finally, Section 7 recalls the main contributions and outlines perspectives for future work.

2 Fuzzy Vocabulary

Fuzzy set theory was introduced by Zadeh [10] for modeling classes or sets whose boundaries are not clear-cut. For such objects, the transition between full membership and full mismatch is gradual rather than crisp. Typical examples of such fuzzy classes are those described using adjectives of the natural language, such as *young, cheap, fast*, etc. Formally, a fuzzy set F on a referential U is characterized by a membership function $\mu_F : U \to [0, 1]$ where $\mu_F(u)$ denotes the grade of membership of u in F. In particular, $\mu_F(u) = 1$ reflects full membership of u in F, while $\mu_F(u) = 0$ expresses absolute non-membership. When $0 < \mu_F(u) < 1$, one speaks of partial membership.

In the approach we propose, it is assumed that the user specifies a vocabulary defined by means of fuzzy partitions. Let R be a relation defined on a set $\mathcal{A}$ of q categorical or numerical attributes $\{A_1, A_2, \ldots, A_q\}$. A fuzzy vocabulary on R is defined by means of fuzzy partitions of the q domains. A partition $\mathcal{P}_i$ associated with the domain of attribute A_i is composed of m_i fuzzy predicates $\{P_{i,1}, P_{i,2}, \ldots, P_{i,m_i}\}$, such that for all $x \in domain(A_i)$:

$$\sum_{j=1}^{m_i} \mu_{P_{ij}}(x) = 1$$

where $\mu_{P_{ij}}(x)$ denotes the degree of membership of x to the fuzzy set P_{ij}. More precisely, we consider partitions for numerical attributes (Fig. 1) composed of fuzzy sets, where a set, say P_i, can only overlap with its predecessor P_{i-1} or/and its successor P_{i+1} (when they exist). For categorical attributes, we simply impose that for

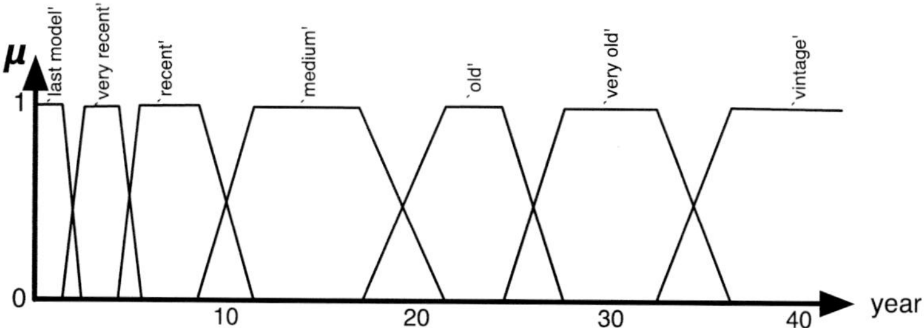

Fig. 1 A partition over the domain of the attribute *year*

each domain value the sum of the satisfaction degrees on all elements of a partition is equal to 1. Each $\mathcal{P}_i$ is associated with a set of linguistic labels $\{L_{i,1}, L_{i,2}, \ldots, L_{i,m_i}\}$.

As an example, let us consider a database containing ads about second hand cars and a view named *secondHandCars* of schema (*id*, *model*, *description*, *year*, *mileage*, *price*, *make*, *length*, *height*, *nbseats*, *consumption*, *acceleration*, *co2emission*) as the result of a join-query over the database. A common sense partition and labelling of the domain of the attribute *year* is illustrated in Fig. 1. Table 1 shows a possible common sense partition and labelling of the domain of the categorical attribute *make*.

Table 1 A partition over the domain of the attribute *make*

	make																	
	Dodge	Jeep	...	Honda	...	Nissan	Renault	Peugeot	Dacia	...	ARO	Oltcit	...	VW	Lamborghini	Skoda	...
American	1	1	...	0	...	0	0	0	0	...	0	0	...	0	0	0	...
Asian	0	0	...	1	...	0.6	0	0	0	...	0	0	...	0	0	0	...
...	...	...	...	...	...	...	...	...	...	...	...	...	...	...	...	...	...
French	0	0	...	0	...	0.4	1	1	0.4	...	0	0	...	0	0	0	...
East-European	0	0	...	0	...	0	0	0	0.6	...	1	1	...	0	0	0	...
German	0	0	...	0	...	0	0	0	0	...	0	0	...	1	0.5	0.6	...
...	...	...	...	...	...	...	...	...	...	...	...	...	...	...	...	...	...

3 Principle of the Approach

Let us denote by R the relation concerned by the selection-projection query Q considered. $\mathcal{A}$ being the set of attributes of R, let us denote by $\mathcal{A}_\pi$ the subset of $\mathcal{A}$ made of the attributes onto which R is projected (i.e., the attributes of the resulting relation), by $\mathcal{A}_\sigma$ the subset of $\mathcal{A}$ concerned by the selection condition, and let us denote $\mathcal{A}_\omega = \mathcal{A} \backslash (\mathcal{A}_\pi \cup \mathcal{A}_\sigma)$. The three main steps are:

1. **detection** of the clusters of answers: identifying them with a clustering algorithm on the attributes from $\mathcal{A}_\pi$ (Subsection 3.1);

2. **description** of the clusters of answers: projecting them on the vocabulary defined on the domains of the attributes from $\mathcal{A}_\pi$ (Subsection 3.2);
3. **characterization** of each cluster in terms of the vocabulary defined on the domains of the attributes from $\mathcal{A}_\omega$ (Subsection 3.3).

The first step consists in using a clustering algorithm to group the answers into several subsets. Step 2 is about using a fuzzy vocabulary to describe each one of these clusters. Step 3 aims at providing one or several characterizations for each of these clusters in the form of a conjunction of modalities (*i.e.* fuzzy labels) from the vocabulary – considered as additional information as it concerns attributes that do not appear in the query. The objective is to find properties that will permit to characterize the clusters beside the data used to produce them.

A characterization is related to the set of clusters built in step 1. It is made of a set of linguistic descriptions, one for each cluster. Let us denote by $C = \{C_1, \ldots, C_n\}$ the set of clusters obtained.

Definition 1. *A characterization $E(c_i)$ attached to a cluster c_i is a conjunction of couples (attribute, label) of the form*

$$\begin{aligned} E(c_i) = \{(A_i,\ L_i) \mid\ & A_i \in \mathcal{A}_\omega \textit{ and} \\ & L_i \textit{ is a linguistic label from the partition of the domain of } A_i, \\ & \textit{or a disjunction of such labels}\}. \end{aligned}$$

Example 1. Let us consider a set of second-hand cars, which could be classified into two sets based on a clustering on the attributes *year* and *mileage* so that:

- Cluster 1 is described by: "*year is recent and mileage is small*";
  A characterization could be: "*price is expensive*".
- Cluster 2 is described by: "*year is old and mileage is high*";
  A characterization could be: "*price is affordable*".

To be informative, a characterization should satisfy two properties: specificity and minimality.

Property 1. Specificity: a characterization must identify and characterize a single cluster, *i.e.* only the elements belonging to the concerned cluster.

In other words, a characterization $E(c_i)$ should satisfy the following equivalence:

$$\forall x \in R,\ Q(x) \wedge E(c_i)(x) \Leftrightarrow x \in c_i, \tag{1}$$

i.e. every element x belongs to a strengthening of Q formed by the conjunction $Q \wedge E(c_i)$ iff x belongs to the cluster c_i. This equivalence guarantees the specificity of $E(c_i)$. As a consequence we get:

$$\nexists x' \in c_j,\ (j \neq i) \text{ such that } E(c_i)(x') \tag{2}$$

Property 2. Minimality: viewing a characterization as a conjunction of predicates (or disjunctions of predicates), then one says that $E(c_i)$ is a minimal characterization of the cluster c_i iff $\nexists E'(c_i) \subset E(c_i)$ so that $E'(c_i)$ characterizes c_i. For instance, if we get back to houses to let (cf. Introduction), and identify a subset of answers whose characterization is E = (*price is expensive*) $\wedge$ (*swimming pool* = *yes*) $\wedge$ (*garden is big*), there should not exist a characterization *e.g.* E' = (*price is expensive*) $\wedge$ (*swimming pool* = *yes*) also characterizing this cluster only.

To obtain the minimality property we need to add the following condition:

$$\nexists E'(c_i) \subset E(c_i) \text{ such that } \forall x \in R,\ Q(x) \wedge E'(c_i)(x) \Leftrightarrow x \in c_i \qquad (3)$$

with (3) guaranteeing that there does not exist any characterization E' more specific than E that also characterizes c_i.

3.1 Detecting Clusters of Answers (Detection Step)

In our approach, we propose to make use of

- predefined partitions of the domains of the attributes, (cf. Subsection 2), and
- a clustering technique to identify different groups of answers.

Concerning the choice of the clustering algorithm, we need an algorithm that does not imply to know in advance the number of clusters to obtain. We propose to use the *l-cmed-select* algorithm, a crisp variant of the *l-fcmed-select* technique proposed in [6], which belongs to the framework of incremental clustering and combines relational clustering and medoid-based methods. *l-fcmed-select* is an extension of the linearised fuzzy c-medoids clustering algorithm [5], *l-fcmed*. The *l-cmed-select* algorithm possesses the following three main characteristics: i) it exploits a medoid definition of the cluster center, and it can therefore be applied to heterogeneous data; ii) it does not require a precise number of clusters to operate, simply an over-estimation; iii) it exploits a linear approximation scheme to update the cluster medoid, looking for the new medoid in the vicinity of its previous position. This approximation alleviates computational costs.

The clustering algorithm is applied to the data restricted to the attributes of $\mathcal{A}_\pi$. For each attribute A_i with a vocabulary defined, let us denote D_i its domain, $|A_i|$ the number of modalities and $v_{i,1}, \ldots, v_{i,|A_i|}$ the sets associated.

3.2 Cluster Projection on the Attributes (Description Step)

Once the clusters are formed, they are projected on the vocabulary in order to provide the user with a description of the answers in natural language. When a cluster satisfies several modalities for a given attribute, a simple way to project it is to return the disjunction of the associated labels. A cluster c_i can be "boxed up" with $2 * p$ points $(x_{min}, y_{min}), (x_{max}, y_{max})$ (one pair for each of the p dimensions) so that these points indicate which fuzzy labels the cluster satisfies (to a degree > 0.5). For instance in Figure 2, cluster 2 satisfies labels 2 and 3 of attribute 1, so the disjunction of these two labels should be considered. As to cluster 1, it only satisfies label 1. Regarding attribute 2, cluster 2 satisfies label b only and cluster 1 satisfies label a. Label b is not satisfied by cluster 1 because its degree is below 0.5.

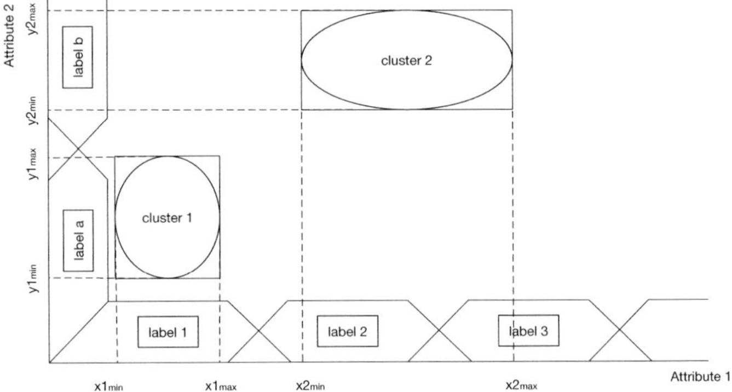

Fig. 2 Projection onto $\mathcal{A}_\pi$

3.3 *Discovering Characterizations (Characterization Step)*

When considering attributes that were not involved in the clustering process, i.e., attributes from $\mathcal{A}_\omega$, the value distributions will in general be closer to the one in Figure 3, since the values of a given cluster are not linked by any similarity relationship anymore. In this case it is not very wise to "box up" one cluster altogether as we did above, because all adjacent labels may not be satisfied by a given cluster. For instance in Figure 3, cluster 2 satisfies labels 1 and 3, but not label 2. One must then check which labels each individual cluster element satisfies.

The first step to discovering characterizations (in the sense of Definition 1) lies in filling a table associating each cluster with its projection on the attributes of $\mathcal{A}_\omega$. For every A_i ($i \in [p+1, q]$) in $\mathcal{A}_\omega$, we indicate which modality L^i_j, $j \in [1, |A_i|]$ (or disjunction of modalities) is satisfied by each cluster.

Table 2 Correspondance between modalities and clusters

	A_{p+1}	A_{p+2}	$\cdots$	A_q
C_1	$(L_2^{p+1} \vee L_3^{p+1})$	L_3^{p+2}	$\cdots$	L_8^n
C_2	L_4^{p+1}	L_9^{p+2}	$\cdots$	L_6^n
$\vdots$	$\vdots$	$\vdots$		$\vdots$
C_k	L_5^{p+1}	$(L_3^{p+2} \vee L_4^{p+2})$	$\cdots$	L_8^n

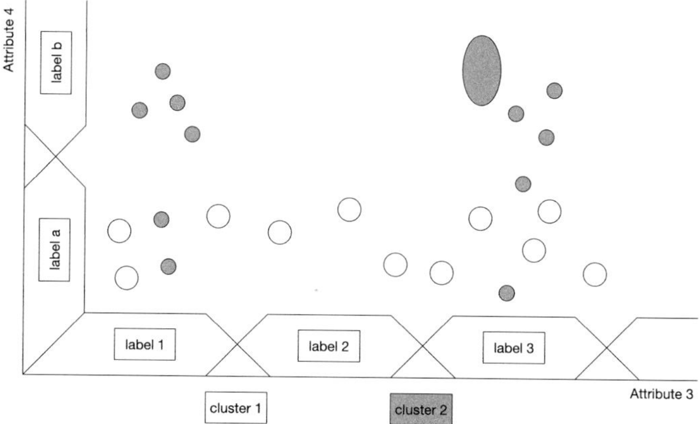

Fig. 3 Projection onto $\mathcal{A}_\omega$

Based on the elements from Table 2, we use Algorithm 1 to find characterizations for the clusters. This algorithm takes as input the number of clusters as well as the data from Table 2. For each cluster c_i (line 2) we look for characterizations of every "size" (line 4) — starting with a single couple (attribute, label), then with two of them, then three, ... — while checking that they are all indeed specific in the sense of Eq. (1) (the candidate characterizations are all minimal, as narrowed down in line 5) with Algorithm 2 (line 6). Those which are specific are then added to the set of characterizations of the cluster considered (line 7).

With Algorithm 2, we check whether a characterization (conjunction of conditions) is specific for a given cluster. To do this, the characterization is confronted to every other cluster (lines 8 and 9). For each conjunct of the characterization, the corresponding labels associated with the clusters and attributes from Table 2 are compared to it in order to check whether the two overlap or not (line 12) — two conjuncts are said to overlap if they are the same or if they have a disjunct in common. If, at the end of the loops, no overlapping has been found, then the characterization really is specific to the cluster.

4 Experimentation

In order to check the effectiveness of the approach, we performed a preliminary experimentation using a synthetic dataset with houses to let as in [2]. The attributes considered were *price*, *surface*, *garden area*, and *swimming pool*. The dataset was generated with the objective to obtain two subgroups, hence the convenient distribution of the data. The algorithm used was *l-cmed-select* (a crisp version of the *l-fcmed-select* algorithm presented in [6]), and 4 was the (over-estimated) specified number of clusters. The distance measure (4) was used to compare numerical attributes, and identity for categorical ones.

```
Input: n clusters c ; q − p attributes/values for each cluster ;
Output: a set of characterizations for each cluster
begin
    foreach cluster c_i do
        Charact(c_i) ← ∅;
        for j ← 1 to q − p do
            for every possible characterization E of size j that is not a superset of any
            element of Charact(c_i) do
                if E is_specific(i,E) then
                    Charact(c_i) ← Charact(c_i) ⋃ E
                end
            end
        end
    end
end
```

Algorithm 1. Characterizations Finder

```
Input: int i, condition cc
Output: bool spec
Let cc be a conjunction of conditions, cc_g the condition of cc on attribute A_g
Let T[i, j] be the condition describing cluster c_i on attribute j (in Table 2)
n clusters c ; q − p attributes/values for each cluster ;
function is_specific () return
begin
    spec ← true;
    l ← 1;
    while l ⩽ n  &  spec do
        if l ≠ i then
            h ← 1;
            while h ⩽ q − p  &  spec do
                if cc_h overlaps with T[l,h] then
                    spec ← false
                end
                h ← h + 1
            end
        end
        l ← l + 1
    end
    result ← spec
end
```

Algorithm 2. Specificity Checker

$$dist(x,\ y) = \frac{|x - y|}{\max(x,\ y)} \tag{4}$$

Two other parameters were required for the clustering algorithm: minimal cluster size *minSize* and maximal cluster compactness *maxCompactness*. Empirically chosen, these values were $minSize = 10$ and $maxCompactness = 0.75$. The results of the clustering algorithm over the *price* and *surface* attributes are in Figure 4a. After processing Algorithm 1 on the data, several characterizations were found.

- Cluster 0 was described as: *price is cheap and surface is small*;
  The following characterizations were found:
  - *garden area is small*;
  - *swimming pool = no*.
- Cluster 1 was described as: *price is expensive and surface is big*;
  The following characterizations were found:
  - *garden area is large*;
  - *swimming pool = yes*.

Here, each cluster was associated with one label for each attribute. The first two ones $\mathcal{A}_\pi = \{price, \ surface\}$ were the ones on which the clustering process was carried out, while the other two $\mathcal{A}_\omega = \{swimming\text{-}pool, garden\text{-}area\}$ each provided a characterization for each cluster, both specific and minimal.

With a real dataset, data is usually not as sparse as in Figure 4a but closer to that of Figure 4b. Real data from second hand cars ads were used here, with attributes (*price*, *mileage*, *year*, *option level*, *security level*, *comfort level*, *brand*, *model*). Figure 4b is a representation of the data with the query looking for the prices and mileage of cars that cost between 25,000 and 30,000€ or below 10,000€. In this case, some outliers are present and the border between clusters is not as clear-cut as in the former case, making it difficult (if not impossible) to find labels characterizing only one cluster. This leads us to defining more flexible variants of our approach. Due to space limitation, we can only outline the principle of these relaxed variants (cf. Section 5).

To assess the efficiency of this approach, we used another synthetic dataset on a Macbook pro with a 3GHz Intel Core i7 processor and 16Gb RAM. We checked the impact of two parameters on the execution time: the cardinality of the dataset and the number of attributes in $\mathcal{A}_\omega$. $|\mathcal{A}_\pi|$ was set to 3 for both experimentations. The results of the first experiment are presented in Figure 5a. $|\mathcal{A}_\omega|$ was set to 10. The clustering part execution times are acceptable under 10,000 tuples of data. Let us emphasize that the clustering step is performed on a *set of answers*, not on a base relation, and one may consider that 10,000 already corresponds to a rather large answer set. Execution times for the explanation process (description and characterization) are below one second. In the second experiment, we set the number of tuples to 10,000. The results, presented in Figure 5b, show that the excution time remains negligible as long as $|\mathcal{A}_\omega|$ is under 19.

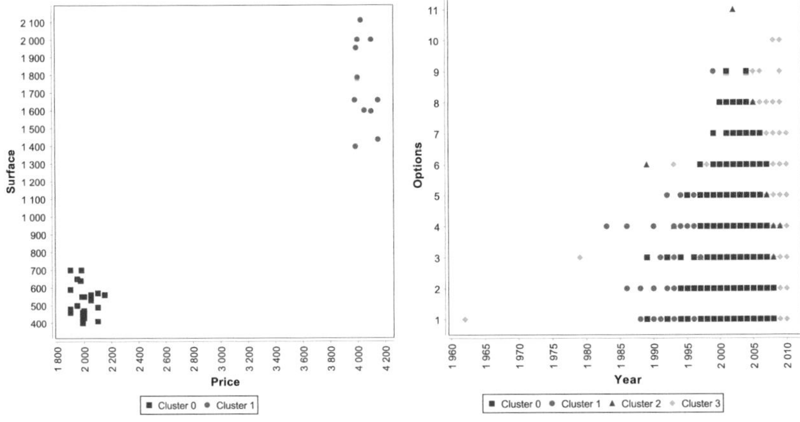

(a) Cluster of housing data over the attributes *price* and *surface*

(b) Cluster of second hand cars over the attributes *year* and *option level*

Fig. 4 Different clustering results

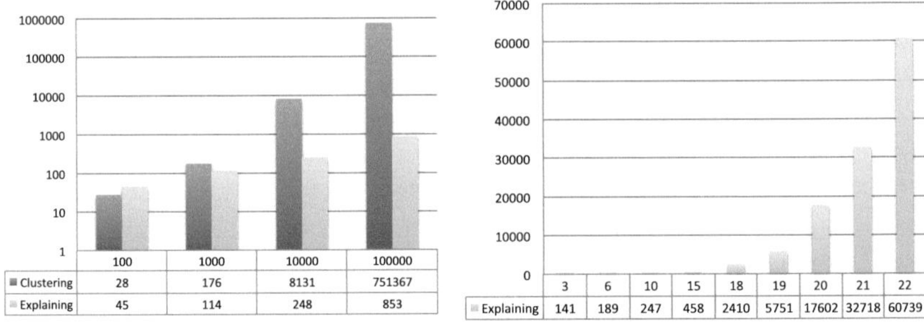

(a) Execution times (in ms, log scale) depending on the number of tuples processed for the clustering and explanation parts

(b) Execution times (in ms) depending on the number of attributes in $\mathcal{A}_\omega$ for the explanation part

Fig. 5 Execution times in milliseconds.

5 Towards a Flexible Approach

In the approach described above, the definition of specificity (Formula 1) is very drastic, and in case of real-world datasets where clusters are not clear-cut, the risk is high that no characterization may be found. Two solutions may be thought of, both based on the same incremental clustering technique as before.

First, one may take into account the representativity of a point in a given cluster, so as to ignore the points that are far away from the medoid when checking the specificity of a candidate characterization for this cluster. Practically, when no characterization is found for a cluster C_i in its entirety, one may check whether there exist characterizations for the 90% most representative elements of C_i (i.e., the closest to its medoid), or — if it is not the case —, for the 80% most representative elements and so on (not going further than 50%, for instance, since one still wants to characterize the majority of the elements of C_i). Then, the definition of specificity does not change (it remains Boolean), but since smaller clusters are taken into account, the chances to find characterizations are increased.

A second solution is to view specificity as a gradual concept and to modify the way a cluster is projected onto the vocabulary so as to take into account the weight of a label in the description of a cluster. Then, instead of a disjunction, the projection of a cluster on a given dimension would be a fuzzy set of labels, where the degree attached to a label would depend on the proportion of elements of the cluster that rewrite into that label. In other words, it would be the (normalized) sum of the membership degrees of the elements of the clusters relatively to the fuzzy label considered. It is then necessary to define the extent to which two such fuzzy sets are distinct (since it will be used in the algorithm for assessing the specificity degree of a candidate characterization). Let us first consider a characterization involving a single attribute. Let E_1 and E_2 be the respective projections of the clusters C_1 and C_2 onto an attribute A_k of $\mathcal{A}_\omega$, whose associated fuzzy partition is denoted by P_k. One may define:

$$\mu_{distinct}(E_1,\ E_2) = 1 - \max_{L_i \in P_k} \min(\mu_{C_1}(L_i),\ \mu_{C_2}(L_i)). \tag{5}$$

When several attributes are involved, two characterizations are globally distinct if they are so on at least one attribute and we get:

$$\mu_{distinct}(E_1,\ E_2) = \max_{A_j \in E} (1 - \max_{L_i \in P_j} \min(\mu_{C_1}(L_i),\ \mu_{C_2}(L_i))). \tag{6}$$

Finally, the specificity degree attached to a candidate characterization associated with a given cluster C may be defined as:

$$\mu_{spec}(E) = \min_{C' \neq C} \mu_{distinct}(E,\ E') \tag{7}$$

where E' denotes the projection of C' onto the attributes present in E.

In the algorithm, one may then consider that E is a valid characterization for C if its specificity degree is over a specified threshold.

6 Related Work

Fuzzy approaches to answer explanations have been previously proposed in [1, 2]. In [1], the answers to a fuzzy query are ranked according to an overall aggregation function and additional information (positive and negative) is provided about the different results.

Case-based reasoning is at the heart of [2], as the authors study the similarities between situations and their resulting outcomes. To do so, queries with a single output attribute are considered and the result is presented in the form of

- a possibility distribution reflecting the values taken by this attribute, and
- a function giving the number of cases supporting a particular outcome attribute value.

The fact that a single attribute is considered makes it relatively easy to detect clusters of answers (they correspond to distinct peaks of the distribution) by looking at the associated curve. However, the authors do not give any detail about how this detection process could be generalized and automated (which we do by using a clustering technique). To find explanations for a given distribution, they propose to look for attribute values that are shared by elements in one peak and different in the others, through the use of fuzzy sets and membership functions.

In [9], explanations based on causality and provenance are defined. The objective of the authors is different from ours insofar as they do not provide any insight regarding the structure of the results of the queries but rather illustrate causality with "intervention", *i.e.* removing tuples from the database and assessing how the results are modified. A different but very close research direction deals with "why not" answers in [4], looking for explanations for missing elements in an answer set. Causality and provenance are here the keys to figuring out which tuples and which conditions prevented some tuples from being part of the result.

7 Conclusion

In this paper, we have presented an approach aimed to characterize subsets of answers to database queries, using three steps: i) detection: the answers are grouped by means of a clustering algorithm; ii) description: the clusters obtained are described in terms of a fuzzy vocabulary; iii) characterization: other attributes (not involved in the clustering process) are used to highlight the particular properties of each cluster.

Preliminary experimental results show that the approach is indeed effective in finding characterizations in the presence of well-defined clusters. Perspectives include adding flexibility to the characterization process so that overlapping dense clusters can be processed (in the spirit of the two solutions outlined in Section 5).

References

1. Amgoud, L., Prade, H., Serrut, M.: Flexible querying with argued answers. In: Proc. of the 14th IEEE International Conference on Fuzzy Systems (FUZZ-IEEE 2005), Reno, Nevada, USA, pp. 573–578(2005)
2. de Calmès, M., Dubois, D., Hüllermeier, E., Prade, H., Sedes, F.: Flexibility and fuzzy case-based evaluation in querying: An illustration in an experimental setting. International Journal of Uncertainty, Fuzziness and Knowledge-Based Systems **11**(1), 43–66 (2003)

3. Gaasterland, T., Godfrey, P., Minker, J.: An overview of cooperative answering. J. Intell. Inf. Syst. **1**(2), 123–157 (1992)
4. Herschel, M.: Wondering why data are missing from query results?: ask conseil why-not. In: He, Q., Iyengar, A., Nejdl, W., Pei, J., Rastogi, R. (eds.) CIKM, pp. 2213–2218. ACM (2013)
5. Krishnapuram, R., Joshi, A., Nasraoui, O., Yi, L.: Low-complexity fuzzy relational clustering algorithms for web mining. IEEE T. Fuzzy Systems **9**(4), 595–607 (2001)
6. Lesot, M.-J., d'Allonnes, A.R.: Credit-card fraud profiling using a hybrid incremental clustering methodology. In: Hüllermeier, E., Link, S., Fober, T., Seeger, B. (eds.) SUM 2012. LNCS, vol. 7520, pp. 325–336. Springer, Heidelberg (2012)
7. Meliou, A., Gatterbauer, W., Halpern, J.Y., Koch, C., Moore, K.F., Suciu, D.: Causality in databases. IEEE Data Eng. Bull. **33**(3), 59–67 (2010)
8. Pivert, O., Prade, H.: Detecting suspect answers in the presence of inconsistent information. In: Lukasiewicz, T., Sali, A. (eds.) FoIKS 2012. LNCS, vol. 7153, pp. 278–297. Springer, Heidelberg (2012)
9. Roy, S., Suciu, D.: A formal approach to finding explanations for database queries. In: Proceedings of the 2014 ACM SIGMOD International Conference on Management of Data, SIGMOD 2014, pp. 1579–1590. ACM, New York (2014)
10. Zadeh, L.: Fuzzy sets. Information and Control **8**, 338–353 (1965)

Literature Review of Arabic Question-Answering: Modeling, Generation, Experimentation and Performance Analysis

Wided Bakari, Patrice Bellot and Mahmoud Neji

Abstract This study provides a comparative study of Arabic question-answering systems. It presents a review of the main approaches and emphasizes the different experimentations in Arabic. It attempts to describe and detail the recent increase in interest and progress in Arabic question-answering research. It compares already existing question-answering systems for this language. It reviews the various categories of proposed question-answering systems within a broader framework of several languages and Arabic. It projects the actual trends of researches in this area for improving the Arabic. Also, it suggests a performance analysis providing an in-depth review of Arabic question-answering, by first noting the characteristics of, and resources available for, Arabic proposed systems and then reviewing the main Arabic question-answering approaches.

Keywords Question-answering · Approaches · Arabic · Experimentation · Contribution · Limitation · Actual trends · Taxonomy · Performance analysis

1 Introduction

During the last decades, with the accumulation of Arabic information's on the web, it's so difficult to find answers to questions using standard search engines.

W. Bakari(✉) · M. Neji
Faculty of Economics and Management, 3018 Sfax, Tunisia

P. Bellot
Aix-Marseille University, 13397 Marseille Cedex 20, France

W. Bakari · M. Neji
MIR@CL, Sfax, Tunisia

P. Bellot
LSIS, Marseille, France

T. Andreasen et al. (eds.), *Flexible Query Answering Systems 2015*,
Advances in Intelligent Systems and Computing 400,
DOI: 10.1007/978-3-319-26154-6_25

This growing provides more and more tools that can help us to find the precise answers. The regular information retrieval techniques (i.e., search engines) cannot satisfy the immense increasing amounts of Arabic contents and the demand for accurate information.

Actually, there have been a few surveys on Arabic question-answering investigations. For their part, the authors (Shaheen and Ezzeldin, 2012) emphasize that there is no review that covers Arabic question answering systems tools, resources, and test-sets on that date, despite the importance of Arabic. These authors discussed background on systems, resources, tools and general approaches to Arabic question-answering. To our knowledge, his study is the first trend towards identifying the importance of the Arabic and explores the related investigations. It presents Arabic's works on this area. This is the challenging task to provide more reviews to give more importance to the Arabic language in question-answering field.

This study provides a survey of some approaches for various categories of Arabic question-answering systems which will be helpful for new directions of research in this area. We examine the previous works on Arabic question-answering systems. Then we present some classifications of their systems. Special attention is paid to examine a performance analysis of different studies proposed for Arabic.

Our survey is different from the previous one (Ezzeldin and Shaheen, 2012) in many ways. First, it does not explore the analysis of main question answering tasks like question analysis, passage retrieval, and answer extraction. It presents a performance analysis and a comparison between the most proposed systems in Arabic based on several criteria. In summary we suggest that the most approaches in Arabic rely on natural language processing and information retrieval techniques. So, there is not a major distinction between the different investigations.

We begin this study by defining the task of question-answer. We present in Section 3 an efficient analysis of Arabic investigations in question-answering field. Indeed, the several systems available in Arabic are presented in Section 3.1. Next, we dedicate Section 3.2 for a comparative study of the Arabic investigations. The experimental results obtained are presented in Section 3.3. Finally we conclude and give some perspectives (section 4).

2 What's a Question-Answering System?

Question-answering system is a software system that takes a question in natural language in order to find an appropriate answer in a set of documents extracted from data bases, corpus of documents or a web. It's a sophisticated form of information retrieval characterized by information needs that are at least partially expressed as natural language statements or questions, and is one of the most natural forms of human computer interaction (Kolomiyets and Moens, 2011). This software system is a complex task needing effective improvements of different research areas, including information retrieval, natural language processing, database technologies, Semantic Web technologies, human computer interaction, speech processing and computer vision. The basic aim of question-answering system is to provide short and correct answer to the user saving his/her navigation time. The concept of Natural Language Processing plays an important role in developing any question-answering system.

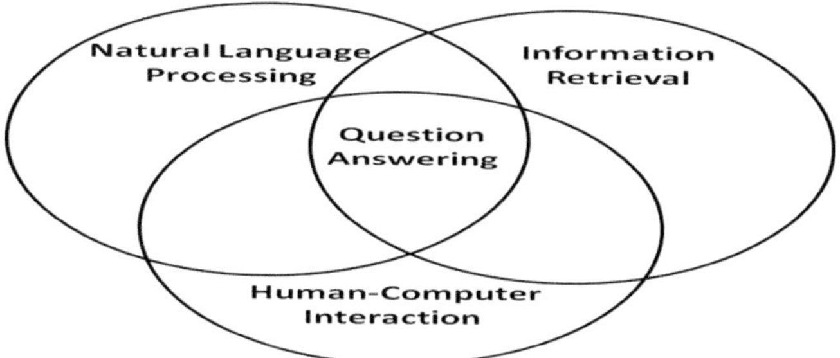

Fig. 1 Intersection of question-answering with different research areas

Table 1 Question-answering components tasks covered by Arabic studies

	Question processing			Document processing				Answer processing		
Components tasks / Question-answering Systems	Question Segmentation	Question Classification	Question Formulation	Sentence Retrieval	Short answer Re-trieval	Passage Retrieval	Paragraph Retrieval	Answer extraction	Answer selection	Anwer validation
AQAS (mohamed et al.,1993)				✓					✓	
QARAB (Hammo et al., 2004)		✓			✓				✓	
ArabiQA (Benajiba et al.,2007)		✓		✓				✓		✓
QASAL(Brini et al.,2009)			✓	✓				✓		
DefArabicQA (Trigui et al.,2010)		✓		✓				✓		
AquASys (Bekhti et al., 2011)	✓				✓			✓		
IDRAAQ (Abouenour et al.,2012)		✓				✓		✓		✓
ALQASIM (Ezzeldin et al., 2013)				✓					✓	✓
System of (NBdour and Gharaibeh, 2013)			✓				✓		✓	
JAWEB (Kurdi et al., 2014)				✓				✓		
Al-Bayan (Abdelnasser et al., 2014)		✓				✓		✓		

Most of the question-answering systems consist of three main modules: question processing, document processing and answer processing. Question processing module plays an important part in component systems processing. If this module doesn't work correctly, it will make problems for other sections. Moreover answer processing module is an emerging topic in question-answering, in which these systems are often required to rank and validate candidate answers. These techniques aiming at discovering the short and precise answers are often based on the semantic classification.

In several areas (information extraction, information retrieval, question-answering…), the accurate answer retrieval is a task that deals with automatically providing an answer of a question posed in natural language. Some attempts were made to reach an acceptable result in the Arabic question answering task.

Question-answering system is an extension of search engines that provides short answers to questions in natural language. In search engines, a user is able to search for information using a set of keywords. So, it can return ranked lists of documents, but they do not deliver answers to the user. However, question-answering system emphases to retrieve a correct answer to the given question posed in natural language from a collection of documents (Grappy and Grau, 2010). For example, search for key actors playing in the movie "Titanic", directed by James Cameron, a possible question would be: Who played the lead roles in the movie Titanic directed by James Cameron ?. In return, the system can meet Leonardo DiCaprio and Kate Winslet.

In spite of the similarity between question-answering systems and search engines (Google and Yahoo, etc.), each of those software's can provide an interface for users and meet the requirements of the user information's (Razmara, 2008), (Pho, 2012). However there is still difference between these two software systems. The most distinguishable one is that search engines rely on keywords to find all the required documents and extract information according to a general theme (El-Ayari, 2007). So, the user scans an often a high number of documents to find the most appropriate answer to his question (Ligozat, 2006).

Table 2 Difference between the search engines and question-answering system (Pho, 2012)

Criteria of comparison	Search Engine	Question-answering system
Task	Respond to user requests.	Answer questions posed in natural language.
Input	Set of keywords.	Question in natural language.
Results	Pages containing the keywords of the query.	1. Precise answer (Named Entities). 2. Short answer: sentences. 3. Long answer: sentences + justification.

3 Arabic Question-Answering: A Performance Analysis

The Arabic question-answering from the last few years reports on global performance of various systems. Few studies have been proposed for Arabic this area. In this mean we present the most relevant investigations in the literature review. At the end, the survey provides an analytical discussion of the proposed question-answering systems in Arabic, along with their main contributions, experimental results, and limitations.

3.1 Literature Review of Question-Answering Systems for Arabic Language

The technology of Arabic question answering has been studied, among others, since the 1970s. In this section we review the studies based on the different techniques used for Arabic question-answering systems and we pinpoint the relevant experimentations achieved by each work.

AQAS: Arabic Question-Answering System
AQAS system presented in (Mohammed et al., 1993), is the first system for Arabic, extracts answers only from structured data. It is knowledge-based and, therefore, extracts answers only from structured data and not from raw text (non structured text written in natural language). AQAS accepts an Arabic sentence (declarative statements or questions to be answered) and generates the appropriate output to the user. It uses frames technique to represent the knowledge base of a radiation domain.

QARAB: Arabic Question Answering System
QARAB (Hammo et al., 2004) provides short answers to questions expressed in the Arabic language. It excludes two types of questions: "ماذا, كيف" (How and Why). This system seeks answers to questions from unstructured documents extracted from the newspaper Al-RAYA. QARAB utilizes techniques from Information Retrieval and Natural Language Processing. It handles the incoming question as a "bag of words" against which the index file is searched to get a list of ranked passages that contains the answer. It does not report any data regarding precision or recall. The answer identified is the whole sentence matching the question keywords. The evaluation process of this system was based on 113 questions and a set of documents collected from the newspaper Al-Raya. QARAB is primary source of knowledge is a collection of Arabic newspaper text extracted from Al-Raya, a newspaper published in Qatar.

ArabiQA: Arabic QA System
ArabiQA (Arabic question answering system), is an Arabic question answering system developed by (Benajiba et al., 2007). This system is based on a generic architecture made up of three modules: a passage retrieval system (JIRS), an Arabic Named Entities Recognition system (NER) and an Answer Extraction (AE)

module. This system is making a precision near 83.3%. ArabiQA is based on the Java Information Retrieval System (JIRS) Passage Retrieval (PR) system and a Named Entities Recognition (NER) module.

QASAL: Question –Answering System for Arabic Language
QASAL (Brini et al., 2009) is a prototype to build an Arabic factual Question Answering system using Nooj platform to identify answers from a set of education books. The Experimentations have been conducted and showed that for a test data of 50 questions the system reached 67.65% as precision, 91% as recall and 72.85% as F-measure.

AQUASYS: Arabic **Qu**estion **A**nswering **Sys**tem
AQUASYS (Arabic Question-Answering System) (Bekhti et al., 2011) is a system focused on named entities and designed to answer factoid questions related that can be of any type: person, location, organization, time, quantity, etc. This system addresses the questions that start with interrogative pronouns (من who, ما what, أين where, متى when, كم العددية how many, كم الكمية how much).

DefArabicQA: Arabic Definition Question Answering System
DefArabicQA is a definitional Question Answering system proposed by Trigui and his associates (Trigui et al., 2010). This system presents an effective and accurate answer to definition questions expressed in Arabic language from Web resources. It is based on an approach that uses a little linguistic analysis and no language understanding capability. DefArabicQA finds candidate definitions by using a set of lexical patterns and categorize these candidate definitions by using heuristic rules and ranks them by using a statistical approach. The system was assessed by an Arabic native speaker. As evaluation metrics, they used MRR. It is a measure used in TREC[1] question-answering section and it is calculated as follows: each question is assigned a score equal to the inverse rank of the first string that is judged to contain a correct answer. If none of the five answer strings contain an answer, the question is assigned a score of zero. The MRR value for the experiment is calculated by taking the average of scores for all the questions (Voorhees, 2001).

ALQASIM: Question Answer Selection and Validation System
In the case of generating multiple answers, the research of (Ezzeldin et al., 2013) exposes ALQASIM system based on the selection and validation of the answer. It seeks to answer multiple choice questions to test QA4MRE @ CLEF 2013. This system uses a new technique by analyzing the reading test comprehension instead of the questions. It provides performance of 0.31 and a precision of 0.36 without using any database collection tests.

---

[1] Text Retrieval Conference http://trec.nist.gov/

ALQASIM is compared with an English system and two works in Arabic in QA4MRE at CLEF 2012. The table below illustrates the performance of systems in term of accuracy and c@1 measures.

IDRAAQ: Information and Data Reasoning for Answering Arabic Questions
The edition of QA4MRE@CLEF has considered for the first time the Arabic language in 2012 with some systems. For example IDRAAQ (Abouenour et al., 2012) system is implemented through a three-level approach to enhance the Passage Retrieval (PR) stage. It Includes the Recognizing Textual Entailment and Answer Validation tasks. Without using the database collections of CLEF and relying on the density of the distance N-gram model, semantic expansion and Arabic WordNet, IDRAAQ achieved a precision of 0, 13 and c@1 equals 0, 21.

System of N Bdour and Gharaibeh
The system of (N Bdour and Gharaibeh, 2013) is an open domain yes/no Arabic question answering system based on paragraph retrieval; it aims to retrieve paragraphs (with variable length) that contain answers to the question. To do this, the authors proposed a constrained semantic representation using an explicit unification framework based on semantic similarities and query expansion (synonyms and antonyms). N Bdour and Gharaibeh used a corpus of 20 Arabic documents, and a collection of 100 different yes/no question. They have generated the correct answers to these questions manually.

JAWEB: Web-Based Arabic Question Answering Application System
JAWEB (Kurdi et al., 2014) is an Arabic web-based question-answering system focused on factoid questions begins with "متى، أين، من، ما، كم الكمية، كم العددية". The questions answered by JAWEB are related to any named entity. It was constructed on the basis of AQUASYS (Bekhti et al., 2011) by providing a user interface as an extension. Thus, the exception of the JAWEB is that it provides a web interface to

Table 3 Example of questions answered by JAWEB

NUMBER	Type	Sentence in Arabic	Sentence in English
1	WHO	من هو محمد طنجة ؟	Who is Muhammad Tangier?
2	WHEN	متى توحدت المملكة العربية السعودية ؟	When was Kingdom of Saudi Arabia united?
3	WHAT	ماهي الأهرامات المصرية ؟	What are Egyptian pyramids?
4	WHERE	أين تقع المملكة العربية السعودية ؟	Where is the Kingdom of Saudi Arabia located?
5	HOW MUCH	كم تبلغ درجة حرارة القشرة الأرضية ؟	How much is the temperature of the Earth's crust?
6	HOW MANY	كم عدد سكان الرياض ؟	How many residents are there in Riyadh?

the system, an additional support for Arabic language presentation in web browsers, an extended corpus as well as an extensive evaluation and testing framework. JAWEB achieved 15-20% recall.

Al-Bayan: An Arabic Question Answering System for the Holy Quran
Al-Bayan (Abdelnasser et al., 2014), an Arabic question-answering system specialized for the Holy Quran. It accepts an Arabic question about the Quran, retrieves the most relevant Quran verses, then extracts the passage that contains the answer from the Quran and its interpretation books (Tafseer). This system achieved 85% accuracy with a collected dataset using the top-3 results. Al-Bayan provides a semantic understanding of the Quran and answering user's questions using reliable Quranic resources. The contribution of this study is summarized in the following points:

- Built a Semantic Information Retrieval module to retrieve the verses semantically related to user's questions.
- Increase the accuracy of a question analysis with application of a highly accurate Arabic tool for morphological analysis and disambiguation and a question classification with Support Vector Machine.
- Extract the ranked answers from the retrieved verses.

In this case, authors exploit the Quran and its interpretation books (Tafseer) of trusted Quranic scholars. Abdelnasser and his associates proposed a new taxonomy for Quranic Named Entities and constructed an Arabic Question Classifier based on state-of-the-art techniques.

In our research we focused on Logical inference based systems, to the best of our knowledge the research on Arabic question-answering have not been covered widely before. It is shown that there is no logical inference approach proposed for Arabic. Thus, there is no question-answering system that covers this type of approaches. This is the challenge to propose a novel method purely logical.

3.2 Comparative Study

Despite the importance of the Arabic language, there have not been many comparative studies of question-answering systems. Our study reviews a comparison of the attempts in Arabic question-answering. We cover almost all the proposed approaches in this area.

Table 4 Comparison between Arabic investigations based on the extracted features and used tools.

	AQAS (mohamed et al.,1993)	QARAB (Hammo et al., 2004)	ArabiQA (Benajiba et al.,2007)	QASAL (Brini et al.,2009)	DefArabicQA (Trigui et al.,2010)	AquASys (Bekhti & Al-Harbi, 2011)
Open/restricted Domain	Restricted radiation	:Open	Open	Open	Open	Open
Language of implementation	Not mentioned	Not mentioned	Java	Not men-tioned	Java	Not mentioned
WorldNet support	-	-	-	-	-	-
Ontology support	-	-	-	-	-	-
Linguistics resources	-	Abuleil's tagger (Abuleil and Evens, 2002),	-	NOOJ platform	-	-
Approach	Used a knowledge-based model Search in structured data	Treats the incoming question as a "bag of words", The Informa-tion Retrieval system module is based on Salton's vector space model, recognize named entities	Classifies the questions into (name, date, quantity, and definition) questions according to the question words and assigns a higher rank for the passages that have a smaller distance between keywords: distance density.	QASAL uses the NooJ platform as a linguistic development environment to answer factual questions.	Identifies candidate definitions by using a set of lexical patterns, filters these candidate definitions by using heuristic rules and ranks them by using a statistical approach	Segmented the question into interrogative noun, question's verb, and keywords
Source	Structured data	Unstructured data(Al-Raya newspaper corpus)	Corpus	Book of education	Web	Corpus
Answer	Sentence	Short passage	Sentence	Sentence	Sentence	Short answer
Question type	Several forms (declarative statements)	Question started by : من،متى، أين، كم	Factual question	Factual question	Definition question	Factual question
Features	Stop words Removal, tokenization	Type and category of expected answer.	Keywords and named entities recognized.	Type of expected answer, focus, key-words.	Question topic, type of expected answer.	Keywords, Type of expected.
Performance	Not mentioned	Precision : 97.3% Recall 97.3 %	:Precision 83.3%	:Not mentioned	MRR: 0.81	Precision : 66.25% Recall : 97.5% F1-Score : 87.89%

Table 4 (*Continued)*

	IDRAAQ (Abouenour et al.,2012)	ALQASIM (Ezzeldin et al., 2013)	System of (NBdour and Gharaibeh, 2013)	JAWEB (Kurdi et al., 2014)	Al-Bayan (Abdelnasser et al., 2014)
Open/restricted Domain	Open	Open	Open	Open	Restricted : Holy Quran
Language of implementation	Java	Not mentioned	Not mentioned	Dreamweaver, Java	Not mentioned
WorldNet support	Arabic WordNet	Arabic WordNet	-	-	-
Ontology support	SUMO ontology, YAGO ontology.	-	-	-	Qurany ontology
Linguistics resources	Al-Khalil parser	MADA+TOKAN (Habash et al., 2009)	-	Arabic Khoja's stemmer	LingPipe tool[2] Lucene[3]
Approach	IDRAAQ uses an enriched version of AWN.	ALQASIM answers the multiple choice questions of QA4MRE @ CLEF 2013 test-set. It uses a novel technique by analyzing the reading test documents instead of the questions	The system is based on a semantic logical representation approach to retrieve paragraph that contain relevant answers to the question.	JAWEB system analyses the questions and extracts the important information to retrieve the most relevant answers from an Arabic corpus. It provides a user interface.	Understands the semantic of the Quran and answers user's questions using the Quran and its interpretation books (Tafseer).
Source	-	QA4MRE @ CLEF 2013 test-set	Corpus	Corpus	Quran and its interpretation books (Tafseer)
Answer	Sentence	Sentence	Paragraph	Sentence	Passage
Question type	QMC	QMC	Yes/ No question	Factual question	Factual question
Features	Keywords, Type of expected answer.	Not mentioned	Stop words	Expected answer type; keywords	Named entity recognition, question type, Expected answer type.
Performance	Accuracy: 0.13, C@1: 0.21 %	Precision : 0.31% C@1: 0.36 %	Not mentioned	Recall: 15-20%	Precision: 85%

3.3 Experimentations, Contributions and Limitations

This section discusses and mentions the results of several works in Arabic that are clearly presented in section (Existing question-answering systems for

[2] http://alias-i.com/lingpipe/index.html

[3] http://lucene.apache.org/

Table 5 Experimentations, contributions and limitations of Arabic studies

System	Question type	Experimentation	Contributions	Limitations
AQAS (mohamed et al.,1993)	Declarative statements	No published evaluation is available for the system	AQAS is considered the first prototype for Arabic question answering system	The morphological analysis uses a limited size dictionary.
QARAB (Hammo et al., 2004)	Question started by : من،متى، اين، كم	Precision : 97.3% Recall 97.3 %	The presented approach is based on linking an Information Retrieval system with an NLP system that performed linguistic analysis.	The system provides answers to factual questions but does not support other types of questions.
ArabiQA (Benajiba et al.,2007)	Factual question	Precision : 83.3%	The presented approach is based on linking an Information Retrieval system with an NLP system that performed linguistic analysis.	The system implementation has not been completed yet.
QASAL(Brini et al.,2009)	Factual question	No published evaluation is available for the system	The presented approach is based on linking an Information Retrieval system with an NLP system that performed linguistic analysis.	No experimental results or performance metrics are published. The overall functionality of the system is limited.
DefArabicQA (Trigui et al.,2010)	Definition question	MRR: 0.81	The first system in Arabic that deals with Definition question.	The test-set contains only 50 organization definition questions and the answers were assessed by only one Arabic native speaker.
AquASys (Bekhti & Al-Harbi, 2011)	Factual question	Precision : 66.25% Recall : 97.5% F1-Score : 87.89%	The system extensively utilizes NLP techniques to analysis questions and retrieves answers.	The test-set is founded on an untagged corpus.
IDRAAQ (Abouenour et al.,2012)	QMC	Accuracy: 0.13, C@1: 0.21 %	The system treats a different type of questions (QMC) compared to existing studies in Arabic that deal with factual questions.	The experiments allowed by IDRAAQ identify the lacks of the system when processing non factoid questions and at the Answer Validation stage.
ALQASIM (Ezzeldin et al., 2013)	QMC	Precision : 0.31% C@1: 0.36 %	The system treats a different type of questions (QMC) compared to existing studies in Arabic that deal with factual questions.	Many incorrectly answered questions are causative and list questions and questions that were incorrectly translated due to erroneous automatic translation.
System of (NBdour and Gharaibeh, 2013)	Yes/ No question	Not mentioned	The first system including Yes or No Questions in Arabic question-answering.	The authors provide a semantic logical representation based approach but it does not cover all the components of the system (only question analysis).
JAWEB (Kurdi et al., 2014)	Factual question	Recall: 15-20%	A Web-based approach that provides an interface to the user.	The system only works for certain types of questions (factual questions) but does not support other types of questions.
Al-Bayan (Abdelnasser et al., 2014)	Factual question	Precision: 85%	The presented approach Builds a Semantic Information Retrieval model	The Authors are manually revised the 1200 concepts and their verses.

Arabic language). In addition we emphasize the main contributions and limitations for each research. Thus, throughout this paper, we provide a thorough explanation and understanding of the meaning behind the experimental results obtained by the Arabic question-answering researches. Finally, all the results and explanations are summarized in tables.

4 Conclusion and Future Works

This study emphasizes a performance analysis of a state-of-the-art of Arabic question-answering systems. Several configurations are examined: first, the definition of question-answering, then, the presentation of previous works on Arabic question-answering systems, and finally, a comparative study of the Arabic investigations, including the experimentation, contributions and limitations of the different investigations in Arabic.

After reviewing the relevant investigations in Arabic, we emphasize that very little amount of work has been done in this language compared to other ones and thus needs improvement in lots of areas. Additionally, among the current researches in Arabic question-answering is to propose logic-based approaches to answer questions. The lack of researches in this area encourages us to propose our novel approach which is based on logical inference to improve the research in Arabic question-answering. We believe that this kind of approaches will have great potential in the coming years.

Acknowledgements I give my sincere thanks for my collaborators Professor Patrice BELLOT (University of Aix Marseille, France) and Mr Mahmoud NEJI (University of Sfax-Tunisia) that I have benefited greatly by working with them.

References

Abdelnasser, H., Mohamed, R., Ragab, M., Mohamed, A., Farouk, B., El-Makky, N., Torki, M.: Al-Bayan: an arabic question answering system for the holy quran. In: ANLP 2014, p. 57 (2014)

Abouenour, L., Bouzoubaa, K., Rosso P.: IDRAAQ: New Arabic Question Answering System Basedon Query Expansion and Passage Retrieval. CLEF (Online Working Notes/Labs/Workshop) (2012)

Abuleil, S., Evens, M.: Extracting an Arabic Lexicon from Arabic Newspaper Text. Computers and the Humanities **36**(3), 191–221 (2002)

Bekhti, S., Rehman, A., AL-Harbi, M., Saba, T.: AQUASYS: an arabic question-answering system based on extensive question analysis and answer relevance scoring. International Journal of Academic Research **3**(4), 45 (2011)

Benajiba, Y., Rosso, P., Lyhyaoui, A.: Implementation of the ArabiQA question answering system's components. In: Proc. Workshop on Arabic Natural Language Processing, 2nd Information Communication Technologies Int. Symposium. ICTIS-2007, Fez, Morroco, April 3–5, 2007

Bhaskar, P., Pakray, P., Banerjee, S., Banerjee, S., Bandyopadhyay, S., Gelbukh, A.: Question answering system for QA4MRE@CLEF 2012. In: CLEF 2012 Workshop on Question Answering for Machine Reading Evaluation, QA4MRE, September 2012

Brini, W., Ellouze, M., Mesfar, S., Belguith, L.H.: An arabic question-answering system for factoid questions. In: IEEE International Conference on Natural Language Procesing and Knowledge Engineering (IEEE NLP-KE 2009), Dalian, China (2009)

El Ayari, S.: Évaluation transparente de systèmes de questions réponses: application au focus. Actes de RéciTAL, 2007 (2007)

Ezzeldin, A.M., Shaheen, M.: A survey of Arabic question answering: challenges, tasks, approaches, tools, and future trends. In: The 13th International Arab Conference on Information Technology ACIT 2012. December 10–13, 2012. ISSN 1812-0857

Ezzeldin, A.M., Kholief, M.H., El-Sonbaty, Y.: ALQASIM: Arabic language question answer selection in machines. In: Forner, P., Müller, H., Paredes, R., Rosso, P., Stein, B. (eds.) CLEF 2013. LNCS, vol. 8138, pp. 100–103. Springer, Heidelberg (2013)

Grappy A., Grau B.: Validation du type de la réponse dans un système de questions réponses. In: CORIA 2010, pp. 131–146 (2010)

Habash, N., Rambow, O., Roth, R.: MADA+TOKAN: A toolkit for Arabic tokenization, diacritization, morphological disambiguation, PoS tagging, stemming and lemmatization. In: Proceedings of the 2nd International Conference on Arabic Language Resources and Tools (MEDAR), Cairo, Egypt, pp. 102–109 (2009)

Hammo, B., Ableil, S., Lytinen, S., Evens, M.: Experimenting with a Question Answering system for the Arabic language. Computers and the Humanities **38**(4), 397–415 (2004)

Kalady, S., Elikkottil, A., Das, R.: Natural language question generation using syntax and keywords. In: Proceedings of QG2010: The Third Workshop on Question Generation, pp. 1–10 (2010)

Kolomiyets, O., Moens, M.F.: A survey on question answering technology from an information retrieval perspective. Information Sciences **181**(24), 5412–5434 (2011)

Kurdi, H., Alkhaider, S., Alfaif, N.: Development and evaluation of a web based question answering system for arabic language. Computer Science & Information Technology (CS & IT) **4**(2), 187–202 (2014)

Ligozat, A-L.: Exploitation et fusion de connaissances locales pour la recherche d'informations précises. Ph.D. thesis, Université Paris XI – Orsay LIMSI-CNRS, décembre 2006 (2006)

Mohammed, F., Nasser, K., Harb, H.: A knowledge based Arabic Question Answering system (AQAS). ACM SIGART Bulletin, pp. 21–33 (1993)

Moldovan, D., Pasca, M., Harabagiu, S., Surdeanu, M.: Performance issues and error analysisin an open-domain question answering system. ACM Transactions on Information Systems **21**(2), 133–154 (2003)

Bdour, W.N., Gharaibeh, N.K.: Development of Yes/No Arabic Question Answering System. International Journal of Artificial Intelligence & Applications (IJAIA) **4**(1) (2013). doi:10.5121/ijaia.2013.410551

Niu, Y., Hirst, G., McArthur, G., Rodriguez-Gianolli, P.: Answering clinical questions with role identification. In: Proceedings of the ACL-2003 Workshop Natural Language Processing in Biomedicine, ACL, Sapporo, Japan, pp. 73–80 (2003)

Pho, V.-M.: Génération de réponses pour un système de questions-réponses. In: CORIA 2012, pp. 449–454 (2012)

Razmara, M.: Answering list and other questions. Ph.D. thesis, Concordia University, August 2008

Rinaldi, F., Dowdall, J., Schneider, G.: Answering questions in the genomics domain. In: Dans Proc. ACL 2004 Workshop on Question Answering in Restricted Domains (2004)

Trigui, O., Belguith, L.H., Rosso, P.: DefArabicQA: Arabic definition question answering system. In: Workshop on Language Resources and Human Language Technologies for Semitic Languages, 7th LREC. Valletta, Malta (2010)

Trigui, O., Belguith, L.H., Rosso, P., Amor, H.B., Gafsaoui, B.: Arabic QA4MRE at CLEF 2012: arabic question answering for machine reading evaluation. In: CLEF 2012 Workshop on Question Answering For Machine Reading Evaluation, QA4MRE, September 2012

Voorhees, E.: Overview of the TREC 2001 question answering track. In: Proceedings of the 10th Text REtrieval Conference, pp. 42–51 (2001)

Wyse, B., Piwek, P.: Generating questions from OpenLearn study units. In: Proceedings of the International Conference on Artificial Intelligence in Education Workshops (AIED) (2009)

Part IV
Querying Temporal, Spatial and Multimedia Databases

LORI: Linguistically Oriented RDF Interface for Querying Fuzzy Temporal Data

Majid RobatJazi, Marek Z. Reformat, Witold Pedrycz and Petr Musilek

Abstract The concept of Semantic Web, introduced by Berners-Lee in 2001, emphasizes importance of expressing semantics of data stored on the web. The introduced data format called Resource Description Framework (RDF) is a meaningful way of expressing and exploring data relations. It provides basics for constructing semantics-oriented data formats. More and more often RDF is used to represent variety of data including N-ary relations and temporal information. On many occasions this results in complex data structures. Their utilization requires a full understating of applied data configurations.

This paper presents a fuzzy-based Linguistically Oriented RDF Interface – LORI – for querying RDF data containing temporal information and using non-trivial data structures. The interface includes specialized built-in predicates suitable for constructing temporal queries and supporting imprecise phrases describing time and data features, and high-level predicates built based on them. A simple case study using LORI interface for imprecise querying about time-based events is presented.

1 Introduction

The important concepts of Semantic Web [1] and Lined Open Data [2] – perceived as the prelude to Web 3.0 – are associated with a data representation format called Resource Description Framework (RDF) [3]. The significance of RDF comes from its ability to represent semantics of data in a form of relations existing between pieces of information. It can be said that RDF data constitutes an

M. RobatJazi · M.Z. Reformat(✉) · W. Pedrycz · P. Musilek
Electrical and Computer Engineering, University of Alberta, Edmonton, Canada
e-mail: {mrobatja,marek.reformat,petr.musilek}@ualberta.ca

W. Pedrycz
Polish Academy of Sciences, Warsaw, Poland
e-mail: wpedrycz@ualberta.ca

T. Andreasen et al. (eds.), *Flexible Query Answering Systems 2015*,
Advances in Intelligent Systems and Computing 400,
DOI: 10.1007/978-3-319-26154-6_26

important step towards creating a foundation for methods and approaches leading to a more intelligent and human-oriented way of processing, analyzing and utilizing any data and information.

There is an increasing trend of representing 'richer' information that contains multifaceted features and relations, as well as temporal information attached to them. The consequence of that is an introduction of not-trivial data structures. At the same time the users' expectations regarding easiness and effectiveness of 'interaction' with data is growing. The users would like to see more human friendly ways of asking for relevant information. It seems very reasonable to say that with an increasing amount of data it is difficult for the users to ask questions with precisely identified quantitative and temporal values of data. Also, RDF that is not inherently suitable for expressing N-ary relations, and the proposed solutions [4] introduce complexity and difficulties in data processing and analysis.

This paper introduces a Linguistically Oriented RDF Interface – LORI – that provides the user with the ability to exploit temporal data as well as data containing complex relations. LORI eliminates needs for an extensive knowledge of details related to the structure of queried data, and allows for using imprecise expressions built with quantitative and time-based terms. The process of designing and developing LORI embraces multiple aspects of dealing with temporal and complex data, and results in the following realizations:

- designing an architecture of LORI based on an idea of two interfaces: 1) low-level one called *ReasonerInterface* which provides necessary rules and predicates to deal with temporal and complex data, built on Jena's RDF/RDFS reasoner; and 2) *UserInterface* composed of high-level predicates and built by data expert based on *ReasonerInterface*, Section 4.1;
- developing temporal predicates as built-in functions of Jena's RDF/RDFS reasoner; these predicates utilize fuzzy terms to express imprecise declarations of time; the predicates are described in Section 4.2.1;
- proposing and developing a data structure for storing query results, with the ability to create sequences of queries, as well as to store, process and merge individual results, Section 4.2.2;
- identifying a flexible approach for mapping high-level queries to low-level ones; the proposed idea is based on a mapping file that allows for dynamic changes and modifications of mapping rules, Section 4.3;
- a case study illustrating how LORI works, Section 5.

2 RDF and Time Representation

2.1 Resource Description Framework – RDF

The concept of Semantic Web [1] has introduced a principal format of representing data called Resource Description Framework – RDF [3]. A building block of RDF is a simple triple: `subject-predicate-object`, where: `subject` identifies an entity the triple is describing; `predicate` defines a type of relation that exists between the `subject` and `object`; and `object` is an entity or a value describing the `subject` via being in relation with it. For example, in the triple

John **travel** Tokyo

`John` is an entity that is being described; `travel` is a relation that exists between the entities; and `Tokyo` is a value of this relationship. In other words we say that `John` travel(s|ed) to Tokyo. Further, a subject of one triple could be a subject of other triples. For example, the two following triples:

John **type** Person;   John **birth** 1996

indicate that `John` is a person, and he was born in 1996. Additionally, an object of one triple could become a subject of another triple, or a subject of one triple can be an object of another. Overall, multiple entities can be involved in different relations and play different roles in these relations. That leads to a highly interconnected network of related entities.

RDF triples can be used to represent any type of information in any domain. There are a number of initiatives focusing on building repositories of predicates called vocabularies [5][6]. Descriptions of all RDF components are contained in RDF Schema [7].

2.2 RDF Representation of N-ary Relations

In its canonical form, RDF is used to represent a binary relation. An RDF predicate links two entities – subject and object. On multiple occasions, it is necessary to represent additional information that is related to a particular RDF predicate. A solution supported by W3C, based on N-ary Relation [8], is to represent a relation as a class. It means that any predicate that needs to be described by additional features, e.g., strength, certainty, is an item 'located' in the middle of the original triple. The idea is presented below. Let us assume that John traveled to Tokyo. We express this in the following way:

John **travel** Tokyo

If we want to provide additional information about the relation `travel` such as: duration, date, stayed-at, we need to create a class **travelC** and a number of properties: `travel_where`, `travel_when`, `travel_stay_at`. Then, the above triple is represented as a bunch of triples:

John **travel_who** **travelC**

travelC **travel_where** Tokyo

travelC **travel_when** 2015-01-15

travelC **travel_stay_at** Hayat

As we can observe, the property **travel** has been replaced by the class **travelC**. This new property class is described using three items (that play the role of objects) in new triples with **travelC** as their object. Please note, that new properties have been introduced to describe the property.

2.3 *Representing Temporal Data*

Many real-world applications require management of temporal data. The notion of representing time has been addressed in a number of different ways. The approach described in [9] represents temporal knowledge as an interval-based temporal logic. This is suitable for expressing qualitative temporal relation between intervals such as before, after, meets and overlaps. Dutta [10] proposes a model that describes time using a set of accurate disjoint intervals of different lengths, while Kim and Oh [11] have introduced a model for the vague representation of qualitative temporal relations between events.

In realty, many temporal concepts are vague and imprecise. Fuzzy theory is a valid solution for handling these imprecise temporal concepts. Dubois and Prade [12] have proposed an approach to represent and process fuzzy temporal knowledge, while Moon has focused on branching time and uncertainty of temporal information [13]. Carinena at el [14] have concentrated on providing a formal definition of a grammar for expressing fuzzy temporal propositions. The definitions of such concepts as date, time extent, and interval, according to the formalism of possibility theory have been presented in [15]. Fuzzy temporal reasoning has been addressed from the point of view of implementation algorithms [16], and its application to decision-making [17] and web [18].

Pan and Hobbs [19] have introduced a model for representing time in an OWL-represented ontology [20]. They offer a comprehensive description of temporal intervals, instants, durations, and calendar terms. They provide a vocabulary to represent facts about topological relations among instants and intervals.

In general, introduction of time into RDF has been accomplished via creating a new extended version of RDF. For example, in [21][22] the authors use labeling and special data structures to introduce time into RDF graphs. However, the proposed query language is almost impossible to use in existing RDF stores. In the attempt to solve the storage problem, the authors of [23] introduce the tGRIN index structure that builds a specialized index for temporal RDF that is physically stored in a rational database. They use temporally annotated RDF triples of the form: `subject-property:annotation-object`. A similar approach – annotated triples – is used in [24]. Here, the authors focus on temporal validity intervals. They propose a temporal version of the SPARQL language for querying such annotated triples.

3 Ontology of Fuzzy and Temporal Data

The fuzzy ontology [25] used in the proposed system contains a number of concepts that are required for representing and storing fuzzy information. They denote basic items needed to express fuzzy data. The most important of them are: *fuzzyVariable*, *fuzzyTerm*, *fuzzyPair*, and *fuzzyMemberFunction*. An important part of the ontology – necessary for "building" fuzzy information – is object properties. The properties define fundamental relations that exist between different fuzzy concepts. The arrangement of both classes and object relations allows us to build a structure needed for expressing fuzzy information. For more details, please see [26].

In order to express temporal fuzzy data we use a simple ontology of basic temporal concepts [26]. In this ontology, time is represented as a single time point, or as a time interval. A time point, or *time instant*, identifies a single occurrence, while a *time interval* is a temporal entity with a beginning time and an ending time. Usually, the beginning and ending are determined with time instants. An important aspect of temporal data is time stamping of information. This is done with so-called *Valid Time*, i.e., the time that information is true in the real world.

We combine both ontologies: ontology for fuzzy data and temporal ontology, and use them as vocabulary for RDF triples. For details and example of fuzzy temporal ontology please see [26].

4 Linguistically Oriented Interface for Querying RDF Data

4.1 Overview

A specialized interface is required to query RDF data using imprecise terms describing temporal aspects of data, as well as to use imprecise quantitative descriptors that impose additional constrains on the results of queries. Such an interface – called Linguistically Oriented Interface (LORI) – is presented here. It is built based on RDF/RDFS reasoner [27] provided by Jena [28] that works with a set of basic entailments rules, Section 4.2. Jena's RDF/RDFS reasoner allows for developing custom built-in predicates. A set of predicates that allow for querying temporal aspects of data, and processing complex data has been developed.

The above-mentioned predicates as well as entailment rules of RDF/RDFS reasoner are used to construct high-level predicates that provide the users with simpler and more straightforward ways of making inquiries. A set of mapping rules is used as a translator between the reasoner's predicates/rules and high-level predicates. A diagram representing architecture of LORI is shown in Fig. 1. The main building blocks of LORI are: *CustomPredicates&RDFSchema* – a component containing developed predicates and required data structures; *ReasonerInterface* – an element providing an access to built-in custom predicates and RDF/RDFS reasoner's entailment rules; *UserInterface* – a set of high-level predicates available to the users; *MappingEngine* – a unit that uses mapping rules to translate queries built by the user to queries offered by *ReasonerInterface*.

4.2 Predicates, Data Structures and Reasoner Interface

A detailed description of LORI should start with an explanation and details regarding the *CustomPredicates&RDFSchema* component. However, before we can do this we should explain what type of input is acceptable for Jena's reasoner. Jena's RDF/RDFS reasoner works with entailment rules [27] in the form of **triple_patterns** [29]. These triples are executed against RDF data and identify RDF triples that match the 'fixed' positions (i.e., positions that contain explicit values) and have arbitrary values on the 'variable' positions (i.e., positions with a '?' character at their beginnings). For example, the **triple_pattern**:

```
(?subject  foaf:first_name  'John')
```

has the 'fixed' positions: `foaf:first_name` and `'John'`, and the variable position `subject`. As the result, we obtain a set of all entities (values of `subject`) that have the first name John, i.e., entities that are connected via the property `foaf:first_name` with `'John'`. The **triple_patterns** use entities defined within the considered data. Essentially, they are SPARQL graph (triple) patterns that are being matched against data graphs [28].

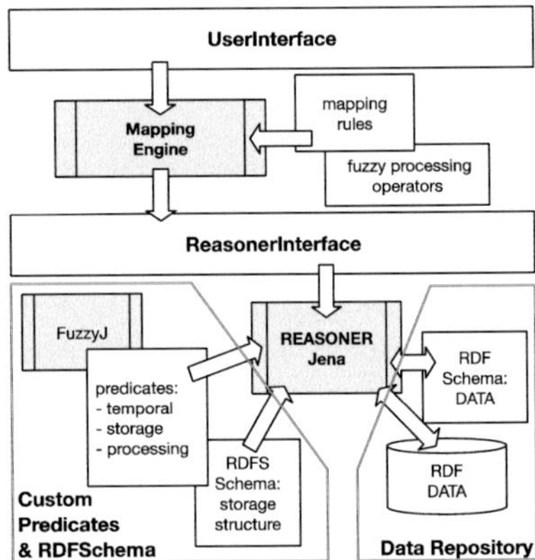

Fig. 1 Configuration of LORI

Processes of constructing queries in the case of temporal and complex data require knowledge and understanding of internal data structures. To address these needs we develop a number of predicates that simplify these processes. These predicates are built according to Jena's syntax and requirements imposed by RDF/RDFS reasoner. As we can see later, these predicates are further mapped into very simple and human-friendly high-level predicates that constitute *UserInterface*.

4.2.1 Fuzzy Temporal Predicates

A library of predicates capable of dealing with temporal instances and intervals is developed. Such a library gives the users the ability to built queries/rules that can express relations between temporal events and infer about them.

An important part required by these predicates is a concept of a time instant:

timeInstant = <specificTime|timeOfAnotherEvent>

The instant is given explicitly via the value of the parameter `specificTime` that can assume the values: NOW, YESTERDAY, LAST_SECOND, LAST_MINUTE, LAST_HOUR, LAST_DAY, or via providing date and time, such as "2012", "2012-05", "2012-05-06", or "2012-05-06 16:23".

The four defined and implemented predicates are:

approx_at_instant(?event[1], ?predicate, ?timeInstant, ?cut)

– to determine a degree of temporal overlapping between an event and an approximated instant in time – it calculates the level of degree to which the statement "something happened approximately around a given time instant" is true.

approx in interval(?event, ?predicate, interval start, interval end, ?cut)

– to determine a degree of overlapping between an event and an interval defined by starting and ending points.

approx_at_instant_before(?event, ?predicate, p, unit_p, ?timeInstant, ?cut)

– to determine a degree of overlapping between an event and an instant that occurs approximately a number of time granules before some instant of time. For example, the predicate `approx_at_instant_before(,?event, $property 2, days, ?time-Instant, ?degree)` calculates the degree of overlapping of an `event` with an instant at about 2 days before `timeInstant`.

approx_in_interval_before(?event, ?predicate, n, unit_n, ?timeInstant, m, unit_m, ?cut)

– to determine a degree of overlapping of an event with an interval that spans across `n` granules of time, and this interval occurs `m` granules of time before `timeInstant`.

The parameters of the predicates are:

`event` – a subject of RDF triple with `property` as its RDF property; if it is not given – nodes are identified by the parameter `predicate`;

`predicate` – an RDF predicate that is the focus of a temporal analysis (this `predicate` should have time as its range);

`timeInstant` – a moment in time defined in multiple ways based on temporal ontology;

[1] In SWRL, the character "?" is used to denote a variable.

interval_start, interval_end – two timeInstant values representing the beginning and the end of a time interval;

p – number of time units;

unit_p – granularity of time units for p;

n – number of time units, "width" of an interval;

unit_n – granularity of time units for n;

m – number of time units 'between' the end of the interval and timeInstant;

unit_m – granularity of time units for m;

cut – a value from the interval <0,1> indicating a desired degree to which the event should 'overlaps' with timeInstant, by default cut is equal to zero.

The predicates – available at *ReasonerInterface* – utilize fuzzy temporal ontology, Section 3, and use the FuzzyJ reasoner [29] suitable to deal with fuzzy data. The predicates are implemented as functions called from within Jena, and they further call FuzzyJ procedures to perform fuzzy calculations and reasoning. The same predicates with a reduced number of parameters are available at *UserInterface*, Section 4.3.

4.2.2 Storage Predicates

The storage predicates increase functionality of LORI by allowing the users to store the results of queries, reuse them, merge them, and perform operations, including fuzzy ones, on them. The results are stored in a form of RDF data. This allows for a full integration of the results with RDF data that the LORI is working with. We have created an RDF Schema that defines classes and properties to build such RDF triples, Fig. 2. Three predicates for dealing with results of queries that use this RDF Schema are developed:

StoreResult(?resName) takes a single parameter resName as a name of RDF graph representing the results; this name is used as identifier of the results;

MergeResults(?resName1, ?resName2, ?resName3) used for merging two RDF graphs containing the results of two different queries, where resName1 is one of them, and resName2 is another, the result of the merge is stored in an RDF graph named resName3; and

ShowResult(?resName) used for displaying an RDF graphs containing results of a single query, where resName is its identifier.

These predicates constitute a very important part of the LORI – they allow for creating sequences of queries where the results of one query can be combined with the results of another query.

```
<rdfs:Class rdf:ID="QueryResult" />
<rdfs:Class rdf:ID="QueryResultNode" />
<rdf:Property rdf:ID="hasNode">
    <rdfs:domain rdf:resource="#QueryResult"/> <rdfs:range rdf:resource="#QueryResultNode"/>
</rdf:Property>
<rdf:Property rdf:ID="hasFuzzyNumber">
    <rdfs:domain rdf:resource="#QueryResultNode"/>
    <rdfs:range rdf:resource="http://www.w3.org/2001/XMLSchema#decimal"/>
</rdf:Property>
<rdf:Property rdf:ID="hasSubject">
    <rdfs:domain rdf:resource="#QueryResultNode"/>
    <rdfs:range rdf:resource="http://www.w3.org/2000/01/rdf-schema#Resource"/>
</rdf:Property>
<rdf:Property rdf:ID="hasPredicate">
    <rdfs:domain rdf:resource="#QueryResultNode"/>
    <rdfs:range rdf:resource="http://www.w3.org/2000/01/rdf-schema#Resource"/>
</rdf:Property>
<rdf:Property rdf:ID="hasObject">
    <rdfs:domain rdf:resource="#QueryResultNode"/>
    <rdfs:range rdf:resource="http://www.w3.org/2000/01/rdf-schema#Resource"/>
</rdf:Property>
```

Fig. 2 The RDF Schema for storing results

4.2.3 Fuzzy Processing Predicates

The introduced data structures for storing results have created an opportunity to design and develop special functions operating on the obtained results. An important set of useful processing utilities includes fuzzy processing predicates. These predicates are implemented as functions inside Jena. They are further used via experts to define processing operators. The definitions are included in a mapping file.

An example of a fuzzy processing predicate is `Most`. It performs fuzzification of data based on a defined fuzzy membership function. Section 4.3 contains details how it is applied to define a fuzzy processing operator, and how it is used together with storage predicates at *UserInterface*. Section 5 illustrates its application.

4.3 UserInterface and MappingEngine

Presented above predicates are still closely related to data, i.e., in order to use them the user has to know details regarding data structures. To avoid this, an expert in the data under consideration builds a set of high-level predicates that 'isolate' the user from data details and its complexity.

The expert constructs high-level predicates and mapping rules. The mapping rules define these predicates and specify a way of 'transforming' them into the predicates and entailment rules offered at *ReasonerInterface*, Section 4.2. *MappingEngine* performs such a translation process. An example of a mapping rule is shown below:

```
highLevel_pred^i(parameter_set)
=> triple_pattern_1|…| triple_pattern_N
```

The high-level predicates **highLevel_pred**^{i} should reflect inquiries frequently made by the users, and be of high importance with meaningful naming and intuitively recognized effects. Their parameters become inputs to individual **triple_pattern**$_{k}$().

In the case of temporal predicates, Section 4.2.1, *UserInterface* offers the same predicates but with a reduced number of parameters. *UserInterface* temporal predicates are:

```
approx_at_instant(?timeInstant, ?cut)
approx_in_interval(?interval_start, interval_end, ?cut)
approx_at_instant_before(?p, unit_p, ?timeInstant, ?cut)
approx_in_interval_before(?n, unit_n, ?timeInstant, m, unit_m, ?cut)
```

Similarity for the storage predicates, *UserInterface* offers the same predicates as the ones defined at the level of *ReasonerInterface*, Section 4.2.2. The predicates **MergeResult** and **ShowResult** are the same, while the third predicate has an additional parameter:

```
StoreResult(?resName,?fuzzy_processing_operator_r)
```

This new parameter fuzzy_processing_operator$_{r}$ is an operator defined by an expert – also in a mapping rules file – in the following way:

```
FProcOperName(fuzzy_processing_operator_r) =>
FuzzyProcessingParam(Most,Subject)
```

It means that, the operator fuzzy_processing_operator$_{r}$ contains the predicate Most, and data entities on which the predicate works. In the case above, these are Subjects of RDF triples representing the stored results.

Fuzzy processing operators defined on processing predicates are very depended on the data under consideration. For specifics see Section 5.

5 Case Study

5.1 Data Structure

This case study shows how to apply LORI to query data containing temporal information and quite complex data structure. The considered RDF data contains triples describing traveling facts: destinations and dates, of a number of individuals. Additionally, the data also includes details regarding their illnesses, i.e., names of diseases, and times when they had it. An example of this data for two individuals X and Y is shown in Fig. 3.

As we can see, Fig. 3, the description of a single trip contains a blank node (*N_* or *P_*). It is needed in order to express two features: destination of the travel, and date when it took place. Information about a single disease also contains a blank node (*M_* and *Q_*) to represent: disease name and a date when a person became ill. Such a structure of data shows an extra complication in data structure – something we would like to 'hide' from the user. At the same time, this gives us an opportunity to show how LORI works and simplifies queries.

5.2 High-level Predicates and Mapping Rules

Complexity of data structure implies a need for high-level predicates. The mapping rules of three predicates that also work as their definitions are shown below:

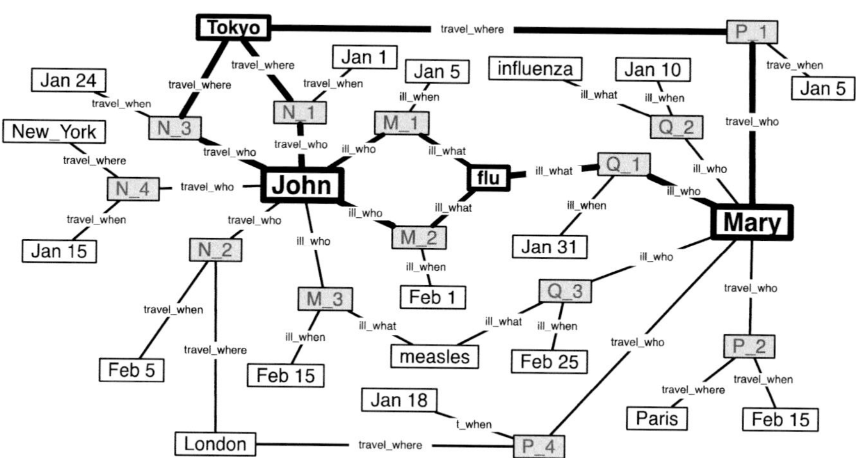

Fig. 3 RDF triples representing data two individuals X and Y (for simplify the properties do not have a prefix `rdfs:get_` used in the text below)

```
WHO_HAS_DISEASE(.*1) =>
    (?Subject rdfs:get_ill_who ?interconnectNode),
    (?interconnectNode rdfs:get_ill_what ?*1)
```

```
WHO_TRAVEL_TO(.*1) =>
      (?Subject rdfs:get_travel_who ?interconnectNode),
      (?interconnectNode rdfs:get_travel_where ?*1)

WHERE_PERSON_TRAVELED(.*1) =>
      (?*1 rdfs:get_travel_who ?interconnectNode),
      (?interconnectNode rdfs:get_travel_where ?Object)
```

The predicate **WHO_HAS_DISEASE**(.*1), where *1 represents an input parameter that should be a person, is mapped into two RDF triple patterns. The first of them returns all Subjects – persons in our case – and interconnectNodes – blank nodes – that are connected via the property rdfs:get_ill_who. The second one takes the identified interconnectNodes and looks for any entities that are connected to it via the property rdfs:get_ill_what. In this way, we obtain a list of all persons – values of Subjects - who had the disease identified by the input parameter *1.

The second predicate **WHO_TRAVEL_TO**(.*1) works in the similar way, but as the result we obtain a list of persons who travelled to the destination identified by *1.

For the predicate **WHERE_PERSON_TRAVELED**(.*1) the first pattern leads to pairs: <*1 interconnectNode> that are connected via rdfs:get_travel_who. This provides all blank nodes connected to a given person – identified by *1. Then we 'follow' the blank nodes and obtain Objects, i.e., places to which the person *1 travelled.

For queries about temporal aspects as well as for storing results, we use the temporal and storage predicates described in Section 4.2. For the case of fuzzy processing operators, we have mentioned that they are data dependent, so in this particular case the expert defines three operators that can be used as an input parameter for the predicate **StoreResult**(?resName,?fuzzy_processing_operator$_r$), they are:

```
FProcOperName(MostWho)=> FuzzyProcessingParam(Most,Subject)
FProcOperName(MostWhere)=> FuzzyProcessingParam(Most,Object)
FProcOperName(Most)=> FuzzyProcessingParam(Most,Both)
```

In this case, fuzzy_processing_operator$_r$ can assume values MostWho, MostWhere, Most. Each of them works on different parts of RDF triples: the first one on subject, the second on object, and the third on both.

At this stage we can also show how the predicate **StoreResult**() with a fuzzy processing operator is translated into triples and rules of *ReasonerInterface*:

```
StoreResult(?resName, MostWho) =>
=> StoreResult(resName_temp)
=> (resName_temp rdf:type queryResult) -> Result(?resName, Most,
Subject)
```

As it can be seen, the predicate is translated into *ReasonerInterface* predicate StoreResult(), and the rule that allows RDF/RDFS reasoner to perform fuzzy processing.

5.3 *Queries and Obtained Results*

In the first query we identify all individuals who traveled most often to Tokyo in mid January 2015, who had flu one week after traveling to Tokyo, over the period of about two weeks. The queries look like this:

```
WHO_TRAVEL_TO('Tokyo'),
approx_in_interval('2015-01-01', '2015-31-01'),
-> StoreResult(ResultT, MostWho)

WHO_HAS_DISEASE('flu'),
approx_in_interval('2015-01-22', '2015-02-11'),
-> StoreResult(ResultD)

MergeResult(ResultT, ResultD, Result)
ShowResult(Result)
```

The *UserInterface* predicate **WHO_TRAVEL_TO**, together with a temporal predicate – `approx_in_interval()` is executed first. This results in a list of individuals – a single entry for each trip – who traveled to Tokyo. The travel takes place approximately between January 1st and 31st. The boundaries of the interval are 'modified' using fuzzy membership functions shown in Fig. 4 (a). The domain of this function is `timeInstant`. For the second predicate, the process is quite similar; it selects individuals who had flu over the period of three weeks – from January 22nd to February 11th, approximately.

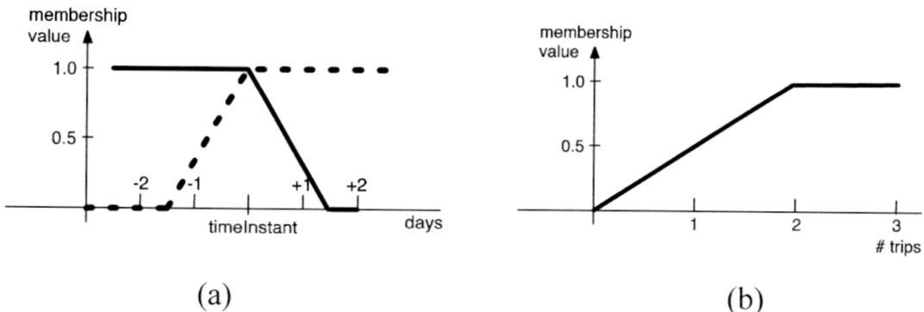

(a) (b)

Fig. 4 Membership functions: (a) modifiers of interval's boundaries to make them `approximate`, and (b) for the predicate `Most`

The results of the first predicate are stored under the name `ResultT`. They are processed using the fuzzy processing operator `MostWho` to identify individuals who traveled to Tokyo most often. In other words, the results indicate how many times a person travelled to Tokyo, and this number is evaluated using the fuzzy membership function associated with the predicate `Most`, Fig. 4 (b). The results of the second high-level predicate are stored with the name `ResultD`. They represent all individuals who had flu approximately between Jan 22nd and Feb 11th, 2015.

The predicate **`MergeResult()`** combines both results. The fusion is based on the individuals – so, the persons who travelled the most to Tokyo and then had flu in the identified period of three weeks. A sample of the final results is shown below:

```
...
ResultT(1): John (Tokyo) -> temp: 1.0 trips: 1.0
ResultT(2): Mary (Tokyo) -> temp: 1.0 trips: 0.5
ResultD(1): John (Flu) -> temp: 1.0
ResultD(2): Mary (Flu) -> temp: 1.0
Result(1): John (Flu/Tokyo) -> temp: 1.0 trips: 1.0
Result(2): Mary (Flu/Tokyo) -> temp: 1.0 trips: 0.5
...
```

As we can see, John who traveled to Tokyo at least 2 times, Fig. 4 (b), and had a flu has been identified as a person who match the query to the highest degree. The part `temp:` indicates degree of satisfaction of the temporal requirement, `trips:` - satisfaction of the fuzzy operator `MostWho`.

The second query should identify individuals who had measles at the beginning of May, and who traveled the most over the second week of May.

```
WHO_HAS_DISEASE('measles'),
approx_at_instant('2015-06-01'),
-> StoreResult(ResultD)

WHERE_PERSON_TRAVELED(),
approx_in_interval('2015-05-07', '2015-05-14',0.75),
-> StoreResult(ResultP, MostWhere)

MergeResult(ResultD, ResultP, Result)
ShowResult(Result)
```

The results, shown below, indicate that the query is not successful. The first predicate provides a list of *persons* who had measles around May 1st, while the second gives a list of *places* visited by individuals – here Susan was in Toronto at least twice (`trips: 1.0`). Both queries have different types of entities as their responses and the result of merge has zero entries.

```
...
ResultD(1): John (Measles) -> (time) 1.0
ResultD(2): Mary (Measles) -> (time) 1.0
ResultD(3): Susan (Measles) -> (time) 1.0
ResultP(1): Toronto (Susan) -> temp: 1.0 trips: 1.0
ResultD(2): Calgary (Paul) -> temp: 1.0 trips: 0.5
...
```

6 Conclusion

Resource Description Framework (RDF) has become a well-know data representation format that finds a lot of applications in variety of domains that require high interconnectivity and semantics-based data formats. Realistic RDF-based applications need to handle time-based and complex relations between pieces of data. There is no doubt it is important to develop methods and techniques capable of storing and utilizing RDF data with temporal information and N-ary relations.

The paper proposes and describes a dedicated interface called LORI – Linguistically Oriented RDF Interface. The key features of this interface allow the users to query temporal data with complex structures suing simple high-level predicates. This predicates are constructed by data experts based on custom predicates developed in the framework of this work. A mapping file that contains definitions of high-level predicates is used for translation of these predicates into 'reasoner-level' predicates. The reasoning processes are preformed via two reasoners: Jena's RDF/RDFS [22] and FuzzyJ [25]. The architecture of LORI is presented, together with a simple case study.

References

1. Berners-Lee, T., Hendler, J., Lassila, O.: The Semantic Web. Scientific American **284**, 34–43 (2001)
2. http://linkeddata.org (accessed July 9th, 2015)
3. http://www.w3.org/TR/2014/NOTE-rdf11-primer-20140225/ (accessed July 9th, 2015)
4. http://blog.iandavis.com/2009/08/representing-time-in-rdf-part-1/ (accessed July 9th, 2015)
5. http://lov.okfn.org/dataset/lov/ (accessed July 9th, 2015)
6. http://www.w3.org/standards/techs/rdfvocabs#w3c_all (accessed July 9th, 2015)
7. http://www.w3.org/TR/rdf-schema/ (accessed July 9th, 2015)
8. http://www.w3.org/TR/swbp-n-aryRelations/ (accessed July 9th, 2015)
9. Allen, J.F.: A General Model of Action and Time. Artificial Intelligence **23**(2), 123–154 (1984)
10. Dutta, S.: An event base fuzzy temporal logic. In: Proc. 18th IEEE Int, Symp. On Multiple-Valued Logic (Palma de Mallorca), pp. 64–71 (1988)
11. Kim, H., Oh, K.: A new representation model in uncertain temporal knowledge. In: Proc. IFSA 1991, pp. 113–116 (1991)
12. Dubois, D., Prade, H.: Processing fuzzy temporal knowledge. IEEE Trans. on Systems Man and Cybernetics **19**, 729–744 (1989)
13. Moon, S.-I., Lee, K.H., Lee, D.: Fuzzy Branching Temporal Logic. IEEE Trans. on Systems, Man, and Cyberntics – Part B: Cybernetics **34**(2), 1045–1055 (2004)
14. Carinena, P., Bugarin, A., Mucientes, M.: A language for expressing expert knowledge using fuzzy temporal rules. In: EUSFLAT-ESTYLF Joint Conference, pp. 171–174 (1999)
15. Barro, S., Marin, R., Mira, J.: A model and a language for the fuzzy representation and handling of time. Fuzzy Sets and Systems **61**(2), 153–174 (1994)
16. Schockaert, S., De Cock, M.: Efficient Algorithms for Fuzzy Qualitative Temporal Reasoning. IEEE Trans. on Fuzzy Systems **17**(4), 794–808 (2009)
17. Lu, Z., Liu, J., Augusto, J.C., Wang, H.: A Many-Valued Temporal Logic and Reasoning Framework for Decision Making. Computational Intelligence in Complex Decision Systems, Atlantis Computational Intelligence Systems **2**, 125–146 (2010)
18. Schockaert, S., De Cock, M., Kerre, E.E.: Reasoning about Fuzzy Temporal Information from the Web Towards Retrieval of Historical Events. Soft Computing **14**(8), 869–886 (2010)
19. Pan, F., Hobbs, J.R.: Time in OWL-S, AAAI Spring Symposium on Semantic Web Services, Stanford University, CA, pp. 29–36 (2004)

20. http://www.w3.org/TR/owl-features/ (accessed July 9th, 2015)
21. Gutierrez, C., Hurtado, C.A., Vaisman, A.A.: Temporal RDF. In: Gómez-Pérez, A., Euzenat, J. (eds.) ESWC 2005. LNCS, vol. 3532, pp. 93–107. Springer, Heidelberg (2005)
22. Gutierrez, C., Hurtado, C., Vaisman, A.: Introducing Time into RDF. IEEE Transaction on Knowledge And Data Engineering (ESWC 2005) **19**(2), 207–218 (2007)
23. Pugliese, A., Udrea, O., Subrahmanian, V.S.: Scaling RDF with time. In: Proc. Second European Semantic Web Conf. (ESWC 2005), pp. 93–107 (2005)
24. Tappolet, J., Bernstein, A.: Applied temporal RDF: efficient temporal querying of RDF data with SPARQL. In: Aroyo, L., Traverso, P., Ciravegna, F., Cimiano, P., Heath, T., Hyvönen, E., Mizoguchi, R., Oren, E., Sabou, M., Simperl, E. (eds.) ESWC 2009. LNCS, vol. 5554, pp. 308–322. Springer, Heidelberg (2009)
25. RobatJazi, M., Reformat, M.Z., Pedrycz, W.: Ontology-based framework for reasoning with fuzzy temporal data. In: IEEE Int. Conf. on Systems, Man, and Cybernetics, Seoul, Republic of Korea, pp. 2030–2035, October 14–17, 2012
26. http://jena.apache.org/documentation/inference/#rdfs (accessed July 9th, 2015)
27. http://jena.apache.org/ (accessed July 9th, 2015)
28. http://www.w3.org/TR/rdf-sparql-query/ (accessed July 9th, 2015)
29. http://rorchard.github.io/FuzzyJ/ (accessed July 9th, 2015)

Spatial Querying Supported by Domain and User Ontologies: An Approach for Web GIS Applications

Khalissa Derbal, Gloria Bordogna, Gabriella Pasi and Zaia Alimazighi

Abstract Geographic data on the Web has intensely increased in recent years. As a consequence, several applications that allow the search of Geographic Information (GI) on the Web have emerged, such as geographic web services, geographic search engines, Web GIS, etc. Till recently, most of research on GI management on the Web was focused on the storage, handling and Web mapping of GI by relying on keywords based query for searching. Nevertheless, GI specific characteristics, and web-user needs, require the development of novel web search methodologies. This paper focuses on spatial web-querying, which, integrated in a Web GIS context, allows retrieving relevant layers of interest that meet user needs. A layer is composed of objects that pertain to the same semantic domain, for example, roads and bridges may compose a transportation network theme. By exploiting the semantic characterization of each layer, in order to retrieve them, we propose in this paper a querying approach empowered by the use of ontologies, which are currently at the heart of many applications of Knowledge Engineering, especially the Semantic Web. A web GIS application prototype, has been developed by using a variety of open GIS software; this prototype is dedicated to the touristic geographic urban domain, and it exemplifies the proposed approach.

Keywords Geographic Information · Spatial web-query · Webmapping · Web GIS · Domain ontology · Open GIS software

K. Derbal(✉) · Z. Alimazighi
LSI Laboratory, Computer Science Department, USTHB, Bab Ezzouar, Algeria
e-mail: {Kderbal,Zalimazighi}@usthb.dz

G. Bordogna
Gloria Bordogna CNR - National Research Council, via Bassini 15, I20133 Milan, Italy
e-mail: gloria.bordogna@idpa.cnr.it

G. Pasi
Gabriella Pasi, Computer Science Department, Degli Studi University Biccoca, Milan, Italy
e-mail: pasi@disco.unimib.it

T. Andreasen et al. (eds.), *Flexible Query Answering Systems 2015*,
Advances in Intelligent Systems and Computing 400,
DOI: 10.1007/978-3-319-26154-6_27

1 Introduction

The data exchange through the Web has changed the Geographic Information System (GIS) paradigm from closed desktop applications to distributed ones [1]. Many Web-based geographic applications were developed in the few last years: Webmapping [2, 3, 4], online GISs or web GISs, geographic Web services such as Web Mapping Services (WMS), Web Feature Services (WFS) and Web Catalogue Services (WCS) [5, 6, 7], and geographic Search Engines (GSE) [8]. All the above applications aim at efficiently serving spatial web queries of different type to retrieve either spatial data (WMS, WFS), metadata or Web pages containing relevant geographic information (GSE). Specifically, Webmapping is defined as "a set of dynamic and interactive mapping applications available on the Web allowing primarily a user to view maps containing more or less geographic information" [7]. Web Feature Service (WFS) allows primarily to view attributes associated with spatial objects on a map. Both WMS and WFS assume that one knows where the map of interest is on the Web, i.e., its URL in order to execute its task. Web Catalogue Services (WCS) allow users to formulate keyword-based queries on a database of metadata of the maps published on the Web by possibly remote and distributed geo servers, in order to discover relevant layers of spatial information. A Web GIS application generally includes both WMS and WFS functionalities, and spatial analysis functionalities for enabling the user to query the content of specific layers listed in a directory, and to evaluate metrics or topologic operators on selected objects in the layers, such as to measure distances of select areas and possibly to construct new maps. [9]. These applications are usually structured in a sequence of tasks; query analysis is generally guided by the user interface, the GI evaluation process returns the requested data, and finally a new map is constructed and delivered. Most of the research works carried out in the field were interested in organizing and handling GI available on the Web [3, 4, 10]. Their main objective is to satisfy Web-user requirements by improving the response time, which constitutes a major constraint in such highly interactive applications [3, 10, 11]. In addition, the delivered maps may be the same for different users (only according to their availability) and may not support their specific needs.

In this paper, we address the task of spatial querying in web GIS by extracting spatial user needs from a textual query in order to identify relevant spatial data layers and build personalized maps. In fact, the complexity of a spatial query processing does not merely come from the difficulty to interpret a NL query. Rather the complexity comes from the particularities of Web GIS applications in which the alphanumerical and geographical components are organized into layers stored in Geographic Data Bases (GDB), which are spatially indexed but do not support textual indexes. To fill this gap, in this paper we propose a textual semantic indexing of the spatial data, so as to enable seeking any explicit or implicit information related to relevant GI. To this purpose, we propose to use the ontology technology. This latter aims to support knowledge management and reasoning on this knowledge, in a perspective of semantic interoperability between humans and/or artificial agents [12]. The following definition of domain ontology [13] motivates our choice: « In the context of computer and information science, an ontology defines a set of representational primitives with which to model a domain of knowledge or discourse.

The representational primitives are typically classes (or sets), attributes (or properties), and relationships (or relations) among class members». A geographic ontology is an ontology organizing objects of a domain in space. Based on a domain geographic ontology we propose both a query analysis approach which integrated in a Web GIS application, it allows to retrieve layers of interest to the user and a carto-user profile ontology, to either filter or integrate the retrieved layers and customize their visualization according to the user preferences. This paper is organized as follows: The next section introduces some research work related to the addressed issue. Section 3 presents a detailed description of the proposed framework. The different steps of our experiments and some results are described in section 4. In section 5, we conclude and present further work proposals.

2 Related Work

It has been estimated that about 30% of the queries submitted to general purpose search engines contain geographic names. This percentage continues to increase with the availability of geographic information and the high growth of the number of Web GIS (Geographical Information System) applications for public use in different contexts of socio-economic activities. Most research in Web-based geographic applications especially in web GIS applications has focused on the storage, handling and dissemination of geographic information by relying on SQL for querying. Many research works have contributed in developing these applications. In [3] a global architecture of web mapping systems is presented, which uses different sources of GI stored in different Geographic Data Bases (raster, vector and metadata).

Web GIS applications on the World Wide Web are usually built in a 3-tier architecture. At the bottom tier there is a data storage system, usually accessible by means of SQL-like queries. The intermediate tier allows the management of geographic data such as their transformation from one geographic projection to another one and the execution of geostatistics operations. At the top tier, there is the user interface (embedded in the Web browser) interacting with the application by HTTP calls. The enabling of a search functionality within a Web GIS is quite limited since Web GISs understand SQL like queries, and thus one must know the names of the stored spatial data layers and attributes to select the layers and to retrieve the data of interest. There is no way to submit a natural language query to retrieve a layer whose content matches the query topic. In the Web context, the popularity of search engines has determined simple keyword based search to become the only widely accepted way for seeking information over the Web. Yet keyword based queries are inherently ambiguous, and this problem of ambiguity is even more evident in spatial queries containing names of geographic entities, like city names. For example the query "Nice" is both the name of a city and of an English word and he query "London" identifies more than forty different cities in the world. Thus, when interpreting spatial queries it is necessary to perform both geo-not geo and geo-geo disambiguation.

This problem has been faced in Geographic Information Retrieval systems (GIR) that make use of geographic ontologies and gazetteers for the recognition of geographic names and their disambiguation [16]. First GIR systems failed to take advantage from the use of geographical information and were outperformed by standard keyword-based systems [14, 15]. More recent, research works that attempted to integrate geographical knowledge and combined textual retrieval with map-based filtering and ranking, have obtained good results. Current approaches to GIR are diversified, ranging from basic IR approaches with no concern for spatial and geographic indexing and reasoning, like generic search engines do, to more specialized approaches applying NLP part- of-speech tagging and geographic name entity recognition for extracting geographic names and spatial relationships from the texts and queries [16].

In our approach we take some ideas from Geographic Information Retrieval (GIR) where textual spatial queries are admitted to extend the query functionality of Web GIS. Nevertheless, we highlight that highly interactive applications such as WCS and GIR systems are, sometimes, confronted with the problem of irrelevance of delivered GI. This may be due to several reasons; (1) the terms of the query are not properly acquired (concerning data entry). The linguistic analyzer is oriented accordingly, towards another context, for example, the use of the term center (commercial center, trade center…), which is highly generic, can affect the search process and (2) the available geographic data do not exactly match the expressed needs. Furthermore, studies have shown that more than 80% of the users would prefer to receive personalized search results instead of the currently generic ones [17]. Query expansion is a technique to assist the user in formulating a better query, by adding additional keywords to the initial request in order to encapsulate its interests there in, as well as to focus the web search output accordingly.

To overcome the above shortcomings, in this paper we propose an ontology-based spatial query analysis approach dedicated to web GIS applications. It allows the end-user to submit her/his spatial query in NL and to receive a result (a visualized map) including the layers that best match user's GI needs without worrying about the underlying process. To do so, we propose to start the search process with a set of terms. The geographic data related to these terms can be then identified by using a geographic ontology that describes concepts of real world related to a specific domain as well as their corresponding GI. For example, road is a concept of our urban ontology that is related to transportation layers in a geographic database. Furthermore, knowing that a well-designed geographic ontology should comprise topological relationships between the represented concepts, the expansion of the terms of the query with the ontology concepts is applied based on the relationships. This expansion can avoid an empty answer but in some cases it can turn in an information overload. The following example illustrates this phenomenon; if the user looks for a hotel in a particular area, implicitly she/he also might seek roads leading to the hotel, of the available transportations and their corresponding stations, etc. This implies the need of several layers to be displayed, thus making the requested map unreadable. As a solution, we propose to integrate in the search process both a carto user profile, defined as an ontology and a filtering module applied after the search process in order to display the preferred layers and visualize them based on the user style preferences. For example a user, searching how to reach a hotel in a city, will be interested in displaying the layers with the public transports and stations at a local map scale. More details of our proposed approach are given in the following sections.

3 Proposed Approach

The overall description of the approach is illustrated in Figure 1. It consists of three modules: (1) *Query analysis module*, which extracts the relevant terms of the query that are then used to explore the geographic ontology. To do so, we developed an algorithm that allows to identify the requested layers of interest by matching the retained terms and the concepts of the ontology (see algorithm 1); (2) *User profile learning module* which aims to associate a cartographic user profile to the web-user. To do so, we propose to capture the user's cartographic orientation, and to construct an ontology that we name "CartoUserOntology" dedicated to the representation of different categories of cartographic users. The users are associated with the cartographic layers that better meet their needs, with respect to semantic and geometric resolutions. We emphasize that the definition of these ontologies must be done with experts in both cartography (as producer) and the addressed domains. In this paper, to the aim of exemplifying the proposed approach, we define some examples of these ontologies to illustrate our idea and to carry out our experimentation and (3) *Filtering module*, which aims to select and integrate the retrieved layers at point 1 that best match the user query requirements based on the "CartoUserOntology".

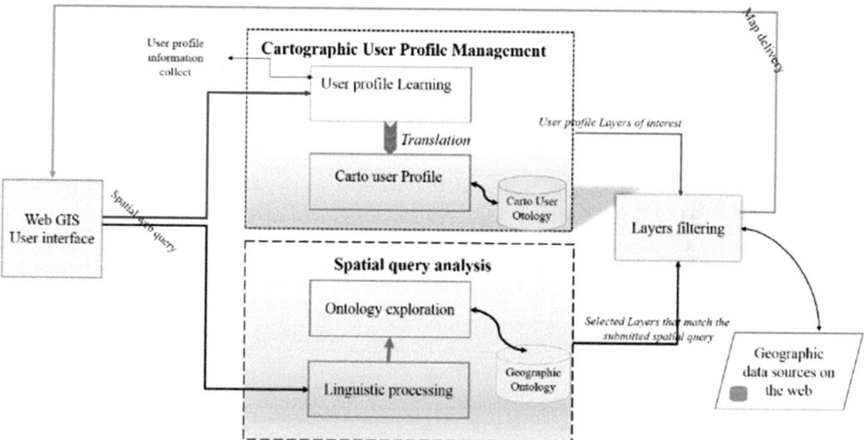

Fig. 1 A high level sketch of the proposed approach

3.1 Linguistic Processing

The first operation carried out by the query analysis module is the lexical analysis of the textual query. We only work with English because of both the availability of several linguistic processing tools, and the existence of the WordNet resource. A spatial web-query in natural language submitted to our Web GIS will undergo a sequence of processing steps, aimed at extracting relevant terms: This is achieved through the following steps: (1) elimination of special characters such as punctuations; (2) tokenization; (3) elimination of stop words; to this purpose we have used a stop list that contains 429 English words and (4) lemmatization that consists of transforming words into a canonical form. The retained terms are used to explore the geographic domain ontology to identify their corresponding cartographic layers. We describe how this is done in the following subsections.

3.2 Design of the Proposed Ontologies

For the definition of our ontologies, we contacted professionals (cartographer, urban planner, etc.), and we have identified an initial set of concepts, attributes, properties and relationships (to be enriched and refined). In the following we provide an example.

Geographic Domain Ontology

The design of our ontology is based on the METHONTOLOGY method developed at the Polytechnic University of Madrid "UPM" [25]. Our choice is motivated by the fact that the reuse of an open access ontology requires its adaptation to both structures and nomenclatures (gazetteers) that comply with the studied geographical area. The example of geographic domain ontology that we have developed, describes a urban domain that is related to the topographic layers (roads, buildings, electricity network, etc.); we call this ontology ''OntoUrbain''. In this ontology three main concepts are identified; Locality and Urban planning, which can be represented by cartographic layers, and Entity, which includes the real-world entities that are not geographically representable but that have relationships with different geographic layers. For example, the entity "car" has a relationship ''roll over'' with the road, or a relationship ''park in'' with the concept cars parking. Figure 2 presents an excerpt of our CartoUrban ontology.

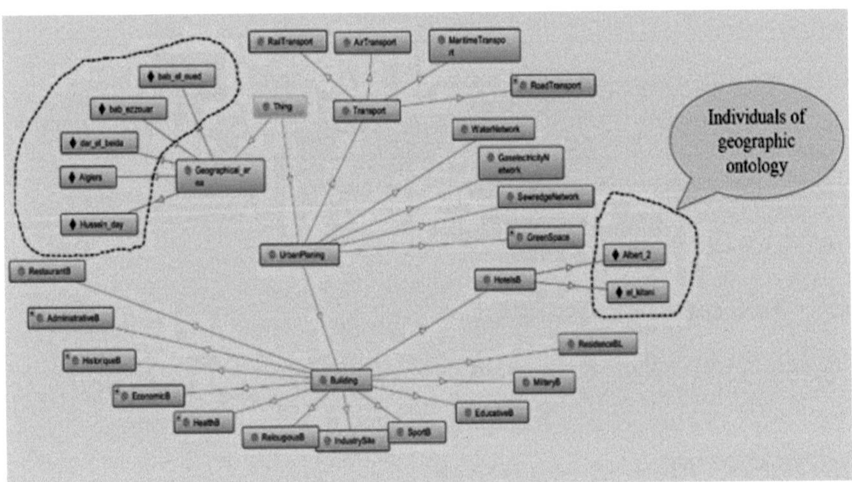

Fig. 2 Excerpt of the implemented geographic domain ontology

The concept "Locality" is related to the administrative boundaries layer. It allows to determine the extent (geographical area) of the requested map. We emphasizes that Toponyms are usually explicitly cited in the queries. They allow to identify the geographic extent and possibly the target geographic objects. Examples of such queries are given in Table1.

Table 1 Examples of spatial queries

Spatial query with extent and target objects (a)	Spatial queries without target objects (b)
Q1: Del parco residence in Milano (Milano)	Museums in Milano
Q2: hotel Alsafir in Algiers (Algiers, Alsafir)	Mosques in Algiers
Q3: Bardo museum in Tunis (Tunis, Bardo)	Cathedrals in Milano

The algorithm aimed at exploring the geographic ontology and consists of finding the ontological concepts that match the terms retained from the first step and hence the different layers of interest. This is achieved by creating a textual index containing all the attribute names extracted from the schema of the layers in the geo-database managed by the Web GIS.

To retrieve more layers the exploration of the geographic ontology can be performed in two ways; Bottom-Top or Top-Down. The first one is recommended when the spatial query contains toponyms like in the queries of the type shown in column (a) of table 1, which constitute the instances of the ontology, while the second strategy is applied in the case of the type in column (b) of table 1. Of course to apply a top-down or bottom-up navigation of the ontology one needs to have a hierarchical ontological structure. Indeed, our developed ontology consists of many concepts related to the cartography domain, such as administrative boundaries road network and building layers, which are in turn composed of sub-layers. For example, a road network can be organized into national roads, departmental and secondary roads layers.

The set of retained terms provided by the first processing step constitutes the entry of our ontology exploration algorithm with a top-down strategy (from general terms to more specific terms). A matching process between these terms and the concepts of the geographic ontology allows to identify the required cartographic layers corresponding to the leaf nodes (Algorithm 1).

CartoUser Ontology

This ontology is defined for categories of users and it aims to propose a formal representation of cartographic profiles with respect to map layers that best meet the user needs to an adequate level of detail. It contains parameters expressing some properties deemed by experts as the most relevant for the displayed maps for the considered profile. We talk about map scale that defines the corresponding Level of detail, style of layers, i.e., colours of legends and icons, preferred background layers, transparency of layers when displayed in overlay mode, etc. We propose two general classes of users: professionals and casual users. The first class includes all potential users of cartography such as urban planner, decision makers in different domains (transportation, urban planning, tourism, etc.), civil protection agents, public authority officers (the police, Border Enforcement, etc.). With each of this category of users, we associate a set of layers (the most relevant ones) as shown in the following example: A rescuer of civil protection (who operates in a specific geographical area) needs specific GI in order to easily reach his destinations: the Road network layer associated with each geographical area in a high Level of Detail (LoD), and the building theme, which contains neighborhoods of the city, hospitals, emergency, etc.

Filtering Layers

To increase the accuracy and to reduce the layers overload that the geographic ontology exploration can provide, the Layers filtering module enables both a selection and an integration of the retrieved layers based on the matching of their information with the user profile. To this end, the user who submits a request will be categorized according to a CartoUser profile ontology and associated to a carto profile by exploiting not only the content of the request but also information the user downloaded from the Web. The CartoUser profile ontology allows associating a cartographic user profile specifying the relevant layers to the web-user associated to a user category. In the case the user cannot be associated to any category it will be identified as a casual user with default needs. The layers resulting from the query evaluation and the layers assigned based on the CartUser profile will then be compared in order to select/integrate the layers that best match the user needs. For example, the evaluation of the query "roads" will retrieve all road layers having different map scales. In her/his CartoUserOntology the user has

Algorithm 1: Ontology Exploration

```
On Entry :
T= {t1,t2,....tn}  // the set of retained terms, result of linguistic processing step
C ={}              // the set of ontological concepts, ci, corresponding to the query terms
I = {}             // the set of individuals(instances) of the
On exit
L={l1,l2,l3,..ln}  // requested cartographic layers, it is initially empty

For each term ti, 0 <i <n do
 Begin
        Browse ontology (CartoUrban,ti)
            if ti corresponds to a concept then
                          Insertconcept(C) // the corresponding concept(class)
                                           // is loaded into C
             else if ti corresponds to a relationship then
                              Insert concept(C) // classes that participate in this
                                                // relationship are loaded into C
                  else if ti corresponds to an instance (leaf Node)
             begin                  get instance I
                                    get superclass ( I, li)
                                    Insertlayer (li, L) // L={}+li
             end
End
// cartographic layers retrieval
For each ci in C do
Begin
     BrowseOntology (CartoUrban,ci,L)  // browsing the ontology until N-1 level of hierarchy in our ontology.
                  get (li,L)           // retrieve the layer li corresponding to ci
 end
  exception: if the concept is an entity it will be replaced by its corresponding concepts in C
 End
// L, represent the set of cartographic layers
```

specified a preferred map scale that will be used to select the corresponding road layer. It can also happen that the preferred layers in the CartoUser ontology are not in the set of the retrieved layers. In this case, they are added to the results in order to provide a meaningful contextual background where to display the retrieved layers in the overlay mode. For example a user looking for "hotels" will also need to display the roads layers as a background contextual information to see how to reach the hotels. We have not yet developed this idea, but it is one of our perspectives in the short-term.

4 Web GIS Implementation Supporting Querying by the Use of Ontologies

In order to demonstrate the feasibility of our approach, we implemented a web GIS application prototype supporting the described query evaluation by the use of ontologies. We made use of well-known open source software tools to implement the components of our prototype. Protégé 2000 was applied to generate the geographic domain ontology, encoded by the W3C standard language for ontologies OWL. The GeoServer package was set up on an Apache Tomcat server as a Web GIS Server to allow deploying on the Web the geographic layers of our application context. Geoserver was selected due to the fact that it supports a variety of data layer formats such as shapefiles, GML files, GEOTIFF, and is compliant with Open Geospatial Consortium (OGC) standard to answer requests for data and layers (WMS and WFS) from the most common web mapping applications implemented as thin clients based on Open layers library functions within common web browsers like Firefox, Explorer and Chrome, (see figure 3). We have also defined, for each layer, an SLD style (Styled Layer Descriptor) to define its visualization properties. In this prototype we have not implemented yet the CartoUser ontology, so we have proposed default preferred properties and layers.

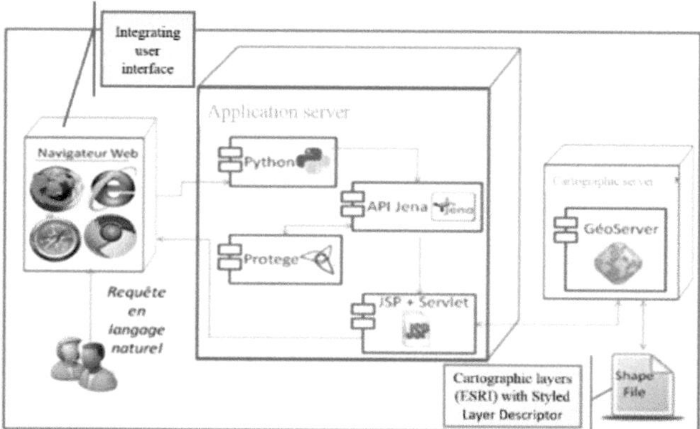

Fig. 3 Software tools and their interactions in our Web GIS prototype

The search interface and query evaluation component performing both the query parsing, lexical analysis and terms expansion based on the geographic domain ontology encoded in OWL and the layers filtering based on the default CartoUSer profile ontology (in the current prototype this latter is equal to all users) were developed by means of the JENA API that proposes a set of predefined classes for analyzing RDFs files like OntClass.

Specifically, the query lexical analysis component that parses the user query and returns the terms to be expanded by exploring the geographic domain ontology was developed in Python programming language, which integrates the Natural Language Toolkit (NLTK) by also exploiting the WordNet thesaurus. A scenario of a running query is described here below, as well as some query processing steps.

Let us consider the following query:

Q1 : "I search some hotels in Babeloued" which is equivalent to
Q2: " Some hotels in Babeloued " or
Q3: "Hotels in Babeloued".

The spatial query is submitted through the search interface of our Web GIS application (see figure 4).

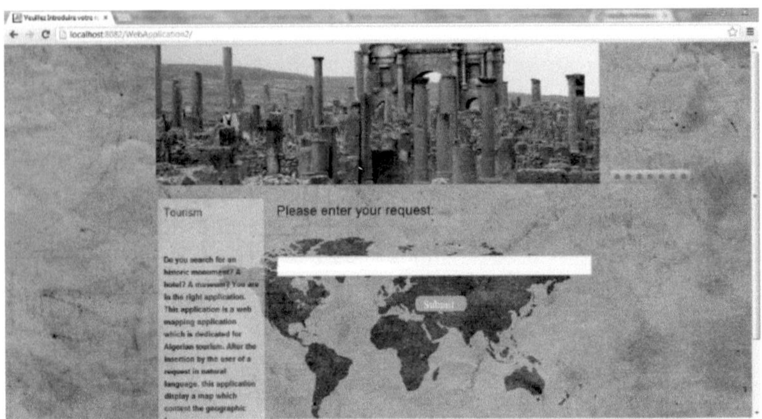

Fig. 4 User interface of the developed web GIS prototype

4.1 Running Scenario

We consider the second query, Q2 : "some hotels in Babeloud".

According to our approach, the linguistic processing module takes in input Q2 and delivers these two terms that are, hotel and Babeloud.

- Babeloued represents a locality in the administrative boundaries layer, it defines the extent of our delivered map.

- "Hotel" is identified as a cartographic layer (see algorithm 1), and all its instances will be retrieved and marked on the map. Nevertheless "Hotel" in the geographic domain ontology is related (has relationships) to many other more specific layers such as restaurants, resort, bed breakfast, family hotels. All these layers will be retrieved.

Nevertheless, in the carto user profile ontology, that at this stage of implementation is the same for all users, the preferred layers are the "road" layer and the "toponyms" layer that are not in the result list of the query Q2. Thus, in order to prevent displaying the retrieved layers on a while background, the filtering module allows integrating the road and toponyms layers as background maps. Figure 5 presents the delivered map, where the required hotels are marked as descriptive information integrating toponyms and -road maps; it is dynamically visualized on the bottom (left) of the visualized map.

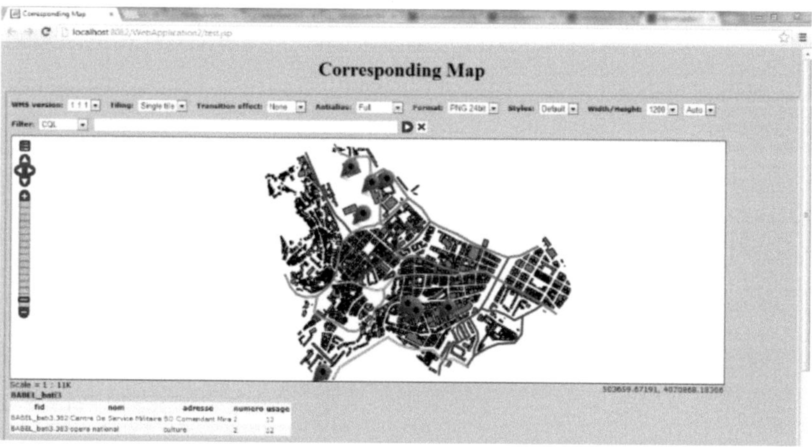

Fig. 5 Delivered result of the Web GIS application

5 Conclusions and Future Issues

This paper has presented a spatial Web-query analyzer integrated into a Web GIS application. Based on the use of domain ontologies this module is able to locate the information relevant to the submitted query, and describing the semantics of the required map (layers of interest). The proposed approach aims to effectively meet GI's user needs by exploiting ontologies. The proposed approach addresses the main methodological and architectural aspects for delivering a readable and relevant map to the user by introducing both a geographic domain ontology to represent the concepts of the application domain and a CartoUser profile ontology to represent the specific cartographic preferences of the categories of users.

As future issues, we aim at addressing:

- The implementation, and assessment, of the carto user profile ontology in the context of spatial query analysis within web GIS application.
- The recognition and evaluation of metrics and topological relationships in queries that allow to target required GI more accurately. For example, the constraint "close to" requires the definition of a rule related to the minimal distance between two geographical objects.
- An extension of the query analysis module to other languages than English.

References

1. Goodchild, M.F.: Geographical information science: fifteen years later. In: Fisher, P.F. (ed.) Classics from IJGIS: Twenty years of the International Journal of Geographical Information Science and Systems, pp. 199–204. CRC Press, Boca Raton (2006)
2. Weibel, R., Burghardt, D.:On-the-fly generalisation. Encyclopedia of GIS. Springer science & Business media (2008). LLC. ISBN: 978-0-387-35973
3. Gaffuri, J.: Toward web mapping with vector data. In: Xiao, N., Kwan, M.-P., Goodchild, M.F., Shekhar, S. (eds.) GIScience 2012. LNCS, vol. 7478, pp. 87–101. Springer, Heidelberg (2012)
4. Derbal, K., Boukhalfa, K., Alimazighi, Z.: A muti-representation and generalisation based webmapping approach using multi-agent system predicates. In: ICWIT, pp. 83–92 (2012)
5. Foerster, T., Lehto, L., Sarjakoski, T., Sarjakoski, L.T., Stoter, J.: Map generalisation and schema transformation of geospatial data combined in a web service context. Computers, Environment and Urban Systems **34**, 79–88 (2010)
6. Burghardt, D., Petzold, I., Bobzien, M.: Relation modeling within multiple representation databases and Generalisation services. The Cartographic Journal **47**(3), 238–249 (2010)
7. Pornon, H., Yalamas, P., Pelegris, E.: Services web géographiques, état de l'art et perspectives. Géomatique Expert **65**, 44–50 (2008)
8. Flora, S.T.: Web-based geographic search engine for location-aware search in Singapore. Expert Systems with Applications **38**, 1011–1016 (2011)
9. Hächler, T.: Online visualization of spatial data. A prototype of an open source internet map server with backend spatial database for the swiss national park. Diploma Thesis, Department of Geography, University of Zürich (2003)
10. Jabeur, N.: A multi-agents system for on-the-fly web map generation and spatial conflict. Ph.D. thesis, Laval University, Quebec, Canada (2006)
11. Vangenot, C.: Multi-representation in spatial databases using the MADS conceptual model. ICA Workshop on Generalisation and Multiple Representations, Leicester, UK (2004)
12. Frédéric, F.: Francky T..Raisonner sur des ontologies lourdes à l'aide de Graphes Conceptuels. LARIA - Laboratoire de Recherche en Informatique d'Amiens (CNRS-FRE 2733) (2004)
13. Gruber, T.R.: The Encyclopedia of Database Systems. In: Liu, L., Özsu, M.T. (eds.) Springer-Verlag (2009)

14. Buscaldi, D., Rosso, P., Arnal, E.S.: Using the WordNet ontology in the GeoCLEF geographical information retrieval task. In: Peters, C., Gey, F.C., Gonzalo, J., Müller, H., Jones, G.J., Kluck, M., Magnini, B., de Rijke, M., Giampiccolo, D. (eds.) CLEF 2005. LNCS, vol. 4022, pp. 939–946. Springer, Heidelberg (2006)
15. Kornai, A.: Evaluating geographic information retrieval. In: Peters, C., Gey, F.C., Gonzalo, J., Müller, H., Jones, G.J., Kluck, M., Magnini, B., de Rijke, M., Giampiccolo, D. (eds.) CLEF 2005. LNCS, vol. 4022, pp. 928–938. Springer, Heidelberg (2006)
16. Bordogna, G., Ghisalberti, G., Psaila, G.: Geographic information retrieval: Modeling uncertainty of user's context. Fuzzy Sets and Systems **196**, 105–124 (2012)
17. Yu, Y., Wenjuan, W., Yiming, S., Tong, L: Search Engine System Based on Ontology of Technological Resources. Journal of Software **6**(9), September 2011
18. Buscaldi, D., Rosso, P., On, A., Petras, V., Santos, D. (eds.): Advances in Multilingual and Multimodal Information Retrieval. In: 8th Workshop of the Cross-Language Evaluation Forum, CLEF 2007, Budapest, Hungary, September 19–21, 2007. Lecture Notes in Computer Science, pp. 815–822. Springer (2008)

Efficient Multimedia Information Retrieval with Query Level Fusion

Saeid Sattari and Adnan Yazici

Abstract Multimedia data particularly digital videos that contain various modalities (visual, audio, and text) are complex and time consuming to deal with. Therefore, managing a large volume of multimedia data reveals the necessity for efficient methods for modeling, processing, storing and retrieving such data. In this study, we investigate how to efficiently manage multimedia data, especially video data. In addition, we discuss various flexible query types including the combination of content as well as concept-based queries that provide users with the ability to perform multimodal query. Furthermore, we introduce a fusion-based approach at the query level to improve query retrieval performance of the multimedia database. Our experimental tests show a significant improvement in the query retrieval performance.

Keywords Multimodal query · Query level fusion · Multimedia database · Cross-modal retrieval

1 Introduction

With an increasing amount of multimedia data production favored by cheap digital devices, such as large capacity and fast accessed media storages, people are exposed to a very large volume of multimedia data in daily life. Storing and searching such a large volume of data in a database becomes a real challenge. Usually most of the users are interested in the semantic contents of videos. Consequently, manual annotation of such a large volume of data in order to prepare them for any content-based search is almost impractical. Thus, dealing with this big amount of digital videos requires automatic methods to extract the semantic contents and efficient ways to store, index and retrieve them.

S. Sattari(✉) · A. Yazici
Department of Computer Engineering,
Middle East Technical University, 06531 Ankara, Turkey
e-mail: {saeid,yazici}@ceng.metu.edu.tr

T. Andreasen et al. (eds.), *Flexible Query Answering Systems 2015*,
Advances in Intelligent Systems and Computing 400,
DOI: 10.1007/978-3-319-26154-6_28

Due to the nature of multimedia data including multidimensionality, multimodality and complexity in automatic concept extraction, manual annotation and conventional database approaches are not as sufficient as expected for managing multimedia content. Therefore, extracting semantic concepts automatically, efficiently storing and indexing them plus handling multimodality as well as efficiently retrieving them continue to be one of the most challenging and fast-growing research areas which have attracted many researchers. Although managing and retrieving textual content (metadata) are easy and straightforward, many researchers are usually interested in the semantic content of the video.

Audio, text, visual objects are considered as the basic components of semantic content of the video data [1]. The process of extracting these entities (annotating semantic information) in the domain of multimedia system research is a challenging task [2], [3]. When we use manual annotation methods for multimedia data information extraction, it is mostly resource consuming, exhausting and slow; therefore, it is an expensive, limited and inefficient process. Thus, an automatic annotation system is required in real life applications. Besides, due to the complexity of multimedia data structure, modeling as well as querying are also complex tasks.

From query level perspective, multimodal data which originated from the same source tend to be correlated [4]. Since the presence of one modal can be used to understand certain semantics of other modals, different modals can take a complementary role in the retrieving multimedia content. Due to the fact that each modal may compensate for weakness of the others, we can benefit from relations among various modals. For instance, a video retrieval system that exploits both audio and visual modals may achieve a better performance in terms of both accuracy and efficiency than a system which exploits either one only [5]. As a result, multimodal correlation analysis has attracted a growing attention in multimedia content analysis researches in the recent years [6,7].

In the literature there are some studies that only focus on the positive correlation among objects of multimedia in the same class while the negative correlation between them is underestimated. From our point of view, both kinds of correlation are important because positive correlation provides the co-occurrence information and negative correlation reflects the exclusive information. For example, in a video shot the "explosion" concept may come from text as well as visual category. Meanwhile, the related section of audio labeled "gun" may have strong positive correlation with that shot while some concepts in audio like "silence" has negative correlation with the same shot. As a result, using both positive and negative correlations would be beneficial. Although combining and propagating correlations along with each modal is proven to improve retrieval performance, the semantic correlations that are propagated among modals throughout the whole dataset are also beneficial. As a result, employing the correlations among the objects and concepts can naturally meet the requirements of multimodal retrieval challenge in which, the retrieval results are of the similar semantics and can be of different media types.

1.1 Motivation

Although various studies exist about multimedia querying, they focus just on one aspect of multimedia for instance, working only single modals or support only query by concept [8]. In addition, there are very few multimedia retrieval approaches that use a fusion at query level [9]. However, to the best of our knowledge, there exists no study in literature that uses three modals (visual, audio, and text) of video data in query level fusion.

The main motivation for this work is the need to handle different query types using multimodal query level fusion. More specifically, we study multimodal content-based and concept-based queries. Compared with the existing studies in literature, the major contributions and advantages of this study are as follow:

1. A client application for query is developed that supports query by concept, query by content and their combination. It also supports multimodal query with logical operations to join various modals (visual, audio and text).
2. We employ a query level fusion that exploits semantic correlation in each modal and among them. That is, we employ term to term correlations to catch relations between modals by looking at co-occurrence information. Then we evaluate them using the vector space retrieval model, which provides some improvement in retrieval performance.

The rest of the paper is organized as follows: in Section 2, previous works and studies regarding to our work are explained. In Section 3, data model for managing video data is described. Section 4 summarizes the various query types supported. Query level fusion is briefly explained in Section 5. In Section 6, experimental results, dataset and retrieval model are described. Lastly, conclusions and some possible future directions are given.

2 Related Works

With a fast technological improvement in cheap and large capacity storage users need to efficiently search among these mass volume of multimedia data. Hence, multimedia information retrieval necessity is fulfilled when some semantic concepts as well as example multimedia objects can be efficiently searched within multimedia data [10,11]. The multimedia objects search is facing a well-known problem that is defined as semantic gap.

A system for multimedia information retrieval was proposed in [12] that utilizes matrix-based mathematical models for content modeling. "Informedia" [13], "Combinformation" [14], "greenstone" [15], "M-Space Browser" [16,17] and "EVIADA" [18] also were proposed as non-web-based multimedia information retrieval systems. They mostly provide searching within particular modal of multimedia data [19]. Since searching inside one modal of multimedia data cannot compensate the requirements of the retrieval of multimedia data [9], searching within multiple modals or multiple knowledge sources like audio, video and text is almost vital. Designing the mixture approaches for multimodal retrieval gain great importance lately in developing effective multimedia systems [20]. This issue has created an important challenge for researchers, as pointed out in [21]:

"To deal effectively with multimedia retrieval, one must be able to handle multiple query and document modalities. In video, for example, moving images, speech, music audio and text (closed captions) can all contribute to effective retrieval. Integrating the different modalities in principled ways is a challenge."

The problem of multimedia modal source combination has been actively investigated in recent years. Westerveld et al. [22] demonstrated how the combination of different modalities can influence the performance of video retrieval. The video retrieval system proposed by Amir et al. [23] applied a query-dependent combination model that the weights are defined based on user experiences. They also utilized a query-independent linear combination model to merge the text/image retrieval systems. Rautiainen et al. [24] used a user-dependent approach to combine the results from text search and visual search. In their system the combinations' scores are predefined by users when the query is submitted. Some of the recent studies have also proven that using a latent relation between modalities increased the performance of query retrieval in multimodal systems [25,26,27].In some studies, multi-modal correlation analysis [28,29] approaches are employed to explore statistic correlations between modalities by analyzing their co-occurrence relationship. For instance, after extracting visual and audio features, correlation can be analyzed between their feature matrices to learn their correlations [30] and then apply a hierarchical manifold space to make the correlations more accurate [31]. However, difficulties still exist due to the heterogeneous feature space of visual or audio modalities.

Rasiwasia et al. [32] utilized the semantic correlation matching to correlate the text and image's low-level features for cross-modal retrieval. Their experimental results demonstrated that using this method outperform state-of-the-art image retrieval systems on a unimodal retrieval. Chuang et al. [33] also used the terms correlation for expanding the query for higher retrieval precision. In the work by Safadi et al. [34] that single level fusion was used to overcome the well-known semantic gap problem, it was demonstrated that the mapping of concepts between text and visual modal have a significant retrieval improvement in their framework.

3 Data Model

Multimedia data contains huge amount of information that exist in complex and compound structures. Being in various forms and the diversity of semantic contents make it difficult to model multimedia data. In this study, multimedia data are categorized as a combination of: visual modal, audio modal and textual modal. In order to prepare an appropriate configuration for testing the query retrieval performance, we integrate various modules proposed in [35,36,37,38]. Preparing this test configuration allows us to query the contents and the concepts of a shot. The details of this integrated system are out of the bound of this paper.

In the data models, method of modeling and storing data related to the video files are defined. Therefore, in such a model that is related to multimedia data, semantic entities and objects should be defined clearly. Besides the hierarchical structure of these entities, since some multimedia data types contain time-specific components, temporal segmentation of such data also should be considered in the data modeling.

In this study, we follow a well-known temporal segmentation for visual data that has time information. The smallest temporal segments, shots, are defined as the minimal group of adjacent frames stating a continuous action and having images from the same area or scene. Therefore, shots contain some common concepts. Video objects are composed of these shots and audio segments are aligned with each shot according to the time overlapping. For the key-frames, image segmentation can be applied for partitioning images into smaller parts. These parts can be mapped directly into objects by some low-level features. For audio concepts, the temporal segmentation is considered as well and the segmented audio parts are classified into some predefined taxonomy.

4 Supported Query Types

This study supports various multimodal content and concept-based queries. We provide a web-based application that allows the users to compose and submit following query types:

- **Query by Concept:** In this type of query, the aim is to retrieve shots that contain concepts provided as query terms. Any combination of modals can take part in query composition process, *(i.e., Retrieve all shots which contain: car, accident, screaming)*.
- **Audio Query by Content:** In this type of the query an example audio file is selected and submitted to the query analyzer. Predefined low-level features of this audio file are extracted. These low-level features are used to find the closest audio segments which are similar to the given audio example. As a result, all shots that their audio segment's low-level features match to the selected audio file are fetched from dataset, *(i.e., Given a sample audio whose content is audio of "car brake", retrieve all the shots that contain the same audio sample (car brake))*.
- **Visual Query by Content:** Handling this type of query is partly equivalent to the method mentioned in audio query by content from mechanism perspective. In this type, an image of an object is selected and sent to the query analyzer unit. The low-level features are extracted and similar shots are fetched from dataset by comparing the low-level features. Another variant of this query is to choose an image that consists of multiple objects to retrieve all the shots that contain the most relevant concepts or objects that exist in the provided sample

image. (*i.e., Given a sample frame or an image of an object (i.e. image of an airplane), retrieve all the shots that contain similar object (airplane)).*

- **Multimodal Query:** This type of query is composed of different concepts and sample audio and an image of an object. The aim is to retrieve shots that contain the concepts provided by query terms as well as the content provided with sample audio and sample image of an object. (*i.e., Retrieve all the shots that includes "disaster" concept, "fire" visual seen, and "explosion" audio segment).*

Using the low level features for the audio tracks and video's frame image, some models are trained beforehand to detect a list of predefined concepts. For a new unlabeled video, these models are used to annotate the video with the learned concepts. These concepts are used for retrieval of shots in the queries.

Furthermore, the extracted low-level features are also stored with the videos. In case of a query by content, the low-level features of newly submitted audio or video file (as a content) are extracted and compared with the stored low-level features to retrieve shots corresponding the content provided by user in the query.

5 Query Level Fusion

Query level fusion is utilized to expand and enrich the query and improve the retrieval performance. As mentioned, each shot is represented by three modals (visual, audio and text). Since each modal explain the same content (single shot) from different view, it is more probable that there are some correlations among these modals. Query fusion exploits these correlations and terms co-occurrence within a modal (intra-modality) and among the modals (inter-modality) [30] to update the weights of the terms in the query.

Intra-Modality

In the intra-modality, Pearson correlation is employed among a single modal's terms. Correlation between two sets of data is a measure of how well they are related. The most common measure of linear correlation is the Pearson Correlation which shows the linear relationship between two sets of data. In simple terms, it answers the question: Can I draw a line to represent the data in which, one set can be used to represent the other set? As we also aim to do so within a single modal. The Pearson correlation is commonly represented by the letter r in such a way that if we have one dataset $\{x_1,\ldots,x_n\}$ containing n values and another dataset $\{y_1,\ldots,y_n\}$ containing n values then that formula for r is:

$$r = r_{xy} = \frac{\sum_{i=1}^{n}(x_i - \bar{x})(y_i - \bar{y})}{\sqrt{\sum_{i=1}^{n}(x_i - \bar{x})^2}\sqrt{\sum_{i=1}^{n}(y_i - \bar{y})^2}}$$

In this formula x and y are two concepts within a modal (i.e. gun and armed-person) in which x_i and y_i are their weights in the manually annotated shots. In this formula n is the number of shots in the dataset. Usually value of r which is in the range of $0.49 < r^2 < 1$ is regarded as a highly correlated coefficient. Therefore, we use this value throughout this study.

For instance, in the visual modal there is a strong correlation between "armed person" and "gun" (r =0.89). When a new query is submitted, if "gun" exists in the query terms and "armed-person" does not exist (therefore its weight in the query vector is zero), we use the weight of the "gun" to update the weight of the "armed person" in the query vector. We repeat this step for all terms in each modal separately to update the weights of the query terms.

Inter-Modality

In the inter-modality, since correlations exist among a set of terms in one modal and another set of terms in the other modal, we should use an approach that quantify the strength of the relationship between two sets of variables (i.e. a set of concepts from visual modal (x) vs. a set of concepts from audio modal(y)).

Selected technique for exploring this relationship between two sets of variables is canonical correlation analysis (CCA). Canonical Correlation Analysis which was proposed by Hotelling [39] finds two sets of basis vectors, one for x and the other for y, such that the correlations between the projections of the vectors (x_i, y_i) onto these basis vectors (w_x^T , w_y^T) are maximized [Fig. 1.]. Moreover, CCA does not stop with the calculation of a single pair of basic vector. Instead, *min* {dimension (x),dimension(y)} number of vectors are calculated for each set. In the context of this study, each set (x, y) is equivalent to a modal. In addition, we assume that our shots are composed of 3 modals: visual, audio and text and each of these modals is represented respectively in an m, n and p dimensional spaces in which each dimension represents a concept.

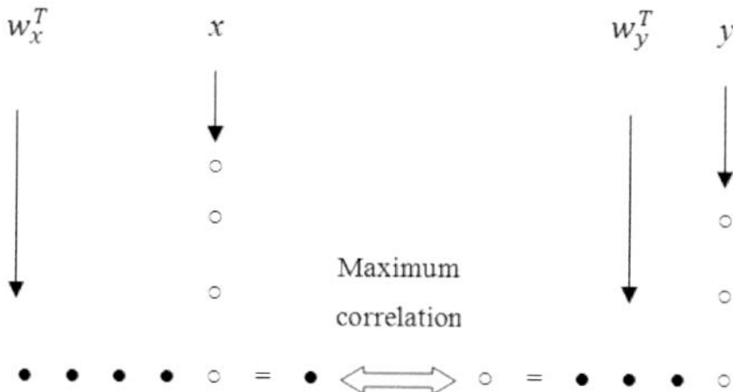

Fig. 1 Finding w_x^T and w_y^T

If we combine the calculated basis vectors we yield two matrices that their dimension is d_i * min{dimension (x),dimension(y)} which each matrix belongs to a modal and d_i is the dimension of that modal. For two modals the related matrices can be used to transform the modals to a new vector space in which the correlation between the modals kept maximized. If we name these matrices $\mathbf{M}_x$ and $\mathbf{M}_y$ and define:

$$S \;= ((x_1,y_1),\ldots,(x_n,y_n)) \text{ of } (x,y)$$

$$S_{x,\mathbf{M}x} = (<\mathbf{M}_x, x_1>,\ldots,<\mathbf{M}_x, x_n>)$$

$$S_{y,\mathbf{M}y} = (<\mathbf{M}_y, y_1>,\ldots,<\mathbf{M}_y, y_n>)$$

Then, the final purpose of the CCA is maximizing *corr* ($S_{x,\mathbf{M}x}$, $S_{y,\mathbf{M}y}$) with regards to the $\mathbf{M}_x$ and $\mathbf{M}_y$.

As an example, if there is an annotated shot that its audio modal is represented as a 17 dimensional vector and visual modal as a 49 dimensional vector, transforming these two vectors to a new vector space would yield two 17 dimensional vectors which the correlation among these two new vectors is maximum. We manually annotate some shots using relevant concepts for each modal. We learn the basis vectors for each modal using these annotated data as a training set.

As mentioned in [40] the geometric interpretation of correlation between two vectors is equivalent to their cosine similarity. Regarding the mentioned explanation, we calculate 2 matrices for each pair of modals. In case of new query following steps are performed. For each modal, top 5 similar vectors in the remaining 2 modals are found by using the cosine similarity. Equivalent vectors in the original spaces of these 2 modals are retrieved for these top 5 similar vectors. At the end, their average is considered as a new vector for the query's correspondent modal. By repeating this process, each modal's terms weight are updated according to the remaining two other modals. We refer to this updated query vector as "fusion query" in the experimental result of this work.

6 Experiments

In this section we describe the dataset and test configuration along with our experimental results. We also explain the query structure and the retrieval model and then evaluate the experimental results.

6.1 *Dataset*

We download 76 news videos (totally five hours) from NTV [41] news archive and categorize them into these categories: accident, military, natural disaster, sport and politic. We also create a concepts list that is a subset of LSCOM [42] concepts. Then the shot boundaries, key frames, visual objects/concepts, audio concepts and subtitle texts are annotated manually for all of the video clips. Briefly saying, each video is splitted into some shots so that each shot represent almost the same scene. Then some Key-frames are extracted for each shot. For each Key-frame, concepts that related to the scene are annotated using the previously created concepts list. Furthermore, for each audio segment a proper audio concept is assigned. At the end, the subtitles of shots are manually extracted and converted to the named entities. In addition, some words are selected from subtitles as the important words. Also, by integrating various modules (the details are out of scope of this paper) and analyzing these shots, we create a dataset that contains automatically extracted concepts from these shots for all three modals.

6.2 Query Structure

In our dataset there are totally 166 different concepts which include audio concepts (17), visual concepts (49) and texts (100). As a consequent, each shot is represented with a 166 dimensional vector:

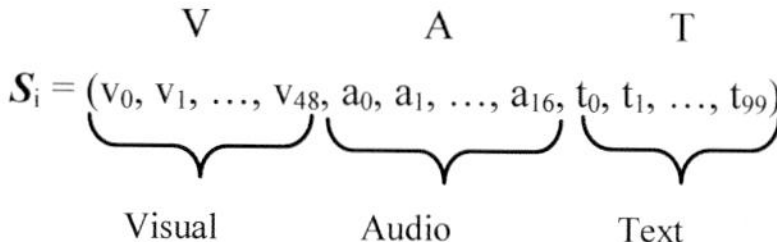

Indices 0 to 48 are assigned to visual concepts and audio concepts are represented form index 49 through 65. For instant if index 53 is for the "alarm" sound and our shot have the "alarm" with score 0.7 in its audio concept list then, 0.7 would be assigned to the value of index 53. If we suppose index 61 is for the "car" sound and the shot doesn't contain the "car" in its audio list then, 0 would be assigned to the value of index 61. We also represent text concept from the indices 100 to 165 but with a minor difference that we put zero or one instead of score (since we don't have score for text). We call this 166 dimensional vector for each shot as "raw result" vector. In this study each query also is defined in a166 dimensional vector. Those terms (concepts) that are indicated in query have one (1) in their related indices value. Remaining indices are assigned zero. Due to the simplicity we entitle this vector as "raw query" vector. In the query fusion processing unit the raw query vector is analyzed and the output is a vector with updated scores that we refer to it as "fused query" vector. The Cosine similarity is used as a similarity measure between the query vector and shot vectors to answer the supported queries in the vector space model.

6.3 Experimental Results

For the test we measure the retrieval performance of multimodal query retrieval using query level fusion for two different test configurations (one being multimodal querying without query-level fusion as the baseline and the other being multimodal with query-level fusion, QF) and plot the results as a precision-recall diagram for the top 50 results [Fig. 2.].

In the configuration labeled as "Baseline" the queries are triple words combinations in which each word is selected from one modal. So, the total number of queries is 83300 (49*17*100). For each query, the raw query vector is used to retrieve top 50 results according to its cosine similarity to the raw data vector. For all the queries the precision and recall are averaged over the number of queries.

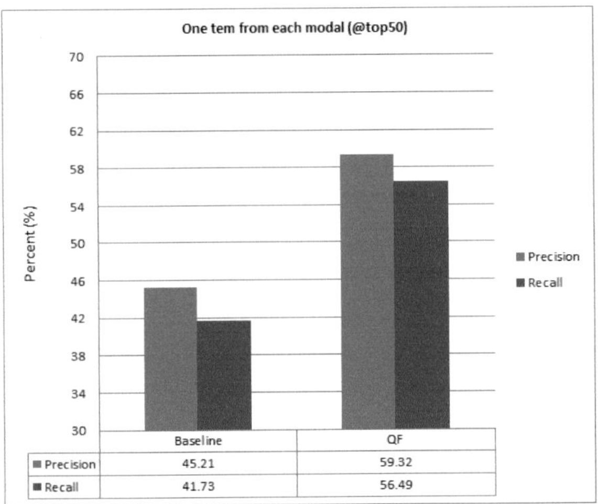

Fig. 2 Retrieval performance between Baseline and QF

In the query-level fusion (QF) test, the approach of selecting the terms and composing the queries is the same as the previous method. The only difference is that we process the raw query vector in query fusion analyzer and then use the fused query vector. Cosine similarity between this vector and raw result vector is used to retrieve the results. As we observe, employing query level fusion improve the retrieval performance in terms of the precision and recall.

7 Conclusion and Future Work

In this study, we introduce an efficient multimodal querying approach using query level fusion to support various query types such as content and concept based queries. Since a typical video file consists of various modals (visual, audio and text), handling all of these modals simultaneously is necessary for acceptable retrieval performance. Therefore, in this study, we support multimodal multimedia queries and show the improvement in retrieval performance.

Despite the fact that we try to address some challenges regarding multimodal multimedia retrieval using correlation, the research in this area still can be improved with several approaches. As possible future works, we identify some approaches that can extend the retrieval performance:

- For calculating correlation between modals a multi-set CCA can be applied to observe if this has any effect on retrieval performance.
- Various clustering methods can be integrated into query level fusion to quick query answering and better performance.

Acknowledgment This work is supported by the research grant from TUBITAK with the grant number "114R0182".

References

1. Yilmaz, T., Yildirim, Y., Yazici, A.: A genetic algorithms based classifier for object classification in images. In: ISCIS, London, pp. 519–525 (2011)
2. Demir, U., Koyuncu, M., Yazici, A., Yilmaz, T., Sert, M.: Flexible content extraction and querying for videos. In: Christiansen, H., De Tré, G., Yazici, A., Zadrozny, S., Andreasen, T., Larsen, H.L. (eds.) FQAS 2011. LNCS, vol. 7022, pp. 460–471. Springer, Heidelberg (2011)
3. Jonker, W., Petkovic, M.: An overview of data models and query languages for content-based video retrieval. In: Advances in Infrastructure for E-Business, Science, and Education on the Internet, Italy (2000)
4. Datta, R., Joshi, D., Li, J., Wang, J.: Image retrieval: Ideas, influences, and trends of the new age. In: ACM Computing Surveys (CSUR), pp. 1–60. ACM (2008)
5. Blei, D.M., Ng, A.Y., Jordan, M.I.: Latent dirichlet allocation. In: JMLR, vol. 3, pp. 993–1022 (2003)
6. Csurka, G., Dance, C., Fan, L., Willamowski, J., Bray, C.: Visual categorization with bags of keypoints. In: Workshop on Statistical Learning in Computer Vision, ECCV, vol. 1, p. 22 (2004)
7. Deerwester, S., Dumais, S., Furnas, G., Landauer, T., Harshman, R.: Indexing by latent semantic analysis. Journal of the American society for Information Science **41**(6), 391–404 (1990)
8. Wei, X.Y., Jiang, Y.G., Ngo, H.: Concept-Driven Multi-Modality Fusion for Video Search. IEEE Transactions on Circuits and Systems for Video Technology **21**(1), 62–73 (2011)
9. Natsev, A., Haubold, A., Tesic, J., Xie, L., Yan., R.: Semantic concept-based query expansion and re-ranking for multimedia retrieval. In: MULTIMEDIA 2007, pp. 991–1000. ACM, NY (2007)
10. Dunckley, L.: Multimedia Databases – An Object-Relational Approach (2003). ISBN # 0 201 78899 3
11. Bhatti, M.A., Rashid, U.: Exploration and management of web based multimedia information resources. In: International Conference on Systems, Computing Sciences and Software Engineering. Bridgeport, USA (2007)
12. Rolleke, T., Tsikrika, T., Kazai, G.: A general matrix framework for modeling information retrieval. In: Proceedings of the ACM SIGIR MF/IR (2003)
13. Sanchez, J.A., Arias, J.A.: Content-based search and annotations in multimedia digital libraries. In: ENC 2003, Fourth Mexican International Conference on Computer Science, pp. 109–117 (2003)
14. Kerne, A., Koh, E., Dworaczyk, B., Mistrot, J.M., Choi, H., Smith, S.M., Graeber, R., Caruso, D.: CombinFormation: a mixed-initiative system for representing collections as compositions of image and text. In: JCDL 2005, ACM/IEEECS Joint Conference on Digital Libraries, pp. 11–20 (2006)
15. Witten, I.H., Bainbridge, D.: Building digital library collections with greenstone. In: JCDL 2005, 5th ACM/IEEECS Joint Conference on Digital Libraries, pp. 425–425 (2005)
16. Wilson, M.L.: Advanced search interfaces considering information retrieval and human computer interaction. In: Agents and Multimedia, Southampton (2007)

17. Schraefel, M.C., Wilson, M., Russel, A., Smith, D.A.: MSPACE: improving information access to multimedia domains with multimodal exploratory search. Communication of the ACM **49**(4), 47–49 (2006). ACM
18. Dunn, W.: EVIADA: ethnomusicological video for instruction and analysis digital archive. In: JCDL 2005, 5th ACM/IEEE-CS Joint Conference on Digital Libraries, p. 407 (2005)
19. Rashid, U., Niaz, I.A., Bhatti, M.A.: Unified multimodal search framework for multimedia information retrieval. In: 4th International Conference on Systems, Computing Sciences and Software Engineering. Springer, Bridgeport (2007)
20. Yan, R.: Probabilistic Models for Combining Diverse Knowledge Sources in Multimedia Retrieval. Ph.D. Thesis. School of Computer Science, Carnegie Mellon University, USA (2006)
21. Manmatha, R.: Multimedia indexing and retrieval. In: Workshop on Challenges in Information Retrieval and Language Modeling (2002)
22. Westerveld, T., Ianeva, T., Boldareva, L., de Vries A.P., Hiemstra, D.: Combining infomation sources for video retrieval. In: NIST TRECVID (2003)
23. Amir, A., Hsu, W., Iyengar, G., Lin, C.Y., Naphade, M., Natsev, A., Neti, C., Nock, H.J., Smith, J. R., Tseng, B.L., Wu, Y., Zhang, D.: IBM research TRECVID- video retrieval system. In: NIST TRECVID (2003)
24. Rautiainen, M., Hosio, M., Hanski, I., Varanka, M., Kortelainen, J., Ojala, T., Seppanen, T.: TRECVID 2004 experiments at mediateam oulu. In: Proc. of TRECVID (2004)
25. Yu, J., Cong, Y., Qin, Z., Wan, T.: Cross-modal topic correlations for multimedia retrieval. In: ICPR, Tsukuba, Japan (2012)
26. Song, Y., Philippe Morency, L., Davis, R.: Multimodal human behavior analysis: learning correlation and interaction across modalities. In: ICMI, Santa Monica, California, USA, pp. 22–26 (2012)
27. Jiang, W., Loui, A.C.: Video concept detection by audio-visual grouplets. Multimedia Information Retrieval **1**, 223–238 (2012)
28. Zeng, J.D., Zheng, H.J., Lu, C., Li, T., Ma, W.: ReCoM: reinforcement clustering of multi-type interrelated data objects. In: ACM SIGIR, pp. 274–281. ACM, Canada (2003)
29. Wang, X.J., Ma, W.Y., Xue, G.R., Xing, L.: Multi-model similarity propagation and its application for web image retrieval. In: ACM Multimedia, pp. 944–951. ACM (2004)
30. Zhang, H., Zhuang, Y.T., Wu, F.: Cross-modal correlation learning for clustering on image-audio dataset. In: ACM Multimedia, pp. 273–276. ACM (2007)
31. Yang, Y., Zuang, Y.T., Wu, F., Pan, Y.H.: Harmonizing hierarchical manifolds for multimedia document semantics understanding and cross-media retrieval. IEEE Transactions, 437–446. (2008)
32. Rasiwasia, N., Pereira, J.C., Coviello, E., Doyle, G., Lanckriet, Gert, R.G., Levy, R., Vasconcelos, N.: A new approach to cross-modal multimedia retrieval. In: Multimedia 2010. ACM, Firenze (2010)
33. Chuang, C.T., Yang, K.H., Lin, Y.L., Wang, J.H.: Combining query terms extension and weight correlative for expert finding. In: International Joint Conferences on Web Intelligence (WI) and Intelligent Agent Technologies (IAT). IEEE (2014)
34. Safadi, B., Sahuguet, M., Benoi, H.: When textual and visual information join forces for multimedia retrieval. In: ICMR 2014. ACM, Glasgow (2014)

35. Aydinlilar, M., Yazici, A.: Semi-automatic semantic video annotation tool. In: ISCIS 2012, Paris, pp. 303–310 (2012)
36. Okuyucu, C., Sert, M., Yazici, A.: Environmental sound classification using spectral. In: IEEE, Turkey (2013)
37. Kucuk, D., Yazici, A.: Exploiting information extraction techniques for automatic semantic video indexing with an application to Turkish news videos. Knowledge-Based Systems **25**(6), 844–857 (2011)
38. Arslan, S., Yazici, A., Sacan, A., Toroslu, I.H., Acar, E.: Comparison of feature-based and image registration-based retrieval of image data using multidimensional data access methods. Data & Knowledge Engineering **86**, 124–145 (2013). Elsevier
39. Hardoon, D.R., Szedmak, S., Taylor, J.S.: Canonical correlation analysis; An overview with application to learning methods. Technical Report CSD-TR-03-02. University of London (2003)
40. Kuss, M., Graepel, T.: The Geometry Of Kernel Canonical Correlation Analysis. Technical Report No. 108. Max Planck Institute (2003)
41. News Channel of Turkey. http://www.ntvmsnbc.com/
42. Naphade, M., Smith, J.R., Tesic, J., Chang, S.-F., Hsu, W., Kennedy, L., Hauptmann, A., Curtis, J.: Large-scale concept ontology for multimedia. IEEE MultiMedia **13**(3), 86–91 (2006). IEEE

A Tire Tread Pattern Detection Based on Fuzzy Logic

Alžbeta Michalíková and Michal Vagač

Abstract Tire tread prints, being common type of evidence at crime scene, contain useful information for the investigator. To determine the brand or manufacturer of a tire related to the tread, database of tire treads is required. It can be built using thousands of tire photographs available in many existing web shops. To build such a database tire photographs must be downloaded and processed automatically.

In this paper we present a method to automatically detect tire tread in provided image from existing database. Our aim is to find ellipses which determine tire tread sample. In the first step we use Hough transform to detect all ellipses from provided image. In the second step the set of ellipses is reduced by fuzzy inference system to select those which describe position of tire tread sample the best and finally at the end we present results of experiments.

Keywords Tire tread print · Image analysis · Image processing · Hough detection · Fuzzy inference system

1 Introduction

Foot prints and tire prints found at crime scene provide reliable evidence of presence of related object (a shoe or a tire). Determination of the brand or manufacturer of the shoe or the tire can be important part in investigation. When brand or manufacturer is known it helps in searching of concrete shoe or tire which can be connected with the crime scene.

Several papers has been published on the topic of footwear or tire prints recognition [1–12]. Main focus of these works is on recognition part of the process. In the

A. Michalíková · M. Vagač(✉)
Department of Computer Science, Faculty of Science, Matej Bel University, Tajovského 40, 97401 Banská Bystrica, Slovakia
e-mail: {alzbeta.michalikova,michal.vagac}@umb.sk
http://www.fpv.umb.sk/

T. Andreasen et al. (eds.), *Flexible Query Answering Systems 2015*,
Advances in Intelligent Systems and Computing 400,
DOI: 10.1007/978-3-319-26154-6_29

most cases authors describe a method which allows to find given foot or tire print in existing database of prints. Common process is to build query image by modifying an image from database (using blur, rotation or crop actions).

However, in practice existence of a database of print samples is one of the biggest problems. Ideally, the samples would be obtained by making print of each existing tire or shoe. In practice, this is not accomplishable, therefore some other solution must be developed. One of the possibilities is to use image (photography) of the tire (shoe), extract tread (sole) from it and then process it to state, which will be comparable with the print from crime scene. Advantage of this solution is presence of thousands of tire (shoe) images on the Internet – in hundreds of web shops which are selling them. These images can be dowloaded automatically by dedicated web crawler. After preparing the database of samples of tire treads (shoe soles), it will be possible to start to develope system for automatic print recognition. This system could replace present manual work, which would lead to better efficiency of crime investigation.

When a tire (shoe) image is downloaded, it must be further processed. First task is to detect part with tire tread (shoe sole) and extract it. Without this step, following sample would be built with unwanted extra information (for example disc wheel in tire image). In this paper, we focus on the step of tire tread detection in one specific type of image – image with whole wheel visible from right side (Fig. 1). In this type of image a tire tread is determined by two half ellipses in the left part of the image. Ellipses detection is achieved using Hough transform technique. Since result of Hough transform may lead to thousands of ellipses a mechanism to choose right ones is needed. This is achieved using fuzzy inference system.

2 Image Preprocessing

In general, there are only a few different kinds of tire photography used in web shops. In this paper we focus on one of them – a photo, where a whole tire (wheel) is visible from the right side (Fig. 1).

Fig. 1 Examples of considered image type

Our aim is to localize part of the wheel with the best possible tire tread sample. In this kind of images the best information we get from the left middle part of the image (Fig. 2).

Fig. 2 Location of the best possible tire tread sample

As it can be seen in Fig. 3b position of desired sample is determined by two half ellipses – borders of the wheel. From experiments we found out that it can be problematic to detect inner (right) half ellipse (in the most cases the reason was improper lighting conditions). Therefore our solution first detects half ellipse describing right border of the wheel (Fig. 3c). Then the second half of this ellipse is used.

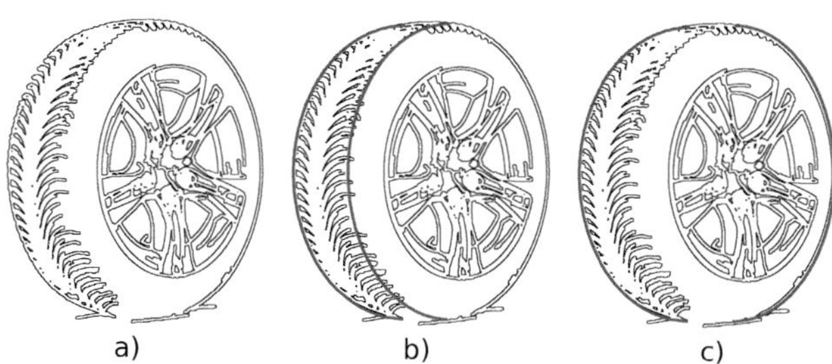

Fig. 3 The location of a tire tread sample is determined using ellipse detection. For better clarification the images are shown after edge detection step. Figure a) depicts the original image, b) shows desired result and c) shows used approach.

As it follows from previous paragraphs, the first task is to detect border of the wheel and describe it analyticaly – as a half ellipse. To detect ellipses in the image, we implemented Hough transform, a technique appropriate for finding analytical shapes in the image. The basic idea of this technique is assumption, that each pixel in the image belongs to a shape with specific parameters. For each shape there is a value in accumulator. This value is incremented for each shape, where the pixel can belong to (according to parameters of the shape). After processing of the whole image the highest accumulator value reveals the parameters of shapes (those defined by the most pixels).

Since Hough transform is ignoring color information (it depends only on foreground/background information), an original image is transformed to binary image using Canny edge detector (Fig. 4).

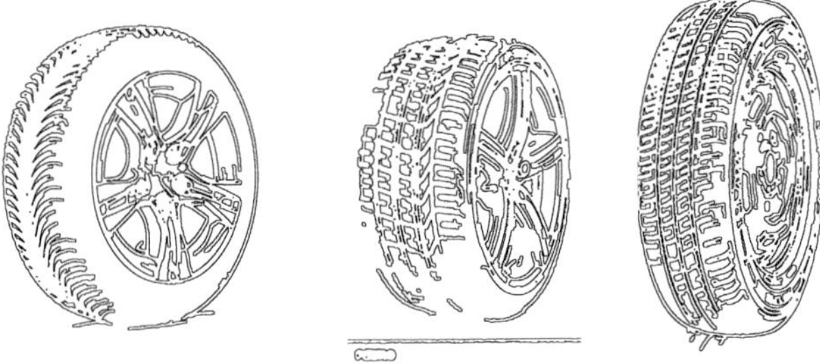

Fig. 4 Examples of binary version of the images as input for Hough transform

Result of Hough transform can contain thousands of ellipses, as can be seen on Fig. 5a. The ellipses of image are processed in sequence and the best of them are choosen.

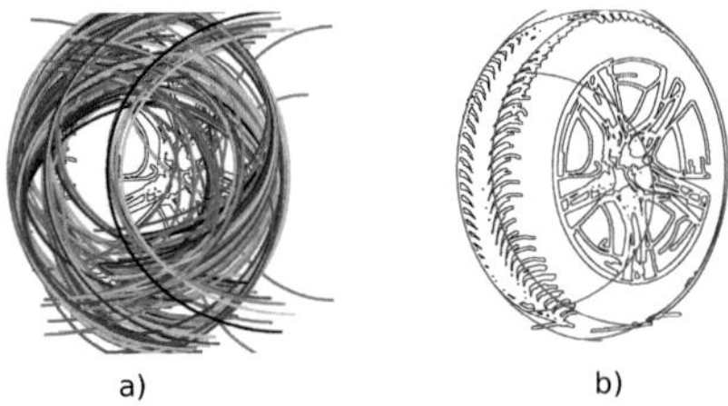

Fig. 5 Example of Hough transform result

3 Fuzzy Inference System

In the next text we will use following notation for the ellipses: S - center of ellipse, A, B - points of the major axis, C, D - points of the minor axis, AS – semi-major axis, CS - semi-minor axis.

Our program detects some ellipses which are described by their centers, lengths of semi-major and semi-minor axis and also by the Hough accumulator values. In general these ellipses have the minor axis parallel with the main axis. All these parameters are normalized into the interval [0, 1]. The next step is to learn the computer to recognize which of found ellipses are suitable for describing tire boundaries. Of course there is a small probability that two different pictures will have the same position of the tire and therefore we could not exactly define the mentioned parameters of ellipse. We could define which values of parameters are better and which are worse. Since the boundary between right and wrong values of parameters is not sharp we use the theory of fuzzy sets to define these parameters. We decide to use Takagi-Sugeno fuzzy inference system. It is working with IF-THEN rules. We choose fuzzy approach because it is suitable for modelling of vague values and its rules are defined in the forms which typically use people in real life. For example

IF Parameters value IS right THEN Ellipse IS suitable.

In the IF part of rule there are inputs/assumptions and in the THEN part there are the outputs/conclusions. Inputs and outputs are usually in the form of linguistic variables what mean that they are non-numeric and their values could be defined by fuzzy sets. We use four input parameters, linguistic variables which are defined by the following way

1. Linguistic variable **X value** represents the location of the points C, D (points of the semi-minor axis). From the Fig. 5b it is obvious that the ellipse is suitable if these points are near the border of the left side or near the border of the right side of the picture. We use three fuzzy sets to model this linguistic variable. We called them low, middle and high (Fig. 6).
2. Linguistic variable **Y value** represents the Y value of the center of the ellipse S. From the Fig. 5b it is obvious that the ellipse is suitable if Y value of the center of ellipse is in the middle of the image therefore we again use three fuzzy sets to model this linguistic variable. We called them low, middle and high (Fig. 7).
3. Linguistic variable **Accumulator value** represents the value in the Hough accumulator for the corresponding ellipse. We use two fuzzy sets to model this linguistic variable low and high (Fig. 8).
4. Linguistic variable **Semi major axis** represents the length of the semi-major axis. The ellipse is preferable if this value is greater. Therefore it is enough to model this variable by two fuzzy sets short and long (Fig. 9).

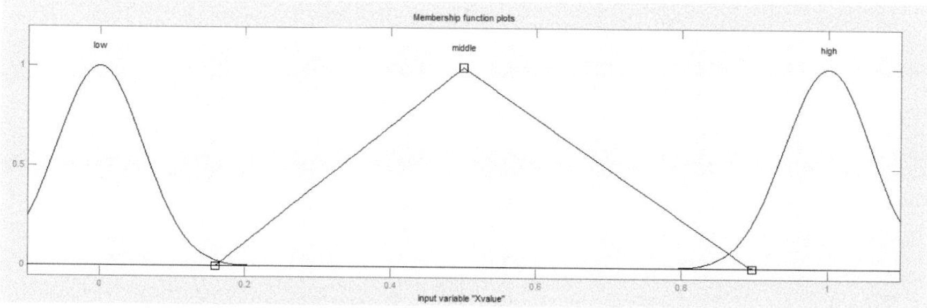

Fig. 6 Fuzzy sets to model linguistic variable X value

The output linguistic variable in our experiments is the suitability of the desired ellipse. In Takagi-Sugeno fuzzy inference system the output could be defined by

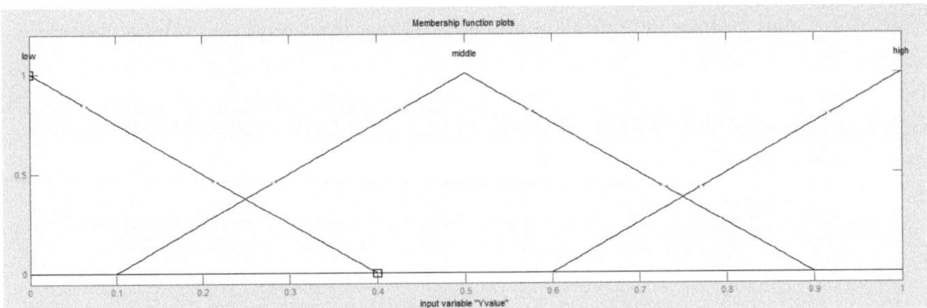

Fig. 7 Fuzzy sets to model linguistic variable Y value

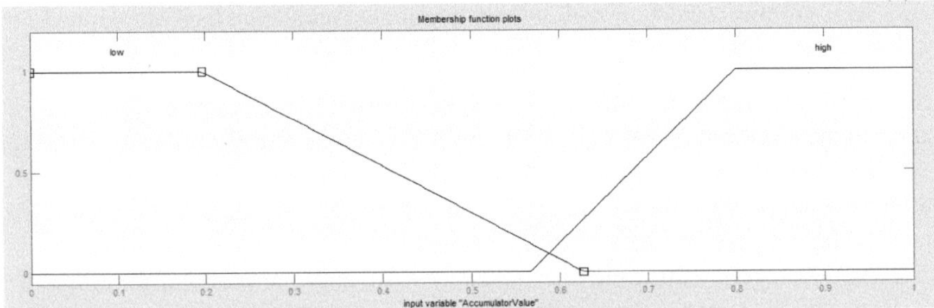

Fig. 8 Fuzzy sets to model linguistic variable Accumulator value

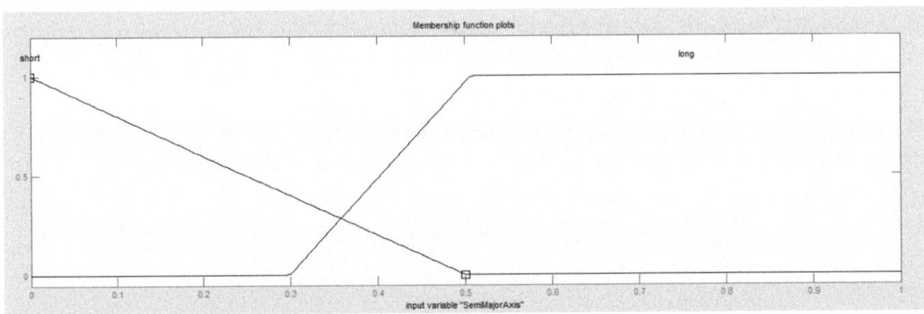

Fig. 9 Fuzzy sets to model linguistic variable Semi major axis

the constant or by the polynomial function. For our experiments we decide to use constant output from the interval [0, 1]. Then the linguistic variable **Suitability of the desired ellipse** we describe by the four values: 0 which means unsuitable; 0.5 hard to say; 0.75 rather suitable and 1 suitable.

Now we are ready to construct the rule base. It contains 36 rules in the following form

IF Xvalue IS low AND Yvalue IS middle AND Accumulator value IS high AND Semi major axis IS long THEN Ellipse IS suitable.

All these parameters are entered into the MATLAB with Fuzzy Logic Toolbox. After this we are able for any 4-tuple of real numbers from unit interval, which represent the parameters of ellipse, assign the value which is also from unit interval and which determine the suitability of select ellipse. The output number closer to one represents more suitable ellipse and number closer to zero represents unsuitable ellipse.

4 Experiments and Results

To evaluate our approach we applied it to a small database of images of tires (over 300 images). Each image was of the kind depicted in Fig. 1. The images were in jpeg format with average resolution 322x460 pixels.

In the first step we converted each image to grayscale and applied Canny edge detector on it. In the next step we used Hough transform technique to detect all possible half ellipses within specific range. In average, there were over 5 million of half ellipses in one image. Detected half ellipses were forwarded into fuzzy inference system, where suitability for each one ellipse was calculated. Finally, we took ten the most suitable half ellipses and the leftmost and the rightmost ones were drawn in the original image.

Evaluation was performed visually – if the drawn half ellipses corresponded with left and right border of the wheel, the result was considered as correct one. From 315 input images 282 were evaluated as correct, what makes 89% success.

The most common problems were with images where border of the wheel doesn't have a shape of plain half ellipse (Fig. 10a). This was related mostly to tires with deep tread. Another common cause of incorrect results was insufficient resolution of input images (images below resolution about 120x180 were in the most cases processed incorrectly). In few cases results were incorrect because of lighting settings when border between background and tire was not clearly defined (Fig. 10b).

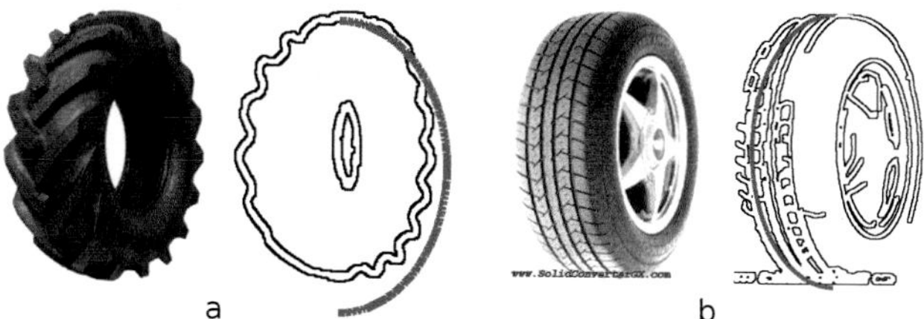

Fig. 10 Two examples of incorrect results

5 Conclusions

In the paper we presented a method which allows to determine a location of tire tread in the image. We created an experiment with real data from the Internet and we confirmed applicability of the method – in the vast majority of cases we have obtained correct results. It prepares ground for an automatic extraction of tire tread.

In the future, we plan to finalize automatic extraction of tire tread for image type described in this paper. Since there are only few different types of tire images used in web shops, we will handle each one of them separately. Then with automatic classification of images of tires we plan to build an automatic system, which will be able to extract tire tread samples from web pages.

Acknowledgments This work was supported by the Slovak Research and Development Agency under the contract No. APVV-0219-12.

References

1. Geradts, Z., Keijzer, J.: The image-database REBEZO for shoeprints with developments on automatic classification of shoe outsole designs. Forensic Science International **82**(1), 21–31 (1996)
2. Thali, M.J., Braun, M., Brschweiler, W., Dirnhofer, R.: Matching Tire Tracks on the Head Using Forensic Photogrammetry. Forensic Science International **113**(1–3), 281–287 (2000)
3. Chazal, P., Flynn, J., Reilly, R.B.: Automated processing of shoeprint images based on the Fourier transform for use in forensic science. IEEE Transactions on Pattern Analysis and Machine Intelligence **27**(3), 341–350 (2005). IEEE Computer Society Washington, DC, USA
4. Zhang, L., Allinson, N.: Automatic shoeprint retrieval system for use in forensic investigations. In: 5th Annual UK Workshop on Computational Intelligence (2005)
5. Pavlou, M., Allinson, N.: Automated Encoding of Footwear Patterns for Fast Indexing. Image and Vision Computing **27**(4), 402–409 (2009). Butterworth-Heinemann, Newton, MA, USA
6. Nibouche, O., Bouridane, A., Crookes, D., Gueham, M., Laadjel, M.: Rotation invariant matching of partial shoeprints. In: Proceedings of the 13th Machine Vision and Image Processing Conference, pp. 94–98. IEEE Computer Society, Washington, DC, USA (2009)
7. Jing, M.Q., Ho, W.J., Chen, L.H.: A novel method for shoeprints recognition and classification. In: International Conference on Machine Learning and Cybernetics, vol. 5, pp. 2846–2851. IEEE (2009)
8. Dardi, F., Cervelli, F., Carrato, S.: A texture based shoe retrieval system for shoe marks of real crime scenes. In: Foggia, P., Sansone, C., Vento, M. (eds.) ICIAP 2009. LNCS, vol. 5716, pp. 384–393. Springer, Heidelberg (2009)
9. Wang, R., Hong, W., Yang, N.: The research on footprint recognition method based on wavelet and fuzzy neural network. In: Ninth International Conference on Hybrid Intelligent Systems, vol. 3, pp. 428–432. IEEE (2009)
10. Patil, P.M., Kulkarni, J.V.: Rotation and Intensity Invariant Shoeprint Matching Using Gabor Transform with Application to Forensic Science. Pattern Recognition **42**(7), 1308–1317 (2009). Elsevier Science Inc., New York, NY, USA
11. Huang, D.-Y., Hu, W.-C., Wang, Y.-W., Chen, C.-I., Cheng, C.-H.: Recognition of tire tread patterns based on gabor wavelets and support vector machine. In: Pan, J.-S., Chen, S.-M., Nguyen, N.T. (eds.) ICCCI 2010, Part III. LNCS, vol. 6423, pp. 92–101. Springer, Heidelberg (2010)
12. Tang, Y., Kasiviswanathan, H., Srihari, S.N.: An efficient clustering-based retrieval framework for real crime scene footwear marks. In: Int. J. Granular Computing Rough Sets and Intelligent Systems, vol. 2, no. 4, pp. 327–360 (2012)

Efficient Bag of Words Based Concept Extraction for Visual Object Retrieval

Hilal Ergun and Mustafa Sert

Abstract Recent burst of multimedia content available on Internet is pushing expectations on multimedia retrieval systems to even higher grounds. Multimedia retrieval systems should offer better performance both in terms of speed and memory consumption while maintaining good accuracy compared to state-of-the-art implementations. In this paper, we discuss alternative implementations of visual object retrieval systems based on popular bag of words model and show optimal selection of processing steps. We demonstrate our offering using both keyword and example-based retrieval queries on three frequently used benchmark databases, namely Oxford, Paris and Pascal VOC 2007. Additionally, we investigate effect of different distance comparison metrics on retrieval accuracy. Results show that, relatively simple but efficient vector quantization can compete with more sophisticated feature encoding schemes together with the adapted inverted index structure.

Keywords Bag of words · Visual Object Retrieval · Distance metrics · SIFT · SVM

1 Introduction

Searching multimedia content and retrieving useful information for the user is becoming a trending research area due to availability of vast amounts of video and image data. Thousands, if not millions, of video content is created every day and uploaded to on-line or cloud communities. In addition to on-line content, much more is kept in local databases. All this data is waiting to be indexed for search and retrieval applications.

Among different indexing approaches, bag-of-visual-words model, which is well known by the information retrieval (IR) community, is the one being used most

H. Ergun · M. Sert(✉)
Department of Computer Engineering, Baskent University, Ankara, Turkey
e-mail: 21020005@mail.baskent.edu.tr, msert@baskent.edu.tr

T. Andreasen et al. (eds.), *Flexible Query Answering Systems 2015*,
Advances in Intelligent Systems and Computing 400,
DOI: 10.1007/978-3-319-26154-6_30

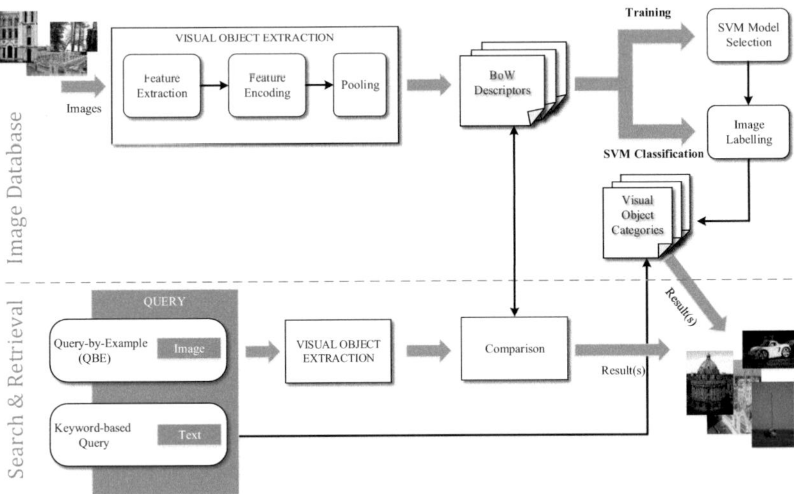

Fig. 1 Overall view of the proposed search and retrieval system

frequently and the one offering most successful results both in terms of precision and query running time performance in visual object retrieval applications. Bag of words or bag of visual words, will be referred as BoW from now on, finds its roots in the document retrieval domain and applied to image domain by Sivic et.al [20]. In very simple terms, BoW approach counts occurrence of local image features and tries to represent higher level image categories using this information. In their very simple form, bag-of-features methods discard all spatial information present in the image and retain only the visual words' visibility frequencies [27]. Lazebnik et al. introduce a novel method of spatial pyramids showing how spatial information can be integrated in BoW pipeline to further improve classification accuracy [11]. Philbin et al. explored image retrieval from a large dataset and showed how complimentary spatial re-ranking can be used to improve retrieval accuracy. [6] increased performance of object retrieval with the introduction of automatic query expansion. Perronnin and Dance applied Fisher kernel encoding to area of image category detection [16]. Jegou et al. introduced Hamming embedding for representing images with binary encodings and for efficient image retrieval against a user query from 1 million images [8]. In order to improve quantization step of BoW, Philbin et al. introduce a method that uses soft assignment of image features to visual codewords [19]. [26] showed how sparse coding can be used in-place of vector quantization to further improve classification accuracy. [24] improved sparse coding with feature-space locality constraints. In [1], the importance of query expansion is proved and also the most important steps which should be taken into consideration for improving object retrieval systems are outlined. Yan et. al. improved tf-idf scheme by learning a similarity matrix from labeled data [25].

In this paper, we are focusing on efficient retrieval of image queries from a mid-to-large scale database with a trade-off between speed and accuracy. We propose a processing pipeline which can be used to issue two different query types to a local database. Block diagram of our proposed pipeline architecture is depicted in Figure 1. One other contribution of this paper is that we evaluated different distance comparison metrics used in literature on 2 different benchmark datasets for the BoW based visual object retrieval systems and we show that improper choice of distance metric and normalization selection can degrade retrieval performance.

The rest of the paper is organized as follows. Section 2 describes our search and retrieval framework. In Section 3, we introduce our keyword- and example-based retrieval architecture by adapting inverted index structure. Comprehensive experimental results and evaluations on three datasets are given in Section 4. Finally we conclude the paper in Section 5.

2 Proposed Search And Retrieval System

In this study, we target two types of querying methodologies against a video/image database, namely keyword based querying and querying by example(QBE). We extract visual objects from all images and create a global representation for the image based on these extracted objects. In the context of QBE, given image representation is directly compared to ones present in the database. This permits user to execute queries like "Find all images which are similar to this one". For keyword querying, we utilize machine learning techniques to learn a single textual representation (also referred to as visual concept) of previously created global image representations. This allows a user to retrieve images from database with a query like "Find all images which is mainly related to *X*" where *X* may be any visual object, event description, or a general concept.

2.1 Visual Object Extraction

As depicted in Figure 1, our visual object extraction scheme consists of three stages. First, we calculate image features for target images. Different local image features can be used at this step, SIFT [13] being one of the most popular choices. Furthermore, more than one type of image features can be extracted. [9] showed that multiple features can be combined to further increase effectiveness of BoW. Color information present in images can be included as well [21].

After image feature extraction comes the encoding step. In this step, we quantize image features into different BoW dictionary bins. Hard or soft assignment may be employed in this step. Vector quantization is the mostly employed hard assignment technique. It can be expanded to include soft-assignment though[18]. Fisher kernel encoding and sparse coding can be used to further increase softness of quantization.

However, in the context of large-scale image retrieval, vector quantization is the mostly adopted technique due to superior run-time performance when compared to more complicated encoding schemes. [5] provides detailed experimental analysis of various encoding schemes as well as their run-time complexity. Next comes the pooling stage. In this step, quantized image features are pooled to create global image representation which constitutes BoW descriptor for the given image. Different encoding schemes may perform better with different pooling techniques. Summation, or average, pooling mostly used in vector quantization type of encodings. Sparse coding performs better with maximum pooling operators. Applying spatial pyramids yields finer-grained pooling regions which boosts classification accuracy [11]. Here we use average pooling technique since we chose vector quantization without any soft-assignment.

2.2 Classifier Design

Different machine learning approaches may be used in this step, support vector machines (SVMs) being most frequently used and successful classifiers in the literature and therefore selected in our study. We make use of the classifier to enable keyword based queries; example based queries (QBE), on the other hand need a slightly modified approach. For QBE, when a query is desired to be executed, BoW descriptor of query image is compared to the ones in database. Instead of SVM based classification, more simple yet powerful distance metrics, such as Euclidean or Manhattan distances, are used to compare images. Then best matches are returned to the user.

SVMs are kernel based classifiers, different non-linear kernels may be employed for classification of different representations. Among non-linear kernels we found χ^2 and histogram intersection kernels to be most useful[5] [11]. One other choice frequently used in the literature is the Earth's mover distance kernel, namely EMD kernel. However, previous work showed that performance of EMD is comparable with χ^2 [27] so we don't use EMD at all. BoW descriptors are nothing but histograms of visual words in a given dictionary; this explains the success of χ^2 and other histogram comparison kernels on BoW data classification. Major drawback of non-linear kernels are their big performance hit. Non-linear SVMs are known to have higher complexity than linear SVMs [26] and they are not preferred for large image databases. On the other hand, when using vector quantization, linear kernels deliver substantially worse results [11] [2] Yang et al. states this is due to high quantization error in encoding step. One alternative is to use of an efficient suitable feature mapping for the data and using linear SVMs in place of non-linear ones. [23] provides a mathematically complete alternative for three of the mostly used histogram kernels and we used their implementation in our work. We investigated different SVM kernels relevant to image retrieval in a previous study and we found χ^2 as the most successful one [19].

2.3 *Distance Metrics*

Distance metrics are used in two different steps of image classification. During BoW image descriptor creation, local image features are compared to dictionary words using a suitable distance metric. While comparing different image BoW descriptors, again a suitable distance comparison is employed. Metric selection for the former is tightly coupled to feature extraction steps used. For SIFT, L2 distance comparison is suggested by the original author[13]. On the other hand, different distance metrics can be employed for the comparison of the latter. We investigate and evaluate the distance metrics, namely L1, L2, Histogram intersection, Hellinger distance, χ^2 and cosine distance, in our study for comparing BoW descriptors.

2.4 *Visual Dictionary Creation*

Vector quantization uses a pre-computed visual words called visual dictionary or codewords. Visual dictionaries can be created using different techniques, however, many papers use simple but effective K-means clustering algorithm and its variants. Target dataset or one another dataset may be used for visual dictionary creation. In this paper, we evaluated visual dictionary creation on the same dataset only.

For creating visual dictionary, all images are processed for local feature extraction and extracted features are clustered into K clusters using K-means. In an average dataset, one might extract 10 millions of image features and clustering of this amount of features may be not tractable. Both processing power and memory resources may be scarce at this step. Hierarchal clustering or approximate k-means clustering algorithms may be employed then. One another technique used is to randomly sub-sample available features before performing clustering.

3 Querying Schemes

While there can be different types of querying, image/video retrieval can be classified into two alternatives: text-based approaches and content-based approaches [12]. Text-based approaches associate each video shot or image frame with single or multiple keywords which permits user to query image or video database with a selection of keywords. Content-based approaches, mostly abbreviated as CBIR, allows user to search media database by supplying an example item.

Text-based approaches require the underlying data to be classified and/or every object contained within is detected prior to keyword query. Mostly attributed to semantic gap phenomenon, classification of images or scenes by computer programs into meaningful categories which can be resolved by humans natively is a non-trivial problem [27]. We describe our methods in the following subsections.

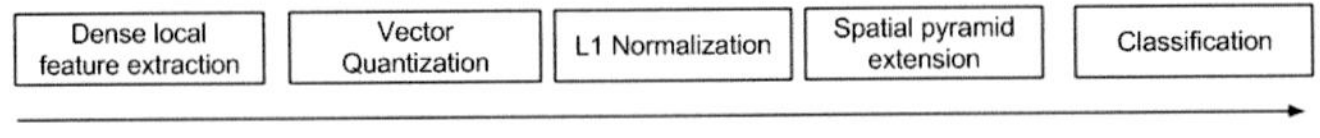

Fig. 2 Classification pipeline overview

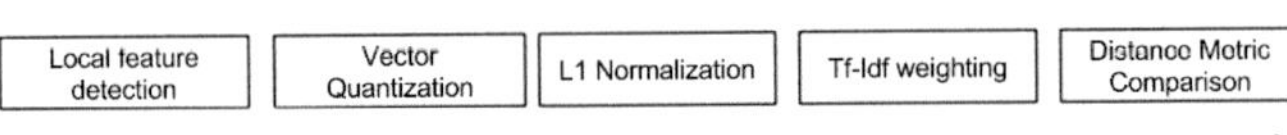

Fig. 3 Retrieval pipeline overview

3.1 Keyword Based Querying

Our image classification pipeline steps employed in keyword based queries are depicted in Figure 2. As a first step, we extract local SIFT features for every image. Local features can be extracted from a given image in one of two ways. One can use an interest point detector, or image can be densely sampled. We use both methods for different type of queries. Many papers show that dense feature extraction outperforms interest-point based extraction on visual classification tasks [11], [10]. For this reason we use dense sampling strategy for keyword type queries. We only extract one type of image feature, namely SIFT features.

After local feature extraction, local features are vector quantized to create image feature histograms. During vector quantization, a previously generated visual word dictionary, codebook, is used for comparison as is described in section 2.4. After vector quantization, created histogram is L1 normalized so that effect of unequal feature cardinalities in different images are neutralized. Then spatial pyramid extension is applied so that spatial layout of images are taken into consideration. At this step, we have the desired BoW descriptor for our query image and we use SVM classifiers to classify image category. Classifier design is detailed in section 2.2.

3.2 Query By Example

Example based retrieval is performed with a slightly modified version of previously described keyword based algorithm. Figure 3 shows a flowchart of this new algorithm. In contrast to classification pipeline, interest point based feature detection works better for exemplar based queries and that's what we have used in our evaluation while performing example based queries. After features are detected and their descriptors are extracted, we apply vector quantization as in classification pipeline. L1 normalization is used once again to get rid of different feature cardinalities. Spatial pyramid extension is skipped in this type of queries due to use of interest point detector. We rely on feature detection methodologies here so that relevant spatial information is represented by detected features. Next step is to insert inverse document frequencies (idf) into created feature histograms so that frequently used visual words are suppressed. After this step, BoW descriptors are ready to be compared.

We evaluate distance metrics described in section 2.3 and provide retrieval accuracies on two different datasets using the metrics.

3.3 Inverted Index

In order to reduce time complexity of underlying retrieval operations, we build inverted index in a way to benefit sparse structure of image descriptors. It keeps a look-up table of all images in database which contains specific codeword. i.e. dictionary entry. Figure 4 shows a graphical interpretation of our inverted index structure.

Let we have a dictionary of size K and a database containing total of N images. Furthermore, to describe sparsity of our descriptors let's also assume average number of SIFT features per image is M. Descriptor of one image consists of K numbers, one for each dictionary word. It can be represented with a K-sized vector as follows:

$$D(w) = (w_1, w_2, w_3, w_4, w_5 \ldots w_K) \tag{1}$$

In the worst case, at most M elements of D is non-zero. In case more than one SIFT feature of image is quantized into the same dictionary word, which is the case in practice, number of non-zero elements will be much smaller which further increases sparsity. If we were going to keep image BoW descriptors as is in our database, both our storage size and image query execution time will be linear both in dictionary size and number of images in our database. Taking into account that K is in range of millions and number of images are tens of thousands, storage requirements tents to increase very quickly, as well retrieval times. For Oxford database, N is 9000 and K is 1 million; so for each query image this results in K * N = 9 billions of basic mathematical operations. In case of more complex distance metrics containing square rooting or natural logarithm retrieval run times will increase dramatically.

On the other hand, inverted index only keeps non-zero elements of a BoW descriptor in its database, as depicted in Figure 4. Since $M \ll K$, this reduces number of basic mathematical operations needed for calculation of query distances to M opera-

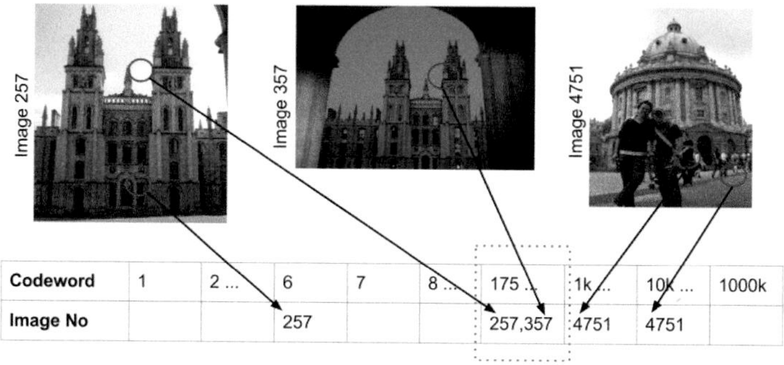

Fig. 4 Structure of the inverted index

Table 1 Summarization of Concepts for Oxford Dataset

Concept	No of Perfect Positives	No of Partial Positives	No of Total Images
All Souls	120	270	390
Ashmolean	60	65	125
Balliol	25	35	60
Bodleian	65	55	120
Christ Church	255	135	390
Cornmarket	25	20	45
Hertford	175	95	270
Keble	30	5	35
Magdalen	65	205	270
Pitt Rivers	15	15	30
Radcliffe	525	580	1105

tions per image. For Oxford database, N is 9000 and M is 3000, which results M*N = 30 million operations. There is a dramatic decrease from the case of no inverting index is used.

One difficulty of using an inverted index is that it is non-trivial to compute distance metrics since all elements of database vectors are not present in the query time. Fortunately, most of the distance metrics, if not the all, can be decomposed into at least two summation terms, one depending on query vector and the other depending on both query and database vectors. Then one can apply distance metrics for each of the query vector and only the non-zero elements of database vector. This results in almost constant time distance comparison for even heavier distance comparison metrics.

4 Experiments and Analysis

In this section, we present our experimental results for QBE and keyword based queries. We used three mostly adopted datasets by researchers in our comparisons. We evaluated our QBE queries on Oxford[17] and Paris[18] datasets, while using Pascal VOC 2007 challenge dataset[7] for keyword based queries.

4.1 Dataset Information

Oxford Buildings Dataset: Oxford dataset is composed of 5062 images containing several pictures of buildings in Oxford, along with false positives. Dataset contains 55 different queries for 11 different buildings. Table 1 summarizes the number of positive examples from each building (concept) type present in the database. In addition to those, there are also 2222 unrelated pictures containing none of the buildings.

Table 2 Summarization of Concepts for Paris Dataset

Concept	Positives	Concept	Positives	Concept	Positives	Concept	Positives
Defense	585	Musee Dorsay	360	Eiffel	1445	Notre Dame	595
Invalides	990	Pantheon	630	Louvre	760	Pompidou	255
Moulin Rouge	1158	SacreCoeur	745	Triomphe	1405		

Table 3 Summarization of Concepts for Pascal Dataset

Concept	Positives	Concept	Positives	Concept	Positives		
aeroplane	238	bus	186	dining table	200	potted plant	245
bicycle	243	car	713	dog	421	sheep	96
bird	330	cat	337	horse	287	sofa	229
bottle	244	chair	445	motorbike	245	train	261
boat	181	cow	141	person	2008	tvmonitor	256

Paris Buildings Dataset: Paris dataset is very similar to Oxford, it contains 6412 images of Paris instead of Oxford. It includes 55 different queries on 12 different Paris buildings. Concepts present in the dataset is shown in Table 2.

Pascal VOC 2007 Challenge Dataset: This dataset is used for keyword-type queries. Pascal dataset contains more than 9000 images for 20 different categories. It is one of the mostly adapted benchmark databases used by image classification community. It is well suited to retrieval tasks because it evaluates query results using retrieval metrics. Dataset contains images of 20 different objects or concepts. For each concept total of 5011 images are selected as either positive or negatives. For each concept, varying number of positives and negatives are provided. Table 3 summarizes numbers for each concept.

During implementation of our algorithms we used various publicly available software packages. OpenCV [3] is used for image processing, feature detection and K-means clustering. [15] is used for Hessian-Affine feature detection and SIFT descriptor creation. VLFeat framework available at [22] is used for homogeneous kernel mapping for SVM classification. LIBSVM [4] library is used for SVM classification. Publicly available software package of [5] is utilized for Pascal VOC 2007 experiments.

4.2 *Experimental Setup*

We evaluate CBIR on Oxford and Paris datasets. We run experiments with aforementioned distance comparison metrics on both datasets. Since different metrics yields different results with different descriptor normalizations, we evaluated all metrics on two types of normalization, L1 and L2. We evaluated our query results using query evaluation software packages by each dataset. Results are presented in mean average precisions. We picked this scheme so that our results are comparable with other studies. In all of our evaluations, we created our own image features, visual dictionaries and BoW descriptors and didn't use any samples supplied by the datasets. To be comparable with other studies we evaluated 1 million words dictionaries for Oxford and Paris datasets. Local image features are detected using Hessian affine local feature detector, on top of that we extracted SIFT feature descriptors. Euclidean, L2, distance is used to compare SIFT features. We didn't apply any query expansion or spatial re-ranking techniques so that raw performance of different distance metrics can be visualized.

In contrast to QBE queries we used 5000 words for Pascal VOC 2007 challenge dataset as increasing dictionary size beyond certain limit does not increase classification precision while increasing computation time. Results are presented using the mAP score like the QBE case. It should be noted that Pascal 2007 dataset was using a slightly different average precision calculation method than Oxford and Paris datasets. As previously noted, we used dense sampling on Pascal dataset instead of using feature detectors. SIFT features are extracted at each 2 pixels. We applied 2 levels of spatial pyramids while creating final image descriptions. χ^2 homogeneous kernel mapping was applied to each image descriptor before using SVM classifiers. This greatly increased classification accuracy compared to linear SVM classification while keeping training and testing times comparable with the linear case. We apply χ^2 expansion with a gamma parameter 1.0 as lower values between [0.1, 1.0] did not yield better results. We cross-validated in training set for best SVM cost parameter and running a grid-search algorithm for cost parameter gave us a cost parameter of 20.

4.3 QBE Retrieval Results

All QBE results are summarized in Table 4. Among all metrics, Hellinger distance performed the best on all configurations. This is consistent with the results presented at [1]. Although Hellinger distance is heavier to compute compared to most of the other distances it was shown that it can be computed very efficiently with an additional normalization step to SIFT descriptor calculation[1]. This normalization step doesn't even need a modification in SIFT extraction routines, it is possible to apply normalization during quantization of image features with dictionary codewords.

Histogram intersection based distance comparison is consistently second after Hellinger distance. This is not a surprising result, it has been shown to be superior for object classification tasks by Lazebnik et al [11]. It should be noted that histogram intersection performs better with L1 on some datasets while it is performing better with L2 normalization on some other datasets. Still, with both normalization techniques are better than other distance metrics excluding Hellinger distance.

L1 and L2 distances should be used with properly normalized descriptors. Cosine distance is normalization agnostic and yields comparable results for both type of normalizations.

Table 4 Comparison of different distance metrics on Oxford and Paris datasets

Metric	Oxford5k	Paris6k	Oxford5k(No Idf)	Oxford5k	Paris6k	Oxford5k(No Idf)
	L1 Normalized			L2 Normalized		
L1	0.6176762	0.62068862	0.60776925	0.038044896	0.033984952	0.038047113
L2	0.075195491	0.043052912	0.075260796	0.61359107	0.60580742	0.59819621
Min	0.61762261	0.6501981	0.60809761	**0.62426698**	**0.64156407**	**0.61962712**
Cos	0.61361635	0.6333003	0.59808475	0.61361593	0.63330024	0.59807926
Hellinger	**0.6378966**	**0.65200478**	**0.6314373**	0.60204697	0.60970277	0.59328997
χ^2	0.59236705	0.62142205	0.58517522	0.55918133	0.5897671	0.55282593

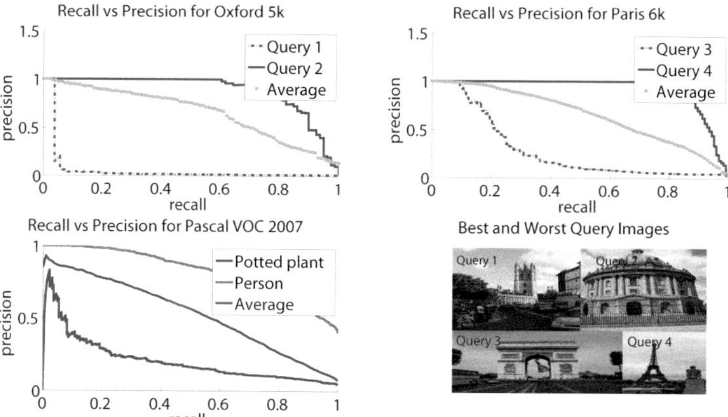

Fig. 5 Precision and recall (PR) curves. The best, average, and worst queries/concepts are considered for each dataset using the proposed scheme: (a) Oxford, (b) Paris, (c) Pascal VoC 2007.

Table 5 Keyword based retrieval results on Pascal VOC 2007 dataset

mAP	aeroplane	bicycle	bird	boat	bottle	bus	car	cat	chair	cow
0.5336	0.6918	0.5594	0.3822	0.6296	0.2409	0.5955	0.7333	0.5548	0.4886	0.3893
dining table	dog	horse	motorbike	person	potted plant	sheep	sofa	train	tvmonitor	
0.4980	0.3627	**0.7507**	0.6346	**0.8109**	0.2372	0.4392	0.4516	0.7381	0.4842	

We also evaluated different metrics on Oxford dataset without IDF weighting scheme applied. Applying IDF weighting adds approximately 2 percent to average precision.

In addition to mAP scores in Table 4, we depicted precision-recall curves for three of the datasets we used in Figure 5. For each dataset, we included best and worst queries/concepts as well average results of each queries. By looking at the PR curves; we can conclude that while some queries achieving more than 95% AP score, some suffer from very low AP scores. Most of these under achievers belong to concepts which relatively have low number of positive samples in our database.

4.4 Keyword Based Retrieval Results

We summarized our results for Pascal VOC 2007 dataset in Table 5. Our mAP score of 53% is a compatible result to Pascal challenge best performers given at [14]. Best performers achieved 59% while average success rate is below our 53% rate. Further taking multiple feature types used by competitors into account this result is a good trade-off between classification run-time performance and accuracy. We should emphasize that it is possible to obtain up to 5% higher accuracies using different encoding schemes than vector quantization. However, using sophisticated encoding schemes greatly increases running time of algorithms. For instance, it is possible to encode one image descriptor under 1 second using vector quantization wheras sparse coding needs 30 seconds and Fisher encodings requires 12 seconds for the same image on a decent CPU[5]. We believe sacrificing query run-time performance for a relatively small increase in accuracy is not optimal for real-world applications.

5 Conclusions

BoW based image classification and object extraction techniques are known to be very successful and efficient on multimedia retrieval tasks. In this paper we represent various components of a multimedia retrieval system which can execute different types of queries on a relatively big image database by adapting inverted index structure to BoW representation. We outlined principal differences between keyword and example based queries with respect to processing pipeline implementation. We showed that with proper choice of implementation parameters, relatively simple but efficient vector quantization can compete with more sophisticated encoding schemes, such as sparse coding or Fisher kernel encoding, in retrieval accuracy maintaining a low processing overhead for database system. Another contribution of our paper was comparison of different distance metric performances while issuing QBE queries. We believe fusion of different distance metrics is a research area which may worth spending some extra time on. We believe integrating not so used correlation information between visual words into distance metric fusion frameworks may further increase of vector quantization retrieval accuracy with a little or at no cost on multimedia database side.

Acknowledgement This work is supported in part by a research grant from The Scientific and Technological Research Council of Turkey (TUBITAK EEEAG) with grant number 109E014. The authors also would like to thank Caglar Akyuz for his valuable supports.

References

1. Arandjelovic, R., Zisserman, A.: Three things everyone should know to improve object retrieval. In: 2012 IEEE Conference on Computer Vision and Pattern Recognition (CVPR), pp. 2911–2918, June 2012
2. Boureau, Y.-L., Bach, F., LeCun, Y., Ponce, J.: Learning mid-level features for recognition. In: 2010 IEEE Conference on Computer Vision and Pattern Recognition (CVPR), pp. 2559–2566, June 2010
3. Bradski, G.: Dr. Dobb's Journal of Software Tools (2000)
4. Chang, C.-C., Lin, C.-J.: LIBSVM: A library for support vector machines. ACM Transactions on Intelligent Systems and Technology **2**, 27:1–27:27 (2011)
5. Chatfield, K., Lempitsky, V.S., Vedaldi, A., Zisserman, A.: The devil is in the details: an evaluation of recent feature encoding methods. In: BMVC, vol. 2, p. 8 (2011)
6. Chum, O., Philbin, J., Sivic, J., Isard, M., Zisserman, A.: Total recall: automatic query expansion with a generative feature model for object retrieval. In: IEEE 11th Int'l Conf. on Computer Vision, ICCV 2007, pp. 1–8. IEEE (2007)
7. Everingham, M. Van Gool, L., Williams, C.K.I., Winn, J., Zisserman, A.: The PASCAL Visual Object Classes Challenge 2007 (VOC2007) Results
8. Jegou, H., Douze, M., Schmid, C.: Hamming embedding and weak geometric consistency for large scale image search. In: Forsyth, D., Torr, P., Zisserman, A. (eds.) ECCV 2008, Part I. LNCS, vol. 5302, pp. 304–317. Springer, Heidelberg (2008)
9. Jiang, Y.-G., Ngo, C.-W., Yang, J.: Towards optimal bag-of-features for object categorization and semantic video retrieval. In: Proceedings of the 6th ACM International Conference on Image and Video Retrieval, pp. 494–501. ACM (2007)
10. Jurie, F., Triggs, B.: Creating efficient codebooks for visual recognition. In: IEEE Int'l Conf. on Computer Vision (ICCV 2005), vol. 1, pp. 604–610, October 2005
11. Lazebnik, S., Schmid, C., Ponce, J.: Beyond bags of features: spatial pyramid matching for recognizing natural scene categories. In: 2006 IEEE Computer Society Conference on Computer Vision and Pattern Recognition, vol. 2, pp. 2169–2178 (2006)
12. Liu, J.: Image retrieval based on bag-of-words model (2013). CoRR, abs/1304.5168
13. Lowe, D.G.: Object recognition from local scale-invariant features. In: The proc. of the 7th IEEE Int'l Conf. on Computer Vision, 1999, vol. 2, pp. 1150–1157. IEEE (1999)
14. Marszałek, M., Schmid, C., Harzallah, H., Van De Weijer, J.: Learning object representations for visual object class recognition. In: Visual Recognition Challange Workshop, in Conjunction with ICCV (2007)
15. Perd'och, M., Chum, O., Matas, J.: Efficient representation of local geometry for large scale object retrieval. In: IEEE Conference on Computer Vision and Pattern Recognition, CVPR 2009, pp. 9–16. IEEE (2009)
16. Perronnin, F., Dance, C.: Fisher kernels on visual vocabularies for image categorization. In: IEEE Conference on Computer Vision and Pattern Recognition, CVPR 2007, pp. 1–8. IEEE (2007)
17. Philbin, J., Chum, O., Isard, M., Sivic, J., Zisserman, A.: Object retrieval with large vocabularies and fast spatial matching. In: IEEE Conference on Computer Vision and Pattern Recognition, CVPR 2007, pp. 1–8. IEEE (2007)
18. Philbin, J., Chum, O., Isard, M., Sivic, J., Zisserman, A.: Lost in quantization: improving particular object retrieval in large scale image databases. In: IEEE Conf. on Computer Vision and Pattern Recognition (CVPR) 2008, pp. 1–8. IEEE (2008)
19. Sert, M., Ergun, H.: Video scene classification using spatial pyramid based features. In: 2014 22nd Signal Processing and Communications Applications Conference (SIU), pp. 1946–1949, April 2014

20. Sivic, J., Zisserman, A.: Video google: a text retrieval approach to object matching in videos. In: Proceedings of the Ninth IEEE International Conference on Computer Vision, 2003, vol. 2, pp. 1470–1477, October 2003
21. Van De Sande, K.E., Gevers, T., Snoek, C.G.: Evaluating color descriptors for object and scene recognition. IEEE Transactions on Pattern Analysis and Machine Intelligence **32**(9), 1582–1596 (2010)
22. Vedaldi, A., Fulkerson, B.: VLFeat: An open and portable library of computer vision algorithms (2008)
23. Vedaldi, A., Zisserman, A.: Efficient additive kernels via explicit feature maps. IEEE Transactions on Pattern Analysis and Machine Intelligence **34**(3), 480–492 (2012)
24. Wang, J., Yang, J., Yu, K., Lv, F., Huang, T., Gong, Y.: Locality-constrained linear coding for image classification. In: 2010 IEEE Conference on Computer Vision and Pattern Recognition (CVPR), pp. 3360–3367, June 2010
25. Yan, Z., Yu, Y.: Sparse similarity matrix learning for visual object retrieval. In: The 2013 Int'l Joint Conf. on Neural Networks (IJCNN), pp. 1–8. IEEE (2013)
26. Yang, J., Yu, K., Gong, Y., Huang, T.: Linear spatial pyramid matching using sparse coding for image classification. In: IEEE Conference on Computer Vision and Pattern Recognition, CVPR 2009, pp. 1794–1801, June 2009
27. Zhang, J., Marszałek, M., Lazebnik, S., Schmid, C.: Local features and kernels for classification of texture and object categories: A comprehensive study. International Journal of Computer Vision **73**(2), 213–238 (2007)

Query Techniques for CBIR

Tatiana Jaworska

Abstract One of the fundamental functionalities of a content-based image retrieval system (CBIR) is answering user queries. The survey of query types and examples of systems using these particular queries are presented here. For our CBIR, we prepared the dedicated GUI to construct user designed query (UDQ). We outline the new search engine which matches images using both local and global image features for a query composed by the user. In our case, the spatial object location is the global feature. Our matching results take into account the kind and number of objects, their spatial layout and object feature vectors. Finally, we compare our matching results to some other search engines.

Keywords Query · Content-based image retrieval · GUI · Composed query · Search engine

1 Introduction

1.1 Query Concept Overview

One of the fundamental functionalities of a content-based image retrieval system (CBIR) is answering user queries. In traditional, text retrieval system of queries has been highly developed, whereas for the image retrieval a content query has not come up to users' expectations. It stems from the fact that content-based searches have important distinctions compared to traditional searches. The limitations of these systems include both the image representations they use and their methods of accessing those representations to find images. Systems based on keyword querying are often non-intuitive and offer little help in understanding why certain images were returned and how to refine the query.

T. Jaworska(✉)
Systems Research Institute, Polish Academy of Sciences,
6 Newelska Street, Warsaw, Poland
e-mail: Tatiana.Jaworska@ibspan.waw.pl

T. Andreasen et al. (eds.), *Flexible Query Answering Systems 2015*,
Advances in Intelligent Systems and Computing 400,
DOI: 10.1007/978-3-319-26154-6_31

Independently of a diversity of methods focused on the retrieval, in particular, search by association, search for a specific image, or category search [25] we can generally divide query methods into (see Fig. 1):

1. Query by keywords [36],
2. Query by example [2, 23],
3. Query by canvas [18, 19],
4. Query by sketches [16],
5. Query by spatial icons [17],
6. Designed query for semantic retrieval [6, 12].

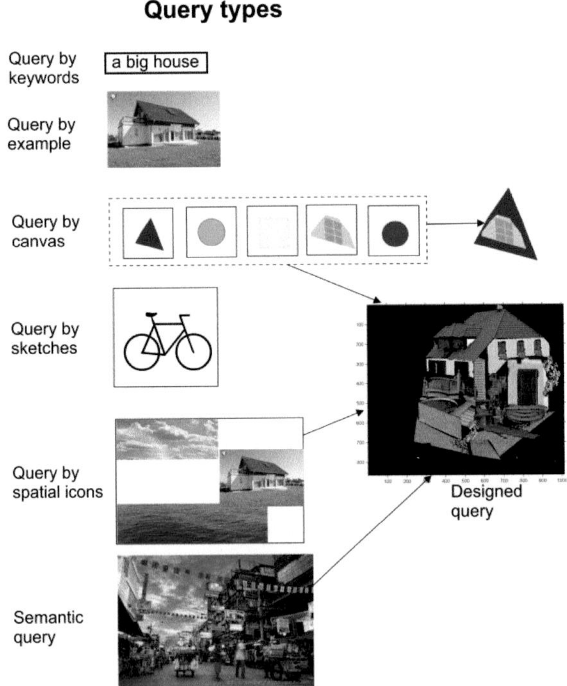

Fig. 1 Query types.

Query by Keywords: The first method for asking the database a question was using keywords as an analog to queries in alpha-numeric DBs. This method requires text annotations for images collected in DB. It is still used in WWW image search engines but it is notorious for incomplete, inconsistent, context sensitive and the ambiguity of meaning of the keywords. In this case [36], the keyword ambiguity is expanded to the selected reference classes most relevant to the query keyword. For example, the keyword 'apple' can mean: 'apple fruit', 'apple computer', 'apple logo', or 'apple tree' from which reference classes are selected.

These attempts try to fill in the semantic gap that exists between the description of an image and the image itself.

Query by Example: At present, most systems use query by example (QBE) whose major advantage is the capability to determine a set of attributes or features that describe the contents of the user's desired image [18, 23]. In a nutshell, QBE provides a full image for the search engine, but its drawback is the fact that the user first has to find an image which he wants to use as a query. In some situations, the most difficult task is to find this one proper image which the user keeps in mind to feed it to the system as a query by example.

Query by Canvas allows the user to compose a visual query using geometrical shapes, colours, and textures. This approach inherently tends to specify objects of interest in an indirect way using primitive features [19]. Moreover, the similarity matching between query and images relies on effective pre-segmentation of regions in the images, which is generally complex and difficult [18].

Query by Sketches enables the user to draw the shape of an object as query but is not popular, perhaps because most users are rather poor at graphic design [23]. For this reason applications have used a query by sketches in a limited form only to images of dominant objects in a uniform background.

Query by Spatial Icons is represented by the visual icons with spatial constraints. It specifies a query using a higher-level visual semantics representation. A query is composed as a spatial arrangement of image segments [17]. The visual query term specifies the region where a salient image region should appear, and a query forming chains of these terms using logical operators. The spatial constraint for regions defines the location and size of the specified visual semantic segments as drawn on a canvas.

Semantic Query contains the associated semantic concept even though is visually dissimilar [21, 36]. Hence, the multiple semantic interpretation results in a retrieval problem because it needs to take into account simultaneously: low-level features, object layout and human associations which an image evokes. For this reason, recently systems have appeared offering designed queries to the user [6, 12].

1.2 Previous Works

The data structure and type of query imply the kind of image information that is retrieved. A CBIR, in turn, should meet the user's diverse requirements depending on the interest domain and a particular need.

Chronologically speaking, the first CBIR systems appeared in the early 90's and used queries by keyword which necessitated the image DBs with annotations. These annotations were more or less relevant to the image context and needed human work to prepare them [29]. Later, QBE appeared which has been the most popular manner used for an image query up to now. Approximately in the late 90's a highly interactive refinement of the search using the sketches or composing images from segments was offered by the system. These three kinds of queries share one disadvantage: it is difficult for the user to prepare, find or design an *ideal* query which they have in mind.

About 10 years later interactive techniques based on feedback information from the user, commonly known as relevance feedback (RF) appeared [31] which, step by step, developed into techniques based on local information from the top retrieved results, commonly known as local feedback or collaborative image retrieval (CIR) [38] which is a powerful tool to narrow down the semantic gap between a low- and high-level retrieval concept [1].

Recently CBIR systems have appeared which enable the user to compose a query from different, previously segmented objects, for instance, a query for face retrieval [37] or a query from composed single images and segments of video frames [6].

In this paper we propose a special, dedicated user's GUI which enables the user to compose their ideal image from the image segments. The data structure and the layout of the GUI reflect the manner of our search engine work. In this paper we present, as a main contribution, the new search engine which takes into account the kind and number of objects, their features, together with different spatial location of the segmented objects in the image.

In order to help the user create the query which they have in mind a special GUI has been prepared to formulate composed queries. An additional contribution is the empirical studies for the proposed search engine consisting in benchmarking it against other known engines.

2 User Designed Query Concept for Semantic Retrieval

Fauqueur and Boujemaa [6] offered the system combining Boolean queries through the region's photometric thesaurus to specify the types of regions which are present and absent in the user's mental/imagined image. The very simple user interaction enables them to combine sophisticated Boolean composition queries with the range query mechanism to adjust the precision of the visual search.

Whereas a front-end module of our CBIR provides a graphical editor which enables the user to compose the image which he/she has in mind from the previously segmented objects. We give the user more tools to design their query than our predecessors.

It is a bitmap editor which allows for the selection of linear prompts in the form of contour sketches (see Fig. 2). These sketches have been generated from images existing in the DB. Next, from the list of object classes the user can select elements to prepare a rough sketch of an imaginary landscape. There are many editing tools available, for instance:

- creating masks to cut off the redundant fragments of a bitmap (see Fig. 3 a) and b));
- changing the bitmap colour (see Fig. 3 c) and d));
- performing basic geometrical transformation, such as: translation, scale, rotation and shear;
- duplicating of repeating fragments;
- reordering bitmaps forward or backward.

This GUI is a prototype, so it is not as well-developed as commercial programs, e.g. CorelDraw. Nevertheless, generally, the user can design an image consisting of as many elements as they need. The only restriction is the number of classes introduced in the DB. At the moment, there are 40 classes but there is no limit for them. Later, the user designed query (UDQ) is sent to the search engine and is matched according to the rules described in sec. 3. However, in the absence of UDQ, the search engine can work with a query consisting of a full image downloaded, for example, from the Internet.

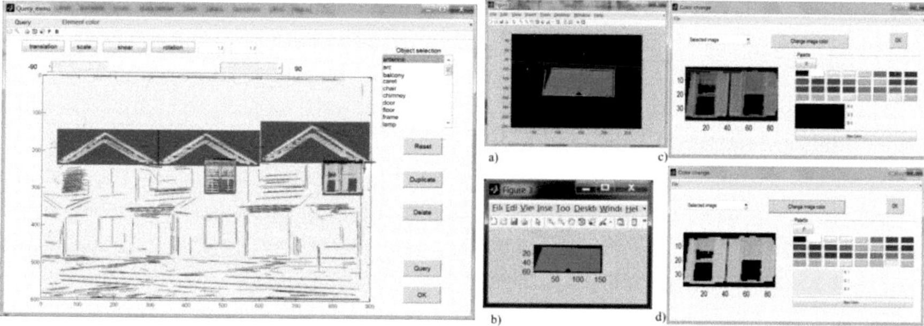

Fig. 2 The main GUI window. An early stage of a terraced house query construction.

Fig. 3 Main components of the GUI. We can draw a contour of the bitmap (see a) and b)) and change the colour of an element c) and d).

3 CBIR Structure

In general, our system consists of four main blocks (Fig. 4): the image preprocessing block [10], the Oracle Database [11], the search engine [13] and the graphical user's interface (GUI). All modules, except the Oracle DBMS, are implemented in Matlab.

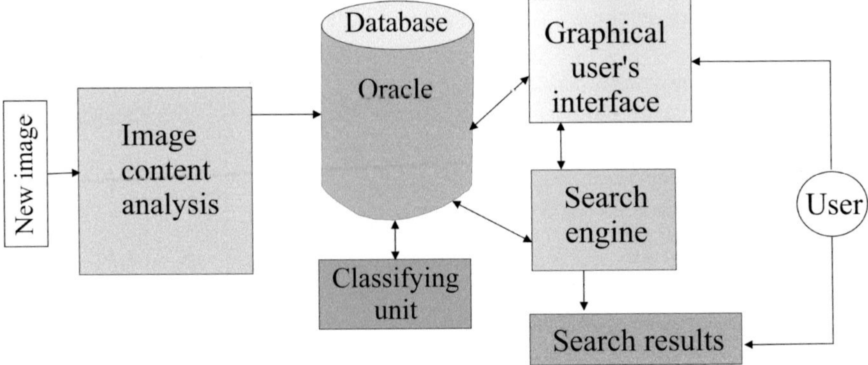

Fig. 4 Block diagram of our content-based image retrieval system.

3.1 Graphical Data Representation and Object Classification

A classical approach to CBIR comprises image feature extraction [30, 39]. Similarly, in our system, at the beginning, the new image (e.g. downloaded from the Internet) is segmented, creating a collection of objects. Each object, selected according to the algorithm presented in detail in [13], is described by some low-level features f_i. We collect $r = 45$ features for each graphical object, for which we construct a feature vector $\mathbf{O} = \{f_1, f_2, \ldots, f_r\}$.

Next, the feature vector **O** is used for object classification. We have to classify objects in order to use them in a spatial object location algorithm and to offer the user a classified group of objects for the semantic selection. So far, four classifiers have been implemented and they are mutually used in our:

- a comparison of features of the classified object with a class pattern;
- decision trees [20, 7];
- the Naïve Bayes classifier [4, 22];
- a fuzzy rule-based classifier (FRBC) [9, 15].

The FRBC is used in order to identify the most ambiguous objects which means these assigned to different classes according the three first classifiers. According to Ishibuchi, the FRBC decides which of the three classes a new element belongs to [14].

3.2 Spatial Object Location

Thanks to taking into account a spatial object location the gap between low-level and high-level features in CBIR has diminished. To describe spatial layout of objects, different methods have been introduced, for example: the spatial pyramid representation in a fixed grid [24], others used the spatial arrangements of regions [28], or the object's spatial orientation relationship [40]. In some approaches, image matching is proposed directly, based on spatial constraints between image regions [35].

Here, spatial object location in an image is used as the global feature [15]. The objects' mutual spatial relationship is calculated based on the centroid locations and angles between vectors connecting them, with an algorithm proposed by Chang and Wu [3] and later modified by Guru and Punitha [8], to determine the first principal component vectors (PCVs).

3.3 Search Engine Construction

Now, we will describe how the similarity between two images is determined and used to answer a query. Let a query be an image I_q, such as $I_q = \{o_{q1}, o_{q2},\ldots, o_{qn}\}$, where o_{ij} are objects. An image in the database is denoted as I_b, $I_b = \{o_{b1}, o_{b2},\ldots, o_{bm}\}$. Let us assume that there are, in total, M classes of the objects recognized in the database, denoted by labels $L_1, L_2, \ldots, L_M$. Then, by the image signature I_i we mean the following vector:

$$\text{Signature}(I_i) = [\text{nobc}_{i1}, \text{nobc}_{i2}, \ldots, \text{nobc}_{iM}] \tag{1}$$

where: nobc_{ik} denotes the number of objects of class L_k present in the representation of an image I_i, i.e. such objects o_{ij}.

In order to answer the query I_q, we compare it with each image I_b from the database. First of all, we determine a similarity measure sim_{sgn} between the signatures of query I_q and image I_b:

$$\text{sim}_{\text{sgn}}(I_q, I_b) = \sum_i (\text{nob}_{qi} - \text{nob}_{bi}) \tag{2}$$

computing as the modified Hamming distance between two vectors of their signatures (cf. (1)), such that $\text{sim}_{\text{sgn}} \geq 0$ and $\max_i (\text{nob}_{qi} - \text{nob}_{bi}) \leq \text{tr}$, tr is a limit of element number of a particular class which I_q and I_b can differ. It means that we prefer images composed of the same classes as the query. Similarity (2) is non-symmetric because if classes in the query are missing from the image the components of (2) can be negative.

Otherwise, we proceed to the next step and we find the spatial similarity sim_{PCV} of images I_q and I_b computing the Euclidean, City block or Mahalanobis distance between their PCVs as:

$$\text{sim}_{\text{PCV}}(I_q, I_b) = 1 - \sqrt{\sum_{i=1}^{3} (PCV_{bi} - PCV_{qi})^2} \tag{3}$$

If the similarity (3) is smaller than the threshold (a parameter of the query), then image I_b is rejected, i.e., not considered further in the process of answering query I_q.

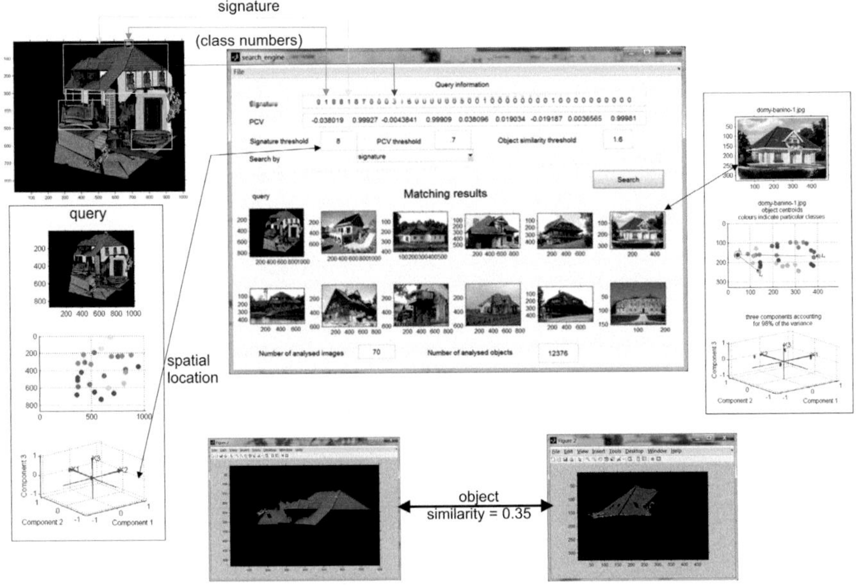

Fig. 5 A graphic concept scheme of our image search engine.

Next, we proceed to the final step, namely, we compare the similarity of the objects representing both images I_q and I_b. For each object o_{qi} present in the representation of the query I_q, we find the most similar object o_{bj} of the same class, sim_{ob} (o_{qi}, o_b) based on the comparison of feature vectors **O** computing the Euclidean distance between o_{qi} and o_b. If there is no object o_{bj} of the class L_{qi}, then sim_{ob} $(o_{qi}, o_b) = 0$.

The process of searching highly similar objects is realized according to the Hungarian algorithm for the assignment problem implemented by Munkres. Thus, we obtain the vector of similarities between query I_q and image I_b:

$$\text{sim}(I_q, I_b) = \begin{bmatrix} \text{sim}_{\text{ob}}(o_{q1}, o_{b1}) \\ \vdots \\ \text{sim}_{\text{ob}}(o_{qn}, o_{bn}) \end{bmatrix} \tag{4}$$

where n is the number of objects present in the representation of I_q. In order to compare images I_b with the query I_q, we compute the sum of sim_{ob} (o_{qi}, o_b) and then use the natural order of the numbers. Therefore, the image I_b is listed as the first in the answer to the query I_q, for which the sum of similarities is the highest.

Fig. 5 presents the main elements of the search engine interface with reference images which are present in the CBIR system. The main (middle) window displays the query signature and PCV, and below the user is able to set threshold values for the signature, PCV and object similarity. The lower half of the window is dedicated to matching results. In the top left of the figure we can see a user designed query comprising elements whose numbers are listed in the signature line. Below the query there is a box with the query miniature, a graph showing the centroids of query components and further below there is a graph with the PCV components. In the bottom centre windows there are two elements of the same class (e.g. a roof) and we calculate their similarity. On the right side there is a box which is an example of PCA for an image from the DB. The user introduces thresholds to calculate each kind of similarity.

4 Results

In this section, we describe experiments conducted on the colour images generated with the help of the UDQ or full images taken from our DB and we will compare our results with the Google image search engine. All images are in the JPG format but in different sizes. In all tables images are ranked according to decreasing similarity determined by our system.

Only in order to roughly compare our system's answer to the query, we used SSIM (Universal image similarity index) proposed by Wang and Bovik [34], being aware that it is not fully adequate to present our search engine ranking. SSIM is based on the computation of three components, namely the luminance, contrast and structural component, which are relatively independent. In case of a big difference of images the components can be negative which may results in a negative index.

A query is generated with the help of the UDQ interface and its size depends on the user's decision as well as the number of objects. The search engine displays a maximum of 11 best matched images from the DB. Although the user designed a query including few details (see tab. 1 query 1 and 2) the search results are quite acceptable.

Applying the UDQ is not obligatory. The user can choose their QBE from among the images of the DB if they find it suitable for their aim. Then the matching results are presented in tab. 1 (column 3).

We also decided to compare our results with the Google image search engine because our system compared with some academic search systems proved significantly more effective. The results obtained are presented in tab. 1 (columns 4 and 5).

Table 1 The matching results and the SSIM (column 1) for UDQ when PCV similarity is calculated based on the City block distance (for thresholds: signature = 17, PCV = 3.5, object = 0.9), (column 2) matches for UDQ when PCV similarity is calculated based on the City block distance (for thresholds: signature = 20, PCV = 4, object = 0.9), (column 3) matches for QBE when PCV similarity is calculated based on the Euclidean distance, (columns 4 and 5) matches for the Google image search engine.

 query1	 query2	 query3	 query2	 query1
 0,1492	 0.1745	 0.2519	 0.3658	 0.0821
 0,1571	 0.1399	 0.2175	 0.0939	 0.1054
 0,1099	 0.0571	 0.1276	 0.0232	 0.1765
 0,1525	 0.1443	 0.3129	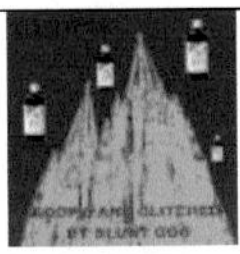 0.0240	 0.2666
 0,1346	 0.1505	 0.2908	 0.3174	 0.1076
 0,0542	 0.0012	 0.1002	 0.2056	 0.0876

Table 1 (*Continued*)

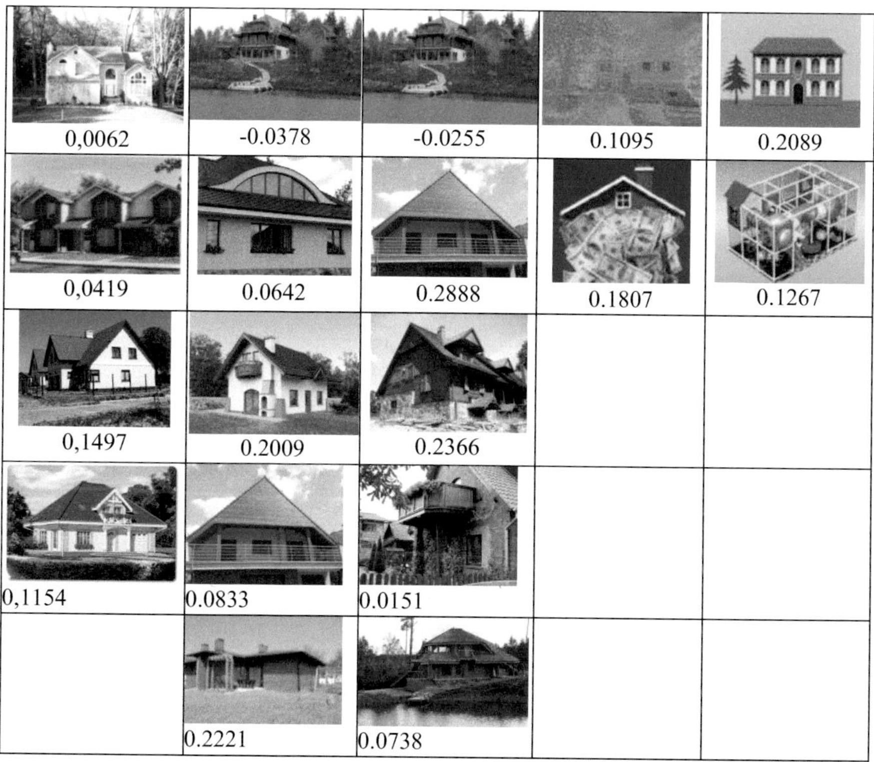

0,0062	-0.0378	-0.0255	0.1095	0.2089
0,0419	0.0642	0.2888	0.1807	0.1267
0,1497	0.2009	0.2366		
0,1154	0.0833	0.0151		
	0.2221	0.0738		

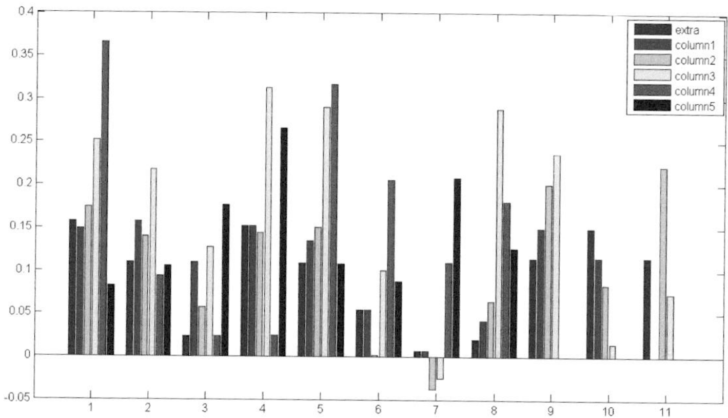

Fig. 6 The comparison of SSIMs for the above-presented results from tab. 1.

In order to better visualise the obtained results, we compare only the SSIMs from tab. 1 as a bar chart (see Fig. 6). There is one extra bar for query1 representing the matches when PCV similarity is calculated based on the Euclidean distance. The other bars refer to the respective columns in tab. 1 according to the legend.

5 Conclusions

We built and described a new image retrieval method based on a three-level search engine. The underlying idea is to mine and interpret the information from the user's interaction in order to understand the user's needs by offering them the GUI. A user-centred, work-task oriented evaluation process demonstrated the value of our technique by comparing it to a traditional CBIR.

Intensive computational experiments are under way in order to determine the optimal choice of model parameters. Furthermore, the results of this study have to be verified in a larger scale evaluation, involving long-term usage of the system for real day-to-day user tasks. However, the preliminary results we have obtained so far, using the simplest configuration, are quite promising.

As for the prospects for future work, the implementation of an on-line version should test the feasibility and effectiveness of our approach. Only experiments on large scale data can verify our strategy. Additionally, a new image similarity index should be prepared to evaluate semantic matches.

References

1. Azimi-Sadjadi, M.R., Salazar, J., Srinivasan, S.: An Adaptable Image Retrieval System With Relevance Feedback Using Kernel Machines and Selective Sampling. IEEE Transactions on Image Processing **18**(7), 1645–1659 (2009)
2. Carson, C., Belongie, S., Greenspan, H., Malik, J.: Blobworld: Image Segmentation Using Expectation-Maximization and Its Application to Image Querying. IEEE Transactions on Pattern Analysis and Machine Intelligence **24**(8), 1026–1038 (2002)
3. Chang, C.-C., Wu, T.-C.: An exact match retrieval scheme based upon principal component analysis. Pattern Recognition Letters **16**, 465–470 (1995)
4. Comer, D.: Ubiquitous B-Tree. ACM Computing Surveys **11**(2), 21–137 (1979)
5. Faloutsos, C., et al.: Efficient and Effective Querying by Image Content. Journal of Intelligent Information Systems **3**, 231–262 (1994)
6. Fauqueur, J., Boujemaa, N.: Mental image search by boolean composition of region categories. Multimed Tools and Applications **31**, 95–117 (2006)
7. Fayyad, U.M., Irani, K.B.: The attribute selection problem in decision tree generation. s.l., s.n., pp. 104–110 (1992)
8. Guru, D.S., Punitha, P.: An invariant scheme for exact match retrieval of symbolic images based upon principal component analysis. Pattern Recognition Letters **25**, 73–86 (2004)

9. Ishibuchi, H., Nojima, Y.: Toward Quantitative Definition of Explanation Ability of Fuzzy Rule-Based Classifiers, Taipei, Taiwan, pp. 549–556. IEEE Society, June 27–39, 2011
10. Jaworska, T.: Object extraction as a basic process for content-based image retrieval (CBIR) system. Opto-Electronics Review **15**(4), 184–195 (2007)
11. Jaworska, T.: Database as a Crucial Element for CBIR Systems, pp. 1983–1986. World Publishing Corporation, Beijing (2008)
12. Jaworska, T.: Multi-criteria object indexing and graphical user query as an aspect of content-based image retrieval system. In: Borzemski, L., Grzech, A., Świątek, J., Wilimowska, Z. (eds.) Information Systems Architecture and Technology, pp. 103–112. Wrocław Technical University Publisher, Wrocław (2009)
13. Jaworska, T.: A search-engine concept based on multi-feature vectors and spatial relationship. In: Christiansen, H., De Tré, G., Yazici, A., Zadrozny, S., Andreasen, T., Larsen, H.L. (eds.) FQAS 2011. LNCS, vol. 7022, pp. 137–148. Springer, Heidelberg (2011)
14. Jaworska, T.: Application of Fuzzy Rule-Based Classifier to CBIR in comparison with other classifiers, Xiamen, China, pp. 1–6. IEEE (2014)
15. Jaworska, T.: Spatial representation of object location for image matching in CBIR. In: Zgrzywa, A., Choroś, K., Siemiński, A. (eds.) New Research in Multimedia and Internet Systems, pp. 25–34. Springer, Wrocław (2014)
16. Lee, Y.J., Zitnick, L.C., Cohen, M.F.: Shadowdraw: real-time user guidance for freehand drawing. ACM Transactions on Graphics (TOG) **30**, 27 (2011)
17. Lim, J.-H., Jin, J.S.: A structured learning framework for content-based image indexing and visual query. Multimedia Systems **10**, 317–331 (2005)
18. Niblack, W., et al.: The QBIC Project: Querying Images by Content Using Colour, Texture and Shape. In: SPIE, vol. 1908, pp. 173–187 (1993)
19. Ogle, V.E., Stonebraker, M.: CHABOT: Retrieval from a Relational Database of Images. IEEE Computer **28**(9), 40–48 (1995)
20. Quinlan, J.R.: Induction of Decision Trees. Machine Learning **1**, 81–106 (1986)
21. Rasiwasia, N., Moreno, P.J., Vasconcelos, N.: Bridging the Gap: Query by Semantic Example. IEEE TransactionS on Multimedia **9**(5), 923–938 (2007)
22. Rish, I.: An empirical study of the naive Bayes classifier. s.l., s.n., pp. 41–46 (2001)
23. Rubner, Y., Tomasi, C., Guibas, L.J.: The Earth Mover's Distance as a Metric for Image Retrieval. International Journal of Computer Vision **40**(2), 99–121 (2000)
24. Sharma, G., Jurie, F.: Learning discriminative spatial representation for image classification. Dundee, s.n., pp. 1–11 (2011)
25. Smeulders, A.W.M., et al.: Content-Based Image Retrieval at the End of the Early Years. IEEE Transactions on Pattern Analysis and Machine Intelligence **22**(12), 1349–1380 (2000)
26. Smith, J.R., Chang, S.-F.: Integrated spatial and feature image query. Multimedia Systems **7**, 129–140 (1999)
27. Srihari, R.K.: Automatic Indexing and Content-Based Retrieval of Captioned Images. Computer, pp. 49–56, September 1995
28. Tuytelaars, T., Mikolajczyk, K.: Local Invariant Feature Detectors: A Survey. Computer Graphics and Vision **3**(3), 177–280 (2007)
29. Urban, J., Jose, J.M., van Rijsbergen, C.J.: An adaptive technique for content-based image retrieval. Multimedial Tools Applied **31**, 1–28 (2006)
30. Wang, Z., Bovik, A.C.: A universal image quality index. Signal Procceing Letter **9**(3), 81–84 (2002)

31. Wang, Z., Bovik, A.C., Sheikh, H.R., Simoncelli, E.P.: Image Qualifty Assessment: From Error Visibility to Structural Similarity. IEEE Transactions on Image Processing **13**(4), 600–612 (2004)
32. Wang, T., Rui, Y., Sun, J.-G.: Constraint Based Region Matching for Image Retrieval. International Journal of Computer Vision **56**(1/2), 37–45 (2004)
33. Wang, X., Liu, K., Tang, X.: Query-Specific Visual Semantic Spaces for Web Image Re-ranking.. s.l., s.n., pp. 1–8 (2011)
34. Xiao, B., Gao, X., Tao, D., Li, X.: Recognition of sketches in photos. In: Lin, W., Tao, D., Kacprzyk, J., Li, Z., Izquierdo, E., Wang, H. (eds.) Multimedia Analysis, Processing and Communications. SCI, vol. 346, pp. 239–262. Springer, Heidelberg (2011)
35. Zhang, L., Wang, L., Lin, W.: Conjunctive patches subspace learning with side information for collaborative image retrieval. IEEE Transactions on Image Processing **21**(8), 3707–3720 (2012)
36. Zhang, Y.-J., Gao, Y., Luo, Y.: Object-based techniques for image retrieval. In: Deb, S. (ed.) Multimedia Systems and Content-Based Image Retrieval, pp. 156–181. IDEA Group Publishing, Hershey, London (2004)
37. Zhou, X.M., Ang, C.H., Ling, T.W.: Image retrieval based on object's orientation spatial relationship. Pattern Recognition Letters **22**, 469–477 (2001)

Dealing with Rounding Errors in Geometry Processing

Jörg Verstraete

Abstract Processing geometric data on computer systems poses interesting challenges. The limited representation in a computer system, combined with the wide variety of calculations can result in robustness problems. As a result of this, it is for example possible that the exact intersection point of two lines cannot be represented by the computer system and its coordinates get rounded. As a result of this, the test to check whether this point is on either line can fail. The solution for this depends on the application at hand. In this article, a solution developed for our application is presented. The proposed solution for handling the rounding errors is quite general: by reconsidering the 9-intersection matrix and derived operations, and as such the solution may be useful in other applications.

Keywords Geometric precision · Robustness · Spatial data · Geometry rounding issues

1 Introduction

The representation of spatial data in a computer system can be considered in two distinct ways: feature based or field based ([1],[2]). In a feature based representation, real world objects such as areas or roads are represented by means of basic geometric objects these range from features represented by a single point to features consisting of multiple polygons with holes. The position of each vertex of such a feature is defined by its coordinates. The field based approach is quite different, and uses either triangular networks or a raster ([1]). In the latter approach, the raster is overlayed with the region of interest, so that the region of interest is partitioned in a number of cells.

J. Verstraete(✉)
Systems Research Institute, Polish Academy of Sciences,
Ul. Newelska 6, 01-447 Warszawa, Poland
e-mail: jorg.verstraete@ibspan.waw.pl
http://www.ibspan.waw.pl

T. Andreasen et al. (eds.), *Flexible Query Answering Systems 2015*,
Advances in Intelligent Systems and Computing 400,
DOI: 10.1007/978-3-319-26154-6_32

With each of the cells, data can be associated. The raster representation is more commonly used when the data that are represented concern a numerical value spread over a region; this is for example data concerning air pollution or demographic data.

In [3], the author presented an algorithm to solve the map overlay problem using artificial intelligent methods. The map overlay problem occurs when gridded data needs to be remapped onto another grid. Gridded data provide a way for representing numerical data that carries a spatial aspect, by covering the region of interest with a raster, and assigning a representative value with each cell of the raster. On figure 1, a feature based representation is used to show both the road network and the water network of the central Warsaw area, while the map is overlayed with air pollution data that is presented as a grid. Each grid cell has three values associated, represented by a bar chart in each cell. They represent the amount of given pollutants (in the example, from left to right, concentrations of SO2, SO4, PM10 and PM25).

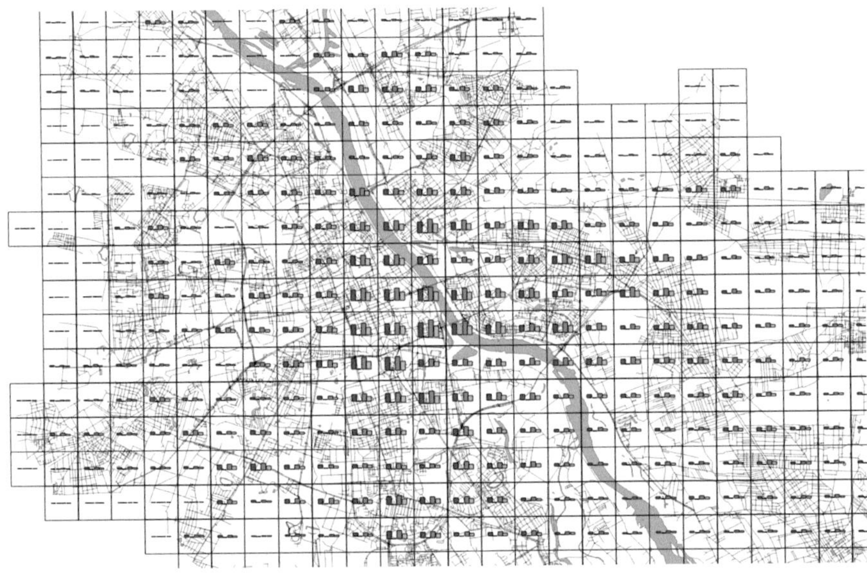

Fig. 1 Example of a geographic data set: a feature based representation is used for roads and water; a grid representation is used for pollution data.

A raster can for instance consist of cells that are e.g. 2km x 2km in size and that have, as associated value, the number of people living inside this square area. Another example would be a raster representing the concentration of an air pollutant, where a representative value is associated with each cell. This representation is commonly used, as it is both an easy format for calculations as well as for representation. At the level of the cells, nothing is known about the underlying distribution: in a cell that has the amount of inhabitants associated, the inhabitants could be uniformly distributed over the cell; or they could all be located in one half. Figure 2 shows some examples on different distributions that yield the same associated value.

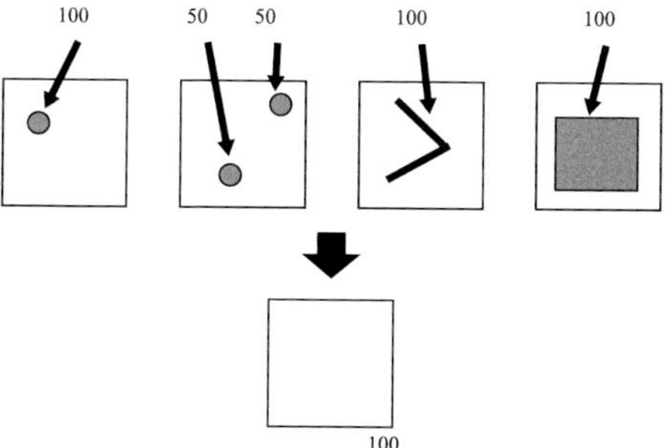

Fig. 2 Example of different underlying distributions that at grid level appear the same. In all cases, the value of the grid cell would be 100, but the underlying distributions are either a single point source, two point sources - each contributes 50, a line source and an area source.

The map overlay problem occurs when incompatible grids need to be combined or compared. Incompatible grids are grids that differ in cells size and/or orientation. Some examples are shown on figure 3. Such situations are quite common, as the grid on which the data are represented is usually related to the way the data were obtained, which can be through mathematical models or real world measurements with data extrapolation. When different grids need to be correlated, the question is which portion of each cell in one grid gets mapped to the cells in the other grids, as there is no one-on-one mapping. Traditionally, areal weighting is used: the percentage of overlap of a cell in one grid and a cell in the other grid serves as the percentage that will be mapped onto the other cell. The problem is that this implicitly assumes that the modelled data is uniformly distributed within each cell. Other approaches are areal smoothing and spatial regression, but they also make assumptions on the underlying data distribution. An overview of the common techniques and problems associated with map overlay can be found in [4]. The method presented in [3] allows for additional data that correlates to the input data, in order to provide information on the underlying distribution.

Fig. 3 Example of different grids that are incompatible. The cases from left to right illustrate grids that are shifted, grids that have a different size of grid cells, grids that are rotated and grids that show a combination of these.

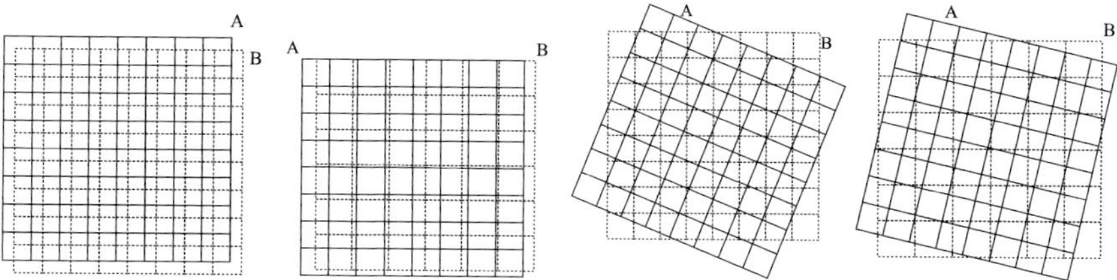

In [3], we presented a technique using a fuzzy inference system to remap gridded spatial data onto a different grid. This is a problem that contains inherent uncertainty; in the presented method, an inference system, with adequately chosen parameters can perform this operation using additional knowledge. The problem concerns gridded data, which for the purposes of the calculations are represented as a collection of gridcells, each in its own defined as a polygon. The additional knowledge can also consist of gridded data, but it can also be a feature based representation. The calculations necessary to implement the application of the fuzzy inference system relies heavily on accurate spatial calculations such as intersections and difference, but also on accurate spatial predicates. The use of the popular JTS library, combined with the fact that the data should be compatible with established data structures, limits the possibilities of the data types. Consequently, the idea was to not interfere with the data structures nor with (intermediate) results of operations, but to derive robust alternatives for the predicates that relate to two geometries.

2 Problem Description

2.1 Robustness Errors in Geometry Calculations

Consider the example shown in figure 4. The line A connects the points (0,0) and (1,1); line B connects the points (0,1) and (1,0). The intersection of both lines is (0.5,0.5). If the representation of the coordinates is limited to integers, the intersection point would get rounded to (1,1). The point would be on one line, but not on the other, even though it was the calculated intersection point. This is a very extreme example, but it illustrates the problem that occurs in general when the number representation for the coordinates has a limited accuracy, as is the case in a standard computer representation. We refer to chapter 4 of [5] for a detailed overview of a classification of related problems and causes.

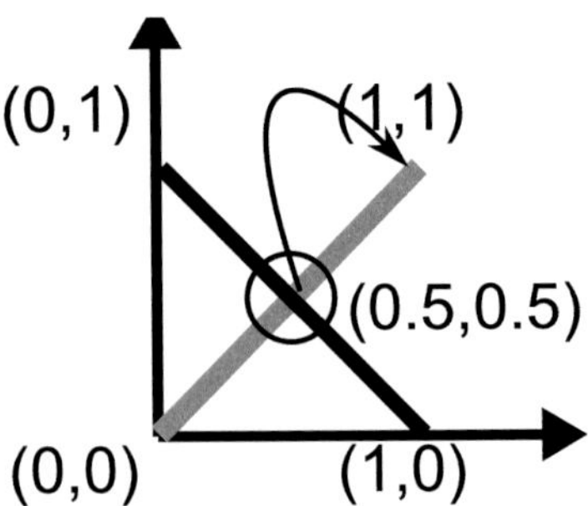

Fig. 4 Example of the problem of limited representation. The intersection point would get rounded to (1,1), which would not longer coincide with the grey line.

Experiments on our algorithm showed that it suffers heavily from such robustness errors, even using the coordinate representation with highest accuracy available in the JTS library.

2.2 *Rulebase Approach for Remapping Data: Spatial Operations*

In order to identify which spatial operations are causing the issues, it is necessary to shortly describe some aspects of the algorithm. The proposed approach for solving the map overlay method starts from finding the intersection between input grid and target grid. The workflow is illustrated in figure 5.

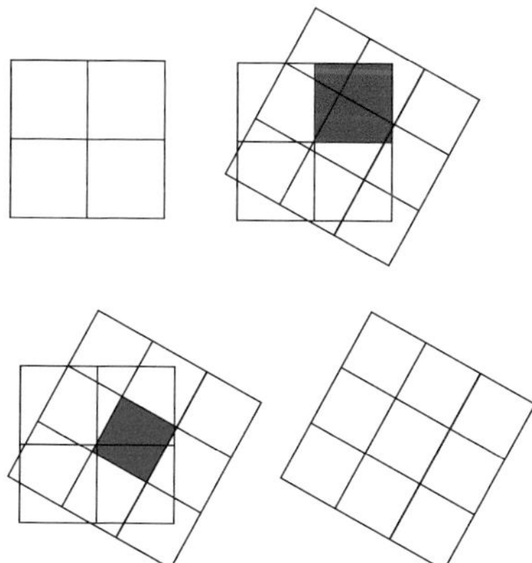

Fig. 5 Steps to remap a grid (top left) to a target (bottom right). The intersection of both grids yields segments, which are calculated (top right, 5 segments forming an input cell are highlighted). The segments are recombined to form the output cell (bottom left, 3 segments forming an output cell are highlighted.

This intersection is an irregular grid, which is such that every cell of the input grid and every cell of the output grid is partitioned in a number of segments (fig 5, top left and bottom right). These segments will be the initial result of the remapping, as they allow to redistribute the data of an input cell and assure that the total value associated with the segments that partition an input cell is the same as the input cell itself. After this, due to their definition, the segments can be recombined to form output cells. For each segment, parameters that relate to the ideal output value are calculated: these parameters usually are defined using predicates that involve topological relations between the segment and the input grid or additional grids that contain data. Examples of parameters are the amount of overlap with a feature or the distance to a high value feature.

The operations that were observed to be prone to the rounding issues are:

- calculating the intersection grid
- determining topological relations between a segment and other geometries

Table 1 Different cases that can be distinguished mathematically, along with the dimension of the intersection.

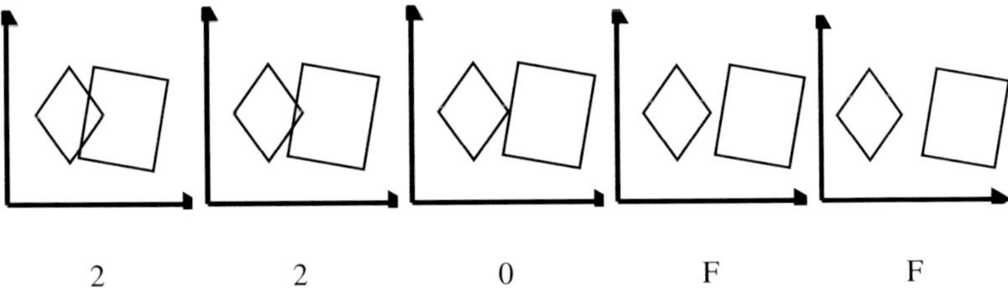

2	2	0	F	F

– calculating the overlap between a segment and cells of another grid (input, additional data or output)
– calculating the overlap between segments and additional non-gridded data (point/lines/polygons)

Each of these four operations involve segments: the first calculates the segments whereas the next three operations involve performing topological checks and calculating intersections to obtain the amount of overlap. The first operation calculates geometries that will be used further on; the next operations use these geometries but do not calculate new geometries for future use. As it is known that it might be impossible to represent the segments correctly, is was determined to not interfere with the calculation of the segments but rather take their possible error into account for the subsequent operations. In the workflow, no new geometries will be calculated from the segments, so there will not be a propagation of the error.

The main issue arises with creating the segments; some possible relative position are shown in table 1. The problem of rounding is illustrated in Figure 6, where two non-intersecting geometries have coordinates rounded to make them touch. This is a simplified example for illustration purposes, but it is possible for such errors, or even bigger errors to occur. In general, the different cases are shown in table 2.

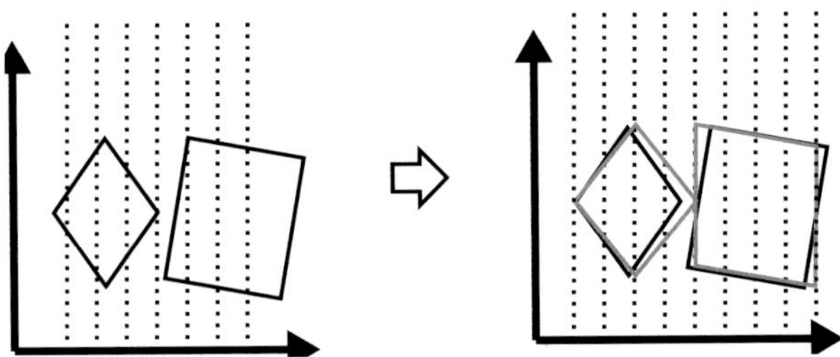

Fig. 6 Example on how the occurrence of rounding modifies geometries and topologies. Two geometries that do not intersect can be represented as such if the grid allows for it (left), but on a different grid the geometries can be rounded which causes them to touch. In general, after rounding, the intersection can have all possible dimensions, depending on the effects of the rounding.

Table 2 The different cases for the dimension of the intersection, along with the possible dimensions of the intersection after rounding of the geometries.

	mathematical dimension	dimension after rounding
case 1	2	2-1-0-F
case 2	1	2-1-0-F
case 3	0	2-1-0-F
case 4	F	2-1-0-F

The segments are also used in combination with additional data, which may be in gridded format or in a feature based representation. As such, the general topological relation of a segments with other geometries also needs to be considered. It is important to realize that due to the rounding of coordinates, information gets lost. It is impossible to determine what the mathematical topological relation is by just looking at the relation between geometries represented by the rounded coordinates. The goal should be to make sure that the errors that occur due to rounding do not cause issues or errors in the algorithms employed.

The biggest problem in our application occurs when a segment is considered to overlap with a cell while this is not the case: this wrongly identifies the cell that contains the segment and may cause wrong distribution of the data. As the main interest between segments and cells of either input or output grid is the percentage they overlap, the case where the rounding error causes a small area (cases 2, 3 and 4 that result an intersection with dimension 2) should be avoided. In general, our application is less sensitive to the loss of a correct small 2-dimensional intersection than it is to the addition of a small 2-dimensional intersection. Intersections between geometries of other dimensions are important for identifying the relation between segments, but the addition of small intersections or the omission off small intersections has little impact on the result. More important is that the small intersections are dealt with in a consistent manner. This is the reason why general rounding error analysis is less suitable: our problem is not symmetrical. False positives in intersections should be avoided at all costs, whereas the occurrence of false negatives are much less of an issue.

3 Operations on Geometries That May not Be Correctly Represented

3.1 Collapsing a Geometry

In order to work with possibly incorrectly represented geometries, a conversion function is introduced that *collapses* a geometry. By collapsing a geometry, we mean converting it to a geometry of lower dimension if the geometry is small enough.

The algorithm is quite simple: if the area of a 2 dimensional geometry is smaller than a given epsilon, the geometry is replaced by its boundary. If the length of a 1 dimensional geometry (which may be the boundary of an already collapsed 2 dimensional geometry) is smaller than a given epsilon, the geometry is replaced by its centroid.

The epsilon value is calculated for a given geometry based on its surface area or its length: the reason is that the scale of the map on which the data are represented is dependent on the data at hand, and this should be taken into account. If all areas are big numbers, the surface area that is considered to be a candidate for collapsing is bigger than if all areas are small numbers. For operations that involve two geometries, the maximum epsilon value of the epsilons for both geometries is chosen.

3.2 9-Intersection Matrix

A common way to determine the topological relations between geometries is to determine the 9-intersection matrix. The 9-intersection matrix contains the intersection of interior, boundary and exterior of 2 regions. To define these concepts, the definition of the neighbourhood of a point is needed.

Definition 1 (Neighbourhood of a Point). *The* neighbourhood *of a point p is a set containing an open set that contains p:*

$$\begin{gathered} A \text{ is a neighbourhood for } p \\ \text{if} \\ p \in A \wedge \exists B \subseteq A : B_x^r = \{y \in \mathbb{R}^2 : |y - x| < r\} \subset B \end{gathered} \tag{1}$$

Definition 2 (Closure of a Set). *The* closure $\overline{A}$ *of a set A is the smallest closed set containing A.*

Using the above concepts, the interior, exterior and boundary of regions can be defined.

Definition 3 (Interior A° of a Region A). *The* interior A° *of a region A is defined as the set of points $p \in A$ such that A contains a neighbourhood of p.*

$$A^\circ = \{p \in A \mid \exists B \subset A, B \text{ is a neighbourhood for } p\} \tag{2}$$

In practice, the interior of a region holds all the points that are inside the region.

Definition 4 (Boundary ∂A of a Region A). *The* boundary ∂A *of a region $A \subset U$ is*

$$\partial A = \overline{A} - A^\circ \tag{3}$$

where $\overline{A}$ is the closure, and A° is the interior of A in U.

Definition 5 (Exterior A^- of a Region A). *The* exterior A^- *of a region A is defined as the complement of A:*

$$A^- = \mathbb{R}^2 - A \tag{4}$$

where $-$ is the notation for set-minus.

In practice, the exterior of a region encompasses all the points that are not part of the region.

To determine the topological relations between two regions, the intersections between the interior, boundary and exterior of both regions are considered: the intersection between the interiors of both A and B, the intersection between the interior of region A and the boundary of region B, and so on. In total, there are *nine* possible combinations; which are grouped in the nine-intersection matrix ([6]). The standard 9-intersection matrix has boolean values: true if the intersection is not empty, false if it is empty. The dimensionally extended matrix holds the dimension of the intersections (false for no intersection, 0 for points or multi-points, 1 for lines or multi-lines and 2 for areas):

$$\begin{pmatrix} dim(A^{\circ} \cap B^{\circ}) & dim(A^{\circ} \cap \partial B) & dim(A^{\circ} \cap B^{-}) \\ dim(\partial A \cap B^{\circ}) & dim(\partial A \cap \partial B) & dim(\partial A \cap B^{-}) \\ dim(A^{-} \cap B^{\circ}) & dim(A^{-} \cap \partial B) & dim(A^{-} \cap B^{-}) \end{pmatrix} \quad (5)$$

The topological relations in the event of geometries that may be wrongly represented need to be corrected for. To allow for this, the calculation of the matrix elements of the 9-intersection matrix will be made more robust. Their meaning and interpretation stays the same, but the determination of the matrix elements accounts for the issues with rounding. For this, it had to be taken into account that in the JTS library, it is not possible to represent an interior or an exterior: the interior of a region has the same representation as the region but carries a different interpretation. To further complicate things, the boundary also uses the same defining coordinates, but uses a different representation (the boundary of an area is a line, the boundary of a line is a set of points).

Intersection of Interiors. The first matrix element, the intersection of the interiors of two geometries g_1 and g_2, is defined by the dimension of the collapsed intersection of the geometries under the following conditions:

- the collapsed intersection is not empty
- the intersection does not equal the collapsed geometry of intersection between the boundary of g_1 and g_2
- the intersection does not equal the collapsed geometry of intersection between g_1 and the boundary of g_2

In all other cases, the matrix element is considered false.

Considering the cases in 2, it is clear that, if the area of the intersection is small enough, it will be collapsed to a line and possibly to a point. If there was no intersection, this will remain. If there was a small intersection that got lost in rounding, it will not be found. One possible remaining problem is when the intersection between g_1 and g_2 is a line that will be collapsed to its centroid which cannot be represented. In this already unlikely scenario, it is very probable that the centroid will result into one of the defining points. However, if it would not, this causes an intersection between interiors as the interior of a point is the point. For the segments, this will not result into a problem, as the intersection has no surface area. When used between segments and other data, this intersection matrix will also not cause issues in our application as the intersection between the other part of the segment and the geometry will carry all the weight.

Intersection of Interior with Boundary. The second matrix element, the intersection of the interior of g_1 with the boundary of g_2 is the dimension of the collapsed intersection between g_1 and the collapsed boundary of g_2 under the following conditions

- the collapsed intersection is not empty
- the collapsed intersection does not equal the collapsed intersection between the collapsed boundaries of both g_1 and g_2

The second conditions effectively checks if the regions are not equal, as in this case there is no intersection between the interior of g_1 and the boundary of g_2. The calculation of the fourth matrix element, the intersection of the boundary of g_1 with the interior of g_2 is similar.

As the boundary cannot be an area, this matrix element is more important when it concerns relating the segment to other non-gridded data. Collapsing the intersection will result in the value being false if the length of the intersection is small enough. This makes an already small intersection even smaller (which leaves the rest of the geometry with the biggest weight), but also minimizes the effect of the intersection if the intersection was introduced due to the rounding.

Intersection of Interior with Exterior. The third matrix element, the intersection of the interior of g_1 with the exterior of g_2, is calculated as the dimension of the difference of g_1 and g_2 if

- the collapsed intersection does not equal the collapsed geometry of g_1

Its symmetrical counterpart, matrix element 7, is calculated similarly.

The condition is to verify that both geometries are not equal, in which case the interior of one does not intersect with the exterior of the other. By collapsing the intersection, the dimension of the intersection will be decreased if the intersection is small enough. This is adequate for our application for all of the rounding cases.

Intersection of Both Boundaries. The fifth matrix element, the intersection of the boundaries of both regions, is defined as the dimension of the collapsed intersection of both boundaries. The collapsing of the intersection most likely has little effect on the end result. This matrix element is mainly of importance when it concerns the relation between segments and additional non-gridded data. Similar to the second matrix element, the behaviour of minimizing the dimension yields the desired result.

Intersection of Boundary with Exterior. The sixth matrix element considers the intersection of the boundary of g_1 with the exterior of g_2. It is defined as the dimension of the difference of the collapsed boundary of g_1 with g_2 if

- the collapsed intersection of the collapsed boundary of g_1 with g_2 does not equal the the collapsed boundary of g_1

As this intersection will never be of dimension 2, it is mainly of importance when segments need to be related to additional non-gridded data. The conclusion is the same as for the above matrix element.

The eight matrix element reverse the roles of the geometries, and is calculated similarly.

Intersection of the Exteriors. As in the library it is not possible to define a geometry that is the size of the universe, the interiors will always intersect with the dimension 2. This is returned as value.

3.3 *Topological Relations*

All topological relations can be expressed as a predicate that involves the 9-intersection matrix. The predicate to verify that two geometries are equal is the matrix

$$\begin{bmatrix} T & * & F \\ * & * & F \\ F & F & * \end{bmatrix} \tag{6}$$

where T indicates that there is an intersection, F indicates there is no intersection and $*$ indicates that this matrix element is not important for the relation. Predicates such as contains, covered by, touches, etc. can all be defined by means of one or more intersection matrices. The documentation of the class IntersectionMatrix of the JTS Topology Suite [7] lists all equivalence relations between predicates and intersection matrices. By using these equivalent definitions while applying the proposed intersection matrix, all predicates circumvent the rounding issues, while adhering to our criteria of minimizing the false detection of intersections with dimension 2.

4 Conclusion

In this article, we presented a way of redefining an implementation of the 9-intersection matrix to solve problems that arise due to the limited computer representation of coordinates. The proposed solution is tailored to the author's application for remapping gridded data, and had specific criteria: false positives in intersections should be avoided at all costs, whereas the occurrence of false negatives are less of an issue. The proposed solution introduces a method to lower the dimension of a geometry based on an automatically calculated scale value. This method is then used in the calculation of the matrix elements of the 9-intersection matrix, which in turn results in satisfactory behaviour for all derived topological predicates.

References

1. Rigaux, P., Scholl, M., Voisard, A.: Spatial databases with applications to GIS. Morgan Kaufman Publishers (2002)
2. Shekhar, S., Chawla, S.: Spatial databases: a tour. Pearson Educations (2003)
3. Verstraete, J.: Solving the map overlay problem with a fuzzy approach. Climatic Change, pp. 1–14 (2014)
4. Gotway, C.A., Young, L.J.: Combining incompatible spatial data. Journal of the American Statistical Association **97**(458), 632–648 (2002)
5. Hoffman, C.M.: Geometric and Solid Modeling (1992). `https://www.cs.purdue.edu/homes/cmh/distribution/books/geo.html`
6. Egenhofer, M.J., Sharma, J.: Topological relations between regions in r^2 and z^2. In: Abel, D., Ooi, B.C. (eds.) Advances in Spatial Databases. Lecture Notes in Computer Science, vol. 692, pp. 316–336. Springer Verlag, Singapore (1993)
7. JTS Topology Suite. `http://sourceforge.net/projects/jts-topo-suite/files/`

Part V
Knowledge-Based Approaches

In Search for Best Meta-Actions to Boost Businesses Revenue

Jieyan Kuang and Zbigniew W. Raś

Abstract The ultimate goal of our research is to build recommender system driven by action rules and meta-actions for providing proper suggestions to improve revenue of a group of clients (companies) involved with similar businesses. Collected data present answers from 25,000 customers concerning their personal information, general information about the service and customers' feedback to the service. This paper proposes a strategy to classify and organize meta-actions in such a way that they can be applied most efficiently to achieve desired goal. Meta-actions are the triggers that need to be executed for activating action rules. In previous work, the method of mining meta-actions from customers' reviews in text format has been proposed and implemented. Performed experiments have proven its high effectiveness. However, it turns out that the discovered action rules need more than one meta-action to be triggered. The way and the order of executing triggers causes new problems due to the commonness, differential benefit and applicability among sets of meta-actions. Since the applicability of meta-actions should be judged by professionals in the field, our concentration is put on designing a strategy to hierarchically sort and arrange so called meta-nodes (used to represent action rules and their triggers in a tree structure) as well as to compute the effect of each meta-node. Furthermore, users will have more concrete options to consider by following the path in trees built from these meta-nodes.

J. Kuang · Z.W. Raś
Deptartment of Computer Science, University of North Carolina, Charlotte, NC 28223, USA
e-mail: jkuang1@uncc.edu

Z.W. Raś(✉)
Institute of Computer Science, Warsaw University of Technology, 00-665 Warsaw, Poland
e-mail: ras@uncc.edu

T. Andreasen et al. (eds.), *Flexible Query Answering Systems 2015*,
Advances in Intelligent Systems and Computing 400,
DOI: 10.1007/978-3-319-26154-6_33

1 Introduction

The competition between companies dealing with similar businesses is always intensive and full of gunpowder, and all of them are trying their best to attract and keep as many customers as possible. To evaluate and improve the performance of a company's growth engine, NPS (Net Promoter System) has been designed and becomes one of the most important measurements nowadays. Generally speaking, NPS system divides customers into three groups: *Promoter, Passive* and *Detractor*, which represent customers' satisfaction, loyalty and likelihood to recommend the company to their friends in a descending order [9].

In our project, we intend to build a hierarchical recommender system driven by action rules and meta-actions for providing proper suggestions to improve clients' NPS rating. There are 34 clients (companies) in our dataset and they deal with similar businesses all over US and south Canada. To collect the research data, over 25,000 customers have been randomly selected to answer a questionnaire that asks customers' personal information, general information about the service and customers' feedback to the service. In terms of giving feedback about the service by customers, they can assign numeric scores ranging from 0 to 10 to express their satisfaction (higher the score is, more satisfied the customer is), and they are also welcomed to give more details which are recorded in text format. Based on the numeric values, customers are categorized into three NPS statuses: 9-10 is promoter, 7-8 is passive and 0-6 is detractor. Furthermore, the NPS rating of a client can be calculated as the percentage difference between customers that are promoter and customers that are detractor.

In previous work, a hierarchical dendrogram has been generated by applying agglomerative clustering algorithm to the semantic distance matrix covering 34 clients [3], which demonstrates the similarity among clients concerning the knowledge hidden in datasets and is the foundation of our project. With the dendrogram and the fact that clients can learn from each other by exchanging their hidden knowledge, a strategy called HAMIS (Hierarchically Agglomerative Method for Improving NPS) has been designed to maximally extend individual dataset for each client by merging it with datasets of clients who are semantically similar, have better NPS rating and better classification results [4]. Then action rules, that show what minimum changes are needed for each client to move its ratings to the Promoter's group, are generated from their datasets enlarged by HAMIS. Meta-actions are the ultimate triggers that will be executed by clients to make action rules effective. Strategy for generating meta-actions is proposed and experiments have been conducted to prove its effectiveness and efficiency.

Clearly, each action rule can be triggered by different groups of meta-actions. In the ideal scenario, all discovered actions rules should be triggered. So the problem of finding the smallest sets of meta-actions triggering all action rules seems important to be solved. However each meta-action has a number of parameters assign to it like its cost or benefit. We may sort all meta-actions in a certain way and then apply them one by one to the discovered action rules only if the increment in NPS by triggering these rules is above some threshold value. Saying another words, some meta-actions may not be executed at all. The same some action rules will not get triggered because the benefit of doing that will be too small.

The main contribution in this paper lays on designing a strategy to classify and organize meta-actions hierarchically for providing clients with an efficient way of picking right choices out of an the entire mass. Traditional influence matrix is transformed into a more straightforward structure named *advanced matrix*. Meta-node, introduced in this paper, is used to store a set of action rules and their triggers. So each meta-node is seen as a possible option to be recommended by our system to a client. When processing meta-actions from the most common one to the least one (number of action rules it is associated with), a new meta-node is built or updated if a new action rule can be triggered. Some meta-nodes are connected and form a "tree". Lower a meta-node is located in a tree, greater is its effect. Our strategy will result in a forest of trees. Details concerning the process of generating these trees is illustrated thoroughly in this paper.

2 Action Rules

The concept of an action rule was firstly proposed by Ras and Wieczorkowska in [6] and investigated further in [2], [5], [1] and [8]. Action rules indicate possibly the smallest sets of changes that should be made to achieve the desirable effect under certain possibilities. An action rule usually consists of three types of features: stable attribute, flexible attribute and decision attribute. Stable attributes refer to these features of which values can not be changed, while flexible attributes denote the features of which values can be changed. Decision attribute is a flexible attribute and the most distinguished one because it contains the decisional values that we aim to transition from or to.

Before explaining the definition of action rules, let's first recall the definition of an information system (dataset). By an information system, we mean a triple $S = (X, A, V)$, where:

1. X is a nonempty, finite set of objects
2. A is a nonempty, finite set of attributes, i.e.
   $a : U \longrightarrow V_a$ is a function (can be partial function) for any $a \in A$, where V_a is called the domain of a
3. $V = \bigcup\{V_a : a \in A\}$.

Based on the partition of attributes in an action rule, we assume that $A = A_{St} \cup A_{Fl}$, where A_{St} and A_{Fl} denote *stable* attributes and *flexible* attributes respectively. Additionally, $A_{Fl} = A_{fl} \cup A_d$, where A_d and A_{fl} represent *decision* attribute and flexible attributes other than decision attribute respectively. Information system S is a decision system if decision attribute defined.

An action rule is composed of *atomic action sets*. By an *atomic action set* we mean a singleton set containing an expression $(a, a_1 \rightarrow a_2)$ called atomic action, where a is an attribute and $a_1, a_2 \in V_a$. If $a_1 = a_2$, then a is called stable on a_1.

Instead of $(a, a_1 \rightarrow a_1)$, we usually write (a, a_1) for any $a_1 \in V_a$. By *Action Sets* we mean a smallest collection of sets such that:

1. If t is an atomic action set, then t is an action set.
2. If t_1, t_2 are action sets, then $t_1 \cup t_2$ is a candidate action set.
3. If t is a candidate action set and for any two atomic actions $(a, a_1 \rightarrow a_2)$, $(b, b_1 \rightarrow b_2)$ contained in t we have $a \neq b$, then t is an action set. Here b is another attribute $(b \in A)$, and $b_1, b_2 \in V_b$.

By the domain of an action set t, denoted by $Dom(t)$, we mean the set of all attribute names listed in t. By an *action rule* we mean any expression $r = [t_1 \Rightarrow t_2]$, where t_1 and t_2 are action sets. Additionally, we assume that $Dom(t_2) \cup Dom(t_1) \subseteq A$ and $Dom(t_2) \cap Dom(t_1) = \emptyset$. The domain of action rule r is defined as $Dom(t_1) \cup Dom(t_2)$.

Now we give an example assuming that a, b and d are stable attribute, flexible attribute and decision attribute respectively in S. Expressions (a, a_2), $(b, b_1 \rightarrow b_2)$, $(d, d_1 \rightarrow d_2)$ are examples of atomic actions. Expression (a, a_2) means that the value a_2 of attribute a remains unchanged. Expression $(b, b_1 \rightarrow b_2)$ means that the value of attribute b is changed from b_1 to b_2. Expression $r = [\{(a, a_2), (b, b_1 \rightarrow b_2)\} \Rightarrow \{(d, d_1 \rightarrow d_2)\}]$ is an example of an action rule saying that if value a_2 of a remains unchanged and value of b will change from b_1 to b_2, then the value of d will be expected to transition from d_1 to d_2.

In early research, action rules can be constructed from two classification rules [7]. Taking the example of action rule r shown above for instance, r can be seen as the composition of two association rules r_1 and r_2, where $r_1 = [\{(a, a_2), (b, b_1)\} \rightarrow (d, d_1)]$ and $r_2 = [\{(a, a_2), (b, b_2)\} \rightarrow (d, d_2)]$. In [11], the definition of support and confidence of action rule r has been proposed. However, there is a slight difference in defining the support in our case. The definition of support and confidence of r, used in this paper, is given below:

- $sup(r) = sup(r_1)$
- $conf(r) = conf(r_1) \cdot conf(r_2)$

The confidence of an action rule is still the multiplication of the confidences of two involved association rules. But the support of action rule becomes the support of association rule r_1. In our example, it is the number of objects x $(x \in X)$ which values of attributes a, b and d are equal to a_2, b_1 and d_1 respectively. By defining support of action rules this way, the number of objects x in S that potentially can be affected by applying this action can be tracked and used to evaluate its performance.

3 Meta-Actions

As an action rule can be seen as a set of atomic actions that need to be made happen for achieving the expected result, meta-actions are the actual solutions that should be executed to trigger the corresponding atomic actions, Table 1 below shows an example of influence matrix which describes the relationships between the meta-actions and atomic actions influenced by them.

Table 1 Meta-actions Influence Matrix for S

	a	b	d
$\{M_1, M_2, M_3\}$		$b_1 \rightarrow b_2$	$d_1 \rightarrow d_2$
$\{M_1, M_3, M_4\}$	a_2	$b_2 \rightarrow b_3$	
$\{M_5\}$	a_1	$b_2 \rightarrow b_1$	$d_2 \rightarrow d_1$
$\{M_2, M_4\}$		$b_2 \rightarrow b_3$	$d_1 \rightarrow d_2$
$\{M_1, M_5, M_6\}$		$b_1 \rightarrow b_3$	$d_1 \rightarrow d_2$

In Table 1, a, b and d are same attributes in decision system S as mentioned in previous section, and $\{M_1, M_2, M_3, M_4, M_5, M_6\}$ is a set of meta-actions which hypothetically trigger action rules generated from S. Each cell in a row shows an atomic action that can be invoked by the set of meta-actions listed in the first column of that row. For instance, the first row shows that the atomic actions $(b_1 \rightarrow b_2)$ and $(d_1 \rightarrow d_2)$ can be activated by executing meta-actions M_1, M_2 and M_3 together. Unlike the traditional influence matrix in [11] and [10] which involves only one single meta-action in each row, here the transaction of atomic actions in our domain relies on one or more meta-actions.

Clearly, an action rule can be triggered with the set of meta-actions listed in one single row of an influence matrix as long as all of its atomic actions are listed in that row. Otherwise, selecting a proper set of meta-actions combined from multiple rows becomes necessary. If we would like to activate action rule r given in previous section, it is quite obvious that the combination of meta-actions listed in the first and second row of Table 1 could trigger r, as meta-actions $\{M_1, M_2, M_3, M_4\}$ cover all required atomic actions (a, a_2), $(b, b_1 \rightarrow b_2)$ and $(d, d_1 \rightarrow d_2)$ in r. On the other hand, one set of meta-actions could possibly trigger multiple action rules, like the meta-action set $\{M_1, M_2, M_3, M_4\}$ triggers not only r as mentioned but also action rule $[\{(a, a_2), (b, b_2 \rightarrow b_3)\} \Rightarrow \{(d, d_1 \rightarrow d_2)\}]$ by following the second and forth row in Table 1, if such action rule exists in S.

Suppose a set of meta-actions $M = \{M_1, M_2, ..., M_n : n > 0\}$ can trigger a set of action rules $\{r_1, r_2, ..., r_m : m > 0\}$ that covers certain objects in S. The coverage or support of M is the summation of support of all covered action rules as shown below, which is the total number of objects in the initial state that are going to be affected by M. The confidence of M is calculated by averaging the confidence of all covered action rules. Furthermore, the effect of applying M is defined as the product of its support and confidence $(sup(M) \cdot conf(M))$. It represents the number of objects in the system that are going to be improved by applying M which also is used for calculating the increment of NPS rating caused by it. Therefore, greater is the effect of M, larger increment of NPS rating will be produced.

- $sup(M) = \sum_{i=1}^{m} sup(r_i)$
- $conf(M) = \frac{\sum_{i=1}^{m} sup(r_i) \cdot conf(r_i)}{\sum_{i=1}^{m} sup(r_i)}$

4 Process of the Strategy

4.1 Background

The importance of a well organized and informative result is mentioned briefly earlier. Now, the reasons for pursuing that are explained thoroughly below:

1. The commonness between sets of meta-actions tells about an enhancement. Although actions rules are triggered by different sets of meta-actions, there are some meta-actions which exist in different sets. The commonness between two sets of meta-actions is defined as the percentage of the common meta-actions both sets occupy in the smaller set. If the commonness between two sets equals to 1, then the smaller set belongs to the larger one. In this case, with the larger set of meta-actions applied, besides the action rules covered by it, the action rules covered by the smaller set are also invoked. So the effect of executing the larger set of meta-actions is expected to be enhanced with the effect of smaller one added.
2. The effect of sets of meta-actions tells about preference. Generally speaking, in contrast to the smaller set of meta-actions, it is understandable that the larger set to which a smaller one belongs is more preferable to clients due to its greater effect. On the other hand, for disjoint sets of meta-actions, the selection preference is no longer built on the commonness, but to some extent directly on their effect. Another words, the ones with more significant effect are certainly more preferable to clients.
3. The applicability of certain meta-actions tells about rejection. In our domain, an action rule will not be effective unless all required meta-actions have been taken. So, clients are encouraged to perform all given meta-actions to achieve the expected improvement. But in real life, we can foresee that some of them will be rejected by clients as it is undeniable that some meta-actions are more acceptable and easier to be executed than other ones because of their cost or other issues. For example, if clients are given a set of meta-actions in which lowering the price is mentioned, clients will not like it, even though they are advised by the system to do it for pleasing the customers who complain about the cost. So meta-actions with poor applicability will lead to the rejections from clients.

Considering these three reasons together, the decision of choosing meta-actions to apply does not solely depend on us any more, as our system can calculate the effect of sets of meta-actions, but we are not the experts in estimating the difficulty or cost of adopting certain meta-actions in the field. The decision must be done by clients and technicians who have professional knowledge about this practice. Also, it should be mentioned that clients probably would ignore a set of meta-actions if its effect and cost are remarkably unbalanced and there is an alternative. So the final decision should be determined by our clients who weight the effect and cost of applying some actions with their experience and expertise, which proves the necessity of the strategy proposed in this paper.

This strategy aims to create actionable paths that are well classified from groups of meta-actions and lead to certain increment of NPS rating. These actionable paths can be seen as a forest of trees where each tree is made of one or more meta-nodes that are hierarchically structured regarding their relation. A meta-node is a choice for client to consider and it represents a set of action rules and their triggers. So, the effect of a meta-node is the effect of the group of meta-actions in it. In a tree (or path), the connection between two nodes is a parent-child relationship and it indicates that the commonness between the groups of meta-actions contained in them equals to 1. So, the node with smaller set of meta-actions is defined as the parent of the node with larger set and it is put on an upper level. The parent-child relationship is a one to many mapping, as one node has at most one parent, but a parent node can have more than one child.

The details concerning the strategy will be explained in following parts, which is the main focus of this paper.

4.2 *Advanced Matrix: Transformation of Influence Matrix*

To find triggers for action rules, influence matrix mentioned earlier is a semi-product for us and it should be transformed into an advanced representation to demonstrate the triggers for each rule straightforwardly. Table 2 is an example of advanced matrix. In the table, there are n meta-actions in total and M_i represents every single meta-action, where $0 < i < n + 1$. There are m action rules and "rule j" denotes every generated action rule, where $0 < j < m + 1$. For each row, each cell corresponding to the meta-action that is responsible for triggering rule j is filled with 1, while other cells are filled with 0.

Table 2 Advanced Matrix: Action Rules and Their Triggers (Meta-actions)

	M_1	M_2	M_3	...	M_n
rule 1	0	1	1	...	1
rule 2	1	0	1	...	1
rule 3	0	0	1	...	1
rule 4	0	1	0	...	1
...	...	...	...	...	...
rule m	0	0	1	...	1

Advanced matrix is transformed from influence matrix. Based on advanced matrix, the association between meta-actions and action rules becomes more apparent and it is easy to sort all the meta-action by their popularity, which is the basis of the following procedures.

4.3 Presentation of the Strategy

As mentioned earlier, the goal of our strategy is to build a forest of trees where each tree is composited by meta-nodes that are linked via a parent-child relationship. Algorithm 1 is designed to organize the advanced matrix in a hierarchical manner and generate a list of meta-nodes T, so it is obvious that an advanced matrix M should be given. At the beginning of Algorithm 1, *meta-list* and *rules* are generated, which are a list of meta-actions descendingly sorted by their popularity in M and a set of all action rules in M respectively. Meanwhile, a map *MetaMap* and a list T are initialized as well. *MetaMap* is created to store the mappings (entries) from sets of meta-actions to their sets of action rules during the entire process. Generally speaking, the content in an entry or mapping in *MetaMap* is identical to a meta-node at the end of the algorithm and each action rule can be involved with only one mapping. The list T will be used to store the continually generated meta-nodes and it will be our final product.

With a sorted list of meta-actions, the algorithm repeatedly runs the main part with one meta-action at a time from the most frequent to least frequent one. Given a meta-action (which is represented as *meta*) in each round, all the action rules are iterated one by one and only the ones associated with *meta* in M will be continued to the following procedures. For each continued action rule (which is denoted as r), the existence of a mapping established in *MetaMap* involving r differs the way of handling it and we can easily tell its existence by attempting to retrieve a mapping E associated with r. If the mapping is valid, it indicates that other triggers for r have been processed and stored in E before *meta*, so a non-empty set of meta-actions (which is denoted as *meta-action-set*) is retrieved along with its *mapped-rules* set from E; Otherwise, a new mapping from *meta* to r should be established instead. In terms of a valid mapping, it is defined as a mapping with a non-empty key (set of meta-actions) in our domain, so an empty entry will be found if a mapping is invalid. For the former situation with a valid existing mapping discovered, it is apparent that a new mapping E' to r has to be built due to the enlargement of its triggers, and the existing mapping E has to be updated accordingly to keep the distinctness of action rules. Building a new mapping in *MetaMap* by creating a new entry E' with *meta* added into the retrieved *meta-action-set* and mapped to r is straightforward, so is the removal of r from *mapped-rules* in the existing entry E. But there is no necessity of keeping E if its action rule set becomes empty after the removal, with regards to the definition of a valid mapping. Similarly, building a new entry E' with only $\{meta\}$ mapped to $\{r\}$ in *MetaMap* for the latter situation is simple as well. However, it is possible that a mapping from *meta* already exists. If this is the case, which implies another set of actions rules that can be triggered by *meta* have already been stored, then r should be added into the existing set of action rules mapped from *meta*, instead of creating a new mapping.

Every time a new mapping (or entry) E' is put into *MetaMap*, no matter if it is from an existing mapping or not, Algorithm 2 will be called on the condition that current action rule r is fulfilled. By r is fulfilled, we mean that the set of meta-actions in the new entry E' is all we need to trigger r according to advanced matrix M. Algorithm 2

Algorithm 1. Hierarchically Organize Triggers Algorithm

```
Input: M: an advanced matrix containing action rules and their corresponding triggers.
Output: T: a list containing all generated metaNodes.
  Generate meta-list: a list of meta-actions that are descendingly sorted by their popularity in M;
  Generate rules: a set of all action rules in M;
  Initialize MetaSets (meta-action-set, mapped-rules);
  Initialize a list T for storing all generated metaNodes;
  for meta ∈ meta-list do
    for r ∈ rules do
      if meta is one of r's triggers then
        retrieve entry E (meta-action-set, mapped-rules) for r ∈ mapped-rules from MetaSets;
        if E is valid then
          add a new entry E'(meta-action-set ∪ {meta}, {r}) to MetaSets;
          mapped-rules = mapped-rules \ {r};
          if mapped-rules is empty then
            remove entry E (meta-action-set, mapped-rules) from MetaSets;
          else
            keep entry E (meta-action-set, mapped-rules) in MetaSets;
          end if
        else
          retrieve entry E' ({meta}, mapped-rules);
          put entry E' ({meta}, mapped-rules ∪ {r}) to MetaSets;
        end if
        if meta-action-set ∪ {meta} trigger r completely then
          T = TreeEditor(meta, r, meta-action-set, T);
        end if
      end if
    end for
  end for
  Sort meta-nodes in T by their effect in a descending order.
  return T;
```

is responsible for two aspects: constructing meta-nodes and building trees by linking meta-nodes regarding their relation, which requires relevant information including current meta-action *meta*, current action rule *r*, *meta-action-set* from *E* and of course *T* for storing meta-nodes. To construct a meta-node for a most recently fulfilled rule *r*, the first step is to check whether there is a meta-node in *T* which contains the same set of meta-actions as *r* does but fulfilling other rules. Since every meta-node can be seen as a mapping in *MetaMap*, the validity of a meta-node is evaluated in a similar way as we did for validating a mapping previously. As shown in Algorithm 2, if the meta-node *N* retrieved for the purpose of validation from *T* is non-empty, then *r* should be added into the action rule set along with other rules represented as "*" that have already been stored in *N*. Otherwise, it means that *N* is empty, so a new entry *(meta-action-set* ∪ *{meta}, {r})* should be assigned to *N* and added into *T* as a new meta-node. With a newborn meta-node *N*, the last but not least step is to set up the parent-child relationship through looking for its parent node. The parent node will

Algorithm 2. TreeEditor

Input: *meta*: a single meta-action.
r: an action rule.
meta-action-set: a set of meta-actions.
T: a tree storing the current added metaNodes and their relationship.
Output: *T*
retrieve metaNode *N (meta-action-set* ∪ *{meta}, *)* from *T*;
if *N* is valid **then**
set *N* as (*meta-action-set* ∪ *{meta}*, * ∪{r});
set *effect(N)* = *effect(N)* + *sup(r)* · *conf(r)*;
if N is someone's parent **then**
update N's children's effect.
end if
else
set *N* as *(meta-action-set* ∪ *{meta}, {r})*;
set *effect(N)* = *sup(r)* · *conf(r)*;
add *N* into *T*;
retrieve metaNode *P (meta-action-set, *)* in *T*;
if *P* is valid **then**
set *P* as *N*'s parent;
set *effect(N)* = *effect(N)* + *effect(P)*;
end if
end if
return *T*;

not exist unless the mapping E' is extended from an existing mapping E in previous procedures in Algorithm 1, in other words, *meta-action-set* is not empty. Therefore, as long as the meta-node P retrieved by looking for a meta-node which has *meta-action-set* is valid, P is N's parent and N is one of P's children. On the other hand, it is unnecessary to set up the relationship for a newly updated meta-node because its parent must be found when it has been made.

Besides organizing sets of meta-nodes hierarchically as a tree, computing the expected effect of each meta-node during the process is another characteristic step in our strategy. It provides clients with concrete clues to evaluate the worth of those meta-actions. Since the effect of triggering a single action rule is defined as the product of its support and confidence, the effect of a meta-node is the summation of the effect of triggering all action rules in it. So whenever a new action rule is added into an existing meta-node, its effect must be added into the meta-node as well. Additionally, if this existing meta-node which has just been updated has any children, its children's effect need to be updated as well. And for a newborn meta-node, besides its own influence computed on the basis of our regulations, its effect is strengthened with its parent's help unless its parent does not exist.

Algorithm 1 will sort the meta-nodes in T by their effect in a descending order and return it after all the meta-actions in *meta-list* have been processed. An example of printed result will be shown in the next section.

5 Experiment

To test its performance, experiments are carried out with the JAVA based implementation of the proposed algorithm. Firstly, we take a small sample of data as an example to show the procedures in Algorithm 1 and the representation of final results. Table 3 is the advanced matrix in our example, and there are six action rules and four meta-actions involved in the sample. These four meta-actions are descendingly sorted by their popularity in *meta-list* as $\{M_3, M_4, M_2, M_1\}$ and they will be checked one by one in such order. Hence, with M_3 as the most frequent meta-action in the list, all the action rules associated with it will be stored in an entry in *MetaMap* at the first round, which is $(\{M_3\}, \{r_1, r_2, r_3, r_4, r_6\})$. There is no action rule that has been fulfilled, so no meta-node will be made yet. When it comes to M_4, the action rules involving it are handled in different ways. Action rules that have already been stored in *MetaMap*, like r_1, r_2, r_3 and r_6, should be moved to another mapping with $\{M_3, M_4\}$, and the previous mapping becomes $(\{M_3\}, \{r_4\})$. For action rules that are new to *MetaMap*, a new entry is built and it is $(\{M_4\}, \{r_5\})$ in this example. In terms of building meta-nodes, when iterating action rules top down, we would find that r_3 is fulfilled with $\{M_3, M_4\}$. Obviously a meta-node $(\{M_3, M_4\}, \{r_3\})$ should be built and its parent who is suppose to be $(\{M_3\}, *)$ doesn't exist, so no parent is affiliated and its effect is $sup(r_3) \cdot conf(r_3)$. The same meta-actions trigger r_6 as well, so the meta-node becomes $(\{M_3, M_4\}, \{r_3, r_6\})$, and its effect is updated as $sup(r_3) \cdot conf(r_3) + sup(r_6) \cdot conf(r_6)$. We should take care of its children's effect as well, but there isn't any, so nothing needs to be done. As all the action rules are appeared in *MetaMap* now, we focus on creating new mappings from existing ones and updating the existing mappings by following the requirements. Along with the same procedures being performed to the rest meta-actions, more meta-nodes will be built, such as $(\{M_3, M_4, M_2\}, \{r_1\})$, $(\{M_2, M_4\}, \{r_4\})$ and so on. Whenever a new meta-node is made, it is necessary to look for its parent. For instance, meta-node $(\{M_3, M_4, M_2\}, \{r_1\})$ is a mapping extended from $\{M_3, M_4\}$, and meta-node $(\{M_3, M_4\}, \{r_3, r_6\})$ does exist, thus these two meta-nodes should be connected with a parent-child relationship as the former is a child of the latter. Meanwhile, the effect of the child's node is its own effect plus its parent's effect.

Table 3 Advanced Matrix for the Sample Data

	M_1	M_2	M_3	M_4
r_1	0	1	1	1
r_2	1	0	1	1
r_3	0	0	1	1
r_4	0	1	0	1
r_5	1	1	1	0
r_6	0	0	1	1

When the algorithm comes to the end, a list of meta-nodes will be printed as Fig. 1 shows. There are five meta-nodes generated and they are listed in an descending order according to their effect, which is basically the priority that clients follow to judge them. Similar as running the program with this sample, we run it with a larger set consisting of 127 action rules and 627 meta-actions, and the program sorts the result in a format as shown in Fig 1, which lists 78 meta-nodes, their relations and their effect.

```
meta-node0= ({m3, m4, m2}, {r1}), its effect is 2+6.15=8.15
meta-node0's parent is ({m3, m4}, {r6, r3})
meta-node1= ({m3, m4, m1}, {r2}), its effect is 1.8+6.15=7.95
meta-node1's parent is ({m3, m4}, {r6, r3})
meta-node2= ({m3, m4}, {r6, r3}), its effect is 3.6+2.55=6.15
meta-node2's parent is not valid
meta-node2's children are: ({m3, m4, m2}, {r1})
meta-node2's children are: ({m3, m4, m1}, {r2})
meta-node3= ({m3, m2, m1}, {r5}), its effect is 4.35
meta-node3's parent is not valid
meta-node4= ({m4, m2}, {r4}), its effect is 1.6.
meta-node4's parent is not valid
```

Fig. 1 Experiment Result with Given Example

6 Conclusion

From the example result shown in Fig.1, it is clear that the strategy helps clients select actions of top priority by considering the predicted effect as a partial reason for the final decision. What's more, the experiment with larger dataset proves the efficiency of organizing the meta-actions in the proposed way. It is justified by the number of generated meta-nodes which is much smaller than the number of action rules or sets of meta-actions triggering them (it is over 30% off the original choices within clients' evaluation). On the other hand, the effect of applying sets of meta-actions is calculated more accurately with the definition of parent-child relationship. Therefore, this strategy provides a more effective way for clients to choose the right set of meta-actions out of the generated suggestions, and conclusively fits our anticipation.

References

1. Hajja, A., Ras, Z.W., Wieczorkowska, A.: Hierarchical object-driven action rules. Journal of Intelligent Information Systems **42**(2), 207–232 (2014). Springer
2. He, Z., Xu, X., Deng, S., Ma, R.: Mining action rules from scratch. Expert Systems with Applications **29**(3), 691–699 (2005). Elsevier
3. Kuang, J., Daniel, A., Johnston, J., Raś, Z.W.: Hierarchically structured recommender system for improving nps of a company. In: Cornelis, C., Kryszkiewicz, M., Ślęzak, D., Ruiz, E.M., Bello, R., Shang, L. (eds.) RSCTC 2014. LNCS, vol. 8536, pp. 347–357. Springer, Heidelberg (2014)
4. Kuang, J., Raś, Z.W., Daniel, A.: Hierarchical agglomerative method for improving nps. In: Kryszkiewicz, M., Bandyopadhyay, S., Rybinski, H., Pal, S.K. (eds.) PReMI 2015. LNCS, vol. 9124, pp. 54–64. Springer, Heidelberg (2015)
5. Paul, R., Hoque, A.S.: Mining irregular association rules based on action and non-action type data. In: Proceedings of the Fifth International Conference on Digital Information Management (ICDIM), pp. 63–68 (2010)
6. Raś, Z.W., Wieczorkowska, A.A.: Action-rules: how to increase profit of a company. In: Zighed, D.A., Komorowski, J., Żytkow, J.M. (eds.) PKDD 2000. LNCS (LNAI), vol. 1910, pp. 587–592. Springer, Heidelberg (2000)
7. Ras, Z., Dardzinska, A.: From data to classification rules and actions. In: Proceedings of the Joint Rough Sets Symposium (JRS07). LNAI, Vol. 4482, pp. 322–329. Springer (2011)
8. Rauch, J., Šimůnek, M.: Action rules and the guha method: preliminary considerations and results. In: Rauch, J., Raś, Z.W., Berka, P., Elomaa, T. (eds.) ISMIS 2009. LNCS, vol. 5722, pp. 76–87. Springer, Heidelberg (2009)
9. Reichheld, F.F.: The one number you need to grow. Harvard Business Review, 1–8, December 2003
10. Touati, H., Kuang, J., Hajja, A., Ras, Z.: Personalized Action Rules for Side Effects Object Grouping. International Journal of Intelligence Science **3**(1A), 24–33 (2013)
11. Tzacheva, A., Ras, Z.W.: Association action rules and action paths triggered by meta-actions. In: Proceedings of 2010 IEEE Conference on Granular Computing, Silicon Valley, CA, pp. 772–776. IEEE Computer Society (2010)

Merging Bottom-Up and Top-Down Knowledge Graphs for Intuitive Knowledge Browsing

Gwendolin Wilke, Sandro Emmenegger, Jonas Lutz, and Michael Kaufmann

Abstract The Lokahi Enterprise Knowledge Browser provides an intuitive and flexible way to query a company's intranet knowledge. In addition to conventional search capabilities, it allows the user to browse through a semi-automatically generated knowledge map that visualizes intranet knowledge as a network/graph structure of semantic relations that are extracted top-down from structured documents, as well as bottom-up from unstructured documents. This paper describes the underlying fuzzy graph data structure, the method for extracting concepts and associations from text documents, and the merging of the resulting data structure with a predefined enterprise ontology.

Keywords Enterprise search · Knowledge browsing · Knowledge map · Knowledge graph · Fuzziness · Association rule mining · Connectivism

G. Wilke(✉)
Institute of Business Information Management IWI, Lucerne University of Applied Sciences and Arts, Zentralstr. 9, 2940, 6002 Lucerne, Switzerland
e-mail: gwendolin.wilke@hslu.ch

S. Emmenegger · J. Lutz
Institute for Information Systems, University of Applied Sciences and Arts Northwestern Switzerland, Riggenbachstr. 16, 4600 Olten, Switzerland
e-mail: {sandro.emmenegger,jonas.lutz}@fhnw.ch

M. Kaufmann
School of Engineering and Architecture, Lucerne University of Applied Sciences and Arts, Technikumstr. 21, 6048 Horw, Switzerland
e-mail: m.kaufmann@hslu.ch

T. Andreasen et al. (eds.), *Flexible Query Answering Systems 2015*,
Advances in Intelligent Systems and Computing 400,
DOI: 10.1007/978-3-319-26154-6_34

1 Introduction

Value creation in companies relies increasingly on knowledge work; that is, creating managing and exchanging information. Intranet search engines are often indispensable tools to make explicitly stored information accessible for knowledge workers. Yet, often, a desired piece of information can only be found if context-specific keywords are known, and, as a result, much of the know-how and the stored information often remains inaccessible. To address this issue, Kaufmann et. al. [9] proposed a conceptual framework for an *enterprise knowledge browser*, which, in addition to conventional search functionality, provides the user with the possibility to visually browse through a company's knowledge landscape, which is represented as a semantic graph structure. A prototype of the enterprise knowledge browser including an intuitive, interactive graphical user interface (GUI) - is currently under development as part of the research project *LOKAHI Inside* in collaboration with the software company FIVE Informatik AG[1], and we call it the *Lokahi Enterprise Knowledge Browser* (LEKB). This paper contributes a more detailed description of the concepts introduced in [9], namely the method for extracting concepts and associations, and the method for merging this extracted information with a predefined enterprise ontology.

The paper is structured as follows: Section 2 gives a brief overview of the components and design principles of the LEKB as far as they are relevant for the underlying Lokahi knowledge graph discussed in this paper. Section 3 gives a brief account of related approaches to networked knowledge representation. Section 4 introduces the Lokahi Knowledge Graph (LKG) as an extension of the graph representation underlying the Entity-Relationship-Model [3]. Sections 5 and 6 describe the Lokahi bottom-up and top-down knowledge extraction approaches, respectively, together with their prototypical implementations and formal representations as Lokahi Knowledge Graphs. The merging of the two graphs is described in section 7, and the result is the graph structure underlying the LEKB. Finally, section 8 concludes the paper and gives an outlook on further work.

2 The Lokahi Enterprise Knowledge Browser

The project Lokahi is currently developing a search interface that is based on an interactive fuzzy graph of keyword associations to enable flexible query answering. A screenshot of the current implementation is presented in Figure 1.

The user can search for keywords, and the system displays (1) a list of associated files and (2) a visualization of a browsable interactive graph that contains associated keywords. If the user enters a search keyword, the system returns not only a standard list of relevant documents, but also a knowledge map that indicates other keywords associated with the search. Thereby, the user can browse and explore the knowledge landscape. By double-clicking on nodes in the graph, the user can add keywords to the filter. Single-clicking on a node navigates the display is to show the selected node

[1] `http://www.fiveinfo.ch`

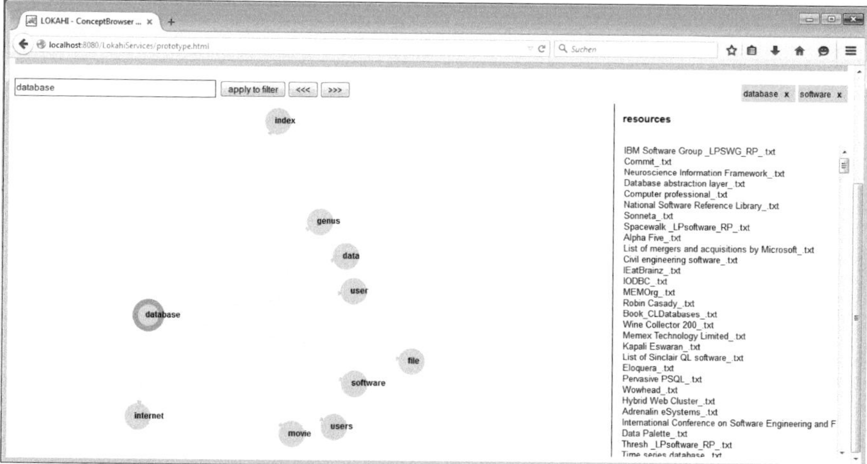

Fig. 1 Screenshot of the current Lokahi Enterprise Knowledge Browser GUI prototype.

as the new central concept. The example in figure 1 is based on bottom-up extraction of keywords and associations from a sample of Wikipedia pages, using the method described in section 5.

The Lokahi Enterprise Knowledge Browser comprises three major components: The first component, knowledge representation, is the *Lokahi Knowledge Graph* (LKG), which is a fuzzy directed labeled multigraph. It is the underlying data structure and conceptual basis for the Lokahi Enterprise Knowledge Browser and provides the formal representation of the visualized semantic network of interconnected keywords.

The second component is the *knowledge extraction* approach, which generates a concrete instantiation of an LKG from data. This approach consists of three steps: In the first step, a so-called *bottom-up Lokahi Knowledge Graph* is generated by automatic information extraction from unstructured documents (text files, wiki pages, emails, websites, etc.). In the second step, structured data, such as data available from enterprise application systems, are enriched with generic knowledge on enterprise architectures to define an enterprise ontology. The result is a so-called *top-down Lokahi Knowledge Graph.* In the third step, both LKGs are merged, and the resulting LKG is the semantic network that is visualized by the Lokahi Enterprise Knowledge Browser.

The third component is the GUI design, which is of paramount importance for user acceptance of the LEKB: A semantic network like the LKG is usually very complex, comprising numerous keywords and semantic connections. As a result, it usually cannot be represented as a planar graph, and, furthermore, it is usually far too big to be displayed as a whole. Therefore, we only display clippings (subgraphs) of the underlying LKG that each represent a local semantic context of a user-selected keyword. Here, it is essential to provide an intuitive way of visually browsing through (and editing) the complete underlying knowledge graph based on single clippings.

In [9], we described the conceptual framework, design principles and general architecture of the Lokahi Enterprise Knowledge Browser. While the third component (GUI design) will be detailed in a separate publication, the present paper elaborates in greater detail the first and second components, namely the underlying graph data structure and the knowledge engineering approach used to enrich it with data, cf. sections 5-7.

2.1 Supporting the Connectivist Learning Paradigm

Connectivism [18] is a relatively new learning theory that was developed with a focus on understanding knowledge in present-day socio-technological interactions. It explains learning in terms of connecting new information within a network of already known concepts. While the classic learning theories (behaviorism, cognitivism and constructivism) adopt the view that learning only occurs inside a person's brain, connectivism additionally addresses learning that occurs in the external world. Knowledge, in that understanding, can be seen as a subclass of information that is *usable* in the sense of Shadbolt [17]. Using technology, connection making can be represented outside the human brain, e.g., in computer systems. Therefore, connectivism can explain organizational learning in terms of making connections in the organizational memory. A knowledge management system that supports the process of connectivist learning can be called a *connective knowledge management system* (CKMS), and it is this idea that underlies the Lokahi Enterprise Knowledge Browser (LEKB). A CKMS empowers individuals and organizations to cope with large amounts of fast changing information by providing them with tools to interactively and iteratively build up a personal or organizational knowledge network structure that embeds single pieces of information in a conceptual context. It thus provides a human-centered approach to knowledge management. The LEKB implements the idea of a CKMS by allowing the user to visually browse (and edit) the organizational knowledge network structure and to access related documents directly in the GUI.

2.2 Top-Down Knowledge Structures Complement Bottom-Up Knowledge Extraction

The knowledge network structure underlying the LEKB is generated using both, bottom-up and top-down knowledge engineering approaches. In the bottom-up approach, automated keyword extraction from unstructured documents (e.g., text files, emails, or wiki-pages) is used together with fuzzy association rule mining to generate a *bottom-up fuzzy knowledge graph*. In the complimentary top-down approach, an RDFS representation of an enterprise ontology is semi-automatically generated from structured documents (e.g., from a customer relationship management system) and an enterprise architecture framework. The result is a *top-down (crisp) knowledge graph*. Both graphs are merged. The merge is driven by the usually much smaller top-down graph, so that the top-down structure (which may be specified by the management) forms the structural skeleton of the LEKB. It also helps linking company internal vocabulary to general terms and technical language, thereby mitigating the problem of intuitive labelling in pure bottom-up approaches.The automatic generation of bottom-up keywords and their association with the top-down skeleton in turn mitigates the problem of expensive manual maintenance present in pure top-down approaches.

2.3 Fuzzy Associations for Intuitive Browsing

The Lokahi knowledge graph is a fuzzy graph. The semantic associations between keywords are represented as fuzzy directed arcs, and the fuzzy membership degree reflects the strength of their semantic connection. In the (crisp) visualization in the GUI, the fuzziness of semantic connections is used to implement a *semantic zooming* functionality that resembles zooming in a geographic map: Zooming out emphasizes terms with stronger semantic connections and blanks terms with weaker connections; as one zooms in, more detail becomes visible. As a consequence, users can carry out interaction with the system on a suitable level of information granularity, which prevents them from getting lost in a flood of detail. Since the top-down graph is crisp, the semantic connections between top-down terms are always visible and provide the visual "skeleton" of the map.

3 Related Approaches

Network structures for knowledge representation have a longstanding history in different application areas and research fields, and we give only a brief account of some prominent approaches that are related to our approach. One of the first to propose network structures for capturing knowledge was C. S. Peirce, who in 1882 proposed *existential graphs* as a form of visual notation for logical expressions [14]. In 1962, M. Masterman [12] introduced *semantic networks* as a form of graph-based notation for linguistic expressions that abstract from the "base language." Based on this idea, J. Sowa [19] devised *conceptual graphs* (CG), which map natural language questions and assertions to a relational database. CGs are bipartite graphs, in which concepts (entities) and associations (relationships) are represented by two different types of nodes.

Another approach is the *semantic networks* of M. R. Quillian [16], which are intended to represent the semantics of English words in the form of a network. In 1976, P. Chen wrote his seminal paper in which he introduced the entity relationship model for knowledge representation [3]. J. D. Novak and D. B. Gowin [13] did research in the domain of human learning and knowledge construction. Their studies led to the development of *concept maps*, which enable users to visualize knowledge that is easily comprehensible by others. Ten years later, [2] introduced *topic maps*, which led to the ISO Standard 13250 of the workgroup ISO JTC1/SC34/WG3[2]. As elaborated in the following section, the knowledge network representation in the

[2] `http://www.topicmaps.org/xtm/`

form of a fuzzy labeled dimultigraph as proposed here is derived from Chen's entity relationship model.

4 The Lokahi Knowledge Graph and Its Relation to the ER-Model

In Chen's entity relationship (ER) model [3], an *entity* is "a 'thing' that can be distinctly identified", and a relationship is "an association among entities." Based on that, a simplified version that omits attributes, an ER-graph can be defined as a directed weighted graph $G = (V, A, W, p, w)$ according to [10, pp.1-2], with sets V, A and W, an *incidence mapping* $p : A \to V^2$ and a *weight function* w. Here, G represents a knowledge network with entity weights and relationships. Every entity is represented by a vertex $v \in V$; every relationship is represented by an arc $a \in A$, and every relationship role is a *symbolic weight* (or *label*) $w \in W$. The incidence mapping p defines two mappings $s, t : A \to V$ by $(s(a), t(a)) := p(a)$, where $s(a)$ is called the *origin (start, head)* of a, and *t(a)* is the *tail* of a. For every arc a, the symbolic weight $w(a)$ represents the role of the entity $t(a)$ in the relationship. Accordingly, arcs represent knowledge of typed associations between entities.

In order to build a knowledge network structure according to the design principles listed in subsections 2.1-2.3, we extend the above notion of a directed weighted (labeled) graph by a fuzzy membership function for arcs.

Definition 1. The Lokahi Knowledge Graph (LKG) *is a fuzzy labeled dimultigraph* $\mathcal{L} = (\Sigma, V, A, \mu_A, \sigma, \tau, \lambda)$, where

- Σ is a finite alphabet.
- $V \subseteq \Sigma^\star$ is the vertex set of $\mathcal{L}$ (where vertices are strings over Σ).
- A is the arc set of $\mathcal{L}$.
- $\mu_A : A \to [0, 1]$ is a fuzzy membership function that indicates the degree of membership of an arc $a \in A$ in $\mathcal{L}$;
- $\sigma : A \to V$ and $\tau : A \to V$ are two maps that indicate the source and target vertex of an arc;
- $\lambda : V \cup A \to \Sigma^\star$ is a labeling function for vertices and arcs.

Notice that the arcs of a Lokahi Knowledge Graph are fuzzy, while the vertex set is crisp.

5 The Bottom-Up Lokahi Knowledge Graph

The bottom-up part of the Lokahi knowledge extraction method automatically extracts keywords from a given corpus of text documents in a company. These keywords and their relations represent *implicit knowledge*, which is usually distributed over a multitude of documents in the form of unstructured and semi-structured data. We assume that there is an intersubjective semantic distance measure between keywords that derives from an organization's collective body of knowledge. For any given keyword b, the set of keywords that are semantically close to b (w.r.t. this distance measure) can be said to establish the *semantic context* of b. It is the goal of the bottom-up part of the Lokahi knowledge extraction to quantify the semantic distance of keywords within a corpus of documents and then visualize their resulting semantic contexts as a browsable visual knowledge graph.

In order to quantify semantic distance, we use the normalized likelihood ratio (NLR) measure, introduced by [8, p.55]. Portmann et. al. [15] have proposed to apply the NLR to text mining as a semiotic and inductive approach to knowledge retrieval. The NLR is based on the *likelihood* measure $L(b_j|b_i)$: Given a document d that contains a keyword of interest b_i, $L(b_j|b_i)$ expresses the strength of evidence that a document d that contains b_i also contains b_j. According to the law of likelihood [5], it is the ratio of likelihoods that is relevant for comparing two hypotheses, not the likelihoods themselves. Instead of the likelihood $L(b_j|b_i)$ itself, Kaufmann uses the normalized likelihood ratio l_{ij} to measure semantic distance of keywords in the corpus:

$$l_{ij} := \frac{L(b_j|b_i)}{L(b_j|b_i) + L(-b_j|b_i)} \in [0, 1]. \tag{1}$$

l_{ij} is a normalized comparison of the hypotheses b_j and $-b_j$ under evidence b_i. Here, $L(b_j|b_i)$ is measured as the number of documents in the corpus that contain both keywords, b_i and b_j, while $L(-b_j|b_i)$ is measured as the number of documents that contain b_i but not b_j. Notice that l_{ij} is not necessarily symmetric, and that the property expresses the fact that an intuitive notion of semantic closeness is not necessarily symmetric as well. For example, if John Doe is a Java programmer in the company, the keyword *Java* is very closely associated with *John Doe*, because it is his main occupation. Yet, *John Doe* may not be the only Java programmer in the company and the company may be specialized in Java-based software development. In this case, the semantic binding of the keyword *John Doe* to *Java* would intuitively be much less close.

The following subsection 5.1 gives a brief outline of a prototypical implementation of the bottom-up knowledge extraction method, and subsection 5.2 provides the formalization of its output as a Lokahi knowledge graph.

5.1 Implementation

Our prototypical implementation of the bottom-up knowledge extraction method consists of three parts:

1. **Keyword Extraction:** In the first step, a set $B \subseteq \Sigma^\star$ of keywords is extracted from the corpus C of documents. Here, $\Sigma^\star$ denotes the set of words over the underlying alphabet Σ (the Unicode character set). We use the Apache Lucene

library MoreLikeThis (MLT)[3] to extract for each document d the set $\kappa(d)$ of top n most relevant keywords of the document and then define the set of bottom up concepts B as the union of all top-n keywords over all documents in the corpus, $B = \bigcup_{d \in C} \kappa(d)$.

2. **Frequent Item Set Mining:** The second step is a pre-processing step that is used to decrease the number of keyword-pairs to be processed in the computationally intensive step three. Using Agrawal's Apriori algorithm [1], we extract from the keyword set B a set F of *two-element subsets of keywords*. F is defined by $F := \{\{b_i, b_j\} | b_i, b_j \in B, b_i \neq b_j, \delta(b_i, b_j) \geq t\}$. This set is called the *frequent keyword set* of C. Here, $\{b_i, b_j\}$ is in F, if b_i and b_j frequently occur in the same document, i.e., if the number of documents $\delta(b_i, b_j)$ that contain both keywords, b_i and b_j, is greater than a given threshold t. In our prototype implementation, the parameter t was set to $t = 0.005$, i.e., 0.5% of all documents must have contained both keywords.
3. **Fuzzy Association Rule Mining:** In the third and last step, every two-keyword set $\{b_i, b_j\} \in F$ is used to generate two fuzzy association rules $l_{ij}/(b_i \rightarrow b_j)$ and $l_{ji}/(b_j \rightarrow b_i)$. Here, l_{ij} is the normalized likelihood ratio defined in (1), and $l_{ij}/(b_i \rightarrow b_j)$ refers to the association rule $(b_i \rightarrow b_j)$ having a fuzzy degree of membership of l_{ij} in the set $R := \{(b_i \rightarrow b_j) | b_i, b_j \in f \in F, b_i \neq b_j\}$ of crisp association rules.

5.2 *Graph Representation*

The output of the Lokahi bottom-up knowledge extraction approach described in the foregoing subsection can be formally represented as the fuzzy set $\mathcal{R} := (R, \mu_R)$ of fuzzy association rules between keywords. Here, R is the set of association rules generated in step 3, and $\mu_R : R \rightarrow [0, 1],\ \mu_R(b_i \rightarrow b_j) = l_{ij}$, is the corresponding fuzzy membership function. Together with the underlying alphabet Σ, $\mathcal{R}$ can be represented as a Lokahi Knowledge Graph:

Definition 2. Let Σ be a finite alphabet, $B \subseteq \Sigma^\star$ and $\mathcal{R} = (R, \mu_R)$ a fuzzy set of association rules $\mu_R/(b_i \rightarrow b_j)$ with $(b_i \rightarrow b_j) \in R$, $b_i, b_j \in B$ and $\mu_R : R \rightarrow [0, 1]$. The *Bottom-Up Lokahi Knowledge Graph* ($\underline{LKG}$) *induced by the pair* $(\Sigma, \mathcal{R})$ is the Lokahi Knowledge Graph $\underline{\mathcal{L}} = (\underline{\Sigma}, \underline{V}, \underline{A}, \underline{\mu_A}, \underline{\sigma}, \underline{\tau}, \underline{\lambda})$ with

- $\underline{\Sigma} := \Sigma$.
- $\underline{V} := head(R) \cup tail(R) \subseteq \underline{\Sigma}^\star$ is the vertex set of $\underline{\mathcal{L}}$. Here, for every association rule $(b_i \rightarrow b_j) \in R$, $head((b_i \rightarrow b_j)) := b_i$ and $tail((b_i \rightarrow b_j)) := b_j$.
- $\underline{A} := \underline{arc}(R)$ is the arc set of $\underline{\mathcal{L}}$. Here, $\underline{arc}$ is a bijection that maps every association rule $(b_i \rightarrow b_j) \in R$ to a unique arc identifier $\underline{a}_{ij} \in \underline{A}$.
- $\underline{\mu_A} := \mu_R \circ \underline{arc}^{-1} : \underline{A} \rightarrow [0, 1]$ is the fuzzy membership function that associates every arc $\underline{a}_{ij} \in \underline{A}$ with the fuzzy membership degree $\mu_A(\underline{a}_{ij}) = \mu_R((b_i \rightarrow b_j))$.

[3] https://lucene.apache.org/core/4_8_0/queries/org/apache/lucene/queries/mlt/MoreLikeThis.html (viewed on 2015-04-25)

– $\underline{\sigma} := head \circ \underline{arc}^{-1} : \underline{A} \to \underline{V}$ and $\underline{\tau} := tail \circ \underline{arc}^{-1} : \underline{A} \to \underline{V}$ indicate source and target vertex of an arc, respectively.
– $\underline{\lambda} : \underline{V} \cup \underline{A} \to \underline{\Sigma}^{\star}$ labels every vertex with itself and every arc with the empty string

$$\underline{\lambda}(x) := \begin{cases} x & \text{if } x \in \underline{V}, \\ \text{'' ''} & \text{if } x \in \underline{A}. \end{cases}$$

Notice that the bottom-up Lokahi Knowledge graph $\underline{\mathcal{L}}$ is not a "true" multigraph because every pair of vertices $(b_i, b_j) \in \underline{V} \times \underline{V}$ can be mapped to exactly one arc $a_{i,j} \in \underline{A}$. Yet, the property is needed later to merge the bottom-up with the top-down knowledge graph, which can have more than one arc between the same vertices.

6 The Top-Down Lokahi Knowledge Graph

The top-down part of the Lokahi knowledge extraction approach consists of two knowledge engineering steps: In the first step, the existing enterprise meta ontology ARCHIMeo[4] is manually extended by an enterprise application ontology that is derived from the company's business context, cf. figure 2(a). Using ARCHIMeo puts the company specific structure in the context of an enterprise architecture framework that serves as a structural "skeleton" in the Lokahi Knowledge Browser[5]. In the second step, the ontology model is semi-automatically enriched with data from the company's application software systems and exported into a simple Java based representation of a Lokahi knowledge graph.

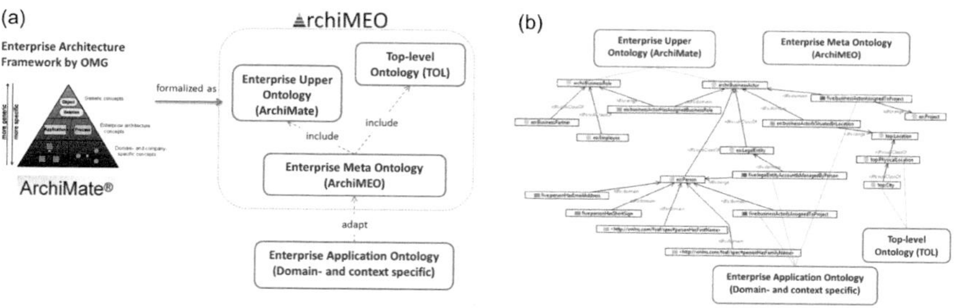

Fig. 2 (a) The parts of the ARCHIMeo meta ontology. (b) Example details of ARCHIMeo after step 1.

[4] ARCHIMeo has been developed at the University of Applied Sciences Northwestern Switzerland (FHNW) as an approach toward relating enterprise ontologies with enterprise architectures [6, 7, 20]. It has been successfully applied in risk management, contract management and software integration support [4, 11, 21].

[5] We also intend to use it for inferring additional information from the merged graph structures in future research.

6.1 Implementation

The existing enterprise ontology ARCHIMeo includes a top-level ontology (including classes such as *PhysicalLocation*), as well as a formalization of the enterprise architecture framework ArchiMate[6] as an enterprise upper ontology (modeling classes such as *BusinessActor*). In the first step, we manually extend it by an enterprise application ontology, cf. figure 2(b). In the second step, the resulting ontology is semi-automatically enriched with data and transformed into a Java-based graph representation. Our prototypical implementation of this step follows a common extract, transform and load (ETL) process with some post-processing:

1. **Extract from Source Systems:** The database schema is manually extracted from existing enterprise systems[7] into a file with a character-separated values (CSV) format. Here, one line represents one instance (*record*) of the source system's object type, such as a specific employee.
2. **Transform and Load Into Enterprise Ontology:** Simple Java-adapters for each data source file read the data records and store them as *instances* of ontology classes in the enterprise ontology. Here, each instance and class is uniquely identified by a URI (uniform resource identifier) that is generated from the respective record's key attribute values as specified in the source system. We also derive *datatype properties* (representing directed relations between instances of classes and RDF literals) and *object properties* (representing directed relations between instances of two classes) from the source systems' record sets.
3. **Infer Human Readable Keywords:** In the ontology, URIs are used as representatives of classes and instances. Yet URIs cannot be used as their representatives in the graphical user interface of the Lokahi knowledge browser, because they do not make sense to a human user. We therefore infer human readable keywords using SPIN[8] (SPARQL[9] Inferencing Notation) rules. We also infer human readable keywords for properties. For instance, persons are labeled with their first and family names (e.g., *John* and *Doe*), and if John Doe is employed by company XY, the respective object property is labeled *employee*.
4. **Transform Into Java-Based Graph Representation:** The relevant parts of the ontology are transformed into a simple in-memory graph representation that is based on the two Java classes *Node* and *Association*. To the *Node* class we map all instances of the ontology, all classes that contain an instance, and all literals. All properties are mapped to the *Association* class. A node is characterized by the *ID* attribute. In the case of a class or instance node, the *ID* is its URI; in the case of a literal node, the *ID* is the literal itself. The *label* attribute contains the keywords inferred in step 3. Every association is attached to a node and is characterized by the *label* attribute, which again captures the inferred keyword, and the *value*

---

[6] www.opengroup.org/subjectareas/enterprise/archimate

[7] In our prototype, these are a human resource (HR) system, a customer relationship management (CRM) system and a project management system.

[8] http://spinrdf.org/

[9] http://www.w3.org/TR/rdf-sparql-query/

attribute, which refers to the *ID* of the instance or literal the association points to.[10]

Notice that we cannot represent the top-down knowledge graph using only labels (keywords) without losing the ability to distinguish two entities with the same label (e.g., two persons with the same first and family names). This is one of the key features of the top-down knowledge structure that can help enhance a pure bottom-up approach. We use a node's *label* attribute only to provide a human-understandable keyword in the GUI instead of URIs. Using *IDs* (URIs) in the knowledge graph allows us to display the same label multiple times in different semantic contexts. For example, one *John Doe* may be working in HR, while another *John Doe* may be employed as a Java programmer in the same company. The user cannot distinguish them based on their label, but he *can* distinguish them based of the label's semantic closeness to the keywords *HR department* and *Java*, respectively. Distinguishing and finding information based on semantic context is the core feature of the Lokahi Knowledge Browser.

Notice also that associations do not have *IDs*, in spite of the fact that two nodes may be connected by more than one association (i.e., "multiple associations" are possible). Here, we assume that association labels are sufficient to distinguish multiple associations. E.g., *John Doe* may be programmer and project leader in *projectXY*, and the two associations may be labeled with *programmer* and *project leader*, respectively.

6.2 Graph Representation

The output of the Lokahi top-down knowledge extraction approach described in the previous subsection can be formally represented by the triple $(\mathcal{N}, label_{\mathcal{N}}, \mathcal{P})$, where $\mathcal{N}$ is the set of node-*IDs* (an *ID* being sufficient for uniquely specifying a node), $label_{\mathcal{N}} : \mathcal{N} \to \Sigma^\star$ the function that assigns to a node the label generated for it via a SPIN rule, and $\mathcal{P}$ is the set of triples of type $id \times value \times label$, each uniquely characterizing an association (i.e., a property). It can formally be described as a Lokahi Knowledge Graph:

Definition 3. Let Σ be a finite alphabet, $\mathcal{N} \subseteq \Sigma^\star$, $label_{\mathcal{N}} : \mathcal{N} \to \Sigma^\star$, and $\mathcal{P} \subseteq \mathcal{N} \times \mathcal{N} \times \Sigma^\star$ a crisp set of labeled associations. The *top-down Lokahi knowledge graph* ($\overline{LKG}$) *induced by the triple* $(\mathcal{N}, label_{\mathcal{N}}, \mathcal{P})$ is the Lokahi knowledge graph $\bar{\mathcal{L}} = (\bar{\Sigma}, \bar{V}, \bar{A}, \bar{\mu}_{\bar{A}}, \bar{\sigma}, \bar{\tau}, \bar{\lambda})$ with

- $\bar{\Sigma} := \Sigma$.
- $\bar{V} := \mathcal{N} \subseteq \Sigma^\star$ is the vertex set of $\bar{\mathcal{L}}$.
- $\bar{A} := \overline{arc}(\mathcal{P})$ is the arc set of $\bar{\mathcal{L}}$, where $\overline{arc}$ is a bijection that maps every $(i, j, k) \in \mathcal{P}$ to a unique arc identifier $\bar{a}_{ijk} \in \bar{A}$.

[10] In our prototypical implementation, we actually use additional attributes to optimize user-friendly GUI interaction. Since these attributes do not affect the underlying graph structure, we do not introduce them here.

- $\bar{\mu}_{\bar{A}} :\equiv 1 : \bar{A} \to [0, 1]$ is the constant function indicating full membership degree of all arcs.
- $\bar{\sigma} := p_1 \circ \overline{arc}^{-1} : \bar{A} \to \bar{V}$ and $\tau = p_2 \circ \overline{arc}^{-1} : \bar{A} \to \bar{V}$ indicate the source and target vertex of an arc, respectively.
- $\bar{\lambda} := \bar{V} \cup \bar{A} \to \Sigma^\star$ is the labelling function for vertices and arcs defined by

$$\lambda(x) := \begin{cases} label_{\mathcal{N}}(x) & \text{if } x \in \bar{V}, \\ p_3(x) & \text{if } x \in \bar{A}. \end{cases}$$

Notice that the top-down Lokahi Knowledge Graph is not a "true" fuzzy graph, since all membership degrees are equal to 1.

7 The Merged Lokahi Knowledge Graph

This section discusses how bottom-up and top-down Lokahi knowledge graphs are merged to form the graph structure that underlies the Lokahi Knowledge Browser.

7.1 Implementation

To build the merged Lokahi knowledge graph, we again use the Java-based in-memory graph representation introduced in subsection 6.1, which is based on the Java classes *Node* and *Association*. The merge is driven by the much smaller top-down graph: For each node of the top-down graph, an identical bottom-up node is looked up in the bottom-up graph. In the case of a match, a single "merged" node is built and labeled with the top-down label[11].

While we replace identical bottom-up and top-down entities (nodes) by a single entity (node), bottom-up and top-down associations between identical node pairs are differentiated, and the merge is implemented by building multiple associations. The rationale for this approach is that bottom-up and top-down nodes can be identical (both represent unique entities in the world), while bottom-up and top-down associations *always* have a different semantic: A bottom-up association expresses the relevance of one entity for another based on likelihood of co-occurence in a document, and is expressed as a fuzzy degee of membership in the entity's context. In contrast, top-down associations are user-modeled crisp associations with a specified semantic (such as "is project leader of").

[11] In practice, class and instance nodes represented by URIs will not realistically be matched with in the bottom-up graph, i.e., the matching focuses on literal notes (values of datatype properties such as a person's first or last name). If a literal node has a match, the corresponding labels also match, since they are identical to the literals, and choosing the top-down label is valid. If indeed a class or instance node, i.e., a URI, should have a match in the bottom-up graph, the top-down label is a human-readable keyword, while the bottom-up label is the URI itself. In this case, it is sensible to choose the top-down label.

7.2 Graph Representation

The merging of the bottom-up and top-down knowledge extraction approaches as described in the foregoing subsection can be formally represented as a "merge" of the corresponding Lokahi Knowledge Graphs:

Definition 4. Let $\underline{\mathcal{L}} = (\underline{\Sigma}, \underline{V}, \underline{A}, \underline{\mu_A}, \underline{\sigma}, \underline{\tau}, \underline{\lambda})$ and $\bar{\mathcal{L}} = (\bar{\Sigma}, \bar{V}, \bar{A}, \bar{\mu}_{\bar{A}}, \bar{\sigma}, \bar{\tau}, \bar{\lambda})$ be a bottom-up and a top-down Lokahi knowledge graph, respectively. We define the *merge of* $\underline{\mathcal{L}}$ *and* $\bar{\mathcal{L}}$ by

- $\Sigma' := \underline{\Sigma} \cup \bar{\Sigma}$.
- $V' = \underline{V} \cup \bar{V}$.
- $A' := \underline{A} \cup arc'(\bar{A})$, where arc' is a bijection that makes $\underline{A}$ and $arc'(\bar{A})$ disjoint.
- $\mu'_{A'} : A' \rightarrow [0, 1]$ with $\mu'_{A'}|_{\underline{A}} := \underline{\mu_A}$ and $\mu'_{A'}|_{\bar{A}} := \bar{\mu}_{\bar{A}}$.
- $\sigma' : A' \rightarrow V'$ with $\sigma'|_{\underline{A}} := \underline{\sigma}$ and $\sigma'|_{\bar{A}} := \bar{\sigma}$.
- $\lambda' := \bar{\lambda}$.

The output of the merging algorithm is again a Lokahi Knowledge Graph:

Corollary 1. *Let* $\underline{\mathcal{L}}$ *and* $\bar{\mathcal{L}}$ *be a bottom-up and a top-down Lokahi knowledge graph, respectively. Then* $\mathcal{L}'$ *as specified in definition 4 is a Lokahi Knowledge Graph, and we call it the* merged Lokahi Knowledge Graph induced by $\underline{\mathcal{L}}$ and $\bar{\mathcal{L}}$.

Notice that $\underline{\mathcal{L}}$ and $\bar{\mathcal{L}}$ are not induced subgraphs of $\mathcal{L}'$, since bottom-up and top-down arcs between identical node pairs are not identified, but included as multiple arcs in order to retain the different semantic of bottom-up and top-down associations.

8 Conclusions and Further Work

We introduced the Lokahi Knowledge Graph as the underlying data structure of the Lokahi Knowledge Browser. We also briefly recapitulated the design principles introduced in [9] for the construction of a concrete instantiation of the Lokahi Knowledge Graph. In accordance with these principles, we proposed three knowledge engineering steps: 1. the construction of a bottom-up knowledge graph from unstructured documents by automatic information extraction based on fuzzy association rule mining, 2. the semi-automatic construction of a top-down knowledge graph from structured data using an enterprise meta ontology, and 3. the merging of the two graphs. In a separate publication, we will elaborate on the GUI design of the Lokahi Knowledge Browser, which is ongoing work. Future work comprises the active use of the top-down ontology structure to enrich the Lokahi knowledge graph with information that is inferred from bottom-up knowledge.

Acknowledgments Part of the presented research has been funded by Five Informatik AG and the Swiss Commission for Technology and Innovation (CTI) as part of the applied research project LOKAHI Inside, CTI-No. 16152.1 PFES-ES.

References

1. Agrawal, R., Imielinski, T., Swami, A.: Mining association rules between sets of items in large databases. In: Jajodia, P.B.S. (ed.) Proceedings of the 1993 ACM SIGMOD International Conference on Management of Data, vol. 2. ACM, New York (1993)
2. Biezunski, M.: Introduction to topic mapping. In: SGML Europe GCA Conference (1997)
3. Chen, P.P.S.: The entity-relationship model - toward a unified view of data. ACM Trans. Database Syst **1**(1), 9–36 (1976)
4. Emmenegger, S., Laurenzini, E., Thönssen, B.: Improving supply-chain-management based on semantically enriched risk descriptions. In: Proceedings of the International Conference on Knowledge Management and Information Sharing, KMIS 2012, pp. 70–80 (2012)
5. Hawthorne, J.: Inductive logic. In: Stanford Encyclopedia of Philosophy. The Metaphysics Research Lab, Stanford University, Stanford (2008)
6. Hinkelmann, K., Merelli, E., Thönssen, B.: The role of content and context in enterprise repositories. In: Proceedings of the 2nd International Workshop on Advanced Enterprise Architecture and Repositories, AER 2010 (2010)
7. Kang, D., Lee, J., Choi, S., Kim, K.: An ontology-based enterprise architecture. Expert Syst. Appl. **37**(2), 1456–1464 (2010). http://dx.doi.org/10.1016/j.eswa.2009.06.073
8. Kaufmann, M.: Inductive Fuzzy Classification in Marketing Analytics. Doctoral Thesis, Université de Fribourg, Fribourg, Switzerland (2012)
9. Kaufmann, M., Wilke, G., Portmann, E., Hinkelmann, K.: Combining bottom-up and top-down generation of interactive knowledge maps for enterprise search. In: Buchmann, R., Kifor, C.V., Yu, J. (eds.) KSEM 2014. LNCS, vol. 8793, pp. 186–197. Springer, Heidelberg (2014)
10. Knauer, U.: Algebraic Graph Theory: Morphisms, Monoids and Matrices, 1st edn. De Gruyter, Berlin (2011)
11. Martin, A., Emmenegger, S., Wilke, G.: Integrating an enterprise architecture ontology in a case-based reasoning approach for project knowledge. In: Proceedings of the First Enterprise Systems Conference (ES 2013) (2013)
12. Masterman, M.: Semantic message detection for machine translation, using an interlingua. In: Proc. 1961 International Conf. on Machine Translation, pp. 438–475 (1961)
13. Novak, J.D., Gowin, D.B.: Learning How to Learn. Cornell University (1984). ISBN: 9780521319263
14. Peirce, C.S.: On junctures and fractures in logic. In: Writings of Charles S. Peirce: 1879–1884, p. 391. Harvard University Press (1882)
15. Portmann, E., Kaufmann, M.A., Graf, C.: A distributed, semiotic-inductive, and human-oriented approach to web-scale knowledge retrieval. In: Proceedings of the 2012 International workshop on Web-scale knowledge Representation, Retrieval and Reasoning, pp. 1–8. ACM, New York (2012)
16. Quillian, M.R.: Semantic Memory. Ph.D. thesis, Carnegie Institute of Technology (now CMU) (1966), abridged version in Minsky, pp. 227–270 (1968)

17. Shadbolt, N.: Knowledge Technologies. Ingenia **8**, 58–61 (2001)
18. Siemens, G.: Connectivism: Learning as network creation (2005). http://www.elearnspace.org/Articles/networks.htm
19. Sowa, J.: Conceptual graphs. In: Handbook on Architectures of Information Systems, pp. 287–311. Springer (1998)
20. Thönssen, B.: An enterprise ontology building the bases for automatic metadata generation. In: Sánchez-Alonso, S., Athanasiadis, I.N. (eds.) MTSR 2010. CCIS, vol. 108, pp. 195–210. Springer, Heidelberg (2010)
21. Thönssen, B., Lutz, J.: Semantically enriched obligation management. In: Proceedings of 4th Conference on Knowledge Management and Information Sharing (KMIS2012) (2012)

A System for Conceptual Pathway Finding and Deductive Querying

Troels Andreasen, Henrik Bulskov, Jørgen Fischer Nilsson, and Per Anker Jensen

Abstract We describe principles and design of a system for knowledge bases applying a natural logic. Natural logics are forms of logic which appear as stylized fragments of natural language sentences. Accordingly, such knowledge base sentences can be read and understood directly by a domain expert. The system applies a graph form computed from the input natural logic sentences. The graph form generalizes the usual partial-order ontological sub-class structures by accommodation of affirmative sentences comprising recursive phrase structures. In this paper we focus on the logical inference rules for extending the concept graph form enabling deductive querying as well as computation of pathways between the concepts mentioned in the sentences.

Keywords Deductive querying of natural-logic knowledge bases · Path finding in knowledge bases · Logical knowledge bases in bio-informatics and medicine

1 Introduction and Background

In a series of papers [1, 2, 4, 5, 6] we have recently developed and described principles and systems design for natural-logic knowledge bases. This work originates in our

T. Andreasen(✉) · H. Bulskov
Computer Science, Roskilde University, Roskilde, Denmark
e-mail: {troels,bulskov}@ruc.dk

J.F. Nilsson
Mathematics and Computer Science, Technical University of Denmark,
Kongens Lyngby, Denmark
e-mail: jfni@dtu.dk

P.A. Jensen
International Business Communication, Copenhagen Business School,
Frederiksberg, Denmark
e-mail: paj.ibc@cbs.dk

T. Andreasen et al. (eds.), *Flexible Query Answering Systems 2015*,
Advances in Intelligent Systems and Computing 400,
DOI: 10.1007/978-3-319-26154-6_35

idea of providing so-called generative ontologies [7]. In our generative ontologies, the concepts are not merely given classes but entire phrases in which the class noun is extended with restrictions for forming subclasses. These restrictive phrases, as in the natural language phrases they reflect and formalize, are endowed with a recursive structure, thereby becoming "generative", in analogy to the well-known notion of generative grammars.

In the above-mentioned more recent papers we go a step further by adopting a simplified form of so-called natural logic [8, 9] as our formal language for stating propositions. Accordingly, a knowledge base (KB) consists of a finite set of affirmative sentences in natural logic. These sentences comprise traditional ontological sub-class relationships as special cases, so there is no separate formal ontology. As discussed in our [1] the natural logic formulations come close to natural language so that the KB can be read by domain experts, for instance, in the bio-sciences. It goes without saying that the formal natural logic dialect cannot accommodate the full meaning content of a text sentence in natural language. As discussed in our [6], it is our contention that semantically the considered natural logic can cover substantial parts of typical textual specifications within the bio-sciences.

In the present paper, we focus on computing conceptual pathways between concept terms stated as a query. In order to achieve this functionality, we have devised a graph form of the knowledge base in which the possibly complex knowledge base sentences are broken down into more elementary ones without essential loss of meaning. As part of this endeavour, we address the deductive querying of natural-logic knowledge bases.

The paper is structured as follows: In section 2 we describe the applied natural logic with the accompanying internal graph form in section 3. In section 4 we describe the inference rules applying to the graph form and brought to bear on pathway querying in section 5.

2 Natural Logic for Knowledge Bases

The knowledge base sentences considered express relationships between classes in an ontology. The applied form of natural logic is meant to be readable for domain experts without background in logic and computer science. At the same time, the considered natural logic dialect constitutes a well-defined logic as discussed in [3, 4, 5].

2.1 Simple Sentences in Natural Logic

The simplest sentence form

Cnoun isa *Cnoun*

expresses class inclusion. *Cnoun*-expressions are common nouns naming introduced classes. The knowledge base ontology is shaped by such sentences, where the class inclusion relationship forms a partial ordering of the classes. As an example, we may have betacell isa cell. Notice that the system is incapable of splitting agglutinated compounds like “betacell” in order to identify a head noun, in casu “cell”.

More generally, the logic admits knowledge base sentences with transitive verbs

Cnoun Verb Cnoun

as in the example betacell produce insulin. Thus, in addition to the strictly ontological class inclusion structure, the knowledge base comprises more general state-of-affairs descriptions.

2.2 Complex Sentences in Natural Logic

Crucially, we further admit compound, recursively structured class terms

Cterm Verb Cterm

as in the sample

cell that produce insulin located:in pancreatic gland,

where the phrase cell that produce insulin denotes a sub-class (complex concept) of the given class cell formed by a restrictive relative clause. Similarly, the adjectival modifier “pancreatic” introduces a subclass of the class “gland”. Generally speaking, the various types of modifiers always act restrictively in the set up.

Restrictive relative clauses may recursively comprise restrictive relative clauses as in the phrase gland that haspart (cell that produce hormone). The parentheses here are for clarification, only. Thus, in principle, an open-ended and unrestricted collection of classes is made available, although in a knowledge base with accompanying queries, obviously, only a finite set would be made explicit. This notion of generative ontologies was launched in a seminal form in [7]. The various suggested language forms are further described in our [1, 4, 5]. Sample knowledge bases are found in our [6].

2.3 The Logical Understanding of Sentences

From a logical point of view, all the knowledge base sentences *Cterm Verb Cterm* are implicitly quantified, namely as

every *Cterm Verb* some *Cterm*

giving for instance every betacell produce some insulin as explicitation of the above betacell produce insulin. As is evident, there are actually four possible quantifier constellations in the above sentence form. However, in this context we only consider the above quantifier form, since it covers substantial parts of the knowledge base information in the considered applications. This is confirmed by the default assumptions applied in natural language concerning this adopted quantifier form.

The natural logics offer an alternative to description logics. Specifically, the natural logics recognize the key role of the main verb in natural language affirmative sentences. This is in contrast to description logics, where sentences come about as extended copula forms, hampering the human comprehension of knowledge bases. For instance, the sample, straightforward sentence betacell produce insulin in description logic becomes the rather incomprehensible betacell $\sqsubseteq$ $\exists$produce.insulin as discussed in [4].

In [1, 4, 5], we discuss further the relationships to syllogistic logic, predicate logic, and description logic. There we also discuss our approach to denials by way of a default assumption amounting to considering classes disjoint unless one is a subclass of the other or they have a common subclass. More generally, we lean towards the closed world assumption, unlike the open world assumption of description logic.

3 The Concept Graph Form

The logical view of sentences supported by inference rules described below affords deductive query facilities. In our system complex sentences are decomposed into simple sentences. The simple sentences are thought of as arcs in a labeled directed graph called the concept graph. Crucially, the decomposition of complex sentences calls for generation of new concept nodes in the graph corresponding to the concept terms as well as any sub-terms in the knowledge base sentences.

The graph view complements the logical view of knowledge base sentences by affording computational path finding between - possibly complex - concepts. We elaborate on the functionalities offered by the graph view in the final sections of this paper.

As an example consider again the sentence cell that produce insulin located:in pancreatic gland. In our system, this given sentence becomes decomposed into the simple sentences:

cell-that-produce-insulin isa cell
cell-that-produce-insulin produce insulin
cell-that-produce-insulin located:in pancreatic-gland

where cell-that-produce-insulin is a system-generated concept term which names a node as illustrated in figure 1. Since adjectival modifications are always taken for

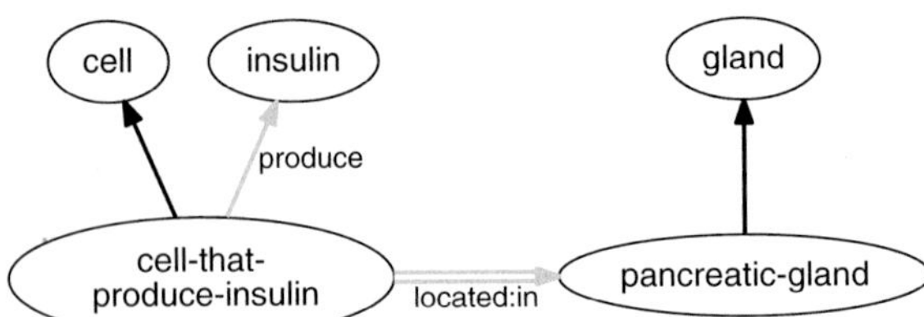

Fig. 1 Graph representation of the sentence "cells that produce insulin are located:in the pancreatic gland"

being restrictive here, the system adds pancreatic-gland isa gland.

In order to ensure that the meaning of a sentence in the knowledge base is properly retained in the graph, we distinguish three different arcs as illustrated in figure 1. The arcs contributing to the definition of a complex concept are drawn as single arrows.

isa-arcs are drawn as black arrows, whereas restrictive contributions to the definition are drawn as grey arrows. The arc stemming from the verb in the main sentence, which creates the proposition, is drawn as a double arrow.

The representation of concepts is assumed to be unique and thus shared across the contributing sentences. Accordingly, the KB sentences give rise to one, usually coherent, graph.

4 Inference Rules

The considered sentences are subject to logical inference rules, that is, inference rules provided for purely logical reasons with reference to the underlying predicate logical explication. These rules admit deductive querying of the knowledge base.

In addition, there may be ad hoc rules supporting introduced relationships cf. the example in section 4.3.

4.1 Logical Inference Rules

First and foremost, the isa relation is made reflexive and transitive, that is, a partial order:

$$\frac{}{C \text{ isa } C}$$

$$\frac{C \text{ isa } X \qquad X \text{ isa } D}{C \text{ isa } D}$$

As a simple example, given the two KB sentences: pancreatic-gland isa endocrine-gland and endocrine-gland produce hormone, we conclude that pancreatic-gland produce hormone using the inheritance rule:

$$\frac{C\ R\ D \qquad C' \text{ isa } C}{C'\ R\ D}$$

Moreover, given that betacell produce insulin and insulin isa hormone we conclude that betacell produce hormone using the rule of property generalization:

$$\frac{C\ R\ D \qquad D \text{ isa } D'}{C\ R\ D'}$$

These two rules are known as monotonicity rules in natural logic. As it appears they express common sense reasoning without appeal to complicated logical inference systems such as resolution and natural deduction.

The transitivity, inheritance and generalisation rules are illustrated in the figures 2 to 4. The inferences drawn by these rules are not materialized in advance in the concept graph. A stated query like betacell produce hormone? is confirmed by appeal

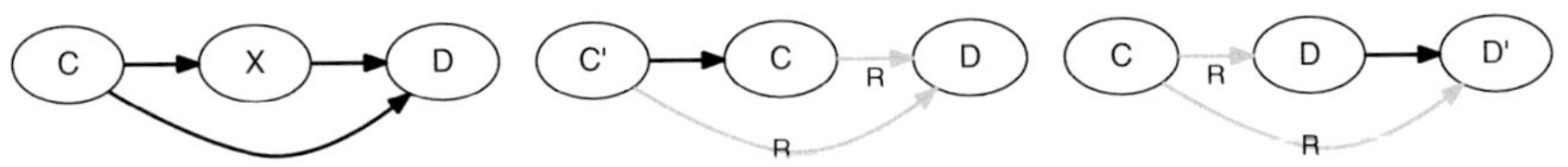

Fig. 2 Transitivity **Fig. 3** Inheritance **Fig. 4** Generalization

to the last of the above inference rules. Thus, derived sentences are not computed in advance.

4.2 *The Subsumption Rule*

The use of decomposed sentences in the KB concept graph calls for a special logical inference rule, termed the subsumption rule. This rule is to ensure that all logically relevant isa class inclusion arcs, less those following by transitivity of isa, become present in the graph.

As an example consider the two concept terms

cell-that-produce-hormone and cell-that-produce-insulin

giving rise to

cell-that-produce-hormone isa cell

cell-that-produce-hormone produce hormone

respectively

cell-that-produce-insulin isa cell

cell-that-produce-insulin produce insulin

and assume that the proposition insulin isa hormone is included ind the KB. In this case, as illustrated in figures 5 and 6, the subsumption rule calculates

cell-that-produce-insulin isa cell-that-produce-hormone

The subsumption pre-computation thus calculates missing class inclusion arcs, and thereby serves to facilitate and crucially speed up subsequent deductive reasoning computations and pathway computations in the concept graph. In some cases, the calculation would have to take regress to inclusion arcs throughout the concept graph. Therefore, we devise the following algorithm for systematically calculating the missing inclusion arcs. The algorithm relies on the principle that all inclusion arcs drawn on in a specific case have already been calculated.

The first step is to rank the concept nodes in the graph according to a depth criterion: Concept nodes which have no non-isa outlet arcs in their definitions are assigned order 0. Concept nodes whose non-isa outlet arcs lead to concept nodes of order 0 are assigned the order 1. Concept nodes whose non-isa outlet arcs lead to concept nodes of order n (and in addition possibly less) are assigned the order $n + 1$. It should be noted that there is no risk of cycles in the definition graph, assuming that there are no cycles in the isa inclusion sub-graph.

The ranking of concept nodes is to ensure that when a pair of concept nodes is checked for subsumption, all the concept nodes pointed to from this pair have already been processed. Accordingly, the algorithm begins with all ranks up to 1 pairs of concept nodes in the entire graph and processes these.

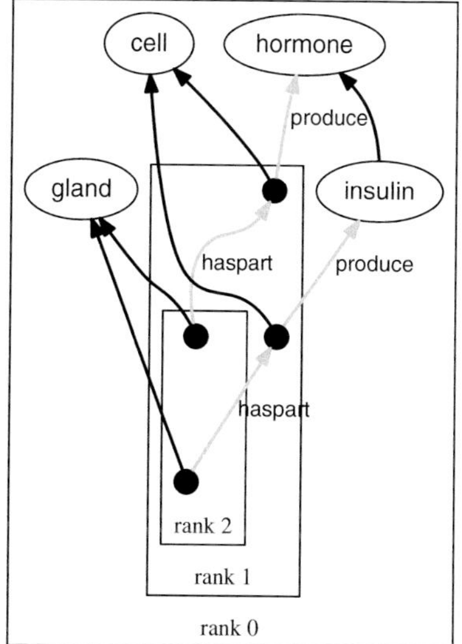

Fig. 5 Before addition

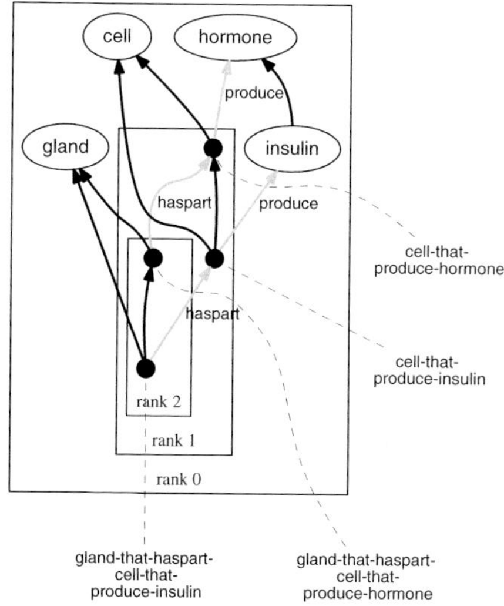

Fig. 6 After addition of subsumption arcs

Consider all pairs of nodes C and C' of rank 1, where

C has arcs $C\ R_i\ D_i$ for $i = 1..m$

and

C' has arcs $C'\ R_i\ D_i'$ for $i = 1..n$

where the sets of arcs $R_i\ D_i$ and $R_i\ D_i'$ may include inherited arcs according to inheritance inference, cf. figure 3. Now, assume that for all $R_i\ D_i'$ there is $R_i\ D_i$ so that D_i isa D_i', either explicitly or by transitivity. In that case, add the arc C isa C'.

The algorithm then proceeds to up to rank 2 pairs of concept nodes (less the pairs having already been processed) and processes these, knowing that the concept nodes pointed to have already been processed. The algorithm continues in this way up to the highest rank being used in the concept graph. An example showing addition of missing arcs at rank 1 as well as rank 2 is illustrated in figures 5 (before) and 6 (after). One should observe that the highest rank is not given statically simply by the syntactic depth nesting of phrases in the original propositions.

4.3 Domain Dependent Inference Rules

As an example of a domain inference rule the has-part relation may be made transitive (cf. [10]) by way of the rule:

$$\frac{C \text{ haspart } X \qquad X \text{ haspart } D}{C \text{ haspart } D}$$

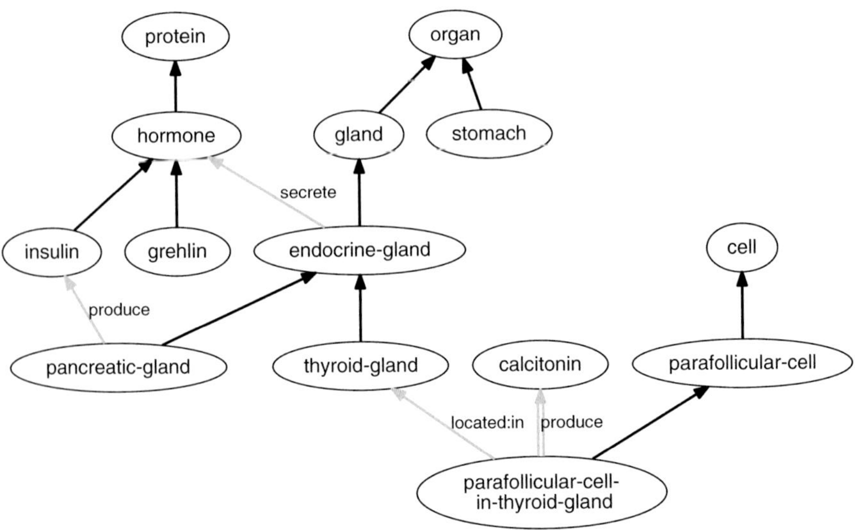

Fig. 7 A fragment of an ontology centered around endocrine gland

Similarly for the complementary part-for relation. Again, these rules are to be activated in the KB system rather than be used for pre-computation of derived relationships.

5 Concept Path Finding

The concept graph is a logical view of the sentences in the corpus from which it has been generated. Sentences are decomposed into simple sentences that correspond to edges in the graph defining concepts and expressing propositions. Thus, a path in the concept graph, a concept path, combines a series of simple sentences and may therefore be rendered in natural language into an explanation of the connection between the end nodes of the path. Concept path finding can thus be applied as a means of knowledge base querying. Given two or more concepts, we can search for natural-language renderings of connections relating these. Given a single concept, we can search for related key concepts. We thus consider queries to reveal connectivity in the graph. Below we mainly consider two-concept queries.

As mentioned above, we assume that the concept graph G is closed wrt subsumption, such that all edges that are inferable by the subsumption rule, are included in G.

Candidate answers to a two-concept query $Q = (C, C')$ are based on paths connecting the two query concepts C and C' or, more specifically, paths connecting C to C'. From any such path we can derive a natural-language rendering corresponding to the connection it provides. For instance, an answer to the query $Q =$ (pancreatic-gland, protein) evaluated on a knowledge base corresponding to the graph in figure 7 involves the path:

(pancreatic-gland produce insulin), (insulin isa hormone), (hormone isa protein)
or more succinctly:
(pancreatic-gland produce insulin isa hormone isa protein)
From this we can derive the natural-language rendering:

pancreatic-gland produce insulin, which is a hormone, which is a protein.

All edges in the concept graph are directed. However, not only directed paths may contribute to answers to two-concept queries. Given a two-concept query $Q = (C, C')$ we consider in principle any undirected path from C to C'. Thus the direction of edges does not influence the paths we are considering, but only the interpretation and thereby the natural language rendering we can apply on these.

5.1 Rendering a Path Into Natural Language

A natural language rendering of a path can be provided as follows. The rendering of an inclusion edge in the beginning of the path X isa Y is "X*, which is a* Y", while an inner edge that continues from a previous node and leads to Z on the path simply adds "*, which is a* Z" to the rendering. Thus the rendering of the path X isa Y isa Z will be "X*, which is a* Y, which is a Z".

When in the beginning of the path, an inclusion edge traversed in the inverse direction, for instance, a path from Z to Y through an edge Y isa Z, can be read as "*some* Z *are* Y", while an inverse inclusion inner edge that continues the path from a previous node and leads to X on the path adds "*, whereof some are* X" to the rendering. Thus, the rendering of the path from Z through Y to X provided by the two edges Y isa Z and X isa Y will be "*some* Z *are* Y*, whereof some are* X".

Semantic relations (i.e. relations other than isa) are named by the main verb in the phrase from which they are extracted, and these may therefore be read "as is". Thus the rendering of the forward direction of an edge $X\ R\ Y$ beginning a path is simply "$X\ R\ Y$" , while an inner edge that continues a path can be read "$R\ Y$". As with the inclusion relation, semantic relations may be traversed in the inverse direction. However, for semantic relations we assume explicitly specified inverse relations. For a relation R the inverse relation is given by $\bar{R} = inv(R)$, where inv is a symmetric mapping given by a domain expert.

When in the beginning of the path, an edge corresponding to the relation R traversed in the inverse direction, for instance, a path from Z to Y through an edge $Y\ R\ Z$, can be read as "*some* Z *are* $inv(R)$ Y", while an inverse semantic inner edge that continues the path from a previous node and leads to X on the path adds "*, whereof some are* $inv(R)$ X" to the rendering. Thus, for instance, the rendering of the path from Z through Y to X provided by the two edges $Y\ R\ Z$ and $X\ R\ Y$ will be "*some* Z *are inv(R) Y, whereof some are inv(R) X.*

As an example, an answer to the query $Q =$ (protein , pancreatic-gland) evaluated on figure 7 based on the path indicated above in inverse direction would lead to the rendering:

> *some protein are hormone, whereof some are insulin, whereof some are produced by pancreatic gland.*

assuming that inv(produce) = produced:by, while an answer to $Q =$ (protein , gland) would lead to:

> *some protein are hormone, whereof some are secreted by endocrine gland, which is a gland.*

assuming that inv(secrete) = secreted:by.

5.2 Reduction

Among the potentially most interesting paths that may be applied to provide answers to a query $Q = (C, C')$, are the shortest paths connecting the two query concepts C and C'. However, due to the fact that a significant number of conceptual edges derivable by the inference rules are not explicitly included in the graph G, we cannot be sure that a shortest path between C and C' in G provides the briefest connection between the two concepts. A path connecting two concepts C and C' may be reduced, replacing edges according to inference, such that premise edges are removed and inferred edges are inserted. Due to the transitivity inference rule, a path or a subpath may be reduced by edge replacement

$$(C \text{ isa } X \text{ isa } D) \qquad \text{replaced by} \qquad (C \text{ isa } D)$$

Similarly we can derive possible replacements from the two monotonicity inference rules. Due to inheritance monotonicity, a path or a subpath may be reduced by replacing edges:

$$(C' \text{ isa } C\ R\ D) \qquad \text{replaced by} \qquad (C'\ R\ D)$$

and due to generalization monotonicity, a path or a subpath may be reduced by:

$$(C\ R\ D \text{ isa } D') \qquad \text{replaced by} \qquad (C\ R\ D')$$

Thus, by applying generalisation twice or transitivity followed by generalisation, we can reduce

(pancreatic-gland produce insulin isa hormone isa protein) to
(pancreatic-gland produce protein)

while by applying inheritance followed by generalization, we can reduce

(pancreatic-gland isa endocrine-gland secrete hormone isa protein) to
(pancreatic-gland secrete protein)

The shortest path in figure 7 connecting *calcitonin* and *protein* (assuming *inv* (produce) = produced:by) is the following:

(calcitonin produced:by parafollicular-cell-in-thyroid-gland
located:in thyroid-gland isa endocrine-gland secrete hormone isa protein)

This path may be reduced to

(calcitonin produced:by parafollicular-cell-in-thyroid-gland
located:in endocrine-gland secrete protein)

Reduction leads to shorter paths and thereby to more succinct natural-language renderings of connectivity. This is obviously at the expense of details which in some cases may provide useful supplementary information. Thus a possibility in a user interface to expand reduced paths to their original form would be a useful feature making a more dynamic interface.

An alternative less coarse-grained reduction principle could also be applied: always retain nodes that have outgoing semantic relation edges (relations other than isa). This would correspond to ignoring inheritance while reducing paths.

5.3 *Query Evaluation Principle*

Evaluating two-concept queries to a concept graph is first of all a matter of finding paths in the graph. The principle indicated above and described in more detail below divides into shortest path computation, reduction and natural-language rendering. The path computation applies a Breadth First Search (BFS) starting from the first query concept.

Given the directed concept graph G and assuming that G is closed wrt subsumption. Let $\bar{G}$ be an undirected version of G and let the query $Q = (C, C')$ be a two-concept query.

1. Derive the set P of all shortest paths from C to C' in $\bar{G}$. Start from C, apply BFS continuously adding all new paths from C to the set B until C' is found, return $P = \{p | p \in B,\ p\ connects\ C\ and\ C'\}$
2. For each path $p \in P$ derive p' by repeatedly reducing subpaths until no further reduction can be performed, set $P = P \cup \{p'\}$
3. Let $\sigma = min(\{l | p \in P,\ l = length\ of\ p\})$
4. Let $\bar{S} = \{p | p \in P, \sigma = length\ of\ p\}$
5. Let S be the set of paths in G corresponding to the paths $\bar{S}$ in $\bar{G}$
6. For all $p \in S$ provide the rendering of p as contribution to the answer to Q

It should be noted that we cannot ensure that all shortest paths will be found by this algorithm. Continuing step 1 until all paths are found may result in additional paths that can be reduced to a shortest path in step 2. There will also be cases where this will lead to a shorter length of the shortest paths found.

6 Summary and Future Work

We have outlined a system for pathfinding in logical knowledge bases. The key component in the system is a concept graph being pre-computed from a given knowledge base which consists of sentences in natural logic. In computing the concept graph we strive – if only heuristically, so far – to strike a balance between "materialized" information in the form of stored arcs versus virtual information deducible by means of the stated inference rules. As the guiding principle we require that all "isa" class inclusion relationships except for those following by transitivity are explicitly recorded. Therefore, the described subsumption algorithm is to be invoked in a pre-computation phase. On the other hand, we refrain from pre-computing the entire transitive closure of the class inclusion as well as what follows from applying the monotonicity rules. Currently we are performing small-scale experiments with a prototype.

References

1. Andreasen, T., Bulskov, H., Nilsson, J.F., Jensen, P.A.: A system for computing conceptual pathways in bio-medical text models. In: Andreasen, T., Christiansen, H., Cubero, J.-C., Raś, Z.W. (eds.) ISMIS 2014. LNCS, vol. 8502, pp. 264–273. Springer, Heidelberg (2014)
2. Andreasen, T., Bulskov, H., Nilsson, J.F., Anker Jensen, P., Lassen, T.: Conceptual pathway querying of natural logic knowledge bases from text bases. In: Larsen, H.L., Martin-Bautista, M.J., Vila, M.A., Andreasen, T., Christiansen, H. (eds.) FQAS 2013. LNCS, vol. 8132, pp. 1–12. Springer, Heidelberg (2013)
3. Nilsson, J.F.: Diagrammatic reasoning with classes and relationships. Moktefi, A., Shin, S.-J. (eds.) Visual Reasoning with Diagrams. Studies in Universal Logic. Birkhäuser, Springer (2013)
4. Nilsson, J.F.: In pursuit of natural logics for ontology-structured knowledge bases. In: The Seventh International Conference on Advanced Cognitive Technologies and Applications, COGNITIVE 2015, Nice, France, March 22–27, 2015. IARIA. ISSN: 2308–4197, ISBN 978-1-61208-390-2
5. Andreasen, T., Nilsson, J.F.: A case for embedded natural logic for ontological knowledge bases. In: Proceedings of the 6th International Conference on Knowledge Engineering and Ontology Development (2014)
6. Andreasen, T., Bulskov, H., Nilsson, J.F., Jensen, P.A.: Computing pathways in bio-models derived from bio-science text sources. In: IWBBIO 2014, pp. 217–226 (2014)
7. Andreasen, T., Nilsson, J.F.: Grammatical Specification of Domain Ontologies in journal. Data & Knowledge Engineering **48**(2), 221–230 (2004)
8. van Benthem, J.: Essays in Logical Semantics. Studies in Linguistics and Philosophy, vol. 29. D. Reidel Publishing Company (1986)
9. van Benthem, J.: Natural logic, past and future. In: Workshop on Natural Logic, Proof Theory, and Computational Semantics 2011. CSLI Stanford (2011). `http://www.stanford.edu/~icard/logic&language/index.html`
10. Smith, B., Rosse, C.: The role of foundational relations in the alignment of biomedical ontologies. In: Fieschi, M. (ed.) MEDINFO 2004 (2004)

Generalized Net Model of an Expert System Dealing with Temporal Hypothesis

Panagiotis Chountas, Krassimir Atanassov, Evdokia Sotirova and Veselina Bureva

Abstract A theory of the Generalized nets was used to construct a generalized net model which describes the possibility to evaluate the time of the occurrence or completion of the events. Twenty predicates for checking the validity of the special circumstances, related to time-moments of two events were introduced. The model can be used to analyze the specific moments in which the facts have happened. In result we can answer questions such as: "Is the fact valid?" "How many facts it contradicts", "How many facts it confirms?"

Keywords Generalized Net · Expert Systems · Temporal reasoning

1 Introduction

Temporal reasoning is especially important for medical systems, as the final diagnosis is often strongly affected by the sequence in which the symptoms develop.

P. Chountas
Faculty of Science and Technology, Department of Computer Science,
University of Westminster, 15 New Cavendish Street, London W1W 6UW, UK
e-mail: p.i.chountas@westminster.ac.uk

K. Atanassov
Bioinformatics and Mathematical Modelling Department,
Institute of Biophysics and Biomedical Engineering,
Bulgarian Academy of Sciences, Acad. G. Bonchev Str. Bl. 105, 1113 Sofia, Bulgaria
e-mail: krat@bas.bg

K. Atanassov · E. Sotirova(✉) · V. Bureva
Intelligent Systems Laboratory, "Prof. Dr. Asen Zlatarov" University,
1 "Prof. Yakimov" Blvd. 8010 Burgas, Bulgaria
e-mail: {esotirova,vbureva}@btu.bg

T. Andreasen et al. (eds.), *Flexible Query Answering Systems 2015*,
Advances in Intelligent Systems and Computing 400,
DOI: 10.1007/978-3-319-26154-6_36

Current research on time structures in artificial intelligence is reviewed, and a temporal GN model based on intuitionistic fuzzy set theory is developed. The proposed model can be used for instance for a simple and natural representation of symptoms, and provides for efficient computation of temporal relationships between symptoms.

Temporal reasoning and querying in medical related environments has been investigated as a subject of research since the end of 1980s. During the 1990s the research community published its proposals and results in computer science and specialised media [1-4]. Temporal representation and reasoning in medicine and, more generally, all research focusing on time-related aspects and medical information have certain peculiarities that distinguish them from other research computer science topics. Like general research on time-related topics in computer science, this research is orthogonal with respect to the required scientific and technical skills. Dealing with time to manage medical information requires that scientific results from various technical fields be shared, such as modal logic, constraint networks, data modeling and querying, fuzzy logic and its extensions, data mining, and others.

Time is an important and underestimated aspect of modelling and managing processes within medical information systems. Despite the current advances in understanding how to effectively model the rich semantics of time within databases, we still are faced with several challenges when devising temporal database methods and techniques to be adopted by decision support, data mining, and business processes. The rapid growth of clinical data repositories and integration techniques require new approaches for temporal representation and querying in database systems.

In a series of research papers were described ten Generalized Net (GN, see [5, 8, 9]) models, collected in book [12]. They describe the way of functioning and the results of the work of different types of Expert Systems (ESs, see, e.g., [6, 11, 17, 18]). Some types of these ESs are introduced for a first time as possible extensions of the ESs, which can be described by the GNs.

The first four from the nine GN-models describe ordinary ESs; the fifth and seventh - ESs with priorities of their Database (DB) facts and Knowledge Base (KB) rules, so, the separate facts and/or rules can be changed at the time of the ES functioning. Sixth GN-model describes an ES containing not only facts but also metafacts that can be represented by rules. Eighth GN-model represents Intuitionistic Fuzzy ES (IFES see [7]; for the intuitionistic fuzziness see [13]). On its base the ninth GN-model is constructed so that it represents functioning of an ES working with temporal facts and answering to the temporal questions [10]. The tenth GN-model (see [14]) represents extension to the later ES. In [16] is described GN-model of ESs with Frame-Type Data Bases (FTDB). It is extended in [15] by adding intuitionistic fuzzy estimations of the frames.

Now, we will discuss the possibility to evaluate the time of the occurring/finishing of a sequence of events. Given the fact that there may be several interacting processes related to disease/syptoms as well as therapy in a patient, medical reasoning is complex by nature. The interactions in the specific situation (additive, subtractive, etc.), and the events causing them, need to be dynamically determined. Understanding and reasoning explicitly with time provides a more accurate representation of reality and may help to exclude impossible scenarios. Our model can be used to analyze the specific moments in which patient related specific facts have happened. In the result we can answer questions such as: ''Is the fact or process occurence valid?'' ''How many facts it denies or contradicts?'', ''How many existing facts it confirms?''

2 A GN-Model

In the beginning we introduce the following 20 predicates that will be used to check the validity of the special questions, related to time-moments of two events:

$V_1(X, Y) = X$ occurred before Y in the first happening of the events,
$V_2(X, Y) = X$ occurred before Y in the last happening of the events,
$V_3(X, Y) = X$ always occurred before Y,
$V_4(X, Y) = X$ sometimes occurred before Y,
$V_5(X, Y) = X$ formerly occurred before Y,
$V_6(X, Y) = X$ occurred together with Y in the first happening of the events,
$V_7(X, Y) = X$ occurred together with Y in the last happening of the events,
$V_8(X, Y) = X$ always occurred together with Y,
$V_9(X, Y) = X$ sometimes occurred together with Y,
$V_{10}(X, Y) = X$ formerly occurred together with Y,
$V_{11}(X, Y) = X$ finished before Y in the first happening of the events,
$V_{12}(X, Y) = X$ finished before Y in the last happening of the events,
$V_{13}(X, Y) = X$ always finished before Y,
$V_{14}(X, Y) = X$ sometimes finished before Y,
$V_{15}(X, Y) = X$ formerly finished before Y,
$V_{16}(X, Y) = X$ finished together with Y in the first happening of the events,
$V_{17}(X, Y) = X$ finished together with Y in the last happening of the events,
$V_{18}(X, Y) = X$ always competes together with Y,
$V_{19}(X, Y) = X$ sometimes finished together with Y,
$V_{20}(X, Y) = X$ formerly finished together with Y.

The GN-model is shown in Fig. 1. It contains 7 transitions and 23 places. Token α_1 enters the net via place l_1. This token has initial characteristic

$$x_0^{\alpha} = \text{“Question (hypothesis) about events } X \text{ and } Y\text{”.}$$

For brevity, below we will omit index i for the tokens $\alpha_1, \alpha_2, \ldots, \alpha_i, \ldots$ and their initial characteristics. β-token enters place b_1 with an initial characteristic

“DataBase”.

It enters place b_2 and permanently stays there with the above characteristic, that in some moments can be updated with a new facts, that are characteristics of some α-tokens from place l_{23}.

The transitions Z_1 has the form$Z_1 = \langle\{l_1, l_{23}, b_1, b_2\}, \{l_2, l_3, l_4, l_5, b_2\},$

	l_2	l_3	l_4	l_5	b_2
l_1	*false*	*false*	*false*	*false*	*true*
l_{23}	*false*	*false*	*false*	*false*	*true* $\rangle$,
b_1	*false*	*false*	*false*	*false*	*true*
b_2	$W_{2,2}$	$W_{2,3}$	$W_{2,4}$	$W_{2,5}$	*true*

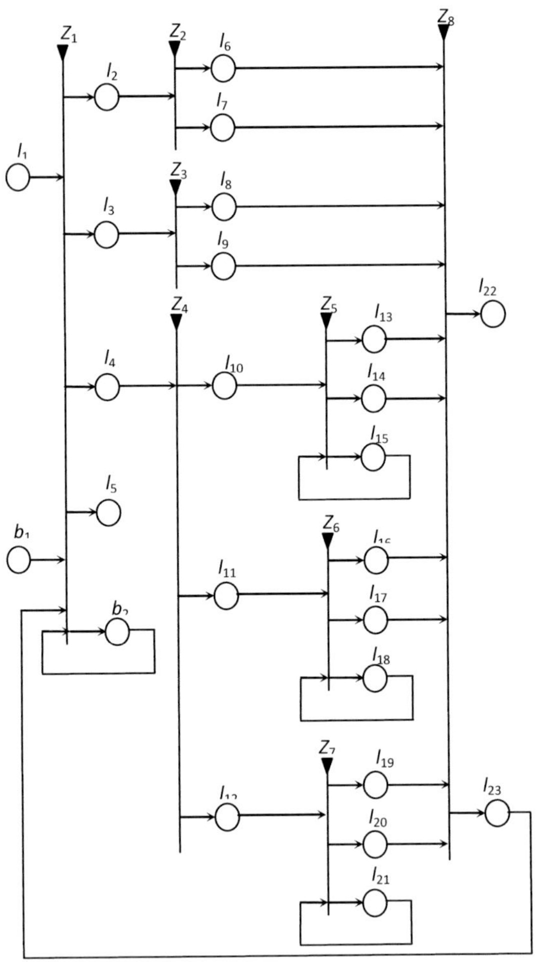

Fig. 1 GN-model of the of an Expert System Dealing with Temporal Hypothesis

where:

$W_{2,2}$ = "The predicate is from V_1-, V_6-, V_{11}- or V_{16}-type",
$W_{2,3}$ = "The predicate is from V_2-, V_7-, V_{12}- or V_{17}-type",
$W_{2,4}$ = "The predicate is from the other types (V_3-, V_4-, V_5-, V_8-, V_9-, V_{10}-, V_{13}-, V_{14}-, V_{15}-, V_{18}-, V_{19}- or V_{20}- type)";
$W_{2,5}$ = "In the DataBase there is an answer to the question in x_0^α".

The α-token, entering place l_2, l_3 or l_4, obtains, respectively, the characteristic:

"The question in x_0^α is a predicate from V_1-, V_6-, V_{11}- or V_{16}-type"

in place l_2,

"The question in x_0^α is a predicate from V_2-, V_7-, V_{12}- or V_{17}-type"

in place l_3,

"The question in x_0^α is a predicate from V_3-, V_4-, V_5-, V_8-, V_9-, V_{10}-, V_{13}-, V_{14}-, V_{15}-, V_{18}-, V_{19}- or V_{20}- type"

in place l_4, and

"An answer to the question (hypothesis) in x_0^α" in place l_5.

The transition Z_2 has the form:

$$Z_2 = \langle \{l_2\}, \{l_6, l_7\}, \begin{array}{c|cc} & l_6 & l_7 \\ \hline l_2 & W_{2,6} & W_{2,7} \end{array} \rangle,$$

where the predicates in the index matrix are the following:

$W_{2,6}$ = "$t_1^X < t_1^Y$",

$W_{2,7} = \neg W_{2,6}$,

where t_1^X and t_1^Y are first time-moments when the events X and Y have happened, respectively, and $\neg P$ is the negation of predicate P.

The α-token, entering place l_6 or l_7 obtains, respectively, the characteristics:

"The hypothesis (predicate) with the form from V_1-, V_6-, V_{11}- or V_{16}-type is true",

in place l_6,

"The hypothesis (predicate) with the form from V_1-, V_6-, V_{11}- or V_{16}-type is false",

in place l_7.

The transition Z_3 has the form:

$$Z_3 = \langle \{l_3\}, \{l_8, l_9\}, \begin{array}{c|cc} & l_8 & l_9 \\ \hline l_3 & W_{3,8} & W_{3,9} \end{array} \rangle,$$

where the predicates in the index matrix are the following:

$W_{3,8}$ = "$t_n^X < t_n^Y$",

$W_{3,9} = \neg W_{3,8}$,

where t_n^X and t_n^Y are last time-moments when the events X and Y were happened.

The α-token, entering place l_8 or l_9 obtains, respectively, the characteristic:

"The hypothesis (predicate) with the form from V_2-, V_7-, V_{12}- or V_{17}-type is true",

in place l_8,

"The hypothesis (predicate) with the form from V_2-, V_7-, V_{12}- or V_{17}-type is false",

in place l_9.

The transition Z_4 has the form:

$$Z_4 = \langle \{l_4\}, \{l_{10}, l_{11}, l_{12}\}, \begin{array}{c|ccc} & l_{10} & l_{11} & l_{12} \\ \hline l_4 & W_{4,10} & W_{4,11} & W_{4,12} \end{array} \rangle,$$

where the predicates in the index matrix are the following:

$W_{4,10}$ = "The predicate is from V_3-, V_8-, V_{13}- or V_{18}-type",

$W_{4,11}$ = "The predicate is from V_4-, V_9-, V_{14}- or V_{19}-type",

$W_{4,12}$ = "The predicate is from V_5-, V_{10}-, V_{15}- or V_{20}-type".

The α-token, entering place l_9, l_{10} or l_{11} obtains respectively, the characteristic

"The hypothesis (predicate) is from V_3-, V_8-, V_{13}- or V_{18}-type"

in place l_{10},

"The hypothesis (predicate) is from V_4-, V_9-, V_{14}- or V_{19}-type"

in place l_{11}, and

"The hypothesis (predicate) is from V_5-, V_{10}-, V_{15}- or V_{20}-type"

in place l_{12}.

The transition Z_5 has the form:

$$Z_5 = \langle \{l_{10}, l_{15}\}, \{l_{13}, l_{14}, l_{15}\}, \begin{array}{c|ccc} & l_{13} & l_{14} & l_{15} \\ \hline l_{10} & false & false & true \\ l_{15} & W_{15,13} & W_{15,14} & W_{15,15} \end{array} \rangle,$$

where the predicates in the index matrix are the following:

$W_{15,13}$ = "$(\forall i \leq k)\ (t_{2i-1}^X < t_{2i-1}^Y)$",

$W_{15,14} = \neg W_{15,13}$,

$W_{15,15}$ = "$(\forall i < k)\ (t_{2i-1}^X < t_{2i-1}^Y)$",

where i is the number of the current check, k is the number of all facts related to the current procedure, where t_{2i-1}^X and t_{2i-1}^Y are the times when the events X and Y were happened.

The α-token, entering place l_{15} does not obtain new characteristic.

The α-token, entering place l_{13} or l_{14} obtains, respectively, the characteristic

"The hypothesis (predicate) with the form from V_3-, V_8-, V_{13}- or V_{18}-type is true",

in place l_{13},

"The hypothesis (predicate) with the form from V_3-, V_8-, V_{13}- or V_{18}-type is false",

in place l_{14}, respectively.

The transition Z_6 has the form:

$$Z_6 = \langle \{l_{11}, l_{18}\}, \{l_{16}, l_{17}, l_{18}\}, \begin{array}{c|ccc} & l_{16} & l_{17} & l_{18} \\ \hline l_{11} & false & false & true \\ l_{18} & W_{18,16} & W_{18,17} & W_{18,18} \end{array} \rangle,$$

where the predicates in the index matrix are the following:

$W_{18,16} =$ "$(\forall i \le k)\ (t^X_{2i-1} < t^Y_{2i-1})$",

$W_{18,17} = \neg W_{18,16,}$

$W_{15,15} =$ "$(\forall i < k)(t^X_{2i-1} < t^Y_{2i-1})$",

were all parameters are as above. The same is valid for the description of the next transition.

The α-token, entering place l_{18} does not obtain new characteristic.

The α-token, entering place l_{16} or l_{17} obtains, respectively, the characteristic

"The hypothesis (predicate) with the form from V_4-, V_9-, V_{14}- or V_{19}-type is true",

in place l_{16},

"The hypothesis (predicate) with the form from V_4-, V_9-, V_{14}- or V_{19}-type is false",

in place l_{17}.

The transition Z_7 has the form:

$$Z_7 = \langle \{l_{12}, l_{21}\}, \{l_{19}, l_{20}, l_{21}\}, \begin{array}{c|ccc} & l_{19} & l_{20} & l_{21} \\ \hline l_{12} & false & false & true \\ l_{21} & W_{21,19} & W_{21,20} & W_{21,21} \end{array} \rangle,$$

where the predicates in the index matrix are the following:

$W_{21,19} =$ "$(\forall i \le k)\ (t^X_{2i-1} < t^Y_{2i-1})$",

$W_{21,20} = \neg\ W_{21,19,}$

$W_{21,21} =$ "$(\forall i < k)(t^X_{2i-1} < t^Y_{2i-1})$".

The α-token, entering place l_{21} does not obtain new characteristic.

The α-token, entering place l_{19} or l_{20} obtains, respectively, the characteristic

"The hypothesis (predicate) with the form from V_5-, V_{10}-, V_{15}- or V_{20}-type is true",

in place l_{19},

"The hypothesis (predicate) with the form from V_5-, V_{10}-, V_{15}- or V_{20}-type is false",

in place l_{20}.

$$Z_8 = \langle \{l_6, l_7, l_8, l_9, l_{13}, l_{14}, l_{16}, l_{17}, l_{19}, l_{20}\}, \{l_{22}, l_{23}\}, \begin{array}{c|cc} & l_{22} & l_{23} \\ \hline l_6 & true & W_{6,23} \\ l_7 & true & false \\ l_8 & true & W_{8,23} \\ l_9 & true & false \\ l_{13} & true & W_{13,23} \\ l_{14} & true & false \\ l_{16} & true & W_{16,23} \\ l_{17} & true & false \\ l_{19} & true & W_{19,23} \\ l_{20} & true & false \end{array} \rangle,$$

where

$W_{6,23} = W_{8,23} = W_{13,23} = W_{16,23} = W_{19,23} =$ "The initial question is valid and must be uploaded in the Data Base".

In all cases, the α-tokens do not obtain new characteristics.

3 Conclusion

In this position paper we have proposed and discussed a few promising subjects of research issues that arise when one is faced with the task of managing the multifaceted temporal aspects of information and knowledge that for instance physicians encounter during their clinical activities.

In the current paper we constructed a new GN-model in the area of ESs. It is an extension of Third, Fifth and Eight GN-models from [8]. In this book there are some other GN-models of other ES-types, that can also be re-written in the form of the above extension. For example, we can construct GN-model of an intuitionistic fuzzy ES with temporal components. All these models show that the concept of an ES can be extended, but in all these extensions it can be described by GNs. Some of the next extensions of the above model will be presented in future.

References

1. Keravnou, E.T.: Medical temporal reasoning (editorial). Artif. Intell. Med. **3**(6), 289–290 (1991)
2. Goodwin, S.D., Hamilton, H.J.: It's about time: an introduction to the special issue on temporal representation and reasoning. Comput. Intell. **12**, 357–358 (1996). Shahar, Y., Combi, C.: Temporal reasoning and temporal data maintenance in medicine: issues and challenges. Time oriented systems in medicine. Comput. Biol. Med. 27(5), 353–68 (1997)
3. Chittaro, L., Montanari, A.: Temporal representation and reasoning in artificial intelligence: issues and approaches. Ann. Math. Artif. Intell. **28**(1–4), 47–106 (2000)

4. Bettini, C., Montanari, A.: Temporal representation and reasoning. Data Knowl. Eng. **44**(2), 139–141 (2003)
5. Alexieva, J., Choy, E., Koycheva, E.: Review and bibloigraphy on generalized nets theory and applications. In: Choy, E., Krawczak, M., Shannon, A., Szmidt, E. (eds.) A Survey of Generalized Nets, Raffles KvB Monograph, No. 10, pp. 207–301 (2007)
6. Alty, J., M. Coombs, Expert Systems. NCC, (1984)
7. Atanassov, K.: Remark on a temporal intuitionistic fuzzy logic. In: Second Scientific Session of the "Mathematical Foundation Artificial Intelligence" Seminar, Sofia, March 30, 1990, Preprint IM-MFAIS-1-90, Sofia, pp. 1–5 (1990)
8. Atanassov, K.: Generalized Nets. World Scientific, Singapore (1991)
9. Atanassov, K.: On Generalized Nets Theory. "Prof. M. Drinov" Academic Publishing House, Sofia (2007)
10. Atanassov, K.: Temporal intuitionistic fuzzy sets. Comptes Rendus de l'Academie Bulgare des Sciences, Tome 44, No. 7, 5–7 (1991)
11. Atanassov, K.: Remark on intuitionistic fuzzy expert systems. BUSEFAL **59**, 71–76 (1994)
12. Atanassov, K. Generalized Nets in Artificial Intelligence. In: Generalized nets and Expert Systems. vol.1. "Prof. M. Drinov" Academic Publishing House, Sofia, 1998
13. Atanassov, K.: Intuitionistic Fuzzy Sets. Springer Physica-Verlag, Berlin (1999)
14. Atanassov, K., Chountas, P., Kolev, B., Sotirova, E.: Generalized Net Model of an Expert System with Temporal Components. Advanced Studies in Contemporary Mathematics **12**(2), 255–289 (2006)
15. Atanassov, K., Peneva, D., Tasseva, V., Sotirova, E., Orozova, D.: Generalized net model of an expert system with frame-type data bases with intuitionistic fuzzy estimations. In: First Int. workshop on Intuitionistic Fuzzy Sets, Generalized Nets and Knowledge Engineering, London, pp. 111–116, September 6–7, 2006
16. Atanassov, K., Sotirova, E., Orozova, D.: Generalized Net Model of Expert Systems with Frame-Type Data Base. Jangjeon Mathematical Society **9**(1), 91–101 (2006)
17. Buckley, J., Siler, W., Tucker, D.: Fuzzy expert systems. Fuzzy Sets and Systems **20**(1), 87–96 (1986)
18. Payne, E., McArthur, R.: Developing Expert Systems. John Wiley & Sons, New York (1990)

A Bipolar View on Medical Diagnosis in OvaExpert System

Anna Stachowiak, Krzysztof Dyczkowski, Andrzej Wójtowicz, Patryk Żywica and Maciej Wygralak

Abstract In the paper we present OvaExpert - a unique tool for supporting gynecologists in the diagnosis of ovarian tumor, combining classical diagnostic scales with modern methods of machine learning and soft computing. A distinguishing feature of the system is its comprehensiveness, which makes it usable at any stage of a diagnostic process. We gather all the results and solutions making up the system, some of which were described in our other publications, to provide an overall picture of OvaExpert and its capabilities. A special attention is paid to a property of supporting uncertainty modeling and processing, that is an essential part of the system.

Keywords Supporting medical diagnosis · Incomplete data · Bipolar information · Uncertainty · Aggregation · Interval-valued fuzzy sets · Atanassov's intuitionistic fuzzy sets

1 Introduction

OvaExpert, an intelligent system for supporting ovarian tumor diagnosis, is being developed by a team of scientists from two universities from Poznan, Poland: Adam Mickiewicz University and Poznan University of Medical Sciences. This interdisciplinary cooperation was motivated by an alarmingly high mortality rate among women caused by ovarian tumor. The correct and early diagnosis of that kind of tumor is still a problem especially for inexperienced gynecologists and in small medical centers that lack specialized equipment and money for medical examinations. Such deficiency implies problems with collecting all the data by a physician during examinations that, in turn, hinders making a final decision.

A. Stachowiak(✉) · K. Dyczkowski · A. Wójtowicz · P. Żywica · M. Wygralak
Faculty of Mathematics and Computer Science, Adam Mickiewicz University in Poznań, Umultowska 87, 61-614 Poznań, Poland
e-mail: min@wmi.amu.edu.pl

T. Andreasen et al. (eds.), *Flexible Query Answering Systems 2015*,
Advances in Intelligent Systems and Computing 400,
DOI: 10.1007/978-3-319-26154-6_37

To support gynecologists in a diagnostic process a wide range of preoperative diagnostic models have been developed, where the goal is to predict the type of malignancy of a tumor. The most common diagnostic scales are based on scoring systems [1, 10] and logistic regressions [12]. Both the sensitivity and specificity of that models rarely exceeds 90% in external evaluation [7, 13]. Another limitation of those models is that they cannot be applied when some of the attributes' values are missing which is a common problem resulted e.g. from technical limitations of the health care unit, high costs of medical examination or high risk for patient's health.

OvaExpert is meant to be an answer to the problem of low-quality diagnosis in a presence of missing data. Its main aim is to equip a physician with a comfortable tool to collect and manage patient's data in a standardised format, to minimise a negative influence of incomplete data on the final diagnosis, to improve the reliability and efficacy of the diagnosis, also when some data is missing, and finally to present the result in a way that gives maximum information to a doctor. The system is easy to use and intuitive, yet it utilizes modern methods mainly from the area of machine learning, soft computing and fuzzy sets theory. In the following we describe the system in details, focusing on its ability to deal with imprecision, incompleteness and uncertainty. We present main features and components of OvaExpert, some of its theoretical background and we discuss how the above mentioned problems were successfully solved and what is still left to be done.

2 Features of OvaExpert

OvaExpert was meant to integrate present knowledge about ovarian tumors (models, scoring systems, reasoning schemes, etc.) into a single computer-aided system. Its modular architecture enables plugging existing and new methods for supporting ovarian tumor diagnosis into the system as modules and then integrate them to increase the reliability of diagnosis. Its prototype version was implemented and is available at `http://ovaexpert.pl/en` as a demo to provide insight into all functions of OvaExpert. It offers convenient access on PCs, tablets and smartphones. The preliminary concept of the system, together with its architecture, was presented in [4]. In the following we discuss in brief the basic features of the system.

At the very beginning of diagnostic process OvaExpert provides physicians with an comfortable and intuitive interface that allows collecting data about patients in a standardized form and then managing data safely. The system gathers knowledge about symptoms, results of medical examinations and final diagnoses for different types of tumors. The design of the interface was carefully consulted with gynecologists to meet the need for ease of use in all conditions, also on mobile devices, especially on smartphones. An example screenshot of OvaExpert interface is depicted in Fig. 1. So far, data was collected by individual doctors using traditional methods, like spreadsheet or notebook, without paying sufficient attention to the quality and format of these data. Thanks to OvaExpert standardized format we initialized building a knowledge base about different medical cases and set up a continuous learning system. This also enables quality assessment of the diagnostic decisions taken by the

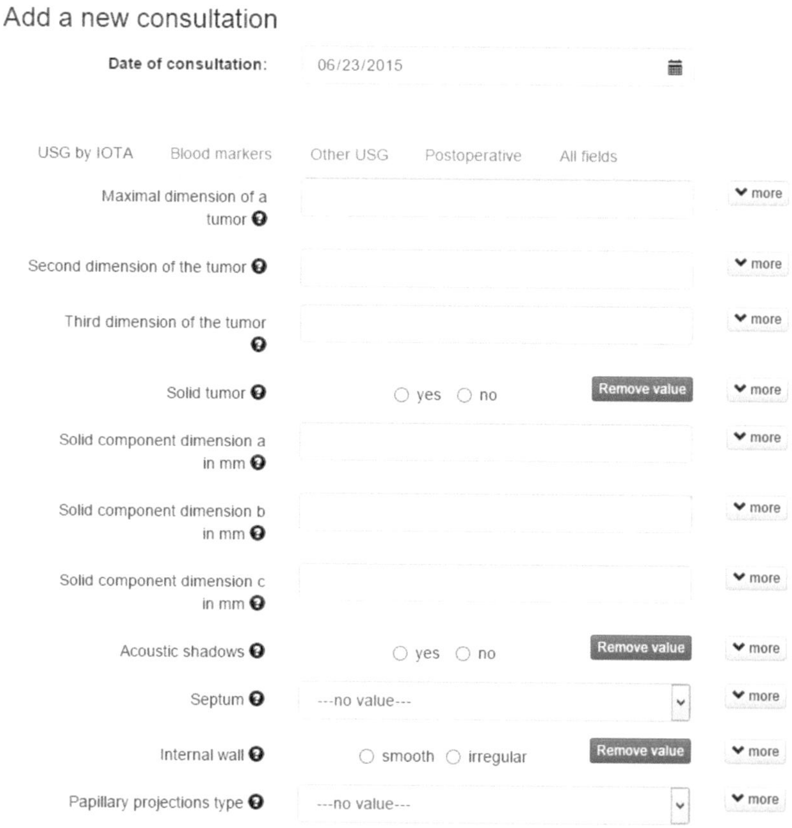

Fig. 1 Adding new consultation using OvaExpert interface

system, performed by specialists from different medical centers, and a collection of data for further scientific research.

At any time, the attending physician can be provided with the history and the visualization of the patient's diagnostic process. During the whole process a gynecologist is accompanied by a system that supports him or her by identifying further research, the execution of which may increase the likelihood of giving accurate diagnosis. Such solution is a great help for inexperienced gynecologists and, moreover, allows to avoid unnecessary examinations and costs related to them.

The main aim of OvaExpert is to support physician in making a final diagnosis, that is in assessing malignancy of the tumor. This complex issue is pursued by many different methods. First of all, OvaExpert implements known prognostic models, including models of IOTA group (International Ovarian Tumor Analysis Group) and RMI. Many gynecologists are familiar with those methods and trust their results. However, those scales had not been prepared to handle incomplete data, while the incompleteness is common in medical practice. For that reason some solutions, for example based on aggregation methods, were proposed and implemented in OvaExpert that allow to make an effective diagnosis in the presence of incomplete or missing diagnostic tests. The details of that solutions are presented in next sections. What is important, OvaExpert system achieves higher efficiency than any of the known models separately.

Finally, OvaExpert presents the result of a diagnostic process in a bipolar way, giving the possibility of diagnosis towards malignant and towards benign together with a degree of impossibility of determining the nature of malignancy. Such presentation informs a physician about the reliability and completeness of a diagnosis.

OvaExpert is a unique tool for many reasons. To the best of our knowledge it is the first time when incompleteness of data was taken into account and incorporated into a system for ovarian tumor diagnosis in a comprehensive way, either at the stage of collecting data about the patient, at the stage of data processing and finally at the stage of presenting the results. This issue is discussed in details in the next section.

3 Uncertainty Handling in OvaExpert

3.1 Uncertainty in Medicine

Uncertainty has attracted increasing attention in health care practice and medical publications as a pervasive and important problem. As studied in [5] there are multiple meanings and varieties of uncertainty in medicine, each of them having unique effects for diagnosis and warrant different courses of action. A lot of forms of uncertainty have been identified, like complexity or ambiguity, that arises from conflicting or incomplete information, as well as from multiple interpretation of some phenomenon, and vagueness that arises from lack of well defined distinctions or imprecise boundaries. Another sorts of uncertainty can be distinguished according to its nature - whether it is objective (arises from a complex or probabilistic nature of a phenomenon), subjective (personal opinion or interpretation) or comes from low quality of information, e.g. incompleteness.

Uncertainty is experienced both by a patient and by a doctor. Functioning in such conditions is an everyday experience in medical practice and is impossible to eliminate completely. However, many tools that support gynecologists, like before mentioned diagnostic scales, neglect that problem and shift the responsibility for good-quality data to a doctor. A different approach is proposed in OvaExpert system. OvaExpert takes into account the uncertainty issue and implements solution for: modeling incomplete data (see Sub. 3.2), reasoning from incomplete data (see Sub. 3.3, 3.5, 3.4) and presenting the final results that incorporates uncertainty factor (see Sub. 3.6).

3.2 Interval Data Modeling

In OvaExpert it is possible to gather over 60 attributes to describe the patient's condition. Some of those attributes are always available for physician, like age or weight, while others are subjective (ultrasound) or may be difficult to obtain (blood markers). Consequently, there is a group of attributes that are imprecise, not well defined, incomplete or not defined.

Formally, in a classical approach, a patient is modelled by a vector $\mathbf{p} = (p_1, p_2, ..., p_n)$ in a space $P = D_1 \times D_2 \times ... \times D_n$, where $D_1, D_2, ..., D_n$ are real closed intervals denoting domains of attributes that describe patients. In OvaExpert we extend this representation by introducing a possibility to model incompletely known data. Each attribute D_i is substituted by its interval version $\hat{D}_i = \mathcal{I}_{D_i}$ and a patient is a vector $\hat{\mathbf{p}} \in \hat{P} = \hat{D}_1 \times \hat{D}_2 \times ... \times \hat{D}_n$. It is thus possible for a doctor to introduce an approximate value of an attribute instead of an exact one, using interval representation. If the value of the attribute is totally unknown than it is calculated as

$$\hat{p}_i = [\underline{p}_i, \overline{p}_i] = [\min_{d \in D_i} d, \max_{d \in D_i} d].$$

3.3 Uncertaintification of Scales

OvaExpert implements six diagnostic scales: SM [10], Alcazar [1], LR1 [12], LR2 [12], Timmerman [11] and RMI1 [6] in two versions: original one and extended, suitable for interval representation. The second version is a novel contribution of OvaExpert into the area of decision-making under incomplete information. It is easy to add another scale to the system if needed, work on ROMA and Adnex indexes is in progress now.

Original diagnostic scale is formalised as a function $m : P \to [0, 1]$. Values returned by a function indicate malignancy of a tumor and are interpreted in the following way:

- $m(\mathbf{p}) > 0.5$ – diagnosis towards malignant;
- $m(\mathbf{p}) < 0.5$ – diagnosis towards benign;
- $m(\mathbf{p}) = 0.5$ – indicates the impossibility of determining the nature of malignancy.

It was crucial to adapt existing diagnostic scales so as they were able to operate on interval-valued representation of a patient described in Sub.3.2. Therefore, an extended diagnostic scale $\hat{m} : \hat{P} \to \mathcal{I}_{[0,1]}$ was constructed as:

$$\hat{m}(\hat{\mathbf{p}}) = \left\{ m(\mathbf{p}) : \forall_{1 \le i \le n} \quad \underline{p}_i \le p_i \le \overline{p}_i \right\} = \left[\min_{\mathbf{p} \in \hat{\mathbf{p}}} m(\mathbf{p}), \max_{\mathbf{p} \in \hat{\mathbf{p}}} m(\mathbf{p}) \right] \quad (1)$$

where by $\mathbf{p} \in \hat{\mathbf{p}}$ we denote that $\mathbf{p}$ is an embedded vector of $\hat{\mathbf{p}}$.

The resultant interval represents all the possible diagnoses that can be made basing on an interval patient description. The more incomplete description, the more uncertain the diagnosis. However, two aspects are worth noting. Firstly, uncertain diagnosis gives more information then no diagnosis at all, which would be the case if original diagnostic scales had been used. Secondly, even such uncertain information could be sufficient to make a proper diagnosis in many cases, since some amount of missing values is acceptable and would not affect the final result significantly.

3.4 *Interval-Valued Aggregation*

The number of different diagnostic scales is large and it is not commonly accepted which one should be used in a particular situation. Moreover, in case of missing data, the result of a single diagnostic scale given by (1) is often uncertain and not so easy to apply by a physician. The biggest challenge was thus to support a physician in making an effective final diagnosis under incomplete information. One of the proposed approaches is to take advantage of the diversity of diagnostic scales and to aggregate their results to benefit from synergy effect. In OvaExpert different interval-valued aggregation methods were implemented and tested. They were divided into groups of aggregation operators based on: arithmetic mean, weighted mean, sum and intersection from the set theory, interval OWA and voting [14]. The conducted evaluation of those methods proved that aggregation is a powerful method to improve the quality of diagnosis as well as to minimise the impact of the lack of data and uncertainty. As can be seen in Fig. 2, even the simplest methods received efficacy which exceed individual diagnostic scales, both in terms of accuracy and the number of diagnosed patients, despite missing data. More details concerning aggregation methods and evaluation methodology can be found in [17].

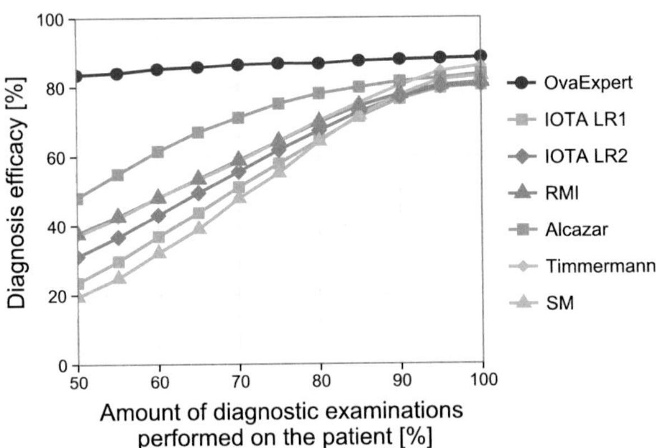

Fig. 2 Efficacy comparison of selected diagnostic models

3.5 *Interval-Valued Fuzzy Classifier*

As a separate module, OvaExpert implements a novel concept of an interval-valued prototype-based fuzzy classifier based on the uncertainty-aware similarity measure. The idea is to preserve full information – including the uncertainty factor – about data during the classification process. The classifier is designed to deal with situations in which both the classified objects and the classes themselves are imprecise, subjective and/or incomplete. In such cases, the resulting classification would also be imprecise or incomplete.

There are two ways to divide patients into classes. A basic, binary classification, discriminates two kinds of tumor: malignant and benign. A multi-class classification allows more sophisticated discrimination into histopathological types of tumor. For each class, one prototype vector which represents the entire class is constructed. We assume that class prototypes as well as objects to be classified (patients) are coded as IVFSs (interval valued fuzzy sets, see [15]). Then, the assignment of patient p_i to classes from $\mathcal{C}$ using the singleton notation can be stated as follows:

$$\widetilde{A}_{p_i} = \sum_{c \in \mathcal{C}} sim_{IF}(\widetilde{iv}(c), \widetilde{iv}(p_i))/c \qquad (2)$$

where sim_{IF} is an uncertainty-aware similarity measure. This approach was discussed in details in [8].

The crucial issue for this approach is the method of constructing prototypes. Prototypes can be formed from data, for example by using clustering algorithms such as k-means, or can result from the application of expert knowledge. Thus the proposed method gives the valuable opportunity to integrate knowledge taken from data and from expert in one tool.

3.6 *Bipolar Presentation of a Medical Diagnosis*

A classical approach to medical diagnostic process involves identifying the most adequate diagnosis. However, it is also possible to follow the criteria that exclude certain diagnoses. It is apparent that in case of doubts regarding the diagnosis, such bipolar - positive and negative - perspective is valuable and carries more information for a doctor.

OvaExpert uses an approach based on Atanassov's intuitionistic fuzzy sets [2, 16] to model bipolarity in the diagnostic process. This concept is innovative in medicine, its use in the diagnosis having only been indicated as a possibility [3, 9]. It is coherent with a basic premise of OvaExpert system that is to accept and to cope with uncertainty. Methods of data modeling and processing presented in previous subsections utilize interval representation to preserve information about the completeness of data. Next, the positive and negative information (diagnosis) is extracted from the obtained results and presented in a bipolar way to a user in a form of a bar chart, as presented in Fig. 3. On the one hand, the patient's condition is described by a degree

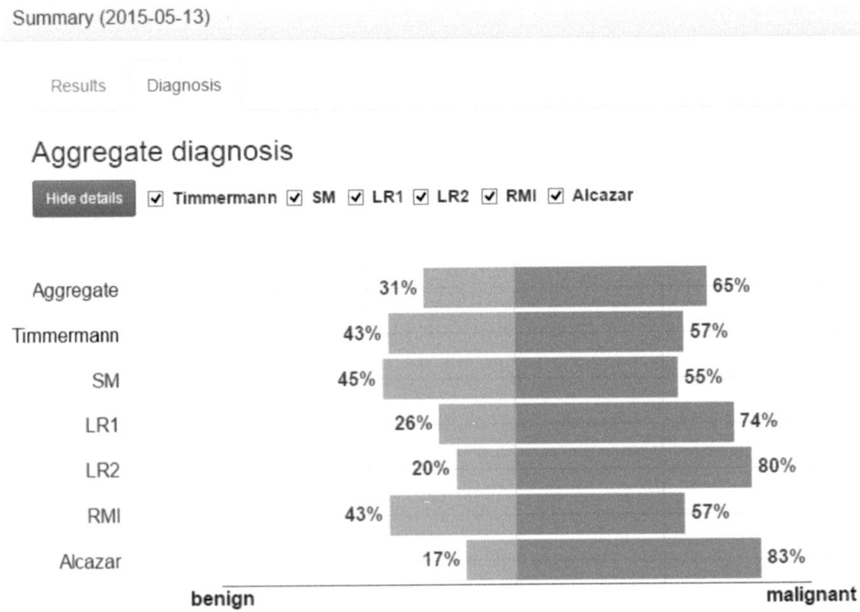

Fig. 3 Bipolar presentation of a diagnosis of ovarian tumor

that indicate a tumor being malignant, and on the other - being benign. Those two degrees sum up to 100% when all the necessary information about patient's attributes is available. Otherwise, the system suggests further examination to increase the reliability and completeness of a diagnosis. The chart may be displayed in aggregated form or with details about diagnostic scales. Apart from the diagnosis itself (whether malignant or benign), the gynecologist is equipped with the additional information about the reliability of that diagnosis, that consists of two factors: a degree of belief towards certain diagnosis (a high advantage over the opposite diagnosis makes it more reliable) and a completeness of a diagnosis (expressed by a length of a bar on the chart); obviously, the diagnosis that was based on incomplete data is less reliable. However, as outlined earlier, our approach makes it possible to make a good-quality diagnosis at the early stage of a treatment process, even if some data is missing, and improve it later, when the examinations will be complemented.

4 Conclusions and Further Work

OvaExpert is an innovative system based on machine learning techniques and computational intelligence. It addresses the need for a tool that not only supports a gynecologist in the final diagnosis, but also assists him or her during the whole diagnostic process, beginning with collecting data about the patient.

The primary advantages of the system are transparency and ease of comprehension of the principles, the ability to take into account knowledge derived both from experts and from data, and the built-in possibility of representation and processing of subjective, imprecise and uncertain information. The system was designed to support less experienced gynecologists and it allows a continuous improvement of the quality of diagnosis.

Moreover, we believe that OvaExpert can connect the medical community in the exchange of experience and verification of knowledge.

In our future work we plan to add new features: ability to integrate expert knowledge into the system, fuzzy rule-based diagnostic module and diagnostic path wizard.

Acknowledgments The project has received the Microsoft Research Award and has been included in the program of Polish Ministry of Higher Education – Innovation Incubator executed by the Poznan Science and Technology Park.

References

1. Alcázar, J.L., Mercé, L.T., et al.: A new scoring system to differentiate benign from malignant adnexal masses. Obstetrical & Gynecological Survey **58**(7), 462–463 (2003)
2. Atanassov, K.T.: Intuitionistic fuzzy sets. Springer (1999)
3. De, S.K., Biswas, R., Roy, A.R.: An application of intuitionistic fuzzy sets in medical diagnosis. Fuzzy Sets and Systems **117**(2), 209–213 (2001)
4. Dyczkowski, K., Wójtowicz, A., Żywica, P., Stachowiak, A., Moszyński, R., Szubert, S.: An intelligent system for computer-aided ovarian tumor diagnosis. In: Filev, D., et al. (eds.) IS 2014, Volume 2: Tools, Architectures, Systems, Applications. AISC, vol. 323, pp. 335–343. Springer, Heidelberg (2015)
5. Han, P.K., Klein, W.M., Arora, N.K.: Varieties of uncertainty in health care a conceptual taxonomy. Medical Decision Making **31**(6), 828–838 (2011)
6. Jacobs, I., Oram, D., et al.: A risk of malignancy index incorporating CA 125, ultrasound and menopausal status for the accurate preoperative diagnosis of ovarian cancer. BJOG: An International Journal of Obstetrics & Gynaecology **97**(10), 922–929 (1990)
7. Moszyński, R., Żywica, P., et al.: Menopausal status strongly influences the utility of predictive models in differential diagnosis of ovarian tumors: An external validation of selected diagnostic tools. Ginekologia Polska **85**(12), 892–899 (2014)
8. Stachowiak, A., Żywica, P., Dyczkowski, K., Wójtowicz, A.: An interval-valued fuzzy classifier based on an uncertainty-aware similarity measure. In: Angelov, P., et al. (eds.) IS 2014, Volume 1: Mathematical Foundations, Theory, Analyses. AISC, vol. 322, pp. 741–751. Springer, Heidelberg (2015)
9. Szmidt, E., Kacprzyk, J.: An intuitionistic fuzzy set based approach to intelligent data analysis: an application to medical diagnosis. In: Abraham, A., Jain, L.C., Kacprzyk, J. (eds.) Recent Advances in Intelligent Paradigms and Applications. STUDFUZZ, vol. 113, pp. 57–70. Springer, Heidelberg (2003)
10. Szpurek, D., Moszyński, R., et al.: An ultrasonographic morphological index for prediction of ovarian tumor malignancy. European Journal of Gynaecological Oncology **26**(1), 51–54 (2005)
11. Timmerman, D., Bourne, T.H., et al.: A comparison of methods for preoperative discrimination between malignant and benign adnexal masses: the development of a new logistic regression model. American Journal of Obstetrics and Gynecology **181**(1), 57–65 (1999)

12. Timmerman, D., Testa, A.C., et al.: Logistic regression model to distinguish between the benign and malignant adnexal mass before surgery: a multicenter study by the International Ovarian Tumor Analysis Group. Journal of Clinical Oncology **23**(34), 8794–8801 (2005)
13. Van Holsbeke, C., Van Calster, B., et al.: External validation of mathematical models to distinguish between benign and malignant adnexal tumors: a multicenter study by the International Ovarian Tumor Analysis Group. Clinical Cancer Research **13**(15), 4440–4447 (2007)
14. Wygralak, M.: Intelligent Counting under Information Imprecision: Applications to Intelligent Systems and Decision Support. Springer (2013)
15. Zadeh, L.: The concept of a linguistic variable and its application to approximate reasoning—i. Information Sciences **8**(3), 199–249 (1975)
16. Zadeh, L.A.: Fuzzy logic and approximate reasoning. Synthese **30**(3–4), 407–428 (1975)
17. Żywica, P., Wójtowicz, A., Stachowiak, A., Dyczkowski, K.: Improving medical decisions under incomplete data using interval-valued fuzzy aggregation. In: Proceedings of the 2015 Conference of the International Fuzzy Systems Association and the European Society for Fuzzy Logic and Technology. Atlantis Press (2015)

Author Index